AF477414

ZDENĚK KOPAL'S BINARY STAR LEGACY

ZDENĚK KOPAL'S BINARY STAR LEGACY

Edited by:

HORST DRECHSEL

Dr. Remeis Observatory, Bamberg, Germany

MILOSLAV ZEJDA

N. Copernicus Observatory and Planetarium Brno, Czech Republik

Reprinted from *Astrophysics and Space Science*
Volume 296, Nos. 1–4, 2005

Springer

Library of Congress Cataloging-in-Publication Data is available

ISBN 1-4020-3131-9

2003055495

Published by Springer,
P.O. Box 17, 3300 AA Dordrecht, The Netherlands.

Printed on acid-free paper

Printed in the Netherlands

TABLE OF CONTENTS

I: INTRODUCTION: REMINISCENCES AND APPRECIATION OF A GREAT ASTRONOMER

II: BINARY STAR MORPHOLOGY AND DYMAMICAL ASPECTS

III: MATHEMATICAL PHYSICS AND NUMERICAL MODELING

III.1: Solution of Eclipse Light Curves

PREFACE

The international conference entitled *"Zdeněk Kopal's Binary Star Legacy"* was held on the occasion of the 90th birthday of the late Professor Kopal from March 31 through April 3, 2004, in his Bohemian home town Litomyšl in the Czech Republic.

This meeting was dedicated to the memory of an outstanding astronomer, who devoted nearly 60 years of his scientific life to the study of close binary systems and is considered as one of the most famous astronomers of the twentieth century.

One of his main achievements was his discovery that close binary components are deformed under the influence of gravitational and centrifugal forces into non-spherical stars – a breakthrough which greatly influenced the development of close binary research until the present days. As an astrophysical application of Édouard Roche's mathematical formulation of the restricted three-body problem, Professor Kopal was the first to represent the shape of close binaries conclusively by means of Roche equipotential surfaces. The Roche lobe concept was suited to explain a wealth of hitherto puzzling observations, evolutionary and interaction effects, and set the stage for the development of whole new branches of modern astrophysics like the proper interpretation of cataclysmic variables, symbiotic stars, or X-ray binaries. Although close binaries always attracted most of his attention and truly were his favourite topic, Professor Kopal had a much wider horizon, with interests in quite different fields like radio astronomy, participating in a lunar mapping project to support NASA's Apollo program, or using his great skill as a numerical mathematician, for instance, in application to kinematical problems and circumstellar particle dynamics, to name only a few. He wrote about 400 scientific papers and 50 books, including his famous monograph *'Close Binary Systems'* of 1959, his textbooks on the method of Fourier decomposition of eclipse light curves entitled *'Language of the Stars'* (1979) and on *'The Roche Problem and its Significance for Double-star Astronomy'* (1989) as well as his autobiography *'Of Stars and Men'* published in 1986.

Professor Kopal was convinced that it is the interaction between two stars bound closely together which provides the best way for a deeper insight into their physics and structure (quite comparable to the disclosure of human characteristics, which are often only revealed by the mutual interaction of individual beings). He liked to speak of the *'Royal Road'* when referring to the determination of elements of binaries by means of light curve analysis. However, he was well aware that there is no real royal road to knowledge in general. This is why he liked to play the self-imposed role of *'Devil's Advocate'* in his frequently expressed critique of science. As sharp a scientist as he was, it was his conviction that one can never be cautious enough with the construction of models and the theoretical interpretation of observational data. *"Observations don't lie!"* and *"don't jump to conclusions!"*

Astrophysics and Space Science **296**: xi–xiii, 2005.
© Springer 2005

belonged to Kopal's favourite aphorisms with which he wanted to raise his voice in warning against the wide-spread inclination of an over-interpretation of data and pushing forward of model-dependent theories without the utmost scepticism.

Such advice is of course also very appropriate for the estimation of the many new results brought forward in the course of this meeting. More than 100 participants from over 20 countries had the pleasure to listen to about 40 talks and study as many posters. The program was set up to reflect Professor Kopal's achievements in close binary research and to enlighten the progress and efforts currently under way to deal with his legacy in this field. In this sense the purpose of contributions was twofold – one part was devoted to the remembrance of late Professor Kopal's trailblazing ideas and also to personal reminiscences conveyed to us by former colleagues and students, while many other presentations reflected the whole spectrum of contemporary research on the structure and evolution of close binaries. It was impressive to follow the modern achievements in 'classical' problems like orbital dynamics of binaries, triple and multiple systems, stellar morphology with due regard to gravitational and radiative interaction, solution of eclipse light curves, evolutionary calculations, and so nice to see the success of new tools for observational recovery and numerical modeling of the distribution and kinematics of circumstellar matter.

The success of the meeting must also be ascribed to the very pleasant surroundings and the warm hospitality offered by the town of Litomyšl, its official representatives, and its citizens. For the Czech people, Professor Zdeněk Kopal's role in astronomy is comparable to that of Václav Havel in politics or that of Bedřich Smetana in the world of music. The town of Litomyšl therefore holds his name in great esteem. He was not only awarded the honorary citizenship in 1991, but on the occasion of his nineteenth birthday the conference participants could witness the unveiling of a unique sculpture at the place of his birth to commemorate Kopal's merits. This artist-made monument symbolizes his crucial discovery of the non-spherical drop-shaped stars revolving around their common centre of mass. The great appreciation of Kopal's achievements was also reflected by the welcoming addresses and concluding remarks of officials as prominent as the Vice-Prime Minister of the Czech Republic, Mr. Petr Mareš, the Governor of the District of Pardubice, Mr. Roman Linek, and the Mayor of Litomyšl, Mr. Jan Janeček, who did us the great honour of their attendance. We are also glad that Professor Kopal's two daughters, Georgina Rudge and Zdenka Smith, both resident in the USA, paid us the honour to attend this memorial conference. We are very much indebted to Dr. Jiří Grygar of the Czech Academy of Sciences, a former graduate student of Professor Kopal, who was one of the most important fathers of the whole venture, and who also acted as a perfect interpreter between Czech and English languages on many occasions. Finally we like to thank Dr. Jan Janik of Masaryk University Brno, who organized the video recording of the conference and the live transmission of major parts to the internet.

We highly appreciate the hospitality of the town of Litomyšl, which offered its splendid Renaissance castle dating from the sixteenth century (which is on the UNESCO list of cultural monuments) as a really exceptional meeting site. We also like to thank the local authorities for a marvellous reception held in the historical premises of the Castle. Special thanks are due to Dr. Milan Skřivánek, a former Director of the Litomyšl archive, who prepared a unique exhibition in the Castle of documents and artefacts preserved in the Kopal archive of Litomyšl. Also two exciting cultural events filled all of us with enthusiasm: Vít Zouhar's chamber opera *'Coronide'*, performed in the charming Castle theatre by the *'Ensemble Damian'*, and a concert of the renowned *'Wallinger String Quartet'*, who enjoyed us with wonderful pieces of the Czech composers Smetana, Martinů, and Dvořák – great events which will be vividly recalled for long.

What remains for us is to acknowledge the precious help and advice provided by our colleagues in the Scientific and Local Organizing Committees, who contributed to the success of such a pleasant meeting.

HORST DRECHSEL
Dr. Remeis Observatory Bamberg

MILOSLAV ZEJDA
N. Copernicus Observatory and Planetarium Brno

COMMITTEES

1. Scientific Organizing Committee

Horst Drechsel (chair), Dr. Remeis Observatory Bamberg, Astronomical Institute of the University of Erlangen-Nürnberg, Sternwartstr. 7, 96049 Bamberg, Germany
drechsel@sternwarte.uni-erlangen.de

Alan Batten, National Research Council of Canada, Herzberg Institute of Astrophysics, Dominion Astrophysical Observatory, 5071, W. Saanich Rd, Victoria, B.C., Canada, V9E 2E7
Alan.Batten@hia-iha.nrc-cnrc.gc.ca

Dmitrij V. Bisikalo, Institute of Astronomy, Russian Academy of Sciences, Moscow, Russia
bisikalo@inasan.rssi.ru

Edwin Budding, Carter Observatory, PO Box 2909, Wellington, New Zealand; Çanakkale Onsekiz Mart University, Faculty of Science & Arts, Physics Department, 17100, Çanakkale, Turkey
ebudding@comu.edu.tr

Osman Demircan, Çanakkale Onsekiz Mart University, Terzioglu Campus, Faculty of Science & Arts, Physics Department, 17100, Çanakkale, Turkey
demircan@comu.edu.tr

Petr Hadrava, Astronomical Institute, Academy of Sciences, 251 65 Ondřejov, Czech Republic
had@sunstel.asu.cas.cz

Pavel Mayer, Astronomical Institute, Charles University, V Holešovičkách 2, 180 00 Praha 8, Czech Republic
Pavel.Mayer@mff.cuni.cz

Zdeněk Mikulášek, Institute of Theoretical Physics and Astrophysics, Masaryk University, Kotlářská 2, 611 37 Brno, Czech Republic
mikulas@ics.muni.cz

Izold Pustylnik, Tartu Observatory, 61602 Estonia
izold@aai.ee

Nikolaj N. Samus, Institute of Astronomy (Russian Acad. Sci.), 48, Pyatnitskaya Str., Moscow 119017, Russia and Sternberg State Astronomical Institute, Universitetskij Prosp. 13, 119992 Moscow, Russia
samus@sai.msu.ru

Astrophysics and Space Science **296**: xv–xvi, 2005.
© Springer 2005

Augustin Skopal, Astronomical Institute, Slovak Academy of Sciences, 059 60 Tatranská Lomnica, Slovakia
astrskop@ta3.sk

Robert E. Wilson, Astronomy Department, University of Florida, Gainesville, FL 32611, USA
wilson@astro.ufl.edu

Marek Wolf, Astronomical Institute, Charles University, V Holešovičkách 2, 180 00 Praha 8, Czech Republic
wolf@beba.cesnet.cz

Miloslav Zejda, N. Copernicus Observatory and Planetarium, Kraví hora 2, 616 00 Brno, Czech Republic
zejda@hvezdarna.cz

2. Local Organizing Committee

Miloslav Zejda (chair)

Miroslav Brož, Petr Hájek, Ondřej Pejcha (BRNO-Variable Star Section)

Hana Mifková, Marek Wolf (Charles University, Prague)

Eva Piknová, Jan Pikna, Michaela Severová (Litomyšl)

Jan Janík, Eva Janouškovcová (Masaryk University, Brno)

1. List of Participants

Francesco **Acerbi**, Via Zoncada 51, Codogno, I-26845 Codogno, Italy, *acerbifr@tin.it*

Berathidin **Albayrak**, Ankara University, Faculty of Science, Dept. of Astronomy and Space Sciences, TR-06100, Tandogan, Ankara, Turkey, *albayrak@astro1.science.ankara.edu.tr*

Volkan **Bakis**, Çanakkale Onsekiz Mart University, Terzioglu Campus, Faculty of Science & Arts, Physics Department, TR-17100, Çanakkale, Turkey, *bakisv@physics.comu.edu.tr*

Carlo **Barani**, Via Molinetto 33, Localita Triulza-Codogno (LO), I-26845, Italy, *cbarani@tiscalinet.it*

Michael **Bauer**, Dr. Remeis Observatory Bamberg, Astronomical Institute of the University of Erlangen-Nürnberg, Sternwartstr. 7, D-96049 Bamberg, Germany, *bauer@sternwarte.uni-erlangen.de*

Dmitrij V. **Bisikalo**, Institute of Astronomy, Russian Academy of Sciences, Moscow, Russia, *bisikalo@inasan.rssi.ru*

Miroslav **Brož**, Observatory and Planetarium, Zámeček 456, CZ-500 08 Hradec Králové, Czech Republic, *miroslav.broz@email.cz*

Edwin **Budding**, Carter Observatory, PO Box 2909, Wellington, New Zealand; Çanakkale Onsekiz Mart University, Terzioglu Campus, Faculty of Science & Arts, Physics Department, TR-17100, Çanakkale, Turkey, *ebudding@comu.edu.tr*

Andrea **Budovičová**, Astronomical Institute, Academy of Sciences, CZ-251 65 Ondřejov, Czech Republic, *andrea@asu.cas.cz*

Marcin **Cikala**, Uniwersytet Mikolaja Kopernika, Centrum Astronomii, ul. Gagarina 11, 87-100 Torun, Poland, *cikala@astri.uni.torun.pl*

Szilárd **Csizmadia**, Konkoly Observatory of the Hungarian Academy of Sciences, H-1525 Budapest, P. O. Box 67., Budapest, Hungary, *csizmadia@konkoly.hu*

Aliakbar **Dariush**, Physics Department, Arsenjan Azad University, Arsenjan, Iran, *dariush75@yahoo.com*

Bert **De Loore**, Astrophysical Institute, Vrije Universiteit Brussel, Pleinlaan 2, B-1050 Brussel, Belgium, *cdeloore@nets.ruca.ua.ac.be*

 *Astrophysics and Space Science* **296**: xvii–xxiii, 2005.
© Springer 2005

Osman **Demircan**, Çanakkale Onsekiz Mart University, Terzioglu Campus, Faculty of Science & Arts, Physics Department, TR-17100, Çanakkale, Turkey, *demircan@comu.edu.tr*

Gojko **Djurašević**, Astronomical Observatory, Volgina 7, 11160 Belgrade, Serbia and Montenegro, *gdjurasevic@aob.bg.ac.yu*

Dijana **Dominis**, Lehrstuhl für Astrophysik, Universität Potsdam, Am Neuen Palais 10, D-14469 Potsdam, Germany, *dijana@astro.physik.uni-potsdam.de*

Horst **Drechsel**, Dr. Remeis Observatory Bamberg, Astronomical Institute of the University of Erlangen-Nürnberg, Sternwartstr. 7, D-96049 Bamberg, Germany, *drechsel@sternwarte.uni-erlangen.de*

Peter **Eggleton**, Lawrence Livermore National Laboratory, 7000 East Ave., Livermore, CA 94551, USA, *ppe@igpp.ucllnl.org*

Michael **Friedjung**, Institut d'Astrophysique, 98 bis Boulevard Arago, F-75014 Paris, France, *fried@iap.fr*

Cezary **Galan**, Uniwersytet Mikolaja Kopernika, Centrum Astronomii, ul. Gagarina 11, 87-100 Torun, Poland, *cgalan@astri.uni.torun.pl*

Kosmas D. **Gazeas**, Department of Astrophysics, Astronomy and Mechanics, National and Kapodistrian University of Athens, GR 157 84, Zografos, Athens, Greece, *kgaze@skiathos.physics.auth.gr*

Constantine L. **Goudas**, Department of Mathematics, University of Patras, Campus at Rio, GR 26 504 Patras, Greece, *stud-mec@otenet.gr,*
goudasconstantine@hotmail.com

Dariusz **Graczyk**, Uniwersytet Mikolaja Kopernika, Centrum Astronomii, ul. Gagarina 11, 87-100 Torun, Poland, *darek.graczyk@astri.uni.torun.pl*

Tomáš **Gráf**, Observatory and Planetarium of Johann Palisa, 17. listopadu 15, CZ-708 33 Ostrava-Poruba, Czech Republic, *tomas.graf@vsb.cz*

Jiří **Grygar**, Institute of Physics, Czech Academy of Sciences, Na Slovance 2, CZ-18221 Praha 8, Czech Republic, *grygar@fzu.cz*

Petr **Hadrava**, Astronomical Institute, Academy of Sciences, CZ-251 65 Ondřejov, Czech Republic, *had@sunstel.asu.cas.cz*

Marcin **Hajduk**, Uniwersytet Mikolaja Kopernika, Centrum Astronomii, ul. Gagarina 11, 87-100 Torun, Poland, *cinek@astri.uni.torun.pl*

Tibor **Hegedüs**, H-6500 Baja, Szegedi ut K.T. 766, Hungary, *hege@electra.bajaobs.hu*

Pavel **Chadima**, Pod zámečkem 291, Hradec Králové, Czech Republic, *pavel.chadima@post.cz*

Anatoly M. **Cherepashchuk**, Sternberg State Astronomical Institute, Universitetskij Prosp. 13, 119992 Moscow, Russia, *cher@sai.msu.ru*

Drahomir **Chochol**, Astronomical Institute, Slovak Academy of Sciences, 059 60 Tatranská Lomnica, Slovakia, *chochol@ta3.sk*

Jan **Janík**, Institute of Theoretical Physics and Astrophysics, Masaryk University, Kotlářská 2, 611 37 Brno, Czech Republic, *honza@physics.muni.cz*

Belinda **Kalomeni**, University of Ege, Faculty of Science, Department of Astronomy and Space Sciences, TR-35100 Bornova, Izmir, Turkey, *belinda@astronomy.sci.ege.edu.tr*

Agata **Karska**, Uniwersytet Mikolaja Kopernika, Centrum Astronomii, ul. Gagarina 11, 87-100 Torun, Poland, *agata@astri.uni.torun.pl*

Nataly **Katysheva**, Sternberg State Astronomical Institute, Universitetskij Prosp. 13, 119992 Moscow, Russia, *nk@sai.msu.ru*

Adela **Kawka**, Astronomical Institute, Academy of Sciences, CZ-251 65 Ondřejov, Czech Republic, *kawka@sunstel.asu.cas.cz*

Masatoshi **Kitamura**, National Astronomical Observatory, Mitaka, Tokyo 181-8588, Japan, *tanikawa@exodus.mtk.nao.ac.jp (subj. Prof. Kitamura)*

Pavel **Koubský**, Astronomical Institute, Academy of Sciences, CZ-251 65 Ondřejov, Czech Republic, *koubsky@sunstel.asu.cas.cz*

Jiří **Krtička**, Institute of Theoretical Physics and Astrophysics, Masaryk University, Kotlářská 2, CZ-611 37 Brno, Czech Republic, *krticka@physics.muni.cz*

Jiří **Kubát**, Astronomical Institute, Academy of Sciences, CZ-251 65 Ondřejov, Czech Republic, *kubat@sunstel.asu.cas.cz*

Blanka **Kučerová**, Institute of Theoretical Physics and Astrophysics, Masaryk University, Kotlářská 2, CZ-611 37 Brno, Czech Republic, *aknalb@physics.muni.cz*

Anna **Litvinchova**, Crimean Astrophysical Observatory, Tavrichesky National University, Crimea, 95007 Ukraine, *pavlenko@crao.crimea.ua*

Tom R. **Marsh**, Department of Physics, University of Warwick, Coventry CV4 7AL, United Kingdom, *t.r.marsh@warwick.ac.uk*

Pavel **Mayer**, Astronomical Institute, Charles University, V Holešovičkách 2, CZ-180 00 Praha 8, Czech Republic, *Pavel.Mayer@mff.cuni.cz*

Ralf **Meyer**, Fürnheim 16, D-91717 Wassertrüdingen, Germany

Zdeněk **Mikulášek**, Institute of Theoretical Physics and Astrophysics, Masaryk University, Kotlářská 2, CZ-611 37 Brno, Czech Republic, *mikulas@ics.muni.cz*

Martin **Navrátil**, Observatory and Planetarium, Zámeček 456, CZ-500 08 Hradec Králové, Czech Republic, *navratil@astrohk.cz*

Stefan **Nesslinger**, Dr. Remeis Observatory Bamberg, Astronomical Institute of the University of Erlangen-Nürnberg, Sternwartstr. 7, D-96049 Bamberg, Germany, *nesslinger@sternwarte.uni-erlangen.de*

Panagiotis G. **Niarchos**, Department of Astrophysics, Astronomy and Mechanics, Faculty of Physics, University of Athens, GR 157 84, Zographou, Athens, Greece, *pniarcho@cc.uoa.gr*

Waldemar **Ogloza**, Mt. Suhora Observatory, ul. Podchorazych 2, 30-084 Krakow, Poland, *ogloza@ap.krakow.pl*

Štefan **Parimucha**, Faculty of Theoretical Physics and Astrophysics, University of P. J. Šafárik, Moyzesova 16, 040 01 Košice, Slovakia, *parimuch@ta3.sk*

Kresimir **Pavlovski**, Department of Physics, University of Zagreb, Bijenicka 32, 10000 Zagreb, Croatia, *pavlovski@phy.hr*

Reza **Pazhouhesh**, Physics Department, Faculty of Sciences, Birjand University, Birjand, Iran, *pazhouhesh@fastmail.ca*

Ondřej **Pejcha**, N. Copernicus Observatory and Planetarium, Kraví hora 2, CZ-616 00 Brno, Czech Republic, *pejcha@astro.sci.muni.cz*

Jiří **Polcar**, Institute of Theoretical Physics and Astrophysics, Masaryk University, Kotlářská 2, CZ-611 37 Brno, Czech Republic, *polcar@physics.muni.cz*

Theodor **Pribulla**, Astronomical Institute, Slovak Academy of Sciences, 059 60 Tatranská Lomnica, Slovakia, *pribulla@ta3.sk*

Andrej **Prša**, University of Ljubljana, Department of Physics, Jadranska 19, 1000 Ljubljana, Slovenia, *andrej.prsa@fmf.uni-lj.si*

Izold **Pustylnik**, Tartu Observatory, 61602 Estonia, *izold@aai.ee*

Eleni **Rovithis-Livaniou**, Department of Astrophysics, Astronomy and Mechanics, Faculty of Physics, University of Athens, GR 15784, Zografos, Athens, Greece, *elivan@cc.uoa.gr*

Krisztina **Ruzsics**, H-6500 Baja, Szegedi ut K.T. 766, Hungary, *kriszta@electra.bajaobs.hu*

Somaya M. **Saad**, Astronomical Institute, Academy of Sciences, CZ-251 65 Ondřejov, Czech Republic, *somaya@sunstel.asu.cas.cz*

Nikolaj N. **Samus**, Institute of Astronomy (Russian Acad. Sci.), 48, Pyatnitskaya Str., Moscow 119017, Russia and Sternberg State Astronomical Institute, Universitetskij Prosp. 13, 119992 Moscow, Russia, *samus@sai.msu.ru*

Selim O. **Selam**, Ankara University, Faculty of Science, Department of Astronomy and Space Sciences, TR-06100 Tandogan, Ankara, Turkey, *selim@astro1.science.ankara.edu.tr*

Sergei **Shugarov**, Sternberg State Astronomical Institute, Universitetskij Prosp. 13, 119992 Moscow, Russia, *sg@sai.msu.ru*

Michal **Siwak**, Astronomical Observatory of the Jagiellonian University, ul. Orla 171, 30-244 Kraków, Poland, *siwak@oa.uj.edu.pl*

Jan **Skalický**, P. Bezruče 440, CZ-563 01 Lanškroun, Czech Republic, *skalicky@physics.muni.cz*

Augustin **Skopal**, Astronomical Institute, Slovak Academy of Sciences, 059 60 Tatranská Lomnica, Slovakia, *astrskop@ta3.sk*

Petr **Sobotka**, Míru 636, Kolín 2, CZ-280 02, Czech Republic, *petr.sobotka@astro.cz*

Júlia **Sokolovičová**, Koleje 17. listopadu, Pátkova 3, CZ-182 00 Praha 8, Czech Republic, *sokolovicova@lycos.com*

Lenka **Šarounová**, Astronomical Institute, Academy of Sciences, CZ-251 65 Ondřejov, Czech Republic, *lenka@asu.cas.cz*

Zdislav **Šíma**, Astronomical Institute, Academy of Sciences, Boční II 1401, CZ-141 31 Praha 4, Czech Republic, *sima@ig.cas.cz*

Petr **Škoda**, Astronomical Institute, Academy of Sciences, CZ-251 65 Ondřejov, Czech Republic, *skoda@sunstel.asu.cas.cz*

Miroslav **Šlechta**, Astronomical Institute, Academy of Sciences, CZ-251 65 Ondřejov, Czech Republic, *slechta@sunstel.asu.cas.cz*

Ladislav **Šmelcer**, Observatory, Vsetínská 78, CZ-757 01 Valašské Meziříčí, Czech Republic, *lsmelcer@astrovm.cz*

Martin **Šolc**, Astronomical Institute, Charles University, V Holešovičkách 2, CZ-180 00 Praha 8, Czech Republic, *martin.solc@mff.cuni.cz*

Stanislav **Štefl**, Astronomical Institute, Academy of Sciences, CZ-251 65 Ondřejov, Czech Republic, *sstefl@pleione.asu.cas.cz*

Dirk **Terrell**, Department of Space Studies, Southwest Research Institute, 1050 Walnut St., Suite 400, Boulder, CO 80302, USA, *terrell@boulder.swri.edu*

Todd **Vaccaro**, Department of Physics and Astronomy, Louisiana State University, 202 Nicholson Hall, TowerDrive, Baton Rouge, Louisiana 70803-4001, USA, *vaccaro@rouge.phys.lsu.edu*

Walter **Van Hamme**, Department of Physics, Florida International University, University Park, CP238, Miami, Florida, FL 33199, USA, *vanhamme@fiu.edu*

Roberto F. **Viotti**, Istituto di Astrofisica Spaziale e Fissica Cosmica, CNR, Via Fosso del Cavaliere 100, I-00133 Roma, Italy, *uvspace@rm.iasf.cnr.it*

Igor M. **Volkov**, Sternberg State Astronomical Institute, Universitetskij Prosp. 13, 119992 Moscow, Russia, *imv@sai.msu.ru*

Irina **Voloshina**, Sternberg State Astronomical Institute, Universitetskij Prosp. 13, 119992 Moscow, Russia, *vib@sai.msu.ru*

Viktor **Votruba**, Institute of Theoretical Physics and Astrophysics, Masaryk University, Kotlářská 2, CZ-611 37 Brno, Czech Republic, *votruba@physics. muni.cz*

Robert E. **Wilson**, Astronomy Department., University of Florida, Gainesville, FL 32611, USA, *wilson@astro.ufl.edu*

Marek **Wolf**, Astronomical Institute, Charles University, V Holešovičkách 2, CZ-180 00 Praha 8, Czech Republic, *wolf@beba.cesnet.cz*

Kadri **Yakut**, University of Ege, Faculty of Science, Department of Astronomy and Space Sciences, TR-35100 Bornova, Izmir, Turkey, *yakut@astronomy. sci.ege.edu.tr*

Atsuma **Yamasaki**, Department of Earth & Ocean Sciences, National Defence Academy, Yokosuka, 239-8686 Japan, *yamasaki@nda.ac.jp*

Petr **Zasche**, Astronomical Institute, Charles University, V Holešovičkách 2, CZ-180 00 Praha 8, Czech Republic, *petr.zasche@email.cz*

Miloslav **Zejda**, N. Copernicus Observatory and Planetarium, Kraví hora 2, CZ-616 00 Brno, Czech Republic, *zejda@hvezdarna.cz*

2. Guests and Accompanying Persons

Marian Denny, USA
Lloyd Denny, USA
Allison Denny, USA
Steve Denny, USA
Irena Fryčáková, Czech Republic
Vanda Goudas, Greece
Julie Gulbransen, USA
Lura Gulbransen, USA
Eva Janouškovcová, Masaryk University, Brno, Czech Republic
(*evina@elanor.sci.muni.cz*)
Libuše Koubská, Czech Republic (*koubska@vydavatelstvimjs.cz*)
Ivana Krejčová, Czech Republic
Hana Mifková, Czech Republic
Georgiana Rudge, USA (*georgiana.rudge@evc.edu*)
Zdenka Smith, USA (*zdenka.smith@noaa.gov*)
Satoko Yamasaki, Japan (*cygnus@a2.ocv.ne.jp*)

I: INTRODUCTION:
REMINISCENCES AND APPRECIATION OF A GREAT ASTRONOMER

DIGGING FOUNDATIONS FOR THE "ROYAL ROAD"

ALAN H. BATTEN

National Research Council of Canada, Herzberg Institute of Astrophysics, Dominion Astrophysical Observatory, Victoria, B.C., Canada; E-mail: batten@dao.nrc.ca

(accepted April 2004)

Abstract. H.N. Russell and Z. Kopal both liked the metaphor of the "Royal Road" to scientific discovery; I discuss which one used it first. I present some personal reminiscences of Professor Kopal and then consider his attitude to the determination of the elements of eclipsing binaries from their observed light changes, comparing it with that of Russell. This leads me to discuss Kopal's work on the evolution of binary stars and his opposition to the prevailing belief in the importance of mass transfer between the components. Kopal's attitudes on these matters puzzle me, but I suggest that at least part of his motivation was to act as a critic of "normal science" within the paradigm that most of us have accepted.

1. The "Royal Road"

Despite all their disagreements, Henry Norris Russell and Zdeněk Kopal shared a fondness for the phrase"the Royal Road". When Ed Budding was preparing his paper for the Çanakkale meeting, nearly two years ago (Budding, 2003), we discussed by e-mail which of the two used the metaphor first. I pointed out that Russell (1948) had entitled the first Russell Lecture (delivered in late 1946) "The Royal Road of Eclipses", while Kopal wrote, it seemed to me as a response, in his monograph, *The Computation of the Elements of Eclipsing Variables* (Kopal, 1950), that there was no royal road to the direct determination of the elements of binaries containing distorted components. Ed Budding was certain that Kopal had used the phrase "Royal Road" earlier. Indeed, in his first monograph, *Introduction to the study of Eclipsing Variables*, Kopal (1946) had already made the same assertion. Characteristically he used several consecutive sentences *verbatim* in both books, and they appeared yet again in the magisterial work *Close Binary Systems* (Kopal, 1959a), where we are introduced, in a rather clever variation of the assonance, to the "equitable avenue" of differential corrections. Russell read the 1946 book in manuscript and provided a foreword. Did this, perhaps unconsciously, influence the choice of title for his eponymous lecture? Apparently not; David DeVorkin (2000a), in his biography of Russell, reveals that the latter had used the phrase "Royal Road" in other astronomical contexts, and first in a lecture delivered early in 1932; before Z. Kopal was eighteen years old and nearly two years before he entered Charles University. The opening sentences of the published version were:

Once in a long time in the history of science, a true "royal road" is found which
leads to the solution of a problem which had previously been unassailable.
There is no better example than the discovery of the spectroscope.

H.N. RUSSELL (1932)

Russell, therefore, seems to have priority for using the phrase in the context of
modern astronomy; he had probably used it in Kopal's hearing many times before
he chose the title of the 1946 lecture. Both DeVorkin (2000b) account and Kopal's
(1986a) own show that, by then, personal relations between the two men had become
very strained. Each man liked the phrase and neither, I suspect, was above using it
deliberately in a way that he knew would needle the other!

Fondness for a particular metaphor may seem to reveal nothing more than the
personal idiosyncrasies of the two men, but I believe that there were deeper parallels between their careers, which, after some personal reminiscences, I would
like to explore by discussing two areas of close binary studies that have developed
considerably in my own professional lifetime: the analysis (and synthesis) of light
curves, and the study of the evolution of binary systems. In the first of these, both
Russell and Kopal played seminal roles or, in the metaphor of my title, dug the
foundations. Digging is the first step in laying foundations, and a Czech colleague
on our Scientific Organizing Committee reminded us, early in our preparations, that
the name "Kopal" means "digger". I think Professor Kopal would have enjoyed the
play on words.

2. Some Personal Reminiscences

Next, of course, to Professor Kopal himself, I claim to be the founding member
of the Manchester School of binary studies. I was not his first student: there were
others one and two years ahead of me, and three contemporary with me, but, for
various reasons, none of them are still active in astronomy. I was the first to come
to Manchester (in the fall of 1955) specifically to study binary stars. Professor
Kopal had been my external examiner in St. Andrews and had helped me with two
undergraduate projects, and I owed my acceptance as his student to my interest in
binary stars and the recommendation of Professor Freundlich, my teacher and his.
Kopal would have been just about forty years old when I first met him, although, even
then, his prematurely grey hair made him look older. He was already a recognized
authority on eclipsing binary systems, as is shown by the two monographs on light
curve analysis that I have cited.

A habit shared by Kopal and Freundlich was using up the clean sides of any
sheet of paper, often as notes to students. Confidential documents, such as Senate
minutes, were not exempt from this form of recycling and, in both my universities,
I learned things not normally in the student curriculum! Kopal once told me that
it was a treat, in his boyhood, to be given a piece of paper with one clean side; no

doubt, that accounted for his habit. In old age, I find myself doing the same thing – I wrote the first draft of this presentation on already used paper. My motive is to save some of our Canadian trees!

The first thing about Professor Kopal that impressed many people was his enthusiasm, directed, of course, especially toward astronomy, but spilling over into many of his other interests. He was a keen and skilled chess-player, for example. During an observing run at the Pic-du-Midi, I played several games with him almost every day. I do not recall that I ever won, although I think I did once or twice hold him to a draw. He acquired that skill very young. I once met W. Jaschek (no relation, so far as I know, to Carlos Jaschek), a Czech astronomer who accompanied Kopal on the 1936 journey by the Trans-Siberian Railway, described in his autobiography (Kopal, 1986b). According to Jaschek, a Russian general on that train wanted a chess partner and young Kopal was the only person willing and able to fill the role. By Jaschek's account (a little different from the autobiography) the Russian general's pleasure soon evaporated when he discovered he never could beat the young upstart! I believe Kopal was also a good tennis player, although I never challenged him at that game. He certainly was at least a competent pianist.

Not all enthusiasts, however, are able to convey and to share their enthusiasms in the way that Kopal could, the way that made him a good teacher. We often disagreed with his lectures, but they were certainly never dull. He was, in particular, a great stimulus to the pioneer staff at Jodrell Bank. Comments made by Hanbury Brown about that role are quoted with obvious pleasure by Kopal (1986c) in his autobiography. Hanbury, and others of the original Jodrell staff, have made similar comments to me in private conversation. It is perhaps difficult for those who did not experience it to understand how great a gulf there was between optical and radio astronomers in those days. Many optical astronomers had ignored the work of Jansky and Reber. Most of the early radio astronomy was radar detection of daytime meteor showers—originally thought to be cosmic-ray events—and did not seem to promise anything fundamentally new in astronomy. I have it on good authority that there were earnest debates in the Council of the Royal Astronomical Society whether or not, *in principle*, papers on "radio astronomy" should be accepted for publication in *Monthly Notices*! On the other hand, those earliest radio astronomers, although brilliant radio technicians, on their own admission did not at first know very much astronomy (Brown, 1988). Kopal played an important role in bridging that gulf by teaching astronomy to the staff of what, then as now, was one of the leading centers of radio astronomy in the world, and in directing their attention to projects likely to be astronomically fruitful.

3. The Analysis of Light Curves

Determining the elements of eclipsing binaries is mathematically more complicated than the corresponding problem for visual or spectroscopic binaries, and has proved more controversial. Henry Norris Russell and Z. Kopal may have been the first to clash on this topic, but they were by no means the last. Let us compare two quotations:

> Least-square solutions should never be attempted until the resources of graphical adjustment have been fully exploited. It is a fair criterion of a good least-squares solution that all the corrections should be of the same order of magnitude as their probable errors, and certainly a criterion of a satisfactory preliminary solution that this should be the case. The computer thus escapes the heart-breaking experience of finding, after the heavy work of least-squares, that the second-order terms were sensible and the whole job must be repeated.
>
> H.N. RUSSELL (1933).

> Before proceeding ... however, it may be advisable to plot l against $\sin^2 \theta$ on a suitable scale, and to inspect the general trend of the light changes which we are going to analyze .... Few investigators will, at this time, resist perhaps the temptation of drawing a continuous *light curve* by free hand to follow the course of observed normals, in order to obtain a more forceful representation of the anticipated light changes. *At no time, however, will there be need to use this curve for anything else but inspection*, or for an *estimate* of preliminary values of certain characteristics of the system (such as depths of the minima, or the moments of inner and outer contacts) which will be improved by subsequent analysis. *No investigator should ever forget that his real task is to interpret the actual observations* (as represented by a discrete set of available normal points) rather than any inferences based upon them. For, irrespective of how closely a light curve could be drawn to follow the course of observed normals, it represents indeed merely a plausible inference which, in some parts, may approach the reality more closely than any normal point but which, in others, may be systematically off; and no matter how small the respective deviations may be, their systematic nature may entail serious cumulative effects.
>
> Z. KOPAL (1959b, italics in the original)

Russell was writing about visual binaries, but he made very similar remarks about eclipsing binaries (Russell, 1942); one could hardly hope for clearer statements of two opposing philosophies of computation. Those two authors were predestined to conflict, especially when one recalls the difference in their ages and cultural backgrounds. Yet each was right in his time and from his own point of view. Russell had a feel for the precision of the observations available to him and a special knack for finding quick and approximate methods, sufficient to do justice to the available data. He regarded such scientific problems as "rather like a glorified crossword puzzle" – but more fun! Kopal had much more concern for mathematical rigor and he and Piotrowski developed their iterative methods for determining the elements of eclipsing binaries just as the earliest photoelectric observations were being published. Of course, even the first photoelectric observations were more precise than older measures, but I suspect that Kopal tended to overestimate their precision. Unlike Russell's methods, however, Kopal's gave not only the elements themselves, but also their uncertainties. I completely agree with Kopal on the importance of quoting uncertainties.

The new methods did not emerge from a vacuum; Kopal himself acknowledged the importance of Takeda's (1934, 1937) basic contributions and of his own collaboration with Stefan Piotrowski (see e.g. Piotrowski, 1947, 1948). I never met Takeda, but I saw at first hand the mutual respect and friendship between Kopal and Piotrowski (and between Kopal and Harlow and Martha Shapley). Naturally, Kopal taught me to use his methods, and I am quite sure that they were an improvement on the old ones but, in the U.S. at least, they were little used. Most investigators remained wedded to the Russell-Merrill methods until the advent of light curve synthesis in the early 1970s. This was partly due to the enormous influence of Russell, exerted after his death through his students, John Merrill and Frank Bradshaw Wood, who both became my friends; but, as DeVorkin argues and Kopal (1986d) himself admitted, a major factor was the time consumed by the calculations required by the new methods, in days when there was nothing faster than an electric desk-top calculator. As an unfortunate result, North-American astronomers tended to underestimate the importance of the work by Kopal and Piotrowski. Kopal himself, at least in Manchester, had the services of a full-time computer (Mrs Gorman – in those days the word "computer" without any adjective meant a human being) whose services I, too, was permitted to use. That made a great deal of difference to how one viewed Kopal's methods, which might well have been the more soundly based, but were definitely more like an equitable avenue than a Royal Road.

In retrospect, I wonder why Kopal did not set me, as a Ph.D. project, the programming of his methods for the Manchester Mark I or Mark II computers. He was rightly concerned that his students should master what was then a new tool and had I undertaken that project, his methods might well have become more widely used. One or two attempts *were* made a little later to program the methods but, as already suggested, the majority of investigators continued with Russell's methods until light curve synthesis became feasible. Instead, Kopal set me on a project which was beyond the capacity of those computers, to say nothing of the student, namely to compute the light variations to be expected during the eclipse of a component filling its Roche lobe. Two decades later, that became possible, but it was not the best thing for a neophyte astronomer to try at that time.

Kopal's own reactions to light curve synthesis are instructive. He conceived his Fourier-transform method in the late 1950s, although I do not recall having discussed it with him in Manchester. Kopal (1986e) says that he raised the matter during the 1958 IAU General Assembly, although neither my memory nor the official account of that meeting (Merrill, 1960) confirms this. I do, however, very clearly remember him discussing the Fourier method three years later at the Berkeley General Assembly, in much the terms in which it is summarized in the official account (Kopal, 1962). The idea was not well received by those members of Russell's group who were present. John Merrill remarked that Russell had looked into this very possibility many years ago and concluded that "the time was not ripe", and he went on to quote the summary by his recently died namesake (and, I believe, distant relative), Paul Merrill, of an attitude that can be traced back to Francis Bacon: "We ask Nature, we do not tell her". In retrospect, neither of these objections seem to me to be very strong. The time might not have been ripe when Russell considered the matter, but by the 1960s the situation was different. Moreover, Kopal could very well have replied (although I do not recall him having done so) that "asking Nature" was precisely what he proposed to do. If he had not just then become heavily involved in the lunar-mapping project, and if a suitable graduate student had been available, we might well have had the Fourier-transform method before the light curve synthesis of Hill and Hutchings (1970). Would so many people, then, have gone on to develop and to use the synthesis methods? One of Kopal's objections to synthesis (more justified when he raised it than now) was the demands it made on computing time – beyond the resources then available in many developing countries. In this, at least, he claimed, the Fourier-transform method was superior. (I do not think that he ever claimed that his new method was the "Royal Road", although he came close to it in Maratea-Kopal 1981 – when he adopted Russell's title for his own introductory lecture to the 1980 NATO Advanced Study Institute.)

Kopal's chief argument against light curve synthesis was that it gave no guarantee that its results were unique, or even correct: he viewed synthesis as "telling Nature". As I said in Çanakkale (Batten, 2003), I believe he was correct in principle but, in practice, the results have been remarkably consistent: self-consistent, consistent with analytical methods, and consistent with our theories of stellar structure and evolution. While the originators of synthesis methods saw themselves as creating physically realistic models of the binaries under investigation, Kopal saw them as making unwarranted assumptions about those systems. He saw a risk that the solution might be constrained to produce values near the adopted ones, even though the system might be very different. To be sure, when Kopal was developing his own methods, attempts at light curve synthesis might well have run the risk of seriously misleading the investigator, but we know very much more about stellar structure and evolution now than was known then; it seems sensible to use our knowledge in tackling what is mathematically a very difficult problem. Most of us see the guarantee of our solutions in the responsible use of that sort of information. To be sure, much of the knowledge (but no longer all of it) came from light curve analysis in the first place, so there is some danger of circularity in our arguments – but it seems strange that the man who insisted that the only "equitable avenue" to the determination of elements of distorted eclipsing systems was an iterative process of differential corrections should have so set his face against iterating between the observed light curves and our increasing understanding of the stars and their atmospheres. Indeed, as I read DeVorkin's account of the strained relations between Russell and Kopal – which I was privileged to see before publication – I was struck by how similar Kopal's attitude towards light curve synthesis, in his later years, was to Russell's attitude towards Kopal's own methods some thirty to forty years earlier. Both men could marshal good arguments to rationalize their cases, but it is hard to avoid the feeling that each felt an undercurrent of resentment at seeing his life's work thrust aside by younger men.

4. Evolution by Mass Transfer

Kopal also contributed to the study of the evolution of close binaries, in which much progress was made during his lifetime. In an obituary notice, Meaburn (1994) claimed that one of Zdeněk Kopal's greatest achievements was the prediction of large-scale mass transfer between the components of close binaries. I disagree and do not believe that Professor Kopal himself would have wished to be remembered for predicting an idea he consistently criticized over the entire period that I knew him. Already, when he arrived in Manchester, Kopal was very sceptical of Otto Struve's interpretations (e.g. Struve, 1949) of spectroscopic observations of systems like U. Cephei in terms of gas streams that could be detected by their effects on the spectra of the components. Kopal's experience of spectroscopic observation was limited and I am not sure that he fully appreciated either its limitations or its possibilities. Struve, on the other hand, had a keen eye for the significant features on a spectrogram (I believe he rarely, if ever, used microphotometer tracings) and a "gut" feeling for what must be happening in a system. I think he had an unusual degree of physical insight, although Kopal disagreed with me about that when we talked in Maratea in 1980. When I worked on U. Cephei, I read and re-read Struve's papers: I was amazed at the detail he had seen on spectrograms inferior to mine, and largely concurred with his conclusions, although I frequently disagreed with the arguments by which he got from one to the other. Perhaps Kopal and I disagreed about Struve's "physical insight" at least partly because we gave different meanings to the term. I meant that Struve had a good idea of what must be going on and brought this to his interpretation of the observations.

In the late 1940s and early 1950s, however, Struve's ideas were highly controversial. Perhaps scepticism about gas streams in binary systems was one of the few things on which Russell and Kopal agreed! Certainly, I was taught to look upon Struve's arguments with a critical eye, although, even then, I found them attractive. Only shortly before I arrived in Manchester, Struve's student, J.A. Crawford (1955) showed that subgiant components in what we now call semi-detached or Algol systems had expanded to the maximum size possible in a close binary. Kopal (1954) reached the same conclusion independently. He later adopted the term "Roche limit" for this maximum size and used it for the rest of his life. Because that term already has a related, but distinct, meaning in discussions of the stability of satellites, I prefer to speak of the "Roche lobe", a phrase that I think I introduced into the literature and seems to have become generally adopted. As you all know, the subgiant components are the less massive in their respective systems, but even in the mid-1950s we knew enough to expect that the more massive component should expand first. The result found by Crawford and Kopal posed a problem, now known as the "Algol Paradox". Crawford went on to suggest that the expanding subgiant had originally been the more massive star, but that it had lost so much mass that it was now the less massive. He was not specific about the fate of the lost mass; either it left the system entirely or it was transferred to the companion. Although he mentioned it only as a possibility, the latter suggestion has become particularly associated with Crawford.

Crawford's paper was published under only his own name, but the mass transfer idea is typical of Otto Struve's bold leaps of speculation, and I have little doubt that it was his. The important point, in our context, is that although Kopal and Crawford independently showed that the subgiant components filled the Roche lobe by expansion, the idea of mass transfer was put forward only in Crawford's paper and, even there, only as one possibility. Kopal consistently favored the other possibility (secular loss of mass from the system) from the beginning of the controversy until the end of his life. Almost everyone – probably everyone here – now accepts the idea of mass transfer, but I checked my memory that Kopal was sceptical from the beginning by re-reading two major papers (Kopal, 1955, 1956), which I read in typescript as soon as I arrived in Manchester. Moreover, in his autobiography, Kopal (1986f) claims equality of priority in the discovery of the expansion of the subgiants, but does not even mention mass transfer. In his 1955 paper Kopal refers to "Crawford's ingenious but not convincing hypothesis". He advanced two objections, both of which he regarded as fundamental: there were no subgiants filling their Roche lobes with masses equal to their companions (which would be systems actually in the process of mass transfer) and there were, as he believed, some subgiants that did not fill their Roche lobes (which was difficult to explain if the mass ratio had been reversed and the subgiants were still expanding). Later, the existence of such undersize subgiants became controversial, while the first objection, as Kopal partly conceded, was answered by the work of Morton (1960) and Smak (1962), who showed that the process of mass transfer would proceed on a Kelvin time scale, i.e. astronomically speaking it would be very rapid. That rapidity, however, strengthened Kopal's more general objections: how could so much mass be transferred without our noticing it, and could the star receiving that amount of mass settle down and look like an ordinary star of its new mass? These were not trivial questions, at least in the late 1950s, and when I left Manchester for Canada, I carried Kopal's scepticism with me. In Victoria, I found that R.M. Petrie, although he enjoyed a much warmer personal relationship with Struve than Kopal did, also took a rather conservative and sceptical view of Struve's ideas on the evolution of binaries.

Another reason for caution about the idea of mass transfer was in the dynamics of the gas streams themselves. Struve's arguments from spectroscopy were very appealing, but they were purely kinematic. Just before I arrived in Manchester, Kopal, with the assistance of R.A. Brooker and V. Hewison, tried to introduce dynamical considerations into the discussion of gas streams. They supposed that the subgiant had expanded to fill its Roche lobe and its rotation, necessarily being slowed in the process, would no longer be exactly synchronous with the orbital rotation. This non-synchronous rotation drove the escape of matter in the upper atmosphere, at or near the inner Lagrangian point. Then they computed trajectories of individual particles for a variety of initial conditions. Kopal published the results, both in the 1956 paper and the 1959 book. Of course, he fully recognized that particle-dynamics computations were at best a crude approximation; hydrodynamic, or even magnetohydrodynamic forces should be taken into account. The application of magnetohydrodynamics to astronomical problems had only just begun, however. The computers of the day were taxed by the particle-dynamics computations, which showed particles circling the more massive star and falling on it, but did not give very good grounds for belief in the kind of streams and semi-stable structures that Struve postulated. For example, the observed rings all rotate in the same sense as the orbital motion, but the computations all led to retrograde rotation – although this, of course, was primarily the effect of the assumption made about the driving force of the initial ejection.

We had, then, several good reasons for scepticism about large-scale mass transfer, even though Morton's and Smak's computations had given some grounds for believing it might be possible. As I have said, I was in the sceptics camp: we *wanted* to find some other explanation for the Algol Paradox. I was converted in Brussels, in the summer of 1966, where Kippenhahn and Weigert (1967) presented the first attempts to calculate what would happen if mass were transferred from one component of a binary to the other. They were not even aware of the Algol Paradox until they had completed their calculations; they were looking for a way to account for the large observed number of white dwarfs. The implications of their calculations were obvious to many of us at that meeting, however, especially to Paczynski and Plavec who, with their respective groups, were not far behind Kippenhahn and Weigert. Mass transfer ceased to be speculation at that meeting and became a hypothesis that could, at least in principle, be tested against observation. I confess, however, that I sometimes wonder whether, if Petrie, who died a few months before that meeting, had still been alive, my conversion would have survived my return to Victoria.

Kopal was not converted at all; although in 1966 he was preoccupied with lunar problems (the first manned landing was then only three years away), it is clear that when he returned to binary star astronomy his basic convictions were unaltered. You can trace his objections to mass transfer as a *major* element in binary star evolution through his papers and books published throughout the years (Kopal, 1955, 1956, 1959a,b, 1978, 1980, 1986a,b). There is a remarkable consistency in his arguments. To be sure, he dropped some as theoretical work showed them not to be as strong as he first thought, and he brought up other, more detailed arguments, as they occurred to him, but the overriding arguments remained, in one dress or another, those just outlined: namely, the process should be at least as conspicuous as a nova outburst, yet, with the possible exception of β Lyrae, we could not point to any system going through it; the amount of mass required to be transferred is very great and it is doubtful, to say the least, that the receiving star can settle down quickly and look like a normal star of its new mass. While some of his arguments were directed towards specific numerical calculations, or interpretations (including my own) of the observations, it is important to understand that basically he was opposed to the whole idea of mass transfer as a *significant* process in binary evolution. He *never* accepted the underlying hypothesis.

Perhaps those first calculations, and even many later ones, were not entirely convincing. Restrictive assumptions had to be made if the computations were to be tractable on the computers then available; the observational data were far from precise, and perhaps sometimes were even misleading. I do not recall a single reliable mass ratio being available for any semi-detached system, even throughout the 1960s and although I tried to do what I could with photographic spectroscopy, really good results were not obtained until solid-state detectors with their high signal-to-noise ratios came into general use. Thus Kopal could justify a sceptical attitude well into the 1970s, although even by then he had become part of small minority, if not a lone voice. Nevertheless, the early calculations were not the unscientific detour that he sometimes seemed to suggest they had been. They did at least show that more realistic models were needed. At a meeting in 1972, Plavec (1973) stated: "I think that *conservative Case A cannot explain the Algols.*" (his italics) and went on to sketch how the computations might be improved. It is interesting that Kopal never cited this paper in his criticisms.

Our inability, during the 1960s, to identify systems in which mass was actually being transferred was, indeed, a serious objection to the whole idea of mass transfer – especially in view of the large number of semi-detached systems known. Even then, we could argue most of the mass would be transferred very quickly, so the process would often escape detection; while, on the other hand, the deep and often total eclipses of the end-product, the semi-detached systems, made them easy targets for the discoverer of variable stars. Now, we can list more objects than just β Lyrae that are possibly in the rapid phase and we recognize a long period of slow mass transfer, after the initial rapid phase. Several semi-detached systems are probably still in this slow phase: I claim some credit myself for helping to show that U. Cephei is one of them. I think that the evidence that mass transfer takes place is now overwhelming and that its importance for the evolution of the systems in which it occurs is irrefutable. I find it very difficult to understand why Kopal fought against, not merely particular computations and their simplifying assumptions, but the very concept of mass transfer as a significant evolutionary mechanism until the end of his days. Once again, there seems to be an element of paradox in his attitude to a subject to which he had himself made seminal contributions.

5. Concluding Thoughts

Professor Kopal often referred to himself as the "Devil's Advocate". He seemed to enjoy playing the role and probably did it to make us think and question our own conclusions. His opposition to the idea that mass transfer can play a significant role in binary star evolution seems, however, to have been genuine. Even so, there was not a great gulf between him and those he criticized. He conceded that *some* mass might be transferred between components, but he certainly believed that loss of mass from the system as a whole was the more important process. On the other hand, proponents of mass transfer are quite ready to take mass loss into account. Whatever may have been the case at the time of the earliest calculations, no-one now believes in *conservative* mass transfer. That was a convenient assumption to make when the physics was less clear than it is now and the capacities of computers were much more limited. The assumption might just have worked; when it did not, at least we had learned something. I do not think that the pioneers of those computations were acting as unscientifically as Kopal sometimes seemed to suggest. John Milton (1608–1674), England's second greatest poet, was no scientist, but he visited Galileo during the latter's house arrest, and was an older contemporary and compatriot of Isaac Newton. He was imbued with something of the same spirit of enquiry that animated those two and drove the scientific renaissance of the sixteenth and seventeenth centuries. I think both Galileo and Newton would have approved of Milton's statement (in his seventeenth-century orthography):

> To be still searching what we know not, by what we know, still closing up truth to truth as we find it (for all her body is *homogeneous* and proportional) this is the golden rule in *Theology* as well as in Arithmetick, . . .
>
> JOHN MILTON, *Areopagitica*, 1644.

As I see it, "searching what we know not, by what we know" is precisely what those of us who have tackled the problems of binary star evolution, both by observational and theoretical means have tried to do. It is the way in which science proceeds. Even Newton did not derive the whole of what we now call "Newtonian mechanics". Apart from his attempt at lunar theory, Newton did not, for example, study very much the theory of perturbations; that was a later development, undertaken mainly by French mathematicians, at least some of whom hoped to prove Newton wrong.

We do, however, need people who make us go back and think again, to check and to re-check our arguments. In 1947 Herbert Dingle delivered his inaugural address as Professor of the History and Philosophy of Science in the University of London (Dingle, 1952), which he entitled "The Missing Factor in Science". The burden of his argument was that literature has a long (and strong) tradition of criticism, which science has lacked. To be sure, we have peer-review (although it was not as fully developed when Dingle wrote now) but that is not quite the sort of criticism that Dingle had in mind. He wanted us to examine the assumptions about the basic notion of science and to look at the fashions of the day, estimating their worth for the advancement of science. The current form of peer-review, on the contrary, tends to perpetuate and even to reinforce fashions. Authors espousing unorthodox opinions often find it hard to get their papers published; consider the way in which those cosmologists who do not subscribe to Big-Bang cosmology have been marginalized – a treatment that is not to the credit of the rest of the astronomical community. In our own subject, evolution of binaries by mass transfer has undoubtedly become a fashion, or, in Kuhn's (1962) terms, a paradigm. I think there are good reasons for adopting this paradigm and am content to do what Kuhn called "normal science" within it. All paradigms, however, sooner or later will be superseded. We are unwise to ignore colleagues who point out insufficiencies in our paradigms, but our own work will not be made useless by changes to new paradigms. Instead, much of that work will be carried over and help to shape the new paradigm.

I do not fully understand why Professor Kopal took so firm a stand against the idea of mass transfer, but perhaps he was motivated by some such consideration as Dingle's. Some of his criticisms were well-merited and I do not think that anyone would deny that there is still much to be learned about the physics of the streams, if they exist, that are transferring the mass from one star to the other. The differences between him and the rest of us are mainly differences of emphasis. If Kopal and Russell laid foundations for the Royal Road of Eclipses, Struve prepared the roadbed and the rest of us have, at most, begun the paving. There is still much work to do before we can travel the motorway in comfort.

References

Batten, A.H.: 2003, in: O. Demircan and E. Budding (eds.), *New Directions for Close Binary Studies*, Çanakkale Onsekiz Mart University, **3**, 289–294.

Brown, R.H.: 1988, *Radio Astronomy and Quantum Optics*, Institute of Physics, Adam Hilger, Bristol, New York and Philadelphia, p. 103.

Budding, E.: 2003, in: O. Demircan and E. Budding (eds.), *New Directions for Close Binary Studies*, Çanakkale Onsekiz Mart University, **3**, 3–18.

Crawford, J.A.: 1955, *Astrophys. J.* **121**, 71–76.

DeVorkin, D.H.: 2000a, *Henry Norris Russell: Dean of American Astronomers*, Princeton University Press, Princeton, New Jersey, p. 276.

DeVorkin, D.H.: 2000b, *ibid*. Chap. 18.

Dingle, H.: 1952, *The Scientific Adventure*, Sir Isaac Pitman and Sons, London, pp. 1–16.

Hill, G. and Hutchings, J.B.: 1970, *Astrophys. J.* **162**, 265–280.

Kippenhahn, R. and Weigert, A.: 1967, *Zs. fuer Astrophys.* **65**, 251–273.

Kopal, Z.: 1946, *Harvard Observatory Monograph* No. 6, Harvard University Press, p. 13.

Kopal, Z.: 1950, *Harvard Observatory Monograph* No. 8, The Observatory, Cambridge, Mass., p. 28.

Kopal, Z.: 1954, *Mem. Soc. Roy. des Sci. de Liege* **15**, 684–685.

Kopal, Z.: 1955, *Ann. d'Astrophys.* **18**, 379–430.

Kopal, Z.: 1956, *Ann. d'Astrophys.* **19**, 298–355.

Kopal, Z.: 1959a, *Close Binary Systems*, Chapman and Hall, London, p. 421.

Kopal, Z.: 1959b, *ibid*. p. 302.

Kopal, Z.: 1962, in: D.H. Sadler (ed.), *Transections of International Astronomical Union*, Vol. XI, Academic Press, London and New York, p. 369.

Kopal, Z.: 1978, *Dynamics of Close Binary Systems*, Reidel, Dordrecht, Holland, Chap. 8.

Kopal, Z.: 1980, in: Z. Kopal and E.B. Carling (eds.), *Photometric and Spectroscopic Binary Systems*, Reidel, Dordrecht, Holland, pp. 1–16.

Kopal, Z.: 1986a, Of Stars and Men: Reminiscences of an Astronomer, Institute of Physics, Adam Hilger, Bristol and Boston, pp. 180–186, 212–213.

Kopal, Z. 1986b, *ibid.* p. 114.

Kopal, Z. 1986c, *ibid.* pp. 245–246.

Kopal, Z. 1986d, *ibid.* p. 424.

Kopal, Z. 1986e, *ibid.* p. 428.

Kopal, Z. 1986f, *ibid.* p. 258.

Kuhn, T.S.: 1962, *The Structure of Scientific Revolutions*, 2nd edn, 1970, University of Chicago Press, Chicago, Chap. II.

Meaburn, J.: 1994, *Quarterly J. Roy. Astron. Soc.* **35**, 229–230.

Merrill, J.E.: 1960, in: D.H. Sadler (ed.) *Transections International Astronomical Union*, Vol. X, Cambridge University Press, Cambridge, pp. 635–638.

Milton, J.: 1644, *Aeropagitica*, reprinted in: C.A. Patrides (ed.), *Milton: Selected Prose*, Penguin Books, Harmondsworth, England, 1974, pp. 235–236.

Morton, D.J.: 1960, *Astrophys. J.* **132**, 146–161.

Piotrowski, S.L.: 1947, *Astrophys. J.* **106**, 472–480.

Piotrowski, S.L.: 1948, *Astrophys. J.* **108**, 36–45, 510–518.

Plavec, M.: 1973, in: A.H. Batten (ed.), *Extended Atmospheres and Circumstellar Matter in Close Binary Systems*, IAU Symposium No. 51, D. Reidel, Dordrecht, Holland, pp. pp. 216–259, see esp. p. 231.

Russell, H.N.: 1932, in *Annual Report of the Smithsonian Institute for 1931*, pp. 199–218.

Russell, H.N.: 1933, *Mon. Not. Roy. Astron. Soc.* **93**, 599–602.

Russell, H.N.: 1942, *Astrophys. J.* **95**, 345–355.

Russell, H.N.: 1948, in: *Harvard Observatory Monograph*, No. 7. The Observatory, Cambridge, Mass., pp. 181–209.

Smak, J.: 1962, *Acta Astron.* **12**, 28–54.

Struve, O.: 1949, *Mon. Not. Roy. Astron. Soc.* **109**, 487–506.

Takeda, S.I.: 1934, *Mem. Coll. Sci. Kyoto Univ.*(Ser. A) **17**, 197–217.

Takeda, S.I.: 1937, *Mem. Coll. Sci. Kyoto Univ.* (Ser. A) **20**, 47–86.

ONE FELLOW'S VIEW OF THE "ROYAL ROAD"

EDWIN BUDDING[1,2]

[1]*Carter Observatory, Wellington, New Zealand; E-mail: ebudding@comu.edu.tr*
[2]*Currently working at: Çannakkale Onsekiz Mart University (COMU), TR 17020, Turkey*

(accepted April 2004)

Abstract. The paper attempts to discern the place and direction of the "Royal Road" of astronomical usage, with the particular perspective that the late Prof. Zdeněk Kopal may have had of it. In seeking further insights into the Kopalian royal road map, we take note of his own "guide stars."

Referring then more closely to the royal road of eclipses, the information content of photometric light curves is discussed, noting in particular, their Fourier representation and the relationship of that to generally believed causes of the light variation.

1. Introduction

I want first to express thanks to our generous hosts. In this setting, of course, we cannot forget our old teacher, the late Prof. Z. Kopal, who provides the theme of the meeting. He also provided essential components, directly or indirectly, to the career paths of many of us. I should again emphasize the sense of great honor I feel in presenting this paper.

Science has two distinct aspects – observational and theoretical. I think Kopal's work can be usefully approached with this in mind. Although not really known for observational work, Kopal was keenly aware of its significance. "Observations don't lie" was a favorite aphorism of his, whose force is in its unspoken insinuation. "Don't jump to conclusions" was also frequently applied with similar cautionary intonation. "The demon speculation" was a bogey to be openly renounced, with due calling to mind examples of where it had tempted scientists, sometimes of the most honorable pedigree, to fall from grace. Since this talk gives a number of personal reflections, it will automatically include various unprovable speculations, and I would ask my hearers, wherever they may be now, to forgive this and rest in peace.

The "Royal Road" was in the title of another recent meeting (Demircan and Budding, 2003) where memories of Kopal were prominent. That meeting was held more or less at the western limit of the old Persian Royal Road, which stretched from the Hellespont to China. In European culture, the expression seems to have a particular association with Aristotle's teaching of Alexander of Macedon. While this image may symbolize an ideal relationship twixt teacher and student, it is also worth keeping in mind what Alan Batten told me was Aristotle's actual message to the young Alexander, i.e. that there is no Royal Road to knowledge. For many, it seems, there are no short-cuts, only hard work and a personal drive to achieve *scientia*. If I try to encapsulate my own – one fellow's – version of this tutorial message, as it came down to me via the mentorship of Prof. Kopal, I could see it as a paradox (the Kopal paradox(?)) thus: although one's own ideas are the main motivator to understanding, they can be the main obstacle to reaching it. But made of such ideas is the strange machine in which we travel along the Royal Road.

Round about the time that Kopal started on his autobiography (Kopal, 1986), he had just enjoyed reading Arthur Koestler's (1981) *The Sleepwalkers* and was sufficiently taken with certain passages that he would sprinkle quotes into his regular Monday afternoon lectures that went from two till tea-time. One Koestlerian image I recall in particular, from Kopal, concerned how human history has wandered along like a herd of buffalo, travelling over a vast plain. From time to time a great bull emerges from the throng and charges forward in some direction, taking a bunch of followers with him. The main stream may then trail along behind. After the energy of that particular episode subsides, it is replaced by another of similar ilk, and similar general aimlessness, when seen against the vast expanse of the plain and timeframe of its background. It is against this background that the importance of the astronomer, as natural philosopher, emerges. For it is the astronomer who is most conscious of this huge plain: its dimensions and its trends. It is he, if any, who can make out its groundrules and chart its ultimate directions.

Kopal's autobiography featured, on its cover, his own set of special guide stars to help him fix his bearings, arranged in the form of the main stars of Orion, that most distinct of constellations. In seeking further insights into the Kopalian roadmap, therefore, we could do worse than take note of these guide stars, and that is what I propose for the next part of this talk. A subsequent part will offer some general thoughts about light curves, Fourier analysis and information.

2. Kopal's Guide Stars

2.1. EDDINGTON

The highest star in our chart (with an outside view of the Celestial Sphere), occupying the position of the giant *Betelgeuse*, is Sir Arthur Eddington, whom Kopal said was generally and justifiably regarded as the greatest astronomer of the first half of the twentieth century. All of Kopal's book-cover guide stars were people of whom he had direct knowledge, and this is perhaps inevitable with primary influences. However, it is also true that Kopal had a deep admiration for scientists of all ages who wrote well, especially those great eighteenth and nineteenth century expositors, mostly French, of classical mechanics. Eddington was the primary contemporary representative, to him, of clear and logical writing in astronomy. Kopal, plenteous author and editor himself, was, of course, not alone in this opinion. Eddington's (1923) *Mathematical Theory of Relativity* was according to Einstein, "the finest presentation of the subject in any language."

Naturally, direct personal influences are important, and the image of Prof. Eddington sitting on the corner of his desk and engaging in meaningful conversation about the progress of work, during his postdoctoral fellowship at Cambridge, seems to have been one of the golden memories of our teacher. That he was able to draw out the great scientist in this way was also no mean achievement for Kopal, since Eddington had a notoriously shy personality. Kopal's autobiography illustrated this with a little story about Eddington's visit to Mt. Wilson in the early 1930s, where he was accompanied, at different times, by two technicians, one of whom apparently found the diffident visitor stuffy and aloof, while the other reported an observant and attentive guest.

The story gave various interesting insights, not least, how differently the same person can be seen through different eyes. But the account was basically about Eddington's humility; a quality that Kopal looked up to.

2.2. RUSSELL

In the position of *Bellatrix* is Henry Norris Russell, recognized as the dean of American astronomy in the pre-War period. In his autobiography, Kopal mentions a series of conferences held at Harvard Observatory in 1939 that became a turning point in his life and brought him into active involvement with close binary systems. There can be little doubt that Russell was the instigator of this, since this field had been a central interest, as Kopal informs, from the beginning of his long career in 1899 to its end in 1956. Russell had been the first to deal systematically with the light curves of eclipsing binaries, with the express purpose of deriving the 'elements,' i.e. the basic physical parameters of stars, such as their sizes, masses and luminosities.

But in his six decades at Princeton, Russell's accomplishments, of course, stretched to much more than eclipsing binaries. He involved himself with a full range of astrophysical subjects, from astrometry to spectral analysis. He was the first to come up with the idea of a stellar evolutionary trend, or 'sequence' as he put it. He determined element abundances and confirmed that stars are composed mostly of hydrogen. This entailed detailed knowledge of atomic physics, where, for example, his work on the interaction of spin and orbital angular momenta of electrons – the Russell-Saunders coupling – has become classical. Russell as a researcher, teacher, writer and advisor, was an enormous force in 20th century astronomy. Kopal's autobiography tends to gloss over this stature and refers to certain negative qualities associated with the somewhat austere role that Russell apparently took on for himself (as may be also inferred from other biographers). As noted before (Batten, 2005), however, Russell introduced the expression "The Royal Road," and whatever differences developed about the best way to travel it, there can be no doubt that Russell first recognized its importance, and Kopal later shared the same conviction.

If Eddington and Russell were the loftier shoulder stars in Kopal's chart, the stars he girt about his middle were scientists of comparable value: on

the right Harold Urey, Richard Prager to the left and Harlow Shapley as the
buckle.

2.3. PRAGER

Richard Prager emerges as the first of a number of kind-hearted and encouraging
senior scientists who were to help in shaping the youthful course of our teacher. We
encounter him in the early pages of *Of Stars and Men* as an organizing force behind
the *Verein der Sternfreunde*: one inspiring meeting of this body in Stuttgart in 1928,
in particular. Talking about those members of Prager's rapidly expanding annual
catalogue of variable stars, the character of whose variability was still unknown,
Dr. Cuno Hoffmeister, in the role of recruiting sergeant, directed his enthusiasm-
filled remarks especially to younger members of the audience. Among other things,
such early exposure to the international nature of astronomical work may have
helped Kopal recognize it as a subject that goes well beyond national boundaries,
and inculcated an attitude of internationalism that was to flourish in his later years
as head of the Manchester department.

Prager's quiet hand could also surely be discerned behind the first three of
Kopal's publications that appeared through the agency of the Czech Astronomical
Society, whilst our author was still at school. These works appear, in fact, to have
been reworkings of the variable stars finding lists, though complete with new charts
and comparison stars. But they were clearly noticed by Dr. Prager – that then *spiritus
agens* of variable star research (Prager, 1935).

2.4. SHAPLEY

Prager was, as it turned out, with Kopal at Harvard Observatory during the War
(WWII), and it is there that we meet the second of the belt stars – Harlow Shapley. Of
the personalities in the constellation, it is this one that may be closest to Kopal, certainly in the manner of human interactions. Kopal declared that Shapley's humor, in
which *Dichtung und Wahrheit* were often intermixed, was "inimitable," yet I would
say that our jovial teacher did incline to copy it, and not infrequently. Shapley used
his roguish sense of humor to good effect in dealing with the strain that developed
between Russell and Kopal, and it was a technology that I saw Kopal employ later
in dealing with departmental problems in Manchester, particularly when they did
not directly involve his own immediate research interests. It is possible that this
impish wit may have sometimes gone too far and too fast for those hearing it to trip
lightly along with. I could think of examples, but it is probably better not to "tell
tales out of school," as Kopal would say.

More to the point, Shapley was essentially the principal agent taking responsibility for Kopal in the United States during those key years from just before to not
long after the War. Shapley's arrival at Princeton just before the First World War
had been, according to Kopal, a godsend to Russell. The latter, then approaching

forty, had just completed his *magnum opus* on the analysis of light curves. Shapley, as his energetic PhD student, was in an excellent position to carry out major new excavations on the royal road. His thesis, published in 1914 as Princeton Observatory Contribution number 3, analysed 90 practical examples of eclipsing binaries, thus creating "at a stroke" the future subject matter of IAU Commission 42.

Shapley was to become much more well known, however, after changing lanes: from eclipses to globular clusters, and thereby discovering the scale and shape of our Galaxy. That these discoveries came as a surprise in the early 1920's can be judged by the 'great debate' that followed, in which Shapley, with Curtis, was at the centre. The debate found a resolution in accordance with the 'principle of mediocrity,' although just how mediocre our Galaxy is, among others, appears still open to question. Frank Shu has commented that "... this debate illustrates forcefully how tricky it is to pick one's way through the treacherous ground that characterizes research at the frontiers of science," a remark that seems to hark back to the demon speculation.

2.5. UREY

At the right of the belt is Harold Urey, the Nobel Prizewinner whose name may be the most widely known among Kopal's stars. His political interventions on nuclear arms issues received ample publicity, while his book (with Ruark) on atoms, molecules and quanta was the standard textbook for many years. Kopal stated that he never met anyone whose work deserved greater attention. He acknowledged that, during the period he was at Manchester, Urey's influence outweighed all others. It was the beckoning finger of Harold Urey that drew Kopal into the Space Age. Here we can see persuasive enough reason why Kopal should pick up his involvement with the Moon, that some have disparaged. "If you are willing to undertake this work I shall help you to raise the money" is a sentence Kopal quotes as indicating the start of a very significant programme at Manchester. And if such a sentence comes to you from the winner of a Congressional Medal of Merit, then, of course, you can take it very seriously.

That there was some recollection of youthful romanticism about this adventure we can see from Kopal's references to that oddly influential old German movie *Die Frau am Mond*. But Kopal's sabbatical trips to Wisconsin were based on his solid record as a mathematician. The work was strongly related to his enduring interests in classical mechanics and positional astronomy. Of course, Urey himself had powerful planetological interests. His systematic treatment in *The Planets: Their Origin and Development* (Urey, 1952) provides scientific underpinning of Kopal's activities in this field.

Of Stars and Men conveys an image of Urey as a force to be reckoned with: an elder statesman in the post-war *realpolitik* of American science, but that there was also a more human side, that no doubt also would have appealed to our teacher, I discerned from another reliable source, this time in New Zealand, some years ago.

This witness reported his astonishment at the vast array of plaudits and recognitive awards adorning the wall of Urey's office. "Oh well," replied Urey, "it's a bit like kissing a nice girl. If the first one goes well, more will surely follow."

2.6. MARTHA SHAPLEY

If the preceding figures were generally in the upper part of the constellation, no less important was the supporting structure that one might associate with the two remaining stars. In the position of the brilliant *Rigel* is Mrs Martha Shapley, from whom sprang much of the benign, behind-the-scenes influence that made Harvard Observatory such a pleasant place to work in. We first meet her as the gifted mathematician and astronomer who, Kopal divined, from his knowledge of both Russell and Shapley, had probably done most of the actual spadework for Princeton Contribution number 3. We then find her at the piano in a faultless private performance of one of the romantic Chopin ballades. Kopal surreptitiously listened in, and was clearly charmed for the rest of his life.

For more illumination on this "First Lady" of the observatory and her gifts, we may look at one of the regular Christmas parties she organized: probably one of the first ones for Kopal. She had decided that there should be a carol from the traditional three wise men: into which roles she cast observer Bart Bok as the gold-bearer, our author as the bringer of priestly incense and Henry Norris Russell as giver of the ointment of prophets. If the casting was perceptive, the music was unfortunately short on harmony.

We saw with Eddington that the same person can be viewed quite differently by different people: with Martha Shapley, we have an example of the contrast between the public and private images of a person. Kopal's book is revealing in the importance it attaches to the latter.

2.7. NEIL ARMSTRONG

Though at the centre of surely one of the most outstanding events in the lives of all now over 35, indeed of the most memorable occasions in human history, Neil Armstrong (as photographed stepping onto the surface of the Moon by Ed Aldrin) is only named in passing in *Of Stars and Men*. And though it is not difficult to learn of the achievements of this heroic pilot, it is surely for what his 'giant leap for mankind' symbolizes that Kopal included Armstrong as the *Saiph* of his constellation. Kopal referred to the day on which the photograph was taken, July 20, 1969, as the greatest day in his life. The section of Kopal's autobiography that deals with the Apollo Program is full of excitement: attending top-level meetings with mission planners, living and socializing in the same residential quarters as astronauts, explaining, in expert panel discussions on live television shows, what was going on to vast audiences in their homes across the USA and abroad. If travel to the Moon was an ancient idea whose time had come, how fantastic it must have been to feel so close to the centre of this huge event. Even if we have had 30 or more years to sober down and drink in the cold water of the critics – and even some astronomers regard the Moon as "nothing but a confounded nuisance" – we would be soulless specimens to fail to sympathize on Kopal's "day of days."

Of course, this excitement was not confined to specialists. Man's journey to the Moon reflects urges that seem close to the core of being human: a drive to explore the unknown that I think almost anyone may recognize (cf. e.g. Whitman, 1900). Kopal's inclusion of Armstrong among his guide stars seems to highlight this sentiment in the psychological motivation for scientific research.

3. Light Curves and Information Analysis

Kopal's (1979) *Language of the Stars* speaks of stars in gravitationally bound systems, especially close pairs, as being like speakers engaged in expository conversation. If we want to find out about them, therefore, we have to 'listen,' and learn how to understand their language.

3.1. FOURIER ANALYSIS OF LIGHT CURVES

One of the most well-used tools to deal with signal deciphering is Fourier decomposition (e.g. Hancock, 1961). When there are well-sampled, periodic data sets, i.e. light curves, coefficients of a Fourier series representation have the important property of being mutually independent and efficiently able to characterize the data's information content. In the regular procedure, data sets are multiplied with relevant frequency components and the coefficients are found from corresponding integrals; although alternative and fast means to derive them, e.g. least squares, are not hard to find.

Close binary system models for the light variations involve the specification of at least 16 independent physical quantities that can be specified separately (cf. e.g. Budding, 1993). The effects these parameters have on the light curve shape are, however, sometimes quite similar to each other, and then, in a deterministic sense, they become interdependent. Budding (2003) argued, referring to Fourier coefficient determinacy, that a typical close system light curve would be insufficient, by itself, to furnish all such physical parameters, although one could often expect to find most, if not all, of the more basic geometric ones. If we seek to use Fourier series coefficients to find these parameters, unfortunately, in general, there seems no very direct method. Early numerical procedures were presented by Mauder (1966) and Kitamura (1967). Kopal (1979) reported that certain linear combinations of a few of the lower coefficients could be related to the geometric elements r_1, r_2, i; but only in the highly idealized circumstances of no photometric proximity effects. In general, incommensurability of eclipse and orbital timescales introduces higher frequency terms. But the greater closeness of the Fourier representation to the actual

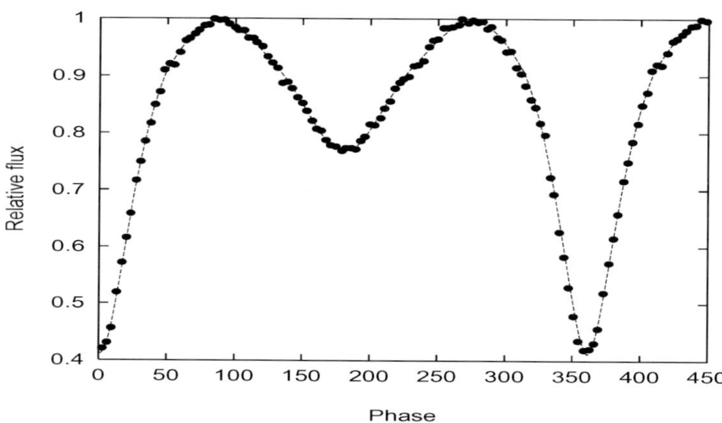

Figure 1. Erdem et al.'s (1993) *B* data of EG Cep and corresponding physical model.

data, which is free of pre-constrained thinking (and, of course, more objective than simply free-hand drawing), could just be telling us to loosen up our minds in approaching light curves. I will try to develop this point by reference to a few examples.

3.1.1. *EG Cep*

Photoelectric *BV* observations of the fairly typical short-period EB-type eclipsing binary EG Cep (HD 194089) were reported by Erdem et al. (1993). Figure 1 shows normalized points, in the *B* filter, and also a theoretical curve that optimally fit these data according to the *CURVEFIT* prescription, with the number of free parameters set sufficient for fitting determinacy (cf. Budding et al., 2004). *CURVEFIT* bases its figure approximation on Legendre polynomial harmonic series for the distorted components, taking into account their density distributions, in the manner originally developed by Kopal (1959).

EG Cep must be a very close system, but a model with a common photospheric envelope ('contact system') is not required. Erdem et al. (2004, in preparation) consider whether a detached pair of near Main Sequence (MS) stars, or the alternative Algol configuration, may give a better representation. We do not go into the details here, but Figure 1 shows a good fitting corresponding to the pair of near MS stars characterized in Table I. Note the relatively wide limits on the (interdependent) error calculation, obtained by inverting the determinacy Hessian around the solution optimum. These fittings permitted at most only 8 parameters to be determined.

Table II presents a Fourier series for the same data. Although the corresponding curve-fitting looks no different to the eye, the significant χ^2 decrease shows that the cosine series is more accurate than the physical model, suggesting still some shortcomings of the latter.

TABLE I

MS model for the EG Cep B light curve

Parameter	Value	Error
m_0	9.584	0.003
L_1	0.948	0.005
L_2	0.052	0.005
r_1	0.442	0.003
r_2	0.301	0.005
T_1	8500	
T_2	6800	
i	$85°8$	$0°2$
M_2/M_1	0.60	0.06
$\Delta\phi_0$	$-0°7$	$0°7$
Δl	0.006	
χ^2/ν	2.03	

TABLE II

Fourier cosine series for the B light curve of EG Cep

Coeff.	Value	Error	Coeff.	Value	Error	Coeff.	Value	Error
a_0	0.8476	0.0009	a_4	-0.0470	0.0013	a_8	-0.0060	0.0013
a_1	-0.0825	0.0013	a_5	-0.0242	0.0013	a_9	-0.0027	0.0013
a_2	-0.1810	0.0013	a_6	-0.0198	0.0013	a_{10}	-0.0007	0.0013
a_3	-0.0584	0.0013	a_7	-0.0089	0.0014	a_{11}	-0.0016	0.0013
$\Delta\phi_0$	$-0°87$		Δl	0.006		χ^2/ν	0.77	

3.1.2. *HN UMa*

Attention has been given to relatively little known systems, like this one, as part of a photometric programme at COMU's new Ulupinar Astrophysics Observatory to follow-up on binary star discoveries reported by Hipparcos. Preliminary information was given in a few IBVS articles appearing in the last year or so (e.g. Demircan et al., 2003). HN UMa was reported (Rucinski et al., 2003) as a low mass-ratio, high space-velocity, A-type contact binary, seen at low orbital inclination. Figure 2, coming from a CURVEFIT application, using a Roche model fitting function, shows that a fair fit to the mean variation of the light curve can be found from a system agreeing with Rucinski et al.'s picture, i.e. an inclination of about 54°. It is clear that the amplitude of this simple quasi-sinusoid, however, is only able to specify one parameter uniquely. Hence, since this amplitude can be matched by a large number of different combinations of system parameters, photometry, at best, can only support a given model. It cannot discriminate between models that differ in sufficient respects to be physically alternative, but are equivalent in determinacy.

E. BUDDING

TABLE III

Fourier series for the V light curve of HN UMa

Coeff.	Value	Error	Coeff.	Value	Error	Coeff.	Value	Error
a_0	0.9206	0.0009	a_4	0.0026	0.0013	b_1	-0.0111	0.0013
a_1	0.0116	0.0013	a_5	0.0016	0.0013	b_2	-0.0029	0.0012
a_2	-0.0556	0.0014	a_6	-0.0002	0.0013	b_3	-0.0034	0.0013
a_3	0.0032	0.0013				b_4	-0.0017	0.0013
$\Delta\phi_0$	0.0		Δl	0.007		χ^2/ν	1.01	

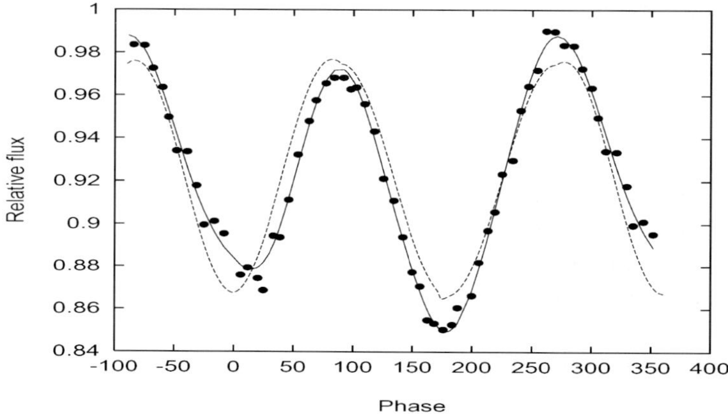

Figure 2. HN UMa light curve with contact model binary having inclination 54° and mass ratio 0.14. A Fourier representation is also shown.

We can also see that although the standard model will fit the mean light cycle, there is some additional variation that is not in the standard model. This type of additional effect could be discussed in terms of other concepts, such as hot spots, cool spots, gas streams or pulsational effects, but we do not go into that here. Instead, we just show, in Figure 2 and Table III, that a Fourier decomposition can accurately characterize the information in the data, independently of model preconceptions. There are here 11 degrees of freedom and 107 data points. Notice how the orthogonal terms spread the accuracy of the specification evenly, so the errors of all the sinusoidal terms are about a tenth of that of an individual point. The main effect of the tidal distortion is measured by the a_2 term, while the scale of the asymmetry is measured by the sine (b_i) terms. Since the Fourier decomposition adequately represents the data, in a χ^2 sense, it would be more unprejudiced to find that a model is consistent with these coefficients, than the coefficients define a particular model.

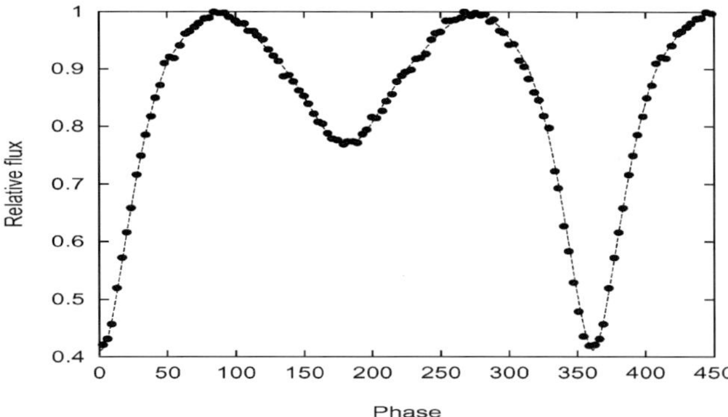

Figure 3. Fourier series representation for the short period RS CVn binary MR Del (Erdem et al., 2002).

3.1.3. *MR Del*

MR Del is another of the new eclipsing binaries discovered by Hipparcos (ESA, 1997) and followed up photometrically. It appears to be one of the 'short-period group' of close binary systems containing active cool stars and may be similar to CG Cyg (cf. Budding and Zeilik, 1987). These stars are of special interest in comparing the Sun with objects that can help us understand its basic properties. A preliminary analysis was given by Erdem et al. (2002).

A Fourier cosine series proved inadequate for fitting the data well at first, but after a small zero-phase displacement was noticed in the light curve, the much better result shown in Figure 3 was obtained. Again, we do not consider physical details, but mention two main points coming from this approach as (a) sensitivity to any small light curve asymmetry and (b) the possibility, suggested in Figure 5, of some additional oscillatory variation in the data. We do not claim that this is established in the present case, but only that Fourier decomposition can provide a useful indication for such behavior if present.

3.1.4. *δ Lib*

This well-known, relatively near Algol system will be discussed in detail in another paper at this meeting (Budding et al., 2005), where optimal fits of a conventional model to various waveband light curves are presented. Again the Fourier series fit, here to the *K* light curve of Lazaro et al. (2002), proves a slightly more accurate rendering of the data than the standard model, and we show the Fourier representation by itself in Figure 4. The depressions on both sides of the out-of-eclipse regions about the primary minimum, which are not as obvious in the raw data, are thought-provoking regarding possible shortcomings of the standard model. There also appear to be similar concave regions around the secondary minimum. Such small-scale departures of real from model light curves are, of course, often pointed out in observational papers.

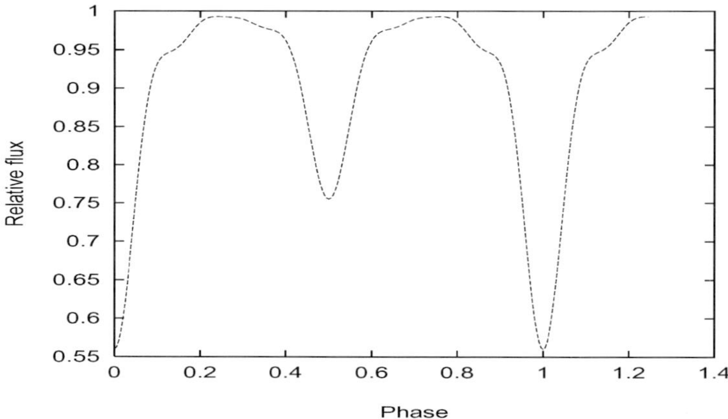

Figure 4. Fourier cosine series fitting for Lazaro et al.'s (2002) K light curve of δ Lib.

3.1.5. *Continuous Transform Operations*

The papers that Kopal published about Fourier analysis of light curves after 1975 (cf. Kopal, 1990) were supported by two other considerations, thus: (i) proximity and eclipse effects might be more readily separated in the frequency than the time domain and (ii) observational noise could also be effectively filtered out as an essentially high frequency component ($\sim 1/\Delta\phi$, where $\Delta\phi$ is the mean phase spacing of the data points). The latter point is clear from the foregoing examples. The first point reflects the Russell-Kopal approach to light curve analysis, i.e. as a royal road to stars *in general*. From this point of view (which differs from modern synthetic approaches), proximity effects are an additional complication, peculiar to the binarity, which must be removed in order to generalize results. The separation in frequency between the lower harmonics of the orbit and the eclipse is not that great, however, so we could expect significant spectral overlap effects in practice. These may become more analyzable from considering continuous, rather than series, Fourier representations.

The real part of the continuous Fourier transform of the same K light curve of δ Lib is shown in Figure 5. This form is reminiscent of the undulatory functions that Kopal became interested in whilst transforming of light losses due to (spherical) eclipses. It has a resemblance to the transform of an inverted triangle function, obtained by subtracting triangle from rectangle transforms ($p[\mathrm{sync}(\omega p/2) - \frac{1}{2}\mathrm{sync}^2(\omega p/4)]$, with period p, angular frequency ω). This function, with its deep minimum at double the orbital frequency, must be showing up the main 'ellipticity' proximity effect. 'Reflection' must do something similar at orbital frequency, while eclipses, viewed as short, deep inverted triangles, will have comparable effects at higher frequencies. This effect of the eclipse explains the lowered maxima at lower frequencies in Figure 5, before they taper down to the expectable sync function form at higher ω.

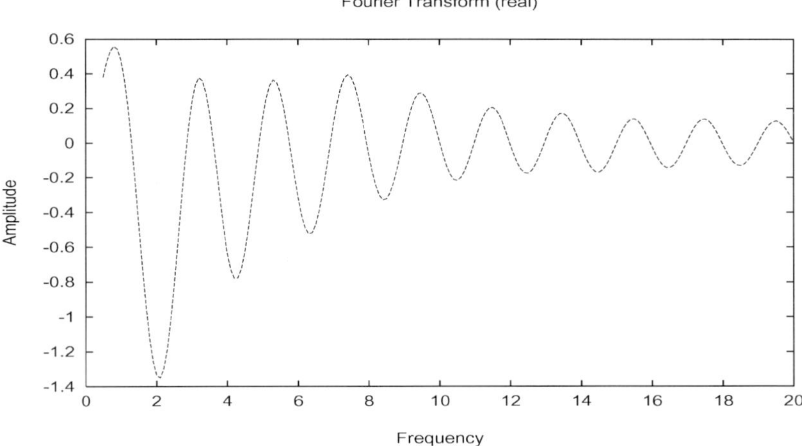

Figure 5. Real component of the Fourier transform of the K light curve of δ Lib.

Using the convolution theorem, Kopal (1979) derived the eclipse transform, for the spherical problem, as a product involving a pair of Bessel functions. The equivalent one-dimensional problem, suggested here, would yield a comparable product of two sync functions, whose arguments, factored by the contact phases, will have their first zeros at correspondingly higher frequencies. Other speakers at this meeting will discuss, in detail, practical procedures for dealing with both the subtraction of the proximity effects and the analysis of the eclipse residues in the frequency domain.

4. Concluding Remarks

This paper has two parts, linked by recollections of Prof. Zdeněk Kopal, his aims and interests. Some main influences from his autobiography were introduced by references to another book by Arthur Koestler. Incidentally, remarks that I read about Koestler have unexpected parallels with comments one might hear about Kopal himself: – *"His work was sometimes out of step with the mainstream. He did not just arrive at different answers to common questions, he tried to bring attention to questions that others were not asking, thus signaling his originality."*
Kopal was not always so intent on bringing attention to questions that others were not asking: his autobiography makes clear that he picked up the royal road from Russell. Having seen its importance and started on its course, he then developed his own methods of travel. And there were, of course, other guide stars apart from Russell. We can see, in these influences, conventionally admirable human qualities. But there was yet a certain restlessness in him, an inclination to break away from the conventional, at times. This may have promoted his interest in analyzing photometric data through Fourier decomposition.

Of course, the choice of how one parameterizes a light curve depends on which type of model one is addressing. Mention has been made of 'standard models,' but, in detail, model testing would be expected to relate to the latest advances of plausible theory. Wilson and Devinney's (1971) numerical integration of the Roche model was an important milestone in such work. With increasing component proximity towards and beyond contact with surfaces of limiting stability the versatility of numerical integration becomes more persuasive, compared with literal expansions, especially in an era of rapidly increasing computational power. It is worth recalling the issue of determinacy, however. This is particularly relevant to just such highly distorted systems, because it is for their quasi-sinusoidal light curves that determinacy must decline, perhaps to only 3 or 4 well-determined parameters. Such light curves can, at best, only corroborate predictions of more detailed models: they cannot provide critical disproof. It is important to keep this in mind against the theoretical complexity of W UMa-type systems. Determinacy is best examined by reference to the parameter error matrix. It should correspond, geometrically, to a closed ellipsoid to ensure formal determinacy of the underlying model. Fast methods of analysis are advantageous for such examination.

In a moment of candor, in our old joint office, Kopal once remarked that he felt the frequency domain approach, that he realized was not getting much attention outside of Manchester, was for a future generation. Whether that generation has arrived yet I do not know. Nor do I know how much I may have misrepresented him or his intentions in this summary. I would beg some accommodation about this point, however. One thing that, in my years at close quarters with Kopal, I could never quite make out were the limits of his sense of humor. A favorite joke of his was about the secret policeman who convicted the Good Soldier Schweik on the grounds that if he had had a camera he could have used it. And if I had had better understanding, I too could have used it here. But I do recall that Prof. was often oddly tickled by misunderstandings.

References

Batten, A.: 2005, *Ap&SS* **296**, 3.

Budding, E.: 1993, *Introduction to Astronomical Photometry*, Cambridge University Press, Cambridge.

Budding, E.: 2003, Publ. Canakkale Onsekiz Mart Univ. *Astroph. Res. Centre* **3**, 3.

Budding, E., Bakis, V., Eredem, A., Demircan, O., Iliev, L., Iliev, I. and Slee, O.B.: 2005, *Ap&SS* **296**, 131.

Budding, E., Zeilik, M.: 1987, *ApJ* **319**, 827.

Budding, E., Rhodes, M., Priestley, J. and Zeilik, M.: 2005, *Astron. Nachr.* **325**, 433.

Demircan, O. and Budding, E.: 2003, Publ. Canakkale Onsekiz Mart University Astrophys. Res. Centre **3**.

Demircan, O. et al.: 2003, *IBVS* 5364.

Eddington, A.S.: 1923, *The Mathematical Theory of Relativity*, Cambridge University Press, Cambridge.

Erdem, A. et al.: 1993, *IBVS* 3915.

Erdem, A. et al.: 2002, *RASNZ Photom. Section Communique*, **2**(6), 6, New Zealand.

ESA: 1997, *Tycho Catalogue* (ESA SP-1200).

Hancock, J.C.: 1961, *An Introduction to the Principles of Communication Theory*, McGraw-Hill.

Kitamura, M.: 1967, *Tables of the characteristic functions of the eclipse and the related delta-functions for solution of light curves of eclipsing binary systems*, Tokyo: University of Tokyo Press.

Kholopov, P.N.: 1959, *General Catalogue of Variable Stars*, Nauk Publishing House, Moscow.

Koestler, A.: 1981, *The Sleepwalkers*, Penguin.

Kopal, Z.: 1959, *Close Binary Systems*, Chapman & Hall, London & New York.

Kopal, Z.: 1979, *The Language of the Stars*, Dordrecht.

Kopal, Z.: 1986, *Of Stars and Men*, Adam Hilger.

Kopal, Z.: 1990, *Mathematical Theory of Stellar Eclipses*, Dordrecht.

Lazaro, C., Arevalo, M.J. and Claret, A.: 2002, *MNRAS* **334**, 542.

Mauder, H.: 1966, *Die Berechnung der Elemente von Bedeckungsveränderlichen durch Fourier-Transformation der Gesamtlichtkurve*, Veröff. Sternwarte Bamberg.

Prager, R.: 1935, *Popular Astron.* **43**, 198.

Rucinski, S.M. et al.: 2003, *AJ* **125**, 3258.

Russell, H.N.: 1939, *ApJ* **90**, 641.

Russell, H.N.: 1948, *Centennial Symposia*, C. Payne-Gaposchkin (ed.), Harvard University Press, Cambridge, Massachusetts.

Urey, H.: 1952, *The Planets, their Origin and Development*, New Haven, CT: Yale University Press.

Whitman, W.: 1900, *Darest Thou*, in *Leaves of Grass*, David McKay, Philadelphia.

Wilson, R.E. and Devinney, E.J.: 1971, *ApJ* **166**, 605.

REMINISCENCES OF A JAPANESE CONTEMPORARY

MASATOSHI KITAMURA

National Astronomical Observatory, Mitaka, 181-8588 Tokyo, Japan;
Fax: +81-422-343690

(accepted April 2004)

Abstract. It was in 1936 when a young Czech student of age 22 came to Japan through Siberia for participating in the solar eclipse expedition. Since then, he had visited us 13 times until 1993 (when he passed away). At each visit his lectures were vivid for us and had strong impact on the audience. Needless to say, it was the late Professor Zdeněk Kopal. In this brief presentation first I reveal my reminiscences of our common time in Manchester and Japan as a contemporary, and second I make some remarks on my activities for the Japanese Official Development Assistance (ODA) for astronomy in developing countries, in a similar way like Professor Kopal contributed in later years to the development of astronomy mainly in middle-east countries.

1. Professor Kopal and Japan

His first contact with Japan was in 1936 when he came to Hokkaido, in the northern part of Japan, for observation of the solar eclipse. He was a young man of age 22, being still a student at Charles University, Prague, Czechoslovakia. For his observations, he brought a 21 cm objective telescope from Prague in his suitcase. The trip took him a week or so with the Trans-Siberian Express (the average speed at that time was about 40 km/h, but it was a comfortable train for a long trip, according to what he wrote in "Of Stars and Men" (Kopal, 1986); subsequently one more week of travel time was necessary to proceed to Hokkaido. His task for the eclipse observation was to photograph the solar corona in white light with his refractor of f/16 focal ratio, with exposure times of 15, 30 and 60 s, respectively. With the help of the Kyoto (ancient capital of Japan) University party, he could obtain three most beautiful photographs of the solar corona successfully.

After the eclipse, he stayed in Japan for another 3 months and traveled a lot within the country, visiting many interesting places. Everything in Japan must have excited his young curiosity. Particularly he was interested in Japanese nature, culture, architecture, castles, gardens and so on. As those were so different from Europe, he took the opportunity to visit many places of his interest. He also learned many Japanese words, and eventually his knowledge of the Japanese language attained such a level that he had no difficulties in basic communication with ordinary Japanese people. I still remember that when he first came to Japan after the World War II and we took dinner together at some Tokyo restaurant he ordered in Japanese

Astrophysics and Space Science **296**: 33–41, 2005.
© Springer 2005

and asked the waitress for "mizu tsumetai" (cold water), "sakana tabetai" (want to eat fish), "sake kudasai" (give me sake wine), etc. Since 1936, he had visited Japan 13 times for various scientific work. We do not know any other Western astronomer who has visited such an isolated Asian country so many times. Figure 1 shows the Himeji castle in the cherry-blossom season in spring which he loved so much. Figure 2 is a picture of the graceful Mt. Fuji, an extinct volcano. With an altitude of 3778 m above sea level, it is the highest one in Japan. During his stay in 1936, young

Figure 1. Himeji castle in spring cherry-blossom season, which Prof. Kopal loved so much.

Figure 2. Mt. Fuji, symbol of Japan, which young Kopal-san climbed up to the summit in 1936.

Kopal-san climbed up to the summit of this mountain. As far as I know, he was the second Western astronomer who had climbed up Mt. Fuji. By the way, the first one was Prof. David Teck Todd of Amhurst College of the United States in 1887, according to the literature (Toshio Sato, 1993). During my stay in Manchester, I was often invited by Prof. Kopal to his home. On such occasions, he showed us photographs taken by himself when he had climbed up this mountain, and talked about his strong impression when he saw the rising sun from the summit. Whenever he came to Japan after the World War II, he seemed to feel strong nostalgia when enjoying the distant view of this beautiful mountain from Tokyo.

2. My First Manchester Time (1960–1962)

The Astronomy Department of the University of Manchester, England, was inaugurated in 1951, and Prof. Kopal came to Manchester as the first astronomy professor according to a recommendation by Prof. S. Chandrasekhar. Before that time, he had been working in the United States and was already well-known for his activities in the field of close binary stars, one of the important fields of modern astronomy. The chance for me came from his recognition of my theoretical papers on the reflection effect (Kitamura, 1954), which lead to similar results like Prof. Kopal's different mathematical expansion approach on the same problem (Kopal, 1954). In August 1960, I was invited by him to take a post-doctoral position at his department.

So far, the knowledge of fundamental astrophysical quantities of stars as their masses, radii, absolute dimensions, and surface temperatures have been best deduced from observations of eclipsing binary stars and analyses of the obtained light curves. Thus, eclipsing binaries have been our principal source of information about the above-mentioned physical quantities. However, the practical methods to analyze the light curves were very complicated at that time, because it was not *a priori* known which one of the two consecutive minima was due to an occultation or a transit[1] eclipse and had in general to be ascertained by trial and error, but also that the method of analysis for the determination of the elements was not the same for partial, total and annular eclipses. Needless to say, no large electronic computer was available at any institute at that time, and the now well-known fine computer code by Wilson and Devinney (1971) did not exist. At that time Prof. Kopal first suggested me to explore any possibility of solving the light curve by a Fourier-transform technique.

In the beginning, I started to try to separate the proximity effect as well as the effects of tidal and rotational distortion and mutual illumination from the eclipse effect in the Fourier domain. In the course of investigation, I found that the distortion effect could not be separated from the eclipse by ordinary Fourier transform, because both effects are mutually related in rather complicated ways apart from the additive

---

[1]If a minimum arises from an eclipse of the smaller star by the larger one we call it an occultation, and if the smaller star is in front, we speak of a transit.

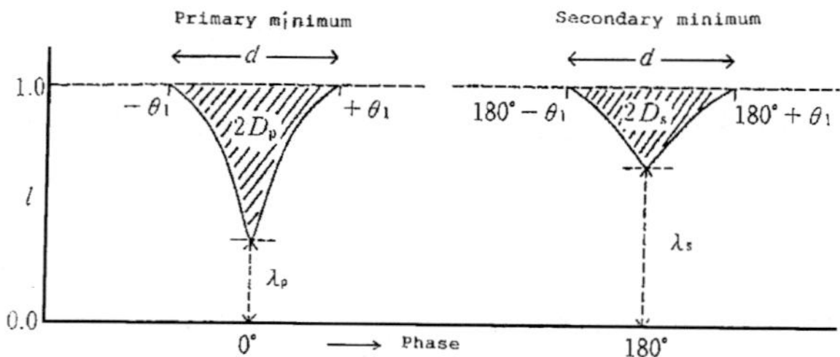

Figure 3. Light curves, in intensity scale, of the primary and secondary minima of an eclipsing binary.

effect, while the reflection effect could be easily separated from the eclipse, because both simply add up to form the light curve. Thus, in the course of deep consideration on what is the best and most reasonable application of the Fourier technique, an idea of "incomplete" Fourier transformation flashed across my mind, like an inspiration (Kitamura, 1965).

This method can provide a number of practical advantages for the analysis, together with the use of the Tables of the Characteristic Functions of the Eclipse (Kitamura, 1967). One of the important merits compared with any other previous method at that time would be that we could know prior to the analysis of the elements, which one of the two consecutive minima is due to an occultation or a transit. In Figure 3, I show two consecutive eclipsing minima in intensity scale schematically. By comparing two quantities:

$$\frac{D}{(1 - \lambda)_{\text{prim}}} \quad \text{and} \quad \frac{D}{(1 - \lambda)_{\text{sec}}}, \tag{1}$$

we can immediately determine, which minimum is due to an occultation and the other due to a transit, because it can be theoretically proved that

$$\frac{D}{1 - \lambda} \quad \text{for transit} \leq \frac{D}{1 - \lambda} \quad \text{for occultation.} \tag{2}$$

It may be mentioned that the theoretical discovery of the above theorem is one of the important products of the incomplete Fourier transform.

In the latter half of my Manchester time, I was engaged, together with Prof. Kopal, in mathematical work on the second-order distortion theory, correctly to squares of the superficial distortion in close binary systems for application to their light and radial velocity changes. The results, which were published as a joint work by Kopal and Kitamura (1968), are practically applicable and correct to r^8 (r: fractional radius of either component).

3. My Second Manchester Time (1968–1970)

At the request of Prof. Kopal I again came to Manchester. The purpose of this visit was to solve the topological problem of the Roche coordinates, which was one of the important problems not only for close binary research but also for celestial dynamics. The old Mercury computer had been replaced by a bigger Atlas computer, even though still far smaller than present supercomputers. The naming of the Roche coordinates was due to Prof. Kopal (1969). In the xy-plane the Roche coordinates are constituted of zero-velocity curves, and the other orthogonal curvelinear ones.

In treating many problems associated to close binaries, Kopal pointed out the importance of the use of these coordinates. As an example, I may show, following Kopal, that the use of these coordinates, instead of the ordinary xy-coordinates, can make the mathematical treatment much simpler. Let us consider the flow equation of gaseous streams around close binary systems described by the well-known Eulerian equation of motion for an inviscid fluid flow:

$$\frac{\partial u}{\partial t} + \frac{1}{\rho} \cdot \text{grad} \quad P = K, \tag{3}$$

where u denotes the velocity of the fluid at any point, P the pressure, ρ its density and K the external force acting upon any element of the fluid. If we consider the case of circular orbits of the component stars and also confine our attention to steady flows in the orbital plane, it is found that the time derivative vanishes in Eq. (3); in Roche coordinates (ξ, η) this equation is simply reduced to

$$K_\xi = \frac{1}{1+q}, \qquad K_\eta = 0, \tag{4}$$

respectively.

Even so, being different from the zero-velocity curves $\xi = \text{const}$, the geometrical features of $\eta = \text{const}$ curves were not disclosed yet.

Adopting our units of length, mass and time as R (separation between the components revolving in circular orbit), $m_1 + m_2$ and ω^{-1}, the zero-velocity curve is given by

$$\xi(x, y) = \frac{1}{r_1} + q\left(\frac{1}{r_2} - x\right) + \frac{1}{2}(1+q)(x^2 + y^2). \tag{5}$$

Here ξ is a dimensionless normalized quantity, and so the orthogonal curve $\eta = \text{const}$ should be given by the differential equation (Kitamura, 1970):

$$\xi_x \eta_x + \xi_y \eta_y = 0. \tag{6}$$

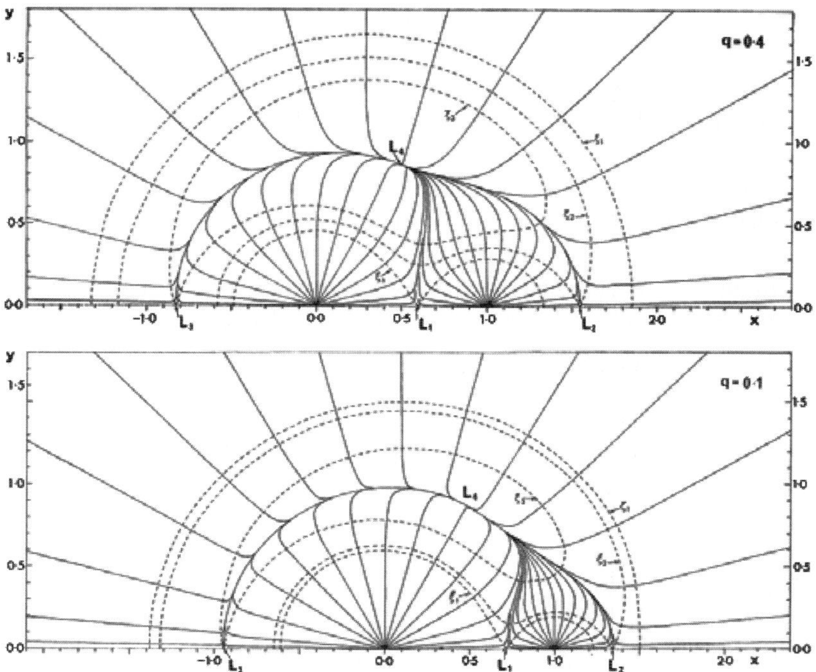

Figure 4. The geometry of one of the Roche coordinates $\eta(x, y) = \text{const}$ in the xy-plane for $y \geq 0$, corresponding to mass ratios $q = 0.4$ and 0.1.

The curve $\eta = \text{const}$ could not be given analytically and can only be generated by solving Eq. (6) numerically.

As examples of the solution, Figure 4 shows the resulting $\eta = \text{const}$ curves for the cases of mass ratios $q = 0.4$ and 0.1. It was disclosed that the resulting orthogonal curves start from the origin (gravitational centre of each component) and converge at the Lagrangian triangular points L_4 and L_5.

As is easily recognized from the results of numerical integrations for the curves $\eta = \text{const}$ in the xy-plane, there should exist some envelopes of the curves $\eta = \text{const}$ whose form should depend on the mass ratio only. Figure 4 reveals that these envelopes pass through the Lagrangian colinear points L_1, L_2, L_3 perpendicular to the x-axis and assemble at the Lagrangian equilateral triangle points $L_{4,5}$. As the envelopes for the curves $\eta = \text{const}$ may be considered as limiting cases of the $\eta = \text{const}$ curves, they should also be governed by the same differential equation as used for the curves $\eta = \text{const}$. However, the initial conditions for numerical integration of the envelopes should be different from those used for the $\eta = \text{const}$ curves, because any envelope under consideration passes through one of the Lagrangian colinear points L_1, L_2, and L_3 perpendicular to the x-axis.

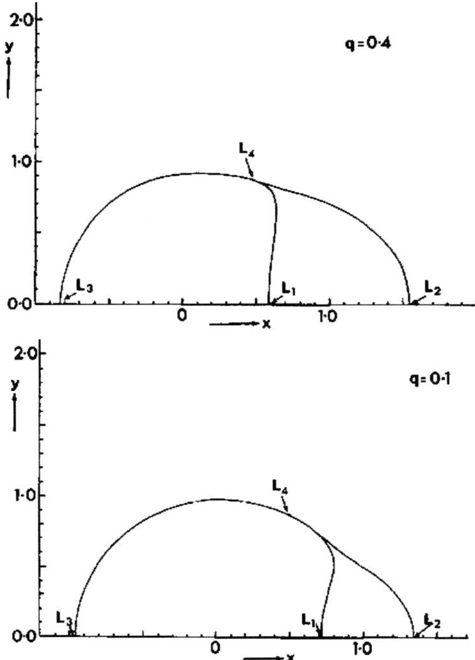

Figure 5. The theoretical envelopes for the curves $\eta(x,y) = $ const in the xy-plane for $y \geq 0$, corresponding to the mass ratios $q = 0.4$ and 0.1. They connect all the Lagrangian singular points.

Figure 6. Another form of Close Binary Systems in a Japanese Garden near Kyoto in 1984.

 M. KITAMURA

TABLE I

Equipment donated by the Japanese Government (ODA) from 1986 to 2003

Year	Reflector size	Receiving institutions/location	Country
Reflecting telescopes and accessories donated			
1987	40 cm	Science Centre	Singapore
1988	45 cm	Bosscha Observatory, Institute of Technology, Bandung	Indonesia
1989	45 cm	Chulalongkorn University, Bangkok	Thailand
1995	45 cm	Arthur C. Clarke, Institute for Modern Technologies, near Colombo	Sri Lanka
1999	45 cm	Asuncin National University	Paraguay
2000	45 cm	PAGASA, Quenzon City, near Manila	Philippines
2001	45 cm	Cerro Caln Astronomical Observatory, University of Chile	Chile
		Planetarium/location	
Planetaria and equipment donated			
1986		Pagoda Cultural Centre, Yangon	Myanmar
1989		Haya Cultural Centre, Amman	Jordan
1989		Space Science Educational Centre, Kuala Lumpur	Malaysia
1990		Auxiliary projectors for the planetarium, Manila	Philippines
1993		Burdwan University, West Bengal	India
1993		Auxiliary projectors for the Planetario de la Ciudad, Buenos Aires	Argentina
1994		Auxiliary projectors for the Planetario de la Ciudad, Montevideo	Uruguay
1998		Ho-Chi Minh Memorial Culture Hall, Vinh City	Viet Nam
1998		Auxiliary projectors for the planetarium, Bangkok	Thailand
1998		Auxiliary projectors for the planetarium	Sri Lanka
1999		Anna Science Centre, Chennai	India
2000		City Park, Tashkent	Uzbekistan
2001		Asuncin National University	Paraguay
2002		Planetario Municipal	Ecuador
2002		Children Museum, San Pedro Sula	Honduras
2002		Plaza de la Cultura, Santo Domigo	Dominican Republic
2003		National Costa Rica University, San Jose	Costa Rica
2004		Laboratorio Central del Instituto Geofisico, Lima	Peru

So, actual numerical integrations for the envelopes in the xy-plane may be carried out from the positions of the Lagrangian colinear points with the same differential Eq. (6). However, the initial conditions for the present case should be taken such that

$$\frac{dx}{dy} = 0 \quad \text{at } x = x_i \quad \text{and } y = 0, \tag{7}$$

where x_i ($i = 1, 2, 3$) stands for the x-coordinate of the respective Lagrangian colinear point. Upon integration, the resulting envelope is found to bend gradually towards the Lagrangian equilateral triangular point L_4 (or L_5), preventing an increase of the independent variable y. Figure 5 shows the results of integrations for the envelopes of $\eta = \text{const}$ curves for the cases of mass ratio $q = 0.4$ and 0.1 as examples. It is found that all Lagrangian singular points are beautifully connected by the envelope curves to $\eta = \text{const}$ ones.

4. Japanese ODA Programme for Astronomy

It is well-known that in his later years, Prof. Kopal contributed much to the development of astronomy, mainly in middle-east countries. Similarly, we have acted for astronomy in developing countries by donating the necessary equipment through the framework of the Japanese Official Development Assistance (ODA). So far, 24 donations of astronomical equipment was made to 19 developing countries untill the end of the Japanese fiscal year 2003. The instruments donated included university-level reflecting telescopes as well as modern planetaria used for educational purposes, together with various accessories. Table I lists all the donated equipment (see Kitamura (2004a,b) for more detail).

Acknowledgements

I am grateful to Prof. H. Drechsel, SOC chairman, for his kind invitation to this memorable and important conference and to Czech LOC people for their various warm help.

References

Kopal, Z.: 1954, *MNRAS* **114**, 101.
Kitamura, M.: 1954, *Publ. Astron. Soc. Japan* **5**, 114 (Part I) and **6**, 217 (Part II).
Kitamura, M.: 1965, *Adv. Astron. Astrophys.* **3**, 27.
Kitamura, M.: 1967, Tables of the Characteristic Functions of the Eclipse, University Tokyo Press.
Kopal, Z. and Kitamura, M.: 1968, *Adv. Astron. Astrophys.* **6**, 125.
Kopal, Z.: 1969, *Ap&SS* **5**, 360.
Kitamura, M.: 1970, *Ap&SS* **7**, 272.
Wilson, R.E. and Devinney, E.J.: 1971, *ApJ* **166**, 605.
Kitamura, M.: 1981, *Ap&SS Library* **91**, 241.
Kopal, Z.: 1986, Of Stars and Men, Adam Hilger, Bristol and Boston.
Kitamura, M.: 2004a, in: H. Haubold (ed.), Seminars of the United Nations Programme on Space Applications, United Nations, New York, pp. 21.
Kitamura, M.: 2004b, *Space Policy* **20**, 31.

RR CrB AND ALGOL – THE FIRST AND LAST STAR IN ZDENĚK KOPAL'S LIFE WITH BINARIES

MARTIN ŠOLC

*Astronomical Institute, Faculty of Mathematics and Physics, Charles University of Prague,
V Holešovičkách 2, CZ-180 00 Praha 8, Czech Republic; E-mail: martin.solc@mff.cuni.cz*

(accepted April 2004)

Abstract. Several important turning points appeared in the personal life and scientific career of Zdeněk Kopal, every time, when he entered a new field of research. Four particular periods of his life will be commented in the present paper: the studies in Prague and early observations in the variable star programme of the Czech Astronomical Society; the first lectures on close binaries at Harvard; the intermezzo of lunar research; and the last period after 1989, when Zdeněk Kopal returned several times to Prague University.

Keywords: variables, eclipsing binaries, lunar cartography, Roche models

When reviewing the rich life of Zdeněk Kopal (1914–1993), one can consult his memoirs *Of Stars and Men* (Kopal, 1986). After school years in Litomyšl and Prague, he enrolled astronomy at Charles University of Prague (1933–1937). The Ernest Denis Fellowship of the Czechoslovak Government allowed him to spend the following academic year at Cambridge with A.S. Eddington. In deciding between the two next offers of fellowships at Lick or Harvard Observatories, the lower travel costs to the East Coast of the USA played the major role. Thus the outburst of World War II reached Kopal at Harvard, working with H. Shapley on studies of eclipsing variables. The feeling of necessity to help led Kopal to MIT where he became involved in the computational work on firing tables. He also acted shortly in the American broadcasting to the occupied Bohemia. After several years spent at Cambridge, MA, and MIT after the War, Kopal followed a recommendation of E. Freundlich and accepted the chair of astronomy at Manchester University in 1952.

Few years later, Harold Urey turned his attention to the beginning of space exploration of the Moon. Kopal started an own programme at the University of Manchester, concentrating on photographic lunar cartography. Sharing his time between teaching duties at different universities throughout the world and the work of a consultant in the lunar programme of NASA, Kopal became more and more involved in space research. This engagement partially diminished parallel with the end of the Apollo programme, and Kopal returned back to the studies of close binaries. The retirement in 1981 allowed him to devote even more time to this topic, which meanwhile took new directions. Zdeněk Kopal wrote, e.g., a basic monograph on the Roche problem, but never fully accepted the idea of mass transfer between the components of close binaries.

Astrophysics and Space Science **296:** 43–54, 2005.
© Springer 2005

The present paper will reveal some details of the history of life of Zdeněk Kopal during the four periods mentioned in the abstract.

1. Variable Star Observer

The first and virtually the most important turning point in the life of Zdeněk Kopal was the starting point, the "kick off" toward the career of a professional astronomer, an event which he commented in his memoirs: "I can specify the day and (almost) the hour when I received the call to become an astronomer, although I cannot recall a particular interest in the subject prior to that fateful 'moment of truth' which decided all my subsequent life."

The young boy was generally interested in natural sciences, he collected beetles and fossils in localities around Prague and consulted the identifications in the National Museum of Prague. On July 31, 1928, when walking through the town and returning home from the Museum, he made a stop at the beginning of the bridge across the Vltava river. At this place, from which there is a nice panoramic view on the Prague Castle, a man offered him to look through his portable telescope, directed this time, however, to the Sun. The curious and inquisitive student could not withstand, and thus he started to observe the sunspots, probably for the first time in his life, full of devotion and astonishment. Immediately after returning home he started to collect and read popular literature on astronomy. During the fall of 1928 he became a frequent visitor of the Štefánik Observatory on Petřín Hill, subscribed the journal Říše hvězd (The Realm of Stars), joined a group of variable star observers and entered the Czech Astronomical Society (CAS).

The 16 years old industrious student soon became the leader of the group and under the patronage of Vladimír Vand and Josef Klepešta began to collect, evaluate and publish the observed data. His great activity is clearly seen from the table summarising the number of observations in 1929–1931 by members of the group for variable stars of which he was the Chairman (Figure 1). His first scientific paper on the variability of RR Coronae Borealis and U Delphini appeared in the journal *Astronomische Nachrichten* in 1932 (Figure 2). The next observations were published in this journal as well, the results appeared however also in publications of the CAS. Besides his observations which concentrated at that time mainly on irregular, semiregular and long-period variables, but only marginally on eclipsing binaries, Kopal prepared charts for the *Atlas d'Etoiles Variables* together with V. Vand and wrote a book on variable stars (in Czech) together with František Kadavý. He also contacted the *American Association of Variable Star Observers* (*AAVSO*) and the *Association Française d'Observateurs d'Étoiles Variables* and sent the observations to the centres of these organisations (Harvard and Lyon Observatories).

Among about 20 papers from his student years listed in ADS, some papers were published in *Monthly Notices*. They deal not only with observations, but there

Pozorovatel:	Místo pozorovací:	Počet pozorování:			
		1929	1930	1931	Celkem
Z. Balík,	Chrudim	50	72	—	122
A. Bláha,	Praha	—	123	130	253
V. Černov,	Krěměnčug, SSSR .	102	332	313	747
K. Goňa,	Praha	—	27	—	27
M. Hylmar,	Ml. Boleslav . . .	—	53	—	53
V. Izera,	Praha	174	40	—	214
F. Kadavý,	Praha	280	2110	4263	6653
Z. Kopal,	Praha, Řmenín . .	553	4465	4817	9835
J. Kraft,	Praha	—	366	—	366
V. Litvan,	Písek	51	—	—	51
B. Macháčková,	Brandýs n. Labem	—	—	291	291
M. Matoušek,	Praha	—	220	1503	1723
V. Nováková,	Praha	24	3	—	27
A. Polanová,	Praha	24	165	—	189
R. Rajchl,	Praha	432	—	—	432
M. Stelčovský,	Praha	—	64	—	64
V. Šedý,	Bohdaneč	114	70	—	184
V. Vand,	Praha	—	124	416	540
		1804	8234	11733	21771

V roce 1931 bylo vykonáno tedy více pozorování, než v letech minulých dohromady. Hlavní částí programu sekce v r. 1931 byla stále pozorování vizuelní, neboť fotografické jsou teprve v začátcích. Mezi neuve...

Figure 1. Summary of variable star observations in 1929–1931 by members of the Czech Astronomical Society (from Kopal, 1932a).

are as well pure theoretical papers on the stability and shape of rotating bodies, on thermodynamics, the two-body problem and the internal structure of close binaries, and on apsidal motion in close pairs. Besides those publications the name of Zdeněk Kopal often appeared in *Říše hvězd* (the volumes of 1932–1938 contain almost two dozens of popular articles on very different topics reaching from cosmology to the origin of life). In the journal *Nature* appeared a common paper with Dr. H. Slouka about the axial rotation of globular star clusters (Slouka and Kopal, 1936).

Soon after Kopal started to study at the University in 1933, V. Vand became the assistant at the Astronomical Institute in 1934. The education lay on the shoulders of Professors František Nušl (positional astronomy) and Vladimír Heinrich, who was the Head of the Institute and taught celestial mechanics, astrophysics and also photometry. The associate Professor Vincenc Nechvíle was responsible for courses on optics and celestial mechanics as well, and the assistants, e.g. Vand and Bohumil Šternberk (who became the director of the Astronomical Institute of the Czechoslovak Academy of Sciences three decades later), had duties in practical work in astronomy.

Thus, Kopal (and to some extent also the other students) had the choice to carry out own observing programmes with telescopes (Figure 3) either at the Štefánik Observatory of CAS (Figure 4) or at the University Observatory (Figure 5). Kopal exclusively preferred the Observatory of the Astronomical Society because of the friendly atmosphere there. The fact that Professor Heinrich had no empathy and

M. ŠOLC

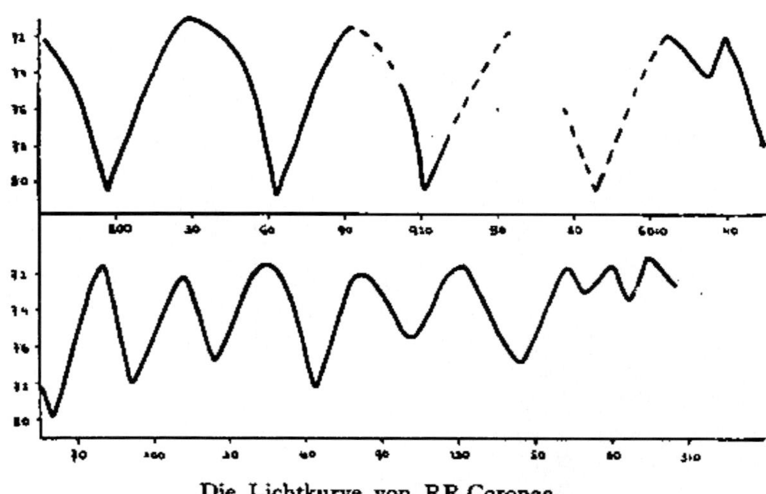

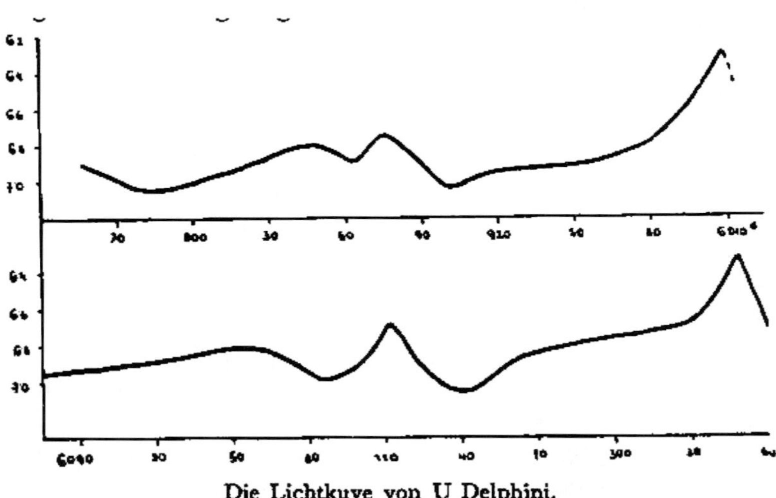

Figure 2. Light curves of RR CrB and U Del (from Kopal, 1932b, pp. 243–248).

understanding for other people – not for young students, but also not for his colleagues – actually prevented an effective operation of the well equipped observatory. Professor Heinrich lost his leading position in 1934 due to some never-cleared events, ranging from bad jokes to damaging of the equipment. On the other side, he continued his teaching and even though he was mainly a theoretician, his lectures covered a broad spectrum from astrophysical spectroscopy to cosmology and general theory of relativity.

Erwin Freundlich spent only a short time in Prague, since he tried to escape from the Nazi regime first by going from Germany to Istanbul in 1933, and later to Prague,

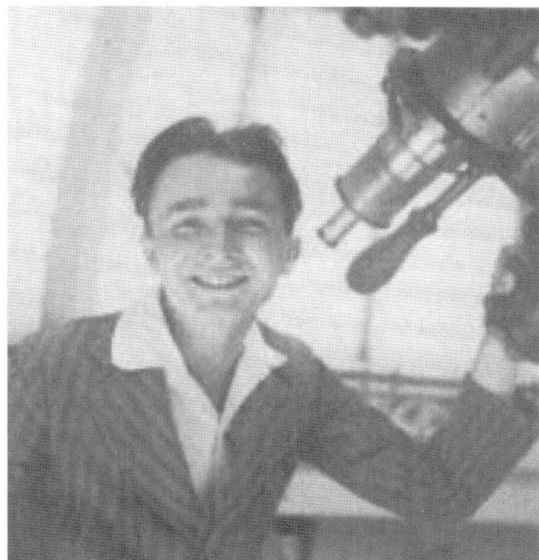 

Figure 3. Left (a): Zdeněk Kopal at the largest telescope of Stefanik Observatory on top of the Petřín Hill in Prague (from Kopal, 1986, p. 62). Right (b): The double refractor of the University Observatory in Švédská Str., on the south slope of the Petřín Hill, in 1933. (Photo from author's archive.)

Figure 4. Observatory of the Czech Astronomical Society on Petřín Hill in Prague, in about 1935. (Photo from author's archive.)

where he arrived in 1936. He lectured at the Astronomical Institute of the German University of Prague and at the Czech Technical University. In the archives of the Czech Astronomical Institute there was an autograph by Professor Freundlich, in which he suggested as a common project a modern astrophysical

Figure 5. University Observatory in Švédská Street, in about 1930. (Photo author's archive.)

Figure 6. Left: Václav Vladimír Heinrich (1884–1965). Right: Erwin Finlay Freundlich (1885–1964). (Photo from author's archive.)

observatory for the Technical and Charles University, situated outside of Prague. Freundlich had experience with building institutes: he succeeded to found the new Institutes in Berlin-Telegraphenberg (with the Einstein Tower) and in Istanbul. Later, when Zdeněk Kopal was appointed Professor at Manchester and Erwin Freundlich at St. Andrew at almost the same time, the friendship of both became very deep (Figure 6).

2. First Lectures on Close Binaries

The next turning points in Kopal's life came in the years 1947–1948, when he hoped to return to Prague and get a chair of astronomy at Charles University. However, things cumulated quickly: while the Communists gained the power in Prague in February 1948, Z. Kopal was appointed Associate Professor at MIT (for numerical analysis) and started at the same time lectures on double stars at Harvard under supervision of H. Shapley.

The origin of the idea that the components of close binaries must fill nonspherical lobes is discussed in other articles of these proceedings, but it dates back to the years around 1940. As can be seen from Kopal´s first notebook with lecture notes on his course, he started with elliptical shapes of components and computed the light curves in case of the eclipses. The notes are written as a mixture of Czech and English sentences or even single words; the heading states that both components are ellipsoids with given eccentricities. A picturesque linguistic colorit and the enthusiasm of the author is demonstrated on one page at the end of the notebook (Figure 7), where the Czech-English comments say (English is in italics) that we finally get a formula for $\gamma_0^{\ 1}$ in which the first term after the minus-sign must be checked and the last parenthesis is "completely correct." The expression for $\gamma_1^{\ 0}$ is not given because it is "very complicated, since it *involves elliptical integrals*," and the comment to $\gamma_1^{\ 1}$ reads verbally:

it is so difficult that we will resign (*involves very many elliptical integrals*) –
(for this moment only) –

In no case we shall give it up.

This lecture notebook was maybe the first one, and the number of Czech words decreases to its end. On the other side, even if the Czech language of Zdeněk Kopal remained excellent after so many years abroad, he sometimes included English words to enrich the dialogue; and he did it always with an unforgettable friendly smile.

3. The Lunar Intermezzo

Zdeněk Kopal step by step attained high respect as a NASA consultant in affairs concerning space research, particularly in the lunar programmes. The turning point here may be found in the discussions with his friend Harold Urey, an enlightened scientist with an unusually broad spectrum of scientific interests, reaching on one side from laboratory experiments on organic synthesis with the aim to search for origins of life, and ending on the other side among many diverse astronomical and astrophysical problems. Zdeněk Kopal started his research by a simple project of lunar cartography, which contributed much to the creation of reliable lunar charts needed for the successful exploration of the Moon by space missions. His colleague from US Air Force, R.W. Carder,

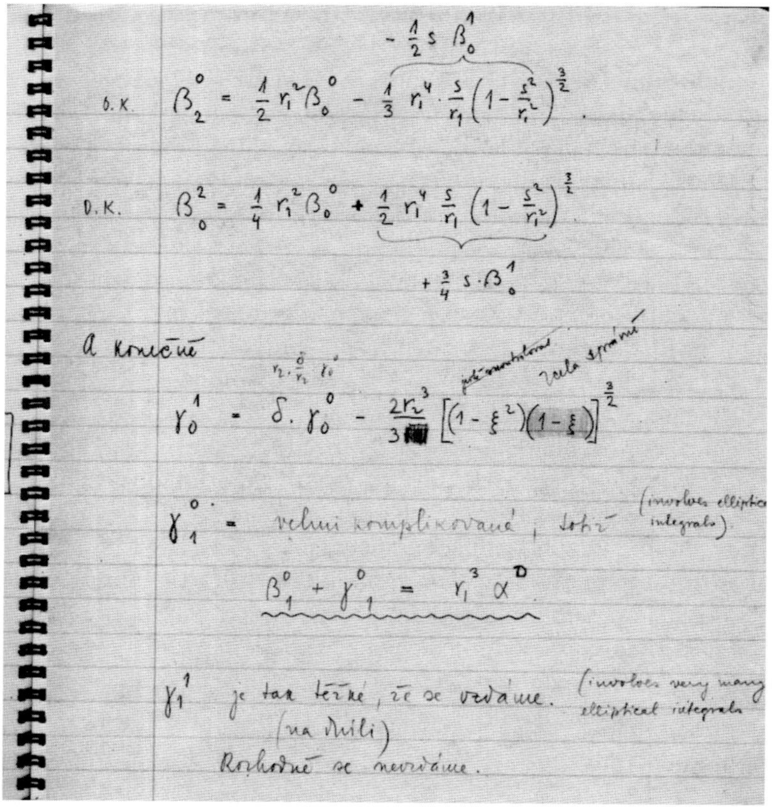

Figure 7. "In no case we shall give it up!" Lecture notes, Pt. I to the course "Orbital revolutions of elliptical components" at Harvard University, 1948. (Photo from author's archive.)

characterized the project undertaken by Kopal by the following words (Rackham, 1962):

Determining heights of lunar mountains by measuring the length of the shadow cast by the peak, dates back to the eighteenth century.[1] But it remained for a small group of students at the University of Manchester, England, working under the direction of Dr. Zdeněk Kopal, to update and refine this technique. The observational method they employed was 35-mm time-lapsed photography taken at Pic-du-Midi of lunar sunrise and sunset over specific parts of the Moon. In order to accomplish this, they modified a 35-mm movie camera so that it would take exposures at a rate of three per min and mounted it on the 24-inch Pic-du-Midi refractor by means of an optical bench attached to the lower end of the telescope. By this method some 12 000 individual

[1] Galilei employed this method already in 1610 by his first telescopic observations of the Moon.

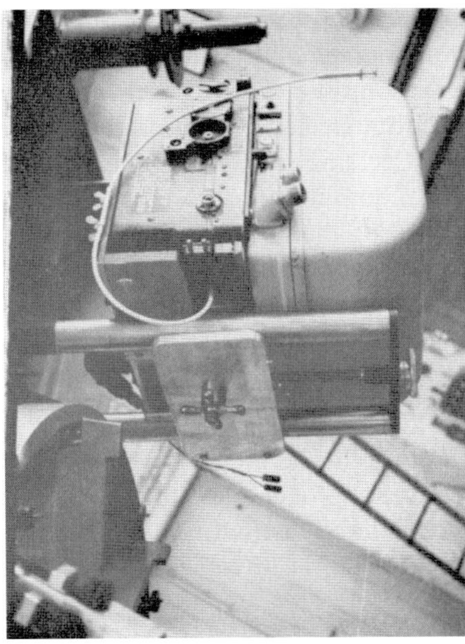

Figure 8. The K-22 camera at Pic du Midi. (Kopal and Rackham, 1962, p. 350.)

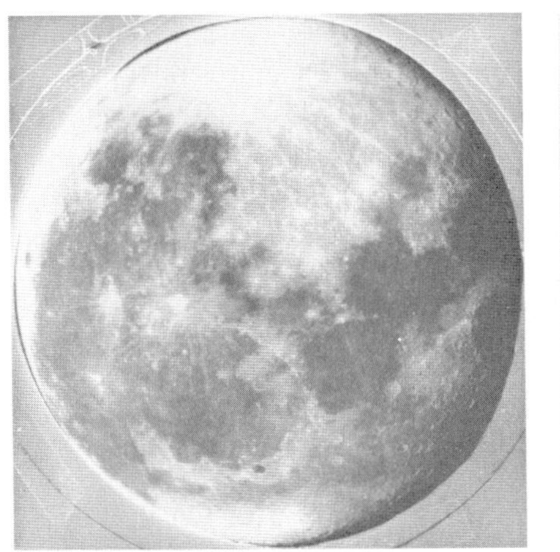
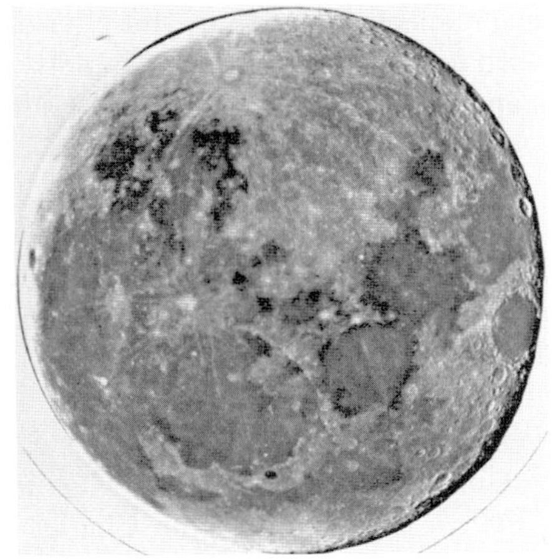

Figure 9. Original photograph taken with the K-22 camera and picture with electronically enhanced contrast and equalised brightness. (Kopal and Rackham, 1962, pp. 357–358.)

photographs were acquired of selected areas of the lunar surface during a 2-year period.

The group adapted the K-22 aerocamera for taking about three pictures per minute and attached it to the 60-cm refractor with a focal length of 18 m of the Pic-du-Midi Observatory (Figures 8 and 9). For the "photometric rectification" they either used the copying of negatives by the "fluorododging method" or the electronic enhancement of contrast in the E.M.I. Research Laboratories in Middlesex in England. The results contributed neatly to the ongoing project of lunar charts with a scale of 1 : 1000000, directed and coordinated by the Aeronautical Chart and Information Center of the US Air Force. The precision of the method was about 1 arcmin in absolute position both in longitude and latitude, and about 10 m in the heights estimated by measuring the geometry of shadows cast on the surrounding surface.

4. Last Visits at Prague University

The last turning point led Zdeněk Kopal to the history of astronomy. He often quoted even in earlier years, that to fulfil the motto he liked so much *Publish or perish!*, then "every astronomer must approach once the moment in his life when he will switch his interests to the history of astronomy, because it is a bottomless dwell of publications." And he supported this "theorem" by many examples of well known astronomers.

He himself included into his memoirs a chapter about the history of astronomy, mainly about discovering and observing the double stars (it is the 7th Chapter in his book *Of Stars and Men*). However, he also knew much about the history of lunar prospecting, charting and physical investigations. For instance, the method he used to rectify the surface geometry by projecting the pictures on a half sphere was applied hundred years ago by Ladislaus Weinek (1848–1913), director of the Prague Clementinum Observatory, who prepared the first photographic atlas of the Moon. Edward Holden, director of the Lick Observatory, provided him with the photographic glass plates and Catherine Wolfe Bruce (1816–1900) financed as Maecenas the whole project. The Lick photographic atlas of the Moon, on 200 printed sheets sized A3, was published in Prague in 10 volumes in 1897–1900, with the contribution of plates taken by the large refractor at the Observatory Paris-Meudon.

When Zdeněk Kopal visited Prague for the first time after the Velvet Revolution, Professor Radim Palouš, Rector Magnificus, conferred upon him the Silver Medal, the highest distinction of the Charles University (Figure 10). On this occasion Zdeněk Kopal also visited the Astronomical Institute in Švédská Street in Prague 5, where he had heard his first lectures almost 60 years ago. The lecture room was so small, that the "students sitting in the first row could write on the blackboard without standing up from the chair" (but there were altogether only two rows). In the late years, Zdeněk Kopal often returned to discussions of the Algol problem, but now also from a historical point of view. But close double stars remained his most beloved favourites, which he compared to human beings (Kopal, 1986):

With the stars, one is almost tempted to say, it is like with human beings. A solitary individual, watched at a distance, will seldom disclose to a distant

Figure 10. Historical building of the Astronomical Institute of Charles University, Švédská Street No. 8, Prague 5. (Photo of 1985 from author's archive.)

observer more than some of his "boundary conditions," insufficient in general to probe his real nature. If, however, two (or more) such individuals are brought together within hailing distance, their mutual interaction will bring out their "internal structure" the more, the closer they come to each other. Similarly, as a star in the sky remains single, there is no way of gauging the detailed properties of its gravitational field, or of the distribution of brightness on its surface. Place, however, another star in its proximity to form a pair bound by mutual attraction: many properties of the combined gravitational field can be inferred at once from the observable characteristics of the components' motion, just as a distribution of surface brightness can be deduced from analysis of the observable light changes.

Acknowledgement

The author acknowledges support by research project J13/98: 113200004 of the Czech Ministry of Education, Youth and Sports.

References

Carder, R.W.: 1962, in: Proceedings of IAU Symposium 14, The Moon, Zdeněk Kopal and Zdenka Kadla Mikhailov (eds.), Academic Press, London and New York, pp. 113–129.
Kopal, Z.: 1932a, *Říše hvězd* **13**, 107–108.

Kopal, Z.: 1932b, *Astron. Nachr.* **242**, 243.
Kopal, Z.: 1986, Of Stars and Men, Adam Hilger, Bristol and Boston.
Kopal, Z. and Rackham, T.W.: 1962, in: Proceedings of IAU Symposium 14, The Moon, Zdeněk
    Kopal and Zdenka Kadla Mikhailov (eds.), Academic Press, London and New York, pp. 350,
    357–358.
Slouka, H. and Kopal, Z.: 1936, *Nature* **137**, 621.

ATMOSPHERIC ECLIPSES IN WR+O BINARIES: FROM KOPAL AND SHAPLEY TO PRESENT DAYS

A.M. CHEREPASHCHUK

Sternberg State Astronomical Institute, Universitetskij Prosp., 13, Moscow 119992, Russia;
E-mail: cher@sai.msu.ru

(accepted April 2004)

Abstract. Kopal and Shapley (1946) were the first to suggest to solve an integral equation for interpretation of atmospheric eclipses in WR+O binary systems. In our studies, this idea was developed in two ways: solution of two integral equations; development of efficient methods of solution of ill-posed problems.

Application of the new method to the interpretation of the light curve of the WN5+O6 eclipsing binary V444 Cyg allowed us to determine the radius and the temperature of the WN5 star core. Some astrophysical applications of these results are presented.

Keywords: atmospheric eclipse, Wolf–Rayet star, ill-posed problem, helium star, black hole, neutron star

1. Introduction

Wolf–Rayet (WR) stars of Population I in the Galaxy are related to:

1. Late stages of evolution of massive stars.
2. Formation of black holes and neutron stars due to collapse of their CO cores.
3. Supernova explosions of Ib and Ic types.
4. Generation of cosmic gamma-ray bursts.

(For recent results, see, e.g., IAU Symposium No. 212, eds. van der Hucht et al., 2003.)

More than 200 WR stars are currently known in the Galaxy (van der Hucht, 2001). About 40% of them are WR+O binaries (Moffat, 1995; van der Hucht, 2001). It is difficult to study basic parameters of isolated WR stars because of their strong optically thick stellar wind ($\dot{M} \simeq 10^{-5}\,\mathrm{M_\odot}$/year, $v \simeq 10^3$ km/sec). Analysis of atmospheric eclipses in WR+O binaries seems to be very promising for determination of the structure of WR extended atmospheres and basic characteristics of WR star "cores" (Kopal and Shapley, 1946; Kopal, 1959). From the analysis of the photoelectric light curve of V444 Cyg (WN5+O6, $P \simeq 4.2$ days) obtained by Kron and Gordon (1943), Kopal (1944) and Russell (1944) demonstrated that the WN5 star had a semi-transparent disc due to the presence of the extended atmosphere. Kopal (1946), for the case of the eclipsing binary ζ Aur, and Kopal and Shapley (1946), for the case of V444 Cyg, were the first to propose a new method of interpretation based on the solution of the integral equation for the light loss during the atmospheric eclipse.

2. Development of Kopal's Idea

Kopal's idea was developed by Cherepashchuk (1966) who suggested that two integral equations, describing both the primary (atmospheric) and the secondary minimum of the light curve of V444 Cyg, should be solved. In the $\cos i < r_{O6}$ case (more than a half of the WR star disc is eclipsed by the O6 star during the conjunction of the components), it is possible to obtain a full solution of the problem: parameters r_{O6}, i and functions $I_c(\xi)$, $I_a(\xi) = 1 - e^{-\tau(\xi)}$, describing brightness and extinction distribution over the WR star disc (ξ is the impact parameter on the WR star disc). The condition $\cos i < r_{O6}$ should be proved before the light curve solution. A sufficient criterion was suggested by Cherepashchuk (1973): if the depth of the secondary minimum of the light curve of the V444 Cyg system is $> \frac{1}{2} L_{WR}$ (L_{WR} is the relative luminosity of the WN5 star), then the condition $\cos i < r_{O6}$ is satisfied. This criterion does not depend on the particular values of the geometric parameters r_{O6}, i. The luminosity of the WN5 star has recently been estimated from high-quality CCD spectra of V444 Cyg (Cherepashchuk et al., 1995). Integral equations describing light loss during eclipses are Fredholm's integral equations of the first kind. The problem of solution of such integral equations is an ill-posed one. Kopal (1946) understood this specific problem quite well and applied a smoothing procedure based on polynomial approximation of the function $\exp(-\tau)$. A very important theorem was proved by Tikhonov (1943): if we select a compact set of functions using physical "a priori" information about the function of interest, an initially ill-posed problem becomes a well-posed one. Efficient algorithms of solving ill-posed problems on compact sets of functions were developed by Goncharsky et al. (1985). The light curve solution for V444 Cyg was obtained at the following compact sets of functions (see Figure 1):

1. The functions of interest $I_a(\xi)$, $I_c(\xi)$ belong to the set of non-negative monotonically decreasing functions (Cherepashchuk, 1975; Cherepashchuk et al., 1984).
2. $I_a(\xi)$ belongs to the set of non-negative concave functions (Antokhin et al., 1997).
3. $I_a(\xi)$, $I_c(\xi)$ belong to the set of non-negative convex-concave functions (Antokhin and Cherepashchuk, 2001).

The set of non-negative convex-concave functions corresponds to the most specific and physically realistic "a priori" information about a WR star: the convex part of the functions $I_c(\xi)$ and $I_a(\xi)$ is related to the opaque core of a WR star, and the concave part of these functions corresponds to its extended atmosphere. It is worthwhile to mention that using such "a priori" information allows us to transform an ill-posed problem into a well-posed one. "A priori" information of

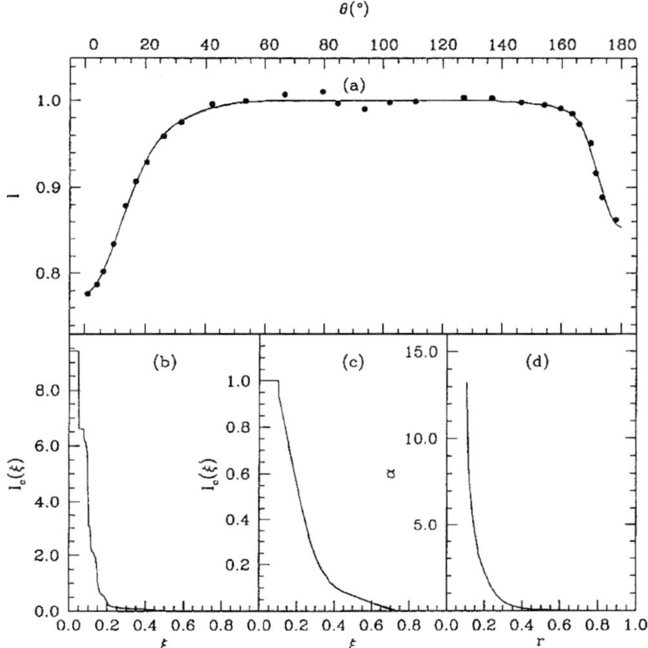

Figure 1. Solution of the rectified light curve of V444 Cyg. (a) observations (dots) versus the theoretical light curve. (b) brightness distribution over the disc of the WN5 star restored on the set of monotonic non-negative functions. (c) extinction distribution over the disc of the WN5 star restored on the set of non-negative concave functions. (d) radial distribution of the extinction coefficient in the extended atmosphere of the WN5 star restored on the set of non-negative concave functions.

this kind seems to be adequate to the physical nature of a WR star and does not depend on a particular physical model of the WR wind. Sphericity of the inner part of the WR wind (and of the corresponding extended atmosphere) is guaranteed by high-velocity spherically-symmetric expansion of the strong, highly supersonic WR wind.

3. Method of the Light Curve Solution

Investigations of supersonic winds in WR+O binaries make it possible to obtain basic characteristics of the interacting stellar wind region as well as of the undisturbed WR wind. The interacting region consists of shocks formed as a result of a wind-wind collision (Cherepashchuk, 1967, 1976, 2000; Prilutsky and Usov, 1976).

As it was pointed out by Kopal and Shapley (1946), it is possible to investigate the proper, undisturbed wind (extended atmosphere) of a WR star analyzing eclipses at optical-continuum frequencies.

The structure of the undisturbed wind was recently studied for the WN5 star in the eclipsing WN5+O6 binary V444 Cyg under the assumption that the functions of interest, $I_a(\xi)$ and $I_c(\xi)$, belong to the compact sets of non-negative concave and convex-concave functions (Antokhin et al., 1997; Antokhin and Cherepashchuk, 2001). The narrow-band ($\Delta\lambda \simeq 70$ Å) continuum ($\lambda 4244$) light curve containing ~ 900 individual photoelectric observations (Cherepashchuk and Khaliullin, 1973) was used. Rectified as well as non-rectified light curves were interpreted. The results of the light curve solutions are close to each other. Our model of the WR+O eclipsing binary is described by a set of integral and algebraic equations. The only model assumption in our light curve solution method is sphericity of the components. Let us write the set of equations describing the light curve of the WR+O eclipsing binary system (Cherepashchuk, 1975).

$$1 - l_1(\theta) = \int_0^{R_a} K_1(\xi, \Delta, r_{O6}) I_o I_a(\xi)\, d\xi \tag{1}$$

$$1 - l_2(\theta) = \int_0^{R_c} K_2(\xi, \Delta, r_{O6}) I_c(\xi)\, d\xi \tag{2}$$

$$I_o \pi r_{O6}^2 \cdot (1 - x/3) + 2\pi \int_0^{R_c} I_c(\xi)\xi\, d\xi = 1 \tag{3}$$

$$I_a(\xi = 0) = 1 \tag{4}$$

$$2\pi \int_0^{R_c} I_c(\xi)\xi\, d\xi = L_{\mathrm{WR}}^{\mathrm{obs}} \tag{5}$$

$$\Delta^2 = \cos^2 i + \sin^2 i \sin^2 \theta \tag{6}$$

Here K_1 and K_2 are the kernels of the integral equations describing the shape of the eclipsed area, θ is the orbital phase angle, Δ is the separation between the centers of the WN5 and O6 star discs, x and I_0 are the limb darkening coefficient and the brightness at the center for the disc of the O6 star. The integral equations (1), (2) describe light losses $1 - l_1(\theta)$, $1 - l_2(\theta)$ at the primary (atmospheric) and the secondary minima of the light curve, respectively. The function of interest, $\tau(\xi)$, is the optical depth in the extended atmosphere of the WN5 star along the line of sight:

$$\tau(\xi) = 2 \int_\xi^{R_a} \alpha(r) \cdot r \cdot dr / \sqrt{r^2 - \xi^2} \tag{7}$$

Here $\alpha(r)$ is the radial distribution of the absorption coefficient in the extended atmosphere of the WN5 star. Equation (4) describes the condition of the existence of an opaque, non-transparent "core" for the WN5 star. Equation (5) describes the equality of the theoretical luminosity of the WN5 star to the observed luminosity $L_{\mathrm{WR}}^{\mathrm{obs}}$ of the WN5 star, estimated from spectrophotometric observations (Cherepashchuk et al., 1995). Four independent unknown values must be determined from equations (1)–(5): two parameters, r_{O6} and i, and two functions, $I_a(\xi) = 1 - e^{-\tau(\xi)}$ and $I_c(\xi)$. In their recent study, Antokhin and Cherepashchuk (2001) used a compact set $\tilde{M} \downarrow$ of non-negative convex-concave functions. After numerical discretization of the functions of interest $z(s) = I_a(\xi), I_c(\xi)$, the compact set $\tilde{M} \downarrow$ is determined by the set of linear inequalities:

$$\tilde{M}\downarrow = \underbrace{z_1 \geq 0, z_1 \geq z_2, z_{i-1} - 2z_i + z_{i+1} \leq 0; i = 1, \ldots, k-1}_{\text{convex part of function}}$$

$$\underbrace{z_{i-1} - 2z_i + z_{i+1} \geq 0, i = k+1, \ldots, n-1; z_{n-1} \geq z_n; 1 < k < n; z_n \geq 0}_{\text{concave part of function}} \quad (8)$$

The position of the transitional point between the convex and concave parts of the function $z(s)$ is a free parameter of the problem.

The following residual functional was used:

$$\eta[I_a(\xi), I_c(\xi), r_{O6}, i] = \left[\frac{\sum_{i=1}^{m} w_i \left(l_i^{\text{obs}} - l_i^{\text{theor}} \right)^2}{\sum_{i=1}^{m} w_i} \right]^{1/2} \quad (9)$$

where l_i^{obs} and l_i^{theor} are the observed and theoretical intensities, m is the number of normal points on the mean light curve, w_i is the weight of the i-th normal point of the light curve. Minimization of the residual functional, η, taking into account the constraints (8) was undertaken using an efficient method of conjugate gradient projection (Goncharsky et al., 1985).

4. Results of the Light Curve Solution

The results of the light curve solution for the eclipsing WN5+O6 binary V444 Cyg are presented in Figure 2. The residual surface $\eta(r_{O6}, i)$ is shown here. The absolute minimum of the residual functional η determines the full solution of our inverse problem: $I_a(\xi), I_c(\xi), r_{O6}, i$. In contrast with the classical light curve synthesis method (Wilson and Devinney, 1971), our method of light curve solution allows us to determine, from an observed high-precision light curve, not only the values of the geometric parameters, but also the functions describing the structure of the disc of the WN5 star. Two solutions were obtained from our analysis (see Figure 2).

1. The residual surface has an absolute minimum $\eta = 0.0067$ for the parameters: $r_{O6} = 0.25a \simeq 9.5\,\text{R}_\odot$, $i = 78°.0$, $L_{\text{WR}} = 0.20$ (determined from the light curve solution by integration of the function $I_c(\xi)$ over the WN5 star's disc), $r_{\text{WR}}^{\text{core}} = (2-3)\,\text{R}_\odot$, $T_{\text{WR}}^{\text{core}} > 70000$ K (determined under the assumption that the effective temperature of the O6 star is $T_{O6} = 40000$ K) (see Figure 3).

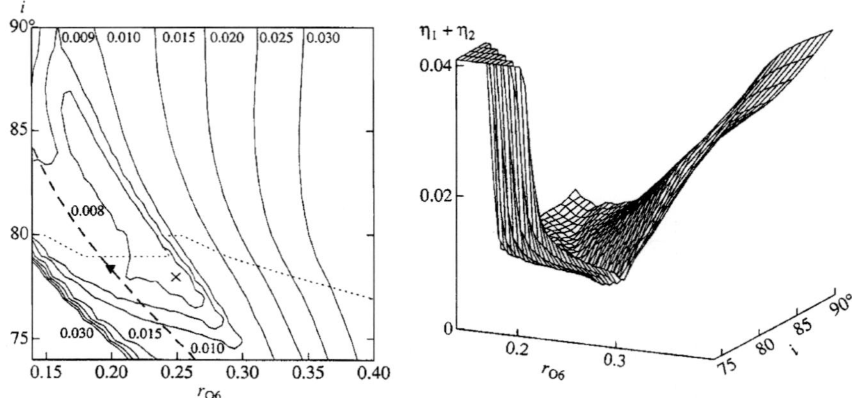

Figure 2. Surface of the total residual $\eta = \eta_1 + \eta_2$ (η_1, η_2 are the residuals in the primary and secondary minima) for the solution of the light curve of V444 Cyg obtained on the set of convex-concave functions. The left panel shows contours of the surface. The cross shows the absolute minimum of the total residuals. The thick dashed line in the left panel corresponds to the assumed luminosity of the WN5 star $L_{\mathrm{WN5}}^{\mathrm{obs}} = 0.38$. The triangle indicates the best-fit solution for this luminosity. The thin dashed line separates the regions where the core of the WN5 star is opaque (below) from the region where it is semi-transparent.

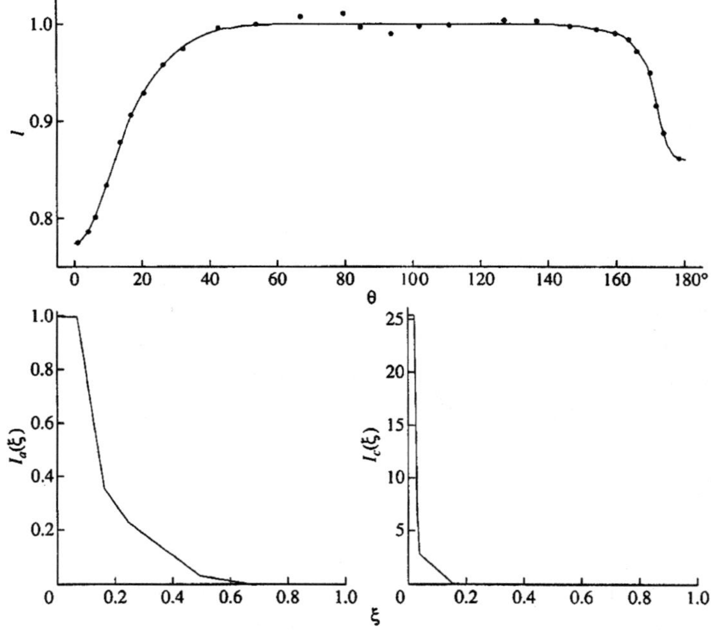

Figure 3. Solution of the light curve of V444 Cyg (upper panel) on the set of non-negative convex-concave functions corresponding to the absolute minimum of the total residual η. Lower panel: extinction and brightness distribution over the disc of the WN5 star.

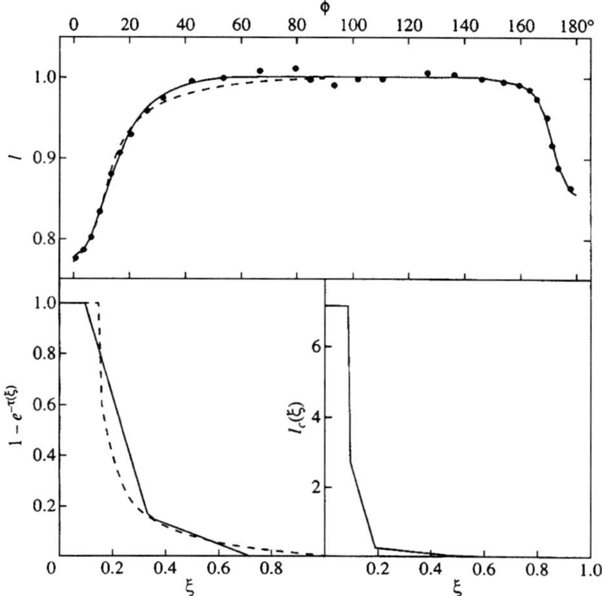

Figure 4. Solution of the light curve of V444 Cyg on the set of non-negative convex-concave functions corresponding to the assumed luminosity of the WN5 star $L_{\mathrm{WN5}}^{\mathrm{obs}} = 0.38$. Lower panel: solid lines – extinction and brightness distribution over the disc of the WN5 star. Dashed lines correspond to Lamers velocity law.

2. The solution corresponding to the assumed luminosity of the WN5 star, $L_{\mathrm{WR}}^{\mathrm{obs}} = 0.38$: $r_{\mathrm{O6}} = 0.20a = 7.6\,\mathrm{R}_{\odot}$, $i = 78°.4$, $r_{\mathrm{WR}}^{\mathrm{core}} \simeq 4\,\mathrm{R}_{\odot}$, $T_{\mathrm{WR}}^{\mathrm{core}} = 52000$ K, $\eta = 0.0087$ (higher by a factor of 1.3 than $\eta = 0.0067$ corresponding to the absolute minimum of the residual functional η). The corresponding theoretical light curve and functions of interest, $I_a(\xi)$ and $I_c(\xi)$, are presented in Figure 4.

The results of the light curve solution for the two cases are close to each other, if we take into account the complicated physical model of the WR+O binary system.

Basic characteristics of the WN5 star are qualitatively the same for both results of the light curve solution: the WN5 star has a compact high-temperature core:

$$r_{\mathrm{WR}}^{\mathrm{core}} = (2-4)\,\mathrm{R}_{\odot}, \qquad T_{\mathrm{WR}}^{\mathrm{core}} \geq 52000\,\mathrm{K}.$$

For the mass of the WN5 star $m_{\mathrm{WN5}} \simeq 10\,\mathrm{M}_{\odot}$, these parameters of the core are compatible with the model of the WN5 star being a helium remnant (bare core) of an initially massive ($m \simeq 30\,\mathrm{M}_{\odot}$) star that lost most of its hydrogen envelope, presumably due to mass exchange in a massive close binary system. Therefore, we can conclude that the WN5 star in the V444 Cyg binary system is at a late stage of its evolution; the collapse of the CO core of this massive helium star can lead to formation of a relativistic object and to a supernova Ib or Ic explosion. From the function $I_a(\xi) = 1 - e^{-\tau(\xi)}$, the density distribution and the corresponding velocity law in the inner part of the wind of the WN5 star was restored (Antokhin et al., 1997; Antokhin and Cherepashchuk, 2001). This velocity law, $v(r)$, suggests accelerated motion of matter in the wind. The acceleration of matter in the wind of the WN5 star was shown to be relatively slow: the acceleration of the wind persists even at large distances from the center of the WN5 star.

The width of the secondary eclipse in V444 Cyg strongly increases with wavelength λ. The characteristic size of the extended atmosphere of the WN5 star increases with λ. It suggests a clumpy structure of the WR wind (Cherepashchuk et al., 1984). Taking into account the clumpy structure of the WR wind allows us to reduce the value of the mass loss rate for the WR star by a factor of ~ 3 (Cherepashchuk, 1990).

5. Confirmations of the Results of the Light Curve Solution

1. An analysis of variable linear polarization at different orbital phases of V444 Cyg yields the core radius for the WN5 star, $r_{\mathrm{WR}}^{\mathrm{core}} < 4\,\mathrm{R}_\odot$ (St-Louis et al., 1993).
2. For the system of V444 Cyg, an investigation of eclipsing effects in the P Cygni absorption component of the N V 4604 emission line gives the radius of the WN5 star's core $r_{\mathrm{WR}}^{\mathrm{core}} < (3-5)\,\mathrm{R}_\odot$ (Moffat and Marchenko, 1996).
3. Cyg X-3, a peculiar short-period ($P \simeq 4^h.8$) X-ray binary system (e.g., Cherepashchuk et al., 1996). The optical companion is a WN3-7 star (van Kerkwijk et al., 1992), it is a massive WR star with $L_{\mathrm{bol}} \simeq 3 \cdot 10^{39}$ erg/s (van Kerkwijk et al., 1992; Cherepashchuk and Moffat, 1994). Its very short orbital period, $P \simeq 4^h.8$, implies the separation between the WR and X-ray components $a \simeq 3.2 - 5.6\,\mathrm{R}_\odot$. Therefore (Cherepashchuk and Moffat, 1994) the parameters of the WN3-7 star's core in Cyg X-3 are the following $r_{\mathrm{WR}}^{\mathrm{core}} < (3-6)\,\mathrm{R}_\odot$, $T_{\mathrm{WR}}^{\mathrm{core}} > 70000-90000$ K.
4. According to Cherepashchuk and Karetnikov (2003) a strong distinction between orbital eccentricity distribution for WR+O and OB+OB binaries suggests formation of WR stars as a result of a strong evolutionary increase of the OB star radius and the mass transfer in binary systems.

6. Some Astrophysical Applications

Up to now, masses of 23 WR stars in WR+O binaries have been measured (e.g., Moffat, 1995). WR stars in binary systems can be considered progenitors of neutron stars and black holes, and thus it is interesting to compare the distributions of masses of relativistic objects to the distribution of masses for CO cores of WR stars at the end of their evolution. Masses of $\sim$40 relativistic objects (18 black holes and $\sim$22 neutron stars) in binary systems are known up to now (Charles, 2001; Thorsett and Chakrabarty, 1999; Cherepashchuk, 2003). The distribution of masses of relativistic objects seems to be bimodal:

$$\langle m_{NS} \rangle = (1.35 \pm 0.15)\,M_\odot, \quad m_{NS} = (1-2)\,M_\odot;$$
$$\langle m_{BH} \rangle = (7 \pm 1)\,M_\odot, \quad m_{BH} = (4-15)\,M_\odot.$$

There is a gap in the $(2-4)\,M_\odot$ range: no neutron star and no black hole were observed in this range. The gap in the $(2-4)\,M_\odot$ range cannot be explained by observational selection effects (Cherepashchuk, 2003). To calculate the final mass of a CO core, we must take into account the strong mass loss by WR stars due to stellar wind ($\dot{M} \simeq 10^{-5}\,M_\odot$/year, dependent on the mass of the WR star). In the case of the WR star wind's clumpy structure, we can reduce the mass loss rate, $\dot{M}_{WR}$, by a factor of ~ 3 (Cherepashchuk, 1990; Nugis et al., 1998). Recently Cherepashchuk (2001) calculated the final masses of CO cores of WR stars taking into account the clumpy nature of the WR winds. The results of the calculations are summarized in Figure 5 where the mass distributions of ~ 40 relativistic objects and of the final CO core masses for 23 WR stars are shown. The distribution of the M_{CO}^f masses is continuous, unlike the mass distribution for the relativistic objects. The range of the mean masses, $\langle M_{CO}^f \rangle = (7.4-10.3)\,M_\odot$, is close to the mean black hole mass, $\langle M_{BH} \rangle = (7 \pm 1)\,M_\odot$. Thus (see Figure 5), the relativistic object mass

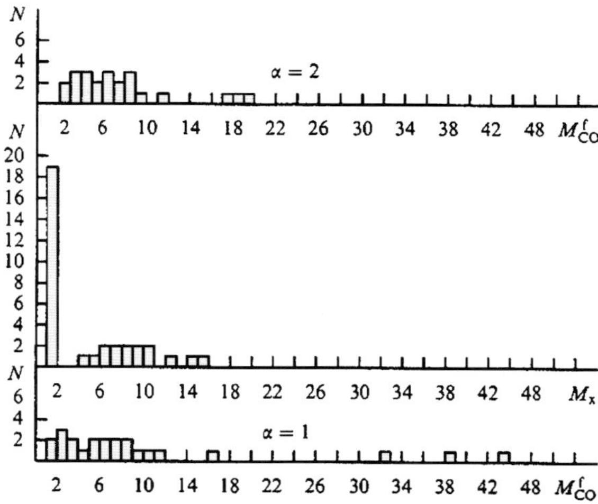

Figure 5. Histogram of distributions of final masses of carbon-oxygen cores M_{CO}^f for 23 WR stars with known masses (the bottom plot corresponds to the case $\alpha = 1$ in the formula $\dot{M}_{WR} = kM_{WR}^\alpha$, the upper plot corresponds to the case $\alpha = 2$). In the middle – the histogram of the M_X – mass distribution for 34 relativistic objects in binary systems is shown. The peak at $1-2\,M_\odot$ corresponds to neutron stars. The distributions of M_{CO}^f are continuous, while the distribution for M_X is bimodal with a gap at $2-4\,M_\odot$. (Cherepashchuk, 2003).

distribution is bimodal in spite of the distribution of masses being continuous for the progenitors. The difference in mass distribution implies that the nature of the relativistic object formed (neutron star, black hole) is determined not only by the progenitor mass, but also by other parameters of the progenitor, such as magnetic field, rotation, etc..

Gershtein (2000), Postnov and Cherepashchuk (2001) suggest the hypothesis that the core collapses of WR stars can be progenitors of γ-ray bursts since, in particular, WR stars have no thick hydrogen envelopes, which facilitates the core collapse energy transformation into the observed γ-ray emission. During the collapse of their CO cores, the WR stars in short-orbital-period binary systems can form rapidly rotating, Kerr black holes (Tutukov and Cherepashchuk, 2004), leading to formation of highly relativistic jets and γ-ray bursts (Paczynski, 1998; MacFadyen and Woosley, 1999). The enormous energy release during γ-ray bursts, ($\sim 10^{51} - 10^{53}$) erg, allows one to suggest that γ-ray bursts, which appear once per day, provide information on late stages of evolution of massive O stars and WR stars at distances corresponding to very large redshifts, $z = 0.4 - 4$.

All these results concerning the nature of WR stars clearly demonstrate the importance of Kopal's pioneer work on the interpretation of atmospheric eclipses in WR+O binary systems. Kopal's idea can also be developed for the interpretation of selective atmospheric eclipses in frequencies of lines (e.g., Khaliullin and Cherepashchuk, 1976).

Acknowledgements

This work was partially supported by the grant 02-02-17524 of Russian Foundation of Basic Researches and Russian President Grant NS-388.2003.3.

References

Antokhin, I.I. and Cherepashchuk, A.M.: 2001, *Astron. Zh.* **78**, 432.

Antokhin, I.I., Cherepashchuk, A.M. and Yagola, A.G.: 1997, *Ap&SS* **254**, 111.

Charles, P.: 2001, in: L. Kaper, E.P.J. van den Heuvel and P.A. Woudt (eds.), *Black Holes in Binaries and Galactic Nuclei: Diagnostic, Demography and Formation*, Vol. 27, Springer, Berlin.

Cherepashchuk, A.M.: 1966, *Astron. Zh.* **43**, 517.

Cherepashchuk, A.M.: 1967, *Variable Stars* **16**, 226.

Cherepashchuk, A.M.: 1973, *Astron. Zh.* **50**, 879.

Cherepashchuk, A.M.: 1975, *Sov. Astron.* **9**, 727.

Cherepashchuk, A.M.: 1976, *Sov. Astron. Lett.* **2**(4), 138.

Cherepashchuk, A.M.: 1990, *Sov. Astron.* **34**, 481.

Cherepashchuk, A.M.: 2000, *Ap&SS* **274**, 159.

Cherepashchuk, A.M.: 2001, *Astron. Rep.* **45**, 120.

Cherepashchuk, A.M.: 2003, *Phys. Uspekhi* **46**, 335.

Cherepashchuk, A.M., Eaton, J.A. and Khaliullin, Kh.F.: 1984, *ApJ* **281**, 774.

Cherepashchuk, A.M. and Karetnikov, V.G.: 2003, *Astron. Zh.* **80**, 42.

Cherepashchuk, A.M., Katysheva, N.A., Khruzina, T.S., Shugarov, S.Yu.: 1996, *Highly Evolved Close Binary Stars: Catalog*, Gordon and Breach, Brusseles, Belgium.

Cherepashchuk, A.M. and Khaliullin, Kh.F.: 1973, *Sov. Astron.* **17**, 330.

Cherepashchuk, A.M., Koenigsberger, G., Marchenko, S.V. and Moffat, A.F.J.: 1995, *A&A* **293**, 142.

Cherepashchuk, A.M. and Moffat, A.F.J.: 1994, *ApJ* **424**, L53.

Gershtein, S.S.: 2000, *Astron. Lett.* **26**, 730.

Goncharsky, A.V., Cherepashchuk, A.M. and Yagola, A.G.: 1985, Ill-posed Problems of Astrophysics, Nauka, Moscow.

Khaliullin, Kh.F. and Cherepashchuk, A.M.: 1976, *Sov. Astron.* **20**, 186.

Kopal, Z.: 1944, *ApJ* **100**, 204.

Kopal, Z.: 1946, *ApJ* **103**, 310.

Kopal, Z.: 1959, Close Binary Systems, Chapman and Hall, London.

Kopal, Z. and Shapley, M.B.: 1946, *ApJ* **104**, 160.

Kron, G.E. and Gordon, K.S.: 1943, *ApJ* **97**, 311.

MacFadyen, A. and Woosley, S.: 1999, *ApJ* **524**, 262.

Moffat, A.F.J.: 1995, in: K.A. van der Hucht and P.M. Williams (eds.), *IAU Symposium 163, Wolf-Rayet Stars: Binaries, Colliding Winds, Evolution*, Vol. 213, Kluwer, Dordrecht, Holland.

Nugis, T., Crowther, P.A. and Willis, A.J.: 1998, *A&A* **333**, 956.

Paczynski, B.: 1998, *ApJ.* **499**, L45.

Postnov, K.A. and Cherepashchuk, A.M.: 2001, *Astron. Rep.* **45**, 517.

Prilutsky, O.F. and Usov, V.V.: 1976, *Sov. Astron.* **20**, 2.

Russell, H.N.: 1944, *ApJ* **100**, 213.

St-Louis, N.: 1993, *ApJ* **410**, 342.

Thorsett, S.E. and Chakrabarty, D.: 1999, *ApJ* **512**, 288.

Tikhonov, A.N.: 1943, *Doklady Acad. Nauk USSR* **39**, 195.

Tutukov, A.V. and Cherepashchuk, A.M.: 2004, *Astron. Zh.* **81**, 43.

van der Hucht, K.A.: 2001, *New Astron. Rev.* **45**, 135.

van der Hucht, K.A., Herrero, A. and Esteban, C. (eds.): 2003, IAU Symposium 212. A Massive Star Odyssey: From Main Sequence to Supernova, San Francisco.

van Kerkwijk, M.H., Charles, P.A. and Geballe, T.R., et al.: 1992, *Nature* **355**, 703.

Wilson, R.E. and Devinney, E.J.: 1971, *ApJ* **166**, 605.

II: BINARY STAR MORPHOLOGY AND DYMAMICAL ASPECTS

RESOLVING THE ALGOL PARADOX AND KOPAL'S CLASSIFICATION OF CLOSE BINARIES WITH EVOLUTIONARY IMPLICATIONS

I.B. PUSTYLNIK

Tartu Observatory, Tóravere 61602, Estonia; E-mail: izold@aai.ee

(accepted April 2004)

Abstract. The Algol paradox posed a serious challenge to the theory of evolution of close binaries. We indicate that it has been resolved by the collective efforts of a whole generation of astronomers rather than by anyone's individual ingenious accomplishment. We discuss the role of illuminating ideas put forward by Zdeněk Kopal in solving the Algol paradox.

Keywords: history – close binaries, evolution

1. Introduction

The amazing progress in understanding of the physics and evolution of close binary systems which we have witnessed in the second half of 20th century raises many intriguing questions with no immediate answers. During the last 30 years the idea of a binary nature of various peculiar objects enabled the elucidation, at least in its basic features of the properties of such strikingly different – in their observable manifestations – double stars as cataclysmic variables, symbiotic and barium stars, X-ray bursters, binary radio pulsars etc. The binary model remains on top of the list of the most productive ideas invoked to explain the nature of elusive and mysterious γ-ray bursts.

What may be equally amazing is the fact that inhabitants of a "cosmic zoo" of binary objects according to proposed evolutionary scenarios all have at least one key element, one crucial episode in their diversified history, which sounds nowadays almost like a cliché – *Roche lobe overflow*. A transparent physical idea of mass transfer driven on a thermal or even dynamical time scale, when a star in course of its nuclear evolution fills in its first critical Roche lobe, revolutionized the whole discipline of close binary research. And this sends us back to the early history of resolving the Algol paradox in an attempt to trace the roots of subsequent spectacular achievements in this field. It would be tempting to look at the whole issue of Algol paradox in a broad historical prospective making use of a rich stock of knowledge accumulated during the later half of the century separating us from pioneering investigations of Struve (1954), Parenago and Massevich (1950), Crawford (1955), Hoyle (1955), and Kopal (1971). Although from time to time I will attempt to place some aspects of the problem of the Algol paradox into a modern prospective, it will be done mostly for illustrative purposes or in order to compare

Astrophysics and Space Science **296**: 69–78, 2005.
© Springer 2005

certain predictions with present realities. The present report does not pretend to be a comprehensive review of this multifold problem. Rather it is an attempt to summarize the most important aspects of the Algol paradox seen in retrospective and to single out some of them which either remained obscured or undeservingly fell to oblivion.

2. Resolving the Algol Paradox

The Algol paradox has been the subject of numerous articles and manuals and it makes no sense to elaborate them in length. It suffices to say that the puzzling feature of Algol-type binaries widely known since Gerard Kuiper's pioneering investigation (Kuiper, 1941) lies in the fact that an early-type primary component (usually of B8-A5 spectral type) with radius and luminosity normal for a main sequence star is accompanied by a low-mass companion (ordinarily the mass ratio is $q \simeq 0.2$–0.3) with characteristics of a subgiant having marked radius and luminosity excesses (as large as 2^m–4^m and even higher).

Although this paradox dates back to the early forties of the 20th century and is associated with the pioneering investigation of Gerard Kuiper, it became especially acute in the early fifties due to advancements of stellar evolution investigations, a rapid progress both in a theory of eclipsing binaries and in stellar photometry. Specifically a proliferation of phototubes and photomultipliers and their low prices made them commercially available even for small and modestly equipped observatories, like Odessa observatory in Ukraine or Tartu Observatory in Estonia where I started my scientific career. I recall how proud was my teacher and supervisor unforgettable Professor V.P. Tsesevich, a former director of Odessa observatory, when he came back from one of his first duty trips behind the iron curtain and ventured to bring in his pocket a small phototube manufactured in UK (in the full swing of the cold war phototubes were proclaimed to be strategic articles prohibited for sale in Soviet block countries).

Application of photomultipliers for the needs of a broad-band photometry of variable stars (following Kron (Kron, 1958) a small telescope equipped with electro-photometer is like a Napoleon, small in size but large in accomplishment) meant a quality change in observations of eclipsing binaries. Introduction of photoelectric photometry (which raised the accuracy of observations up to 0.002^m–0.005^m) enabled us among other things to measure the apsidal motion for a dozen of binaries.

According to the classical theory of apsidal motion in binary systems (first elaborated by Russell in 1928 and later on improved by Chandrasekhar, Stern, Kopal, Martynov and others), the ratio of a full period of apsidal motion U_{aps} to the orbital period P_{orb} obeys in the first approximation a fairly simple relation

$$P_{orb}/U_{aps} = \Sigma_{i=1}^{2} k_i \, r_i^5 (M_{3-i}/M_i) \, F(e),$$

(1)

where M_i, r_i are respectively the mass and radius of the component of a close binary and $F(e)$ is an elementary function of the eccentricity of orbit. Coefficients k_i are determined by the density distribution inside the component stars (to be more specific, by the ratio of the running density of matter to its average value), so that the value of parameter k ranges between 0.75 for a completely homogeneous star and 0 for a point mass, i.e. for the Roche model. Prior to the World War II debates around the possible values of the parameter k still gave no definitive outcome. The measured effect is indeed very subtle, and nature is not quick in cooperating with astronomers to reveal the presence of apsidal motion for many reasons. The apsidal motion manifests itself in the variable displacement of the moment of the secondary minimum from the moment of conjunction, when normally the low-mass component will be eclipsed. But it is not an easy task to measure reliably the effect. Interaction effects between the components (like reflection effect, surface activity, circumstellar matter) considerably influence the shape of the light curve. Also the amplitude of the secondary minimum is small (typically $\leq 0.^{m}05$), and finally the displacement of the moment of minimum is proportional to the product $e \cos \omega$ (e – the eccentricity of orbit and ω being the longitude of the ascending node of the orbit). Thus, we practically miss the systems whose orbits are unfortunately inclined to the earth-bound observer. That is why it took a lot of effort in the mid-fifties analyzing available observational series before for the first time it was firmly established a remarkably high degree of concentration of matter towards center for a dozen of eclipsing variables, yielding the values $k \simeq 10^{-3}$–10^{-2}. This observational result in turn has transformed the Roche model from a beautiful abstract notion into a powerful tool of probing stellar interiors and binary evolution.

Apparently Parenago and Massevich (1950) were the first to gain the fruits of a true statistical approach based on improved observational data. They compiled a catalogue which included quite reliable absolute parameters for a dozen of Algol-type binary systems. Having failed in locating the position of subgiants in the L–M diagram (because of an enormous scatter of the observed points) they inferred that there are no single $L \sim M$ and $R \sim M$ relations valid for the secondary components of Algol-type binaries. But having experimented with two parameter relations they found that $L = f_1(\lambda, M)$, $R = f_2(\lambda, M)$ nicely reproduce the observed points with subgiants forming a distinct sequence and λ being yet some unspecified parameter. Otto Struve, an observational astronomer of great experience endowed with uncommon intuition, guessed that the mysterious parameter λ introduced by Parenago and Massevich is nothing else but the mass ratio of the components q and proved that by plotting the diagram $\Delta M \sim q$ (where ΔM stands for the luminosity excesses observed for subgiants from the theoretical one for main sequence stars) and inserting observational points corresponding to a dozen of well-studied Algol-type binaries (Struve, 1954) (see Figure 1).

And yet, standing only one small step from solving the Algol paradox (because *a posteriori* we know that the diagram $\Delta M \sim q$ first plotted by Struve has direct evolutionary implications, i.e. we should see in it the binary systems caught at

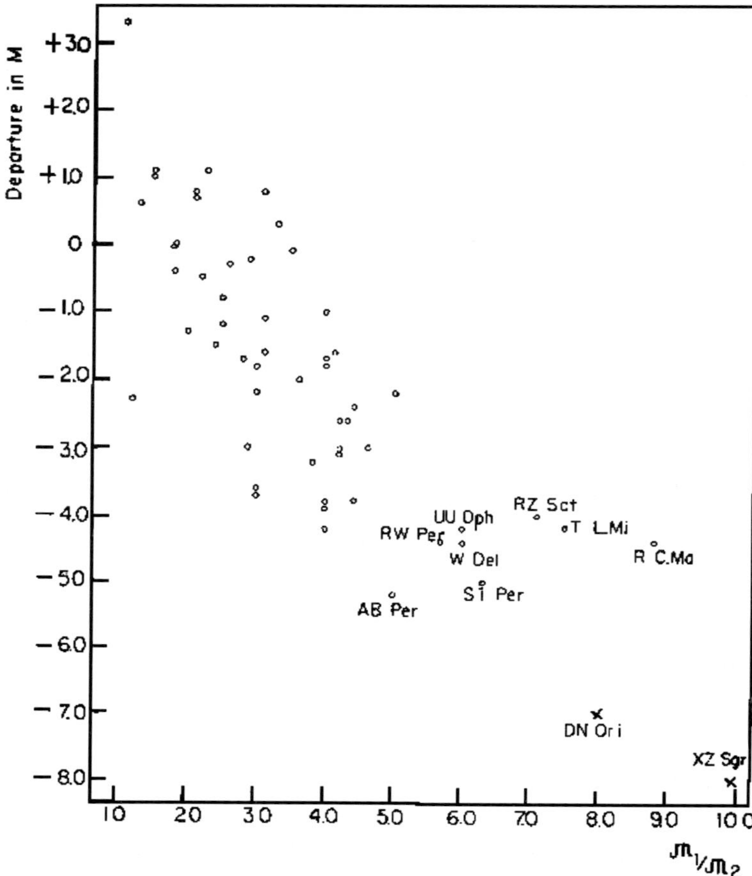

Figure 1. The diagram luminosity excesses ΔM for subgiants relative to main sequence stars in Algol-type binaries versus mass ratio M_1/M_2. This diagram has been constructed for the first time by Struve (1954).

different ages following the crucial mass transfer episode), Struve looked in the wrong direction. He ascribed large luminosity excesses observed in subgiants to initial enrichment by dust particles at the early stage of contraction of "subsidary condensations", in other words the progenitors of subgiants. Being dissatisfied with his own results, yet at the same time being aware of the fundamental implications of the Algol paradox from the evolutionary viewpoint, Struve recommended John Crawford to re-investigate the problem (Crawford, 1955). It would be perhaps too hazardous to speculate over the reasons why Struve overlooked Crawford's explanation of the Algol paradox (after all, Crawford used in his article the same data from Parenago and Massevich and the plot $\Delta M \sim q$, to which he added his own observation that subgiants fill in their respective Roche lobes, a fact certainly known by that time to Kopal (Kopal, 1955).

Independently of Crawford and practically simultaneously with him, Hoyle (Hoyle, 1955) provided a similar evolutionary interpretation of pronounced luminosity and radius excesses observed in subgiant secondary components in Algol-type binaries. For unclear reasons this paper of Hoyle remained unnoticed for many years. It seems to be appropriate to cite at this point a fragment from a private letter to the present author by Prof. Alan Batten (Batten, 1988) without any comment:

> I am myself uncertain how far Hoyle stumbled on the same idea independently. In his book "Frontiers of Astronomy" (published 1955, the same year as Crawford's paper) on pp. 195–202 he gives a very good account of mass-transfer, including a good approximation of the track of the mass-losing component in the H–R diagram. I know that Struve discussed Crawford's work at a Liege symposium a year or two earlier, at which Hoyle was present. I have always assumed that Hoyle had assess to an advance copy of Crawford's paper and was popularizing that. When I recently read his autobiography, however, I realized that his own work on stellar accretion would have got him close to the idea of mass-transfer. Because I have been asked to consider writing another book on binaries, I may write him and ask him whether or not he got the idea independently.

Kopal, the man who has done more than anyone else for investigating the various properties of the Roche model, in 1971 wrote the following lines (for more details see Kopal, 1971):

> The present author pointed out many years ago (Kopal, 1955) that there exists indeed a distinct group of eclipsing variables (which we called the 'semi-detached systems') in which one component has indubitably attained the Roche limit but, unfortunately for our expectations, it is the wrong star! For the most striking feature of such semi-detached systems is the fact that – in every single case known to us so far – it is *the less massive component which appears to be at its Roche limit, while its more massive mate remains well interior to it.* This fact, which has since earned the epithet of an 'evolutionary paradox' has been with us now for more than 15 years and continues to remain a paradox; for in spite of a large amount of work towards its clarification it is not yet satisfactorily understood.

and still further

> other aspects which remain yet to be investigated are so many that a considerable amount of work must be done before more detailed comparisons between theory and observations can possess much meaning.

So what's the point, why the dissident opinion of Kopal on the Algol paradox should deserve a special discussion? The main purpose of my presentation is to convince the reader that in actuality, notwithstanding all rhetoric, Kopal has done, perhaps, more than anyone else for the resolution of the Algol paradox. Pardon my clumsy pun, in my mind one can argue about the paradoxical role of Kopal in the treatment of the Algol paradox. During his long and prolific career Kopal left a profound imprint in studies of a vast amount of interaction effects in close binaries. Tidal distortion of stellar figures, reflection effect, apsidal motion, gravitational darkening, theory of determination of orbital elements, effects of reflection and distortion on the elements of spectroscopic binaries – this is a far from complete list of his accomplishments. Therefore his skeptical attitude towards the way Crawford and Hoyle overcame the Algol paradox is a critical posture of a far-sighted expert, if you wish, of a sober agnostic. If we want to proceed on essence with this discussion let us turn to the original papers on the subject in question by maestro Kopal.

Same year (1955), in the November–December issue of *Annales d'Astrophysique* Kopal publishes an article entitled "The Classification of Close Binary Systems". It seems that the most essential results of this paper are assembled in three tables, Table III, Table IV and Table VIII, with the relative radii in units of semi-major axis, masses of the components, and dimensionless gravitational potentials respectively for stars where both components lie within their critical Roche lobes, with one component filling in its Roche lobe, and with both components very close to their critical Roche lobes. The most startling revelation from inspection of data in Table III is the close equality of gravitational potentials for both component stars constituting a binary. By the way, quite unexpectedly the present author made here his own small "discovery" when recalculating the values of potentials, namely that in all probability Kopal used a slide-rule in his calculations. Comparing our values with the original ones we found almost an ideal coincidence between the values of dimensionless potentials for both components. As we know, Kopal gave the only conceivable physical interpretation for this remarkable coincidence – both stars originated practically in the same time on a cosmic time scale from the same parental body.

In Table IV we notice that unlike the case of detached binaries the values of the gravitational potentials of the components significantly differ and the value of the potential of the less massive component is practically equal to that of the inner Roche lobe. And finally a similar Table VIII constructed for W UMa-type stars convinced Kopal that these systems are in fact contact binaries in view of a close coincidence of their gravitational potentials to that of a critical Roche lobe. Thus an attractive and simple as well as physically substantiated classification of close binary systems composed of three distinctive classes – detached, semi-detached and contact binaries, (which is still in wide use nowadays !) was born. The same conceptually beautiful scheme soon inspired many investigators of close binary systems to combine the advancement of model stellar atmospheres with proper treatment of interaction effects between components to elaborate computer codes for constructing synthetic light curves (for details see, for instance, papers by Hill and Hutchings (1970); Wilson and Devinney (1973)).

Interestingly enough in Table IV we have to do not only with a very limited but also a biased statistics. Not only RS CVn and AR Lac already singled out by Kopal himself as peculiar objects (belonging, as we know, to a separate class of binaries of RS CVn-type), but also u Her and V Pup, which are actually early-type contact rather than semi-detached systems.

As we have seen from a lengthy quotation given above, Kopal has not changed in the least his stance in a debate over the Algol paradox even after 15 years following the publications of the original papers by Crawford and Hoyle. Objections raised in Kopal's paper published in 1971 in "Publications of Astronomical Society of the Pacific" can be summarized as follows. Kopal points to the well-known and generally accepted work by Morton (1960) and Smak (1962), who were the first to indicate that mass transfer due to the Roche lobe overflow and the change of the roles of component stars proceeds on a thermal time scale and so, because of the limited available statistics this event can escape observational detection. Next Kopal concentrates on the following counter arguments:

- the lack of semi-detached systems with practically equal masses of the components (confronted against detached systems where on the contrary, components quite frequently possess almost equal masses),
- while treating the mass transfer between component stars one should take into account in parallel with the changes of orbital period and semi-major axis also the accompanying changes of the dimensions of respective Roche lobes,
- serious discrepancy between the early evolutionary models based on the strict assumption of conservative mass transfer and observational data for low-mass systems like Z Dra or RX Hyd with a total mass of less than 2 $M_\odot$, which requires a substantial mass loss of the progenitors. Otherwise it would be virtually impossible for these systems to attain a semi-detached status within the age of our Galaxy or even in a Hubble time.
- possible tidal interaction effects in binaries involving components which rapidly increase their sizes and according to Kopal could redistribute both the energy and angular momentum and in this way to prevent the expansion of the envelope, when the core devoid of hydrogen will start to shrink as it happens in case of a single star.
- another even more serious problem discussed by Kopal is the viscosity of plasma, notably in the outer regions of an expanding star resulting in asymmetric tidal bulges, giving rise to torques, which transfer angular momentum and kinetic energy as well as potential energy of orbit. The loss of mass, argues Kopal, will be accompanied by the loss and transfer of angular momentum.

Kopal emphatically goes on:

Or consider such wide binary systems in our proximity as those of Sirius or Procyon, in which a typical main sequence star is attended by a less massive white dwarf. The latter obviously could not have attained its far more advanced evolutionary stage than that of its mate if it had been less massive throughout all its astronomical past. The conclusion is, in fact, incontestable that the present

Sirius B or Procyon B must once have been more massive than their principal components – probably very much more so judging from their large disparity on evolutionary stage ... A transfer of mass across so great a gap of space would require extremely special initial conditions of ejection which are most unlikely to be met. Much more probable is the view that in such systems, mass was lost by ejection with velocities well in excess of that of escape from the gravitational field of the binary; and if so, a mechanism which can accomplish this could have done the same to more orthodox binaries whose secondary components are to the right – instead of the left – of the main sequence in the H-R diagram.

To sum up Kopal's statements, as long as the just mentioned uneasy aspects are not addressed in a rigorous way, it is too premature to announce that the Algol paradox is resolved (it is noteworthy that Kopal enters here into polemics with Plavec whose review paper on evolution of close binaries was published in the previous volume of the "Publications of Astronomical Society of the Pacific" (Plavec, 1970).

Among the critical arguments of Kopal only one can be relatively easily dismissed. Concerning the lack of binaries with nearly equal masses of the components, it seems that this is an ideal illustration of a well-known aphorism: "the lack of evidence is not the evidence of the lack". In other words, if we recall such bizarre systems like W Ser with extremely peculiar behavior of the their light curves (see, for instance, Guinan, 1989) and assume that there may be a vast amount of similar systems with rather small inclination angles, will we be able to identify adequately such binaries if they are caught in a phase of rapid mass loss and mass transfer?

Now what is absolutely remarkable in my opinion about other remaining difficult issues of the Algol paradox where Kopal is reluctant to compromise or accept the conventional wisdom of his epoch is the fact that exactly a study, tackling the problems explicitly named in Kopal's article from 1971, became a mainstream of research in physics and evolution of close binaries in the last quarter of the 20th century. In all probability, this will be an intriguing subject of investigation for a historian of astronomy who will clarify one day to what extent the ideas of Kopal fermented discussion of open issues or even predestined subsequent progress in evolutionary studies of close binary systems. My task is more modest, also the format of my presentation leaves little space for in-depth analysis, so my discussion is confined just to several theses:

– in fact by the time of publication of the article in question the combined effect of the mass and angular transfer along with the accompanying changes of the Roche lobes sizes tracing momentary variations in mass ratio was treated properly by a young Soviet researcher L. Sniezhko (Sniezhko, 1967) from Ural University, but as you know, the contacts of Soviet astronomers with their Western colleagues at that time were episodic,

– the complicated problem of mass loss and mass transfer is of course multi-faceted and deserves a dedicated and detailed treatment. For the time being it is suffice to say that in the middle seventies Bogdan Paczynski with his associates invoked an idea of a common envelope, which proved to be instrumental in explaining another paradox of binary evolution, exactly the one concerning the occurrence of white dwarfs in Procyon and Sirius so prominent in critical discourse quoted above in the current report(Paczynski, 1971, 1976). Apropos, Paczynski in his pioneering article dedicated to this problem directly refers to Kopal's article.

– it is symptomatic that at the very end of Chapter 6 of his biographical book "Of Stars and Men" Kopal (1986) writes among other things the following lines:

> . . .it has traditionally been assumed so far that the fractional dimensions of the components of close binary systems remain significantly constant over thousands, or tens of thousands of orbital cycles. But although such components may indeed remain in the state of equilibrium, they may *oscillate* about it with frequency depending on their internal structure; and if so, the ratio of their radii would not remain constant but oscillate with periods comparable with that of their orbit. Moreover, if – as is more than likely – the orbital period of the system is not commensurable with those of free oscillations, photometric beat phenomena will arise which make the shape of the light curve not repetitive from cycle to cycle. . . A theory of the photometric consequences of such oscillations has also been developed to a certain point (cf Kopal, 1982); but its application to practical cases still constitutes a task for the (albeit near) future.

Here we have another example of Kopal's very keen insight into the immediate future trends in close binary research.

Summarizing our brief discussion of the early history of resolving the Algol paradox can one state with confidence that this problem at least by now, at the turn of the second millennium, is behind us? According to Batten (1988)

> it is fair to say that there is a consensus that close binary systems evolve by transfer of mass from one component to another (and out of the system) and that in the systems like RW Tau we *may* be witnessing a late stage of this process, while in β Lyr. . . we perhaps see a somewhat earlier stage. It is also fair to recall that the consensus is challenged (Kopal, 1978). All of us must hope that this argument will be settled during the next 100 years, once and for all. Astronomy, being an observational science, rarely if ever provides us with an *experimentum crucis*; but definite proof that in systems like RW Tau we can see material that has been through the carbon-nitrogen cycle would come close to playing the role of such an experiment.

The most recent studies of the effects of chemical evolution in Algols (see, for instance, Sarna and De Greve, 1996) seem to agree favorably with the theoretical predictions at least for carbon to hydrogen abundances, though much more observational material will be needed before the definitive answer to this tantalizing problem will be obtained.

3. Conclusions

- Algol paradox posed a serious challenge to the theory of evolution of close binaries. It has been resolved by the collective efforts of a whole generation of astronomers rather than by anyone's individual ingenious accomplishment.
- One should not necessarily identify the author with his brainchild. Once the latter is perpetuated, be it a published book, patent or a technological novelty, it starts its own life. If we measure a scientific legacy of an individual scholar not that much by his own words but rather by his real impact on future advancements of research, the fundamental role of Kopal in resolving the Algol paradox can hardly be overestimated.

Acknowledgements

The author is deeply indebted to Prof. Alan Batten for many fruitful discussions. Supporting Grant 5760 from Estonian Science Foundation is gratefully acknowledged.

References

Batten, A.H.: 1988, *Pub. Astron. Soc. Pacific* **100**, 160.

Crawford, J.A.: 1955, *Astrophys. J.* **121**, 71.

Guinan, E.F.: 1989, in: A.H. Batten (ed.), *Algols, Proceedings of the 107th Colloquium of the IAU,* Sidney, B.C., 15–19 August 1988, p. 35.

Hill, G. and Hutchings, J.: 1970, *Astrophys. J.* **162**, 265.

Hoyle, F.: 1955, In *The Frontiers of Astronomy*, Heynemann, London, p. 195.

Kopal, Z.: 1955, *Annales D'Astrophysique* **18**, 379.

Kopal, Z.: 1971, *Pub. Astron. Soc. Pacific* **83**, 521.

Kopal, Z.: 1986, *Of Stars and Men*, IOP Publishing, p. 434.

Kron, G.P.: 1958, in: F.B. Wood (ed.), *The Present and Future of the Telescope of Moderate Size,* University of Pennsylvania Press, p. 129.

Kuiper, G.P.: 1941, *Astrophys. J.* **93**, 133.

Morton, D.C.: 1960, *Astrophys. J.* **132**, 146.

Parenago P.P. and Massevich, A.G.: 1950, *Trudy GAISH* **21**, 81 (in Russian).

Paczynski, B.: 1971, *Ann. Rev. Astron. Astrophys.* **9**, 183.

Paczynski, B.: 1976, in: P. Eggleton, S. Mitton and J. Whelan (eds.), *The Structure and Evolution of Close Binary Systems*, Reidel, Dordrecht, p. 75.

Plavec, M.: 1970, *Pub. Astron. Soc. Pacific* **82**, 957.

Sarna, M.J and De Greve, J.P.: 1996, *Q. J. R. Astron. Soc.* **37**, 11.

Smak, J.: 1962, *Acta. Astr.* **12**, 28.

Sniezhko, L.I.: 1967, *Peremennyje zwiezdy* **16**, 253 (in Russian).

Struve, O.: 1954, *5th Liege Coll. Mem. 8 Soc. Roy. Sci. Liege, 4th Ser.* **14**, 236.

Wilson, R.E. and Devinney, E.J.: 1973, *Astrophys. J.* **182**, 539.

GLOBAL SOLUTION OF DYNAMICAL SYSTEMS – A REPORT TO Z. KOPAL, ON WHAT FOLLOWED SINCE

C.L. GOUDAS[1] and K.E. PAPADAKIS[2]

[1]*Department of Mathematics, University of Patras, Campus at Rio, Patras, Greece*
[2]*Department of Engineering Sciences, University of Patras, Campus at Rio, Patras, Greece;*
E-mail: K.Papadakis@des.upatras.gr

(accepted April 2004)

Abstract. Two basic problems of dynamics, one of which was tackled in the extensive work of Z. Kopal (see e.g. Kopal, 1978, *Dynamics of Close Binary Systems*, D. Reidel Publication, Dordrecht, Holland.), are presented with their approximate general solutions. The 'penetration' into the space of solution of these non-integrable autonomous and conservative systems is achieved by application of '*The Last Geometric Theorem of Poincaré*' (Birkhoff, 1913, *Am. Math. Soc.* (rev. edn. 1966)) and the calculation of sub-sets of 'solutions précieuses' that are covering densely the spaces of all solutions (non-periodic and periodic) of these problems. The treated problems are: 1. The two-dimensional Duffing problem, 2. The restricted problem around the Roche limit. The approximate general solutions are developed by applying known techniques by means of which all solutions re-entering after one, two, three, etc, revolutions are, first, located and then calculated with precision. The properties of these general solutions, such as the morphology of their constituent periodic solutions and their stability for both problems are discussed. Calculations of Poincaré sections verify the presence of chaos, but this does not bear on the computability of the general solutions of the problems treated. The procedure applied seems efficient and sufficient for developing approximate general solutions of conservative and autonomous dynamical systems that fulfil the Poincaré–Birkhoff theorems. The same procedure does not apply to the sub-set of unbounded solutions of these problems.

Keywords: approximate general solution, Duffing problem, Roche limit problem, periodic orbit

1. Introduction

Many of the problems that our memorable Professor tackled, the first being that of binary stars, involves issues related to the most complicated field of dynamical systems. For example, the motions within the Roche limit of a binary star system and the exchange of mass is obviously a dynamical problem. Another one is the recovery of the lunar gravity field expressed in sectorial and tesseral harmonics from lunar orbiter tracking data.

The history of dynamics started with the climax of Newton's solution of the two-body problem in 1687 and the anti-climax of Poincaré who showed the non-existence of other integrals in the restricted three-body problem in 1892 (Poincaré, 1892). Later (1927), G. Birkhoff showed that almost all dynamical systems are not integrable. And the last blow came in 1964, when chaos was found to prevail in large areas of the Poincaré 'Surfaces of Sections'. Ironically, this type of chaos

Astrophysics and Space Science **296**: 79–89, 2005.
© Springer 2005

is called 'deterministic' due to the unique dependence of solutions on their initial conditions. Thus dynamics shifted almost exclusively to 'chaotic dynamics' and many interesting contributions were made, such as the KAM theorem. Some investigators went a step further and named Poincaré 'father of chaos', a title that the great master would not appreciate. The reason is as follows:

Poincaré, after proving the non-integrability of the restricted problem ('Les Méthodes Nouvelles de la Mécanique Céleste', 1892), proceeded to sketch the method of penetration into the solution space of insoluble dynamical systems using 'les solutions précieuses', a name that he gave to the periodic solutions. He conjectured (later, Poincaré (1912) and Birkhoff (1913) proved this conjecture) that periodic solutions are 'dense' everywhere in the set of all bounded solutions. He described this approach as 'the only opening' ('la seule bréche') through which solutions of such insoluble problems can be attempted. Unlike the Poincaré Surfaces of Sections method, his recommendation of *'les solutions précieuses'* as tool for developing a general solution was never thoroughly applied or tested. And this, in spite of the numerous contributions and doctorates awarded for calculations of periodic solutions of many dynamical systems. It is only recently that such an effort was undertaken and reported for some cases (Goudas et al., 2003).

Let us now see the scheme to follow for this endeavor: First, we remind that the solution of a set of differential equations is *general*, if it includes the partial solutions corresponding to any and all possible initial conditions. The above famous conjecture of Poincaré, later named by Birkhoff *'Last Geometric Theorem of Poincaré'*, implies that approximation of the general solution of a system can be made by ignoring the non-periodic solutions of a system fulfilling the conditions of this theorem and secure only a dense set of periodic solutions of it.

This is precisely the 'Report to Z. Kopal' to be presented below, to the man who knew and worked, among many other problems, in this field.

In brief, the objective of this report is to develop an approximate general solution of a non-integrable, autonomous and conservative dynamical system. This presentation covers all cases of such systems with axisymmetric potential and limited to cases where the individual solutions are also axisymmetric and bounded (non-escape). The final product is the direct computation of a satisfactorily dense set of *'solutions précieuses'* (periodic solutions). The density of this set is the parameter pre-assigned ε, which is defined in the 'Last Geometric Theorem of Poincaré' (Birkhoff, 1913).

The procedure to be applied includes four steps:

Step 1. Definition of the suitable space of initial conditions r_0 that contains representative points of each and all solutions of the dynamical system.

Step 2. Development of an algorithm capable to approximate initial conditions of periodic solutions r_0^* such that $|r_0 - r_0^*| < \varepsilon$. The parameter ε is to be kept fixed for all solutions and defines the accuracy of the approximate general solution of the dynamical system.

Step 3. Investigation of the properties of all computed periodic solutions, such as
        stability, morphology, and
Step 4. Compiling an atlas of the general solutions.

This paper includes application and results of this methodology to two well-known dynamical systems:

Case 1. The two-dimensional Duffing problem.
Case 2. The restricted 3-body problem in the neighbourhood of the Roche limit.

A common property of these systems is the axisymmetry of their potential and the symmetry of their periodic solutions. As a consequence the space of initial conditions to be adopted is (x_0, C), where C is the energy constant since $\dot{x}_0 = 0$ and $y_0 = 0$, while $\dot{y}_0$ is uniquely dependent on C.

Remarks:

1. Periodic solutions are grouped in mono-parametric families presentable in (x_0, C) as continuous curves.
2. The stability of periodic solutions is determined by calculating the eigenvalues of their monodromy matrix.
3. The actual space of initial conditions does not include parts where motion is not possible.
4. The approximate general solution cannot cover escape motions. Domains of such solutions, wherever encountered, are marked with an E.

2. Test Case 1: The Duffing Problem

As first example of a problem to be treated we select the two-degree of freedom, non-integrable, conservative dynamical system of Duffing with axisymmetric potential $V(x, y)$.

The two-dimensional Duffing problem is expressed by the potential V, where

$$V = \frac{1}{2}(x^2 + y^2) - \frac{\varepsilon_1}{4}x^4 - \frac{\varepsilon_2}{4}y^4 - \varepsilon_3 x y^2, \tag{1}$$

and is of the axisymmetric type.

The equations of motion are

$$\frac{d^2 x}{dt^2} = x - \varepsilon_1 x^3 - \varepsilon_3 y^2, \quad \frac{d^2 y}{dt^2} = y - \varepsilon_2 y^3 - 2\varepsilon_3 x y. \tag{2}$$

By setting the parameters ε_1 and ε_2 equal to unity, we find the only integral F (energy) of the system to be

$$F = \frac{1}{2}\left[\left(\frac{dx}{dt}\right)^2 + \left(\frac{dy}{dt}\right)^2\right] - \frac{1}{2}(x^2 + y^2) + \frac{1}{4}x^4 + \frac{1}{4}y^4 + xy^2 = C. \tag{3}$$

The limit of motion in the (x, y) plane is defined by inequality

$$\frac{1}{2}(x^2 + y^2) - \frac{1}{4}x^4 - \frac{1}{4}y^4 - xy^2 + C \geq 0. \tag{4}$$

The equilibrium points L_1 through L_7 of the problem (coordinates and energy constant) are:

L_1: $x = 0$	$y = 0$	$C_{L_1} = 0$	unstable
L_2: $x = 1$	$y = 0$	$C_{L_2} = -0.25$	stable
L_3: $x = -1$	$y = 0$	$C_{L_3} = -0.25$	unstable
L_4: $x = 0.3472964$	$y = 0.5526367$	$C_{L_4} = 0.0799888$	stable
L_5: $x = 0.3472964$	$y = -0.5526367$	$C_{L_5} = 0.0799888$	unstable
L_6: $x = -1.879385$	$y = 2.181461$	$C_{L_6} = -4.308606$	stable
L_7: $x = -1.879385$	$y = -2.181461$	$C_{L_7} = -4.308606$	stable

The curves that limit the motion (0-velocity curves) and the positions of the equilibrium points are shown in Figure 1.

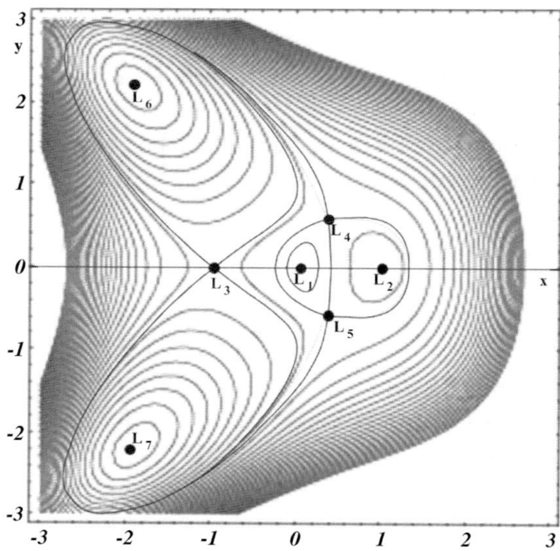

Figure 1. 0-velocity curves and location of the equilibrium points.

The equilibrium points L_2, L_6, and L_7 are stable, and hence the periodic solutions of the overall system, on account of the axisymmetry of the potential (1), appear either as single intersecting normally the x-axis or as symmetrical pairs that do not intersect this axis but encircle the equilibrium points L_6 and L_7.

3. Approximate General Solution of the Duffing Problem

The task of developing an approximation of the general solution of this problem, as already described, will be accomplished by calculating a set of periodic solutions of density ε of the dynamical system. These solutions will be represented by their initial conditions $(x_0, y_0, \dot{x}_0, \dot{y}_0)$. Since all periodic solutions intersecting the x-axis do so with a velocity normal to it (i.e. for $y_0 = 0$, $\dot{x}_0 = 0$), we can make their representation by the phase-space points $\mathbf{r}_0 = (x_0, 0, 0, \dot{y}_0)$ generating them. Selecting at random an initial condition of motion, say $\mathbf{r}_0^*$, leads with probability 1 to a non-periodic solution, and the accuracy of the approximate general solution ε is defined as $\varepsilon = \max[\rho(\mathbf{r}_0^*, \mathbf{r}_0)]$, where $\rho(\mathbf{r}_0^*, \mathbf{r}_0)$ is the distance between any and all $\mathbf{r}_0^*$ and $\mathbf{r}_0$, with $\mathbf{r}_0$ corresponding to any and all periodic solutions of the approximate general solution.

The periodic solutions of the autonomous dynamical system (2) appear in monoparametric families that can be represented in the plane $(x_0, \dot{y}_0)$, or equivalently in the plane (x_0, C), (this latter representation will be used in this paper) with the period serving as parameter of each and all families. Taking into account that the period along the same family varies continuously and hence each and all families constitute a compact one-dimensional manifold imbedded in the two-dimensional space of initial conditions of all solutions (periodic and non-periodic), we can consider each and all families as satisfactorily known (computed) once a sufficient number of members of the family are computed. Essential is, of course, the inclusion in the computed set of the terminal periodic solutions of each and all families.

Due to the oscillatory character of the solutions, the families of periodic orbits will consist of member periodic solutions that re-enter after one complete oscillation in the direction of the y-axis, or after two, or three, etc, oscillations, and so on. Thus families of periodic solutions are distinguished from each other by the number of oscillations required for the re-entrance of their members, so that the terminology to be used is 'family of one oscillation', 'family of two oscillations, etc.

As first approximation of the general solution was selected the one (shown schematically in Figure 2), which includes a set of families of periodic solutions with member solutions of periods corresponding from one to five oscillations per period. The full set of families detected and calculated with these periods and with initial conditions $x \in [0.6, 1.1]$, $C \in [0.5, 1.2]$, is shown in Figure 3. This figure consists of six frames that show the zoom area 1 defined in Figure 2, and illustrate all periodic family curves of solutions re-entering after one oscillation, (top-left frame), after two oscillations (top-middle frame), etc, while the last (bottom-right frame) is

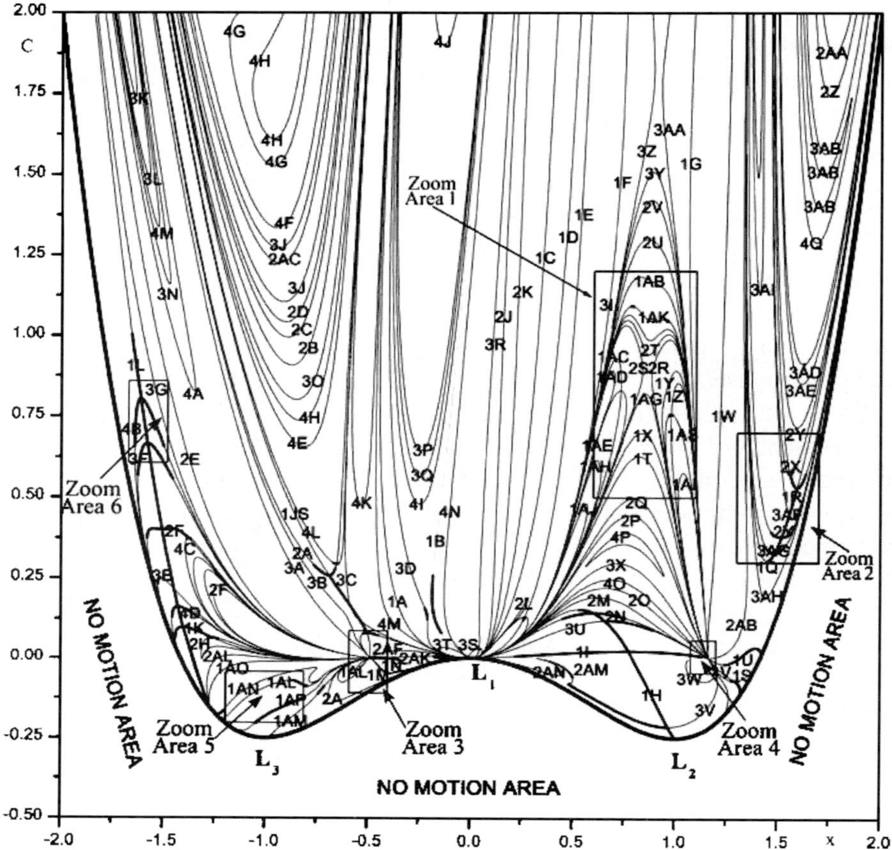

Figure 2. First approximation of the general solution.

a composite presentation of all the families appearing in the previous five frames. This last frame provides evidence of the density of coverage of the space of all solutions (periodic and non-periodic) by the periodic solutions computed. However, the need to develop an approximation of the general solution consisting of at least one periodic orbit in any and all circles of randomly selected centre and of radius ε, for $\varepsilon > 0$, may necessitate calculation of more families using finer steps in the detection procedure and also search for periodic solutions re-entering after a larger number of oscillations. The step used for the approximation appearing in Figure 3 was 10^{-4}, and the search was limited to solutions re-entering after 5 oscillations at the most. In Figure 2, each family curve is labeled by an alphanumeric notation, in which the number indicates the oscillations per period of its member solutions.

Figure 3 shows the approximate general solution within zoom area 6 (see definition of this area in Figure 2), and shows the periodic solution families re-entering

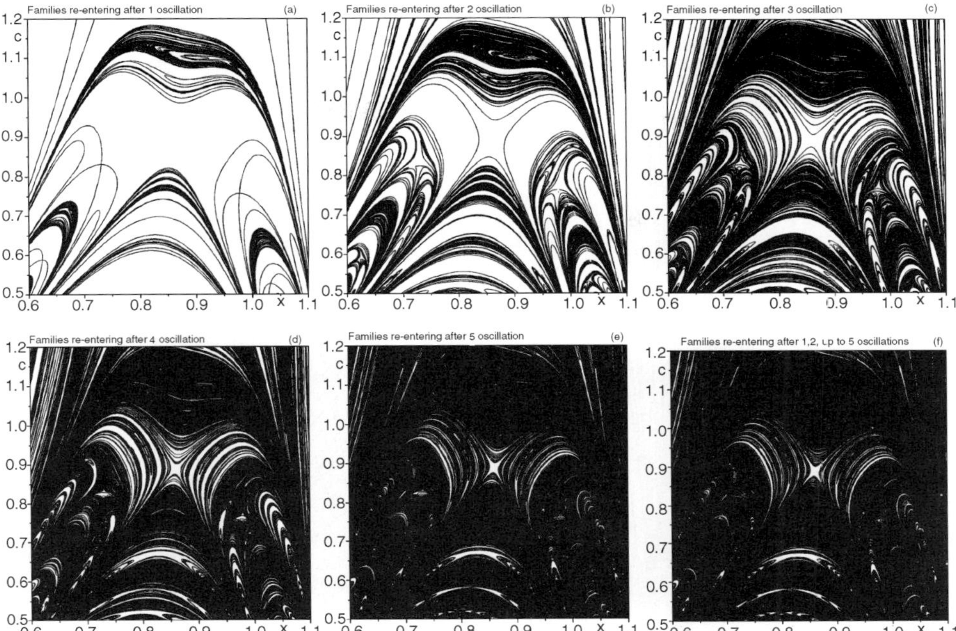

Figure 3. First approximation of the general solution of the Duffing problem inside the zoom area 1, shown in successive steps. The bottom-right frame is the composite of the previous five frames. Blank spaces in the composite frame are limited and mostly provide clear indication of the geometry of the family curves missing in order to achieve complete optical coverage of this limited solution space.

after 1 (a), after 2 (b), after 3 (c), after 4 (d), and after 5 (e), oscillations, whereas (f) shows the composite picture of all previous frames. The accuracy of the general solution inside zoom area 1 is evidently very high.

A second area of the general solution, namely zoom area 6 (see its location in Figure 2) is presented in Figure 4, which shows all periodic family curves re-entering from 1 to 20 oscillations per period. These families cover very densely the saturated part of this figure, but cover inadequately densely the central part. Hence, it is necessary to increase the density of families in the latter part of the space of solutions. This was done by calculating all families of periodic solutions re-entering from 1 to 100 oscillations per period for the sub-area shown as insert in Figure 4. The additional family curves appearing in this insert increased the density of the approximate general solution, although it is evident that achieving densities equal to that of the saturated sub-areas requires raising the re-entrance count to perhaps 1000 oscillations per period. All solution members which are part of the approximate general solution, are saved in an **Atlas of approximate general solution of the two-dimensional Duffing problem**. This atlas can be made available on request.

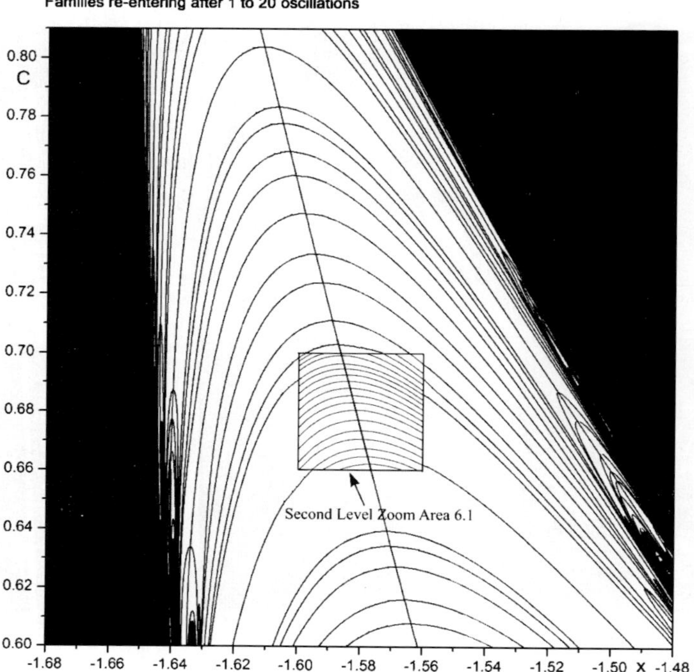

Figure 4. Approximation of the general solution inside zoom area 6 using all families of periodic solutions re-entering after 1–20 oscillations. The insert shows a second approximation inside the low density area using family curves re-entering after 1–100 oscillations.

As already stated the results of this work do not include solutions (and families) that do not cross the x-axis as well as solutions that are not bounded. Yet, no solution is missing from the solution area studied, particularly from the rectangular zoom areas 1 through 6 shown in Figure 2. In Figure 5 we show three periodic solutions of the Duffing problem.

4. Test Case 2: The Restricted Problem Around the Roche Limit

The area of motion in the vicinity of the unstable equilibrium point L_1 of the restricted 3-body problem, defined also as the Roche limit for the dynamics of binary stars, is the next problem to be investigated with the objective to develop an approximate general solution for an area around this limit.

The potential V to be used is

$$V = \frac{1}{2}(x^2 + y^2) + \frac{1 - \mu}{r_1} + \frac{\mu}{r_2}, \tag{5}$$

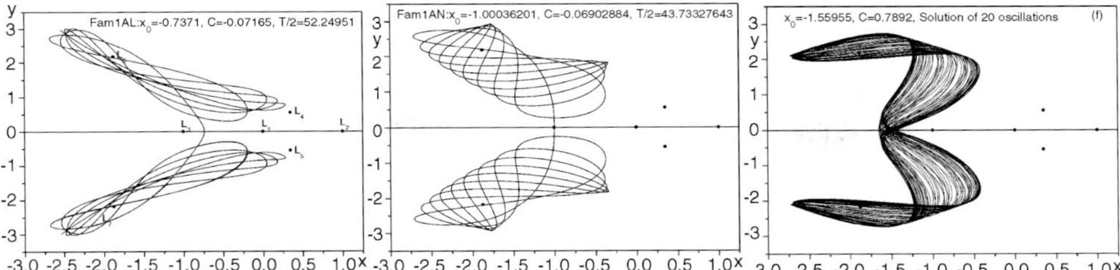

Figure 5. Samples of periodic solutions of the approximate general solution of this problem as an indication of their morphology. Some of these solutions consist of stable, others of unstable ones. In Figure 2, family arcs drawn with heavy lines are stable and with light lines are unstable.

where $r_1^2 = (x + \mu)^2 + y^2$, $r_2^2 = (x + \mu - 1)^2 + y^2$. The corresponding equations of motion are

$$\ddot{x} - 2\dot{y} = x - \frac{(1 - \mu)(x + \mu)}{r_1^3} - \frac{\mu(x + \mu - 1)}{r_2^3}, \tag{6}$$

$$\ddot{y} + 2\dot{x} = y - \frac{(1 - \mu)y}{r_1^3} - \frac{\mu y}{r_2^3},$$

which satisfy the Jacobi integral $2V - (\dot{x}^2 + \dot{y}^2) = C$.

The case treated corresponds to mass parameter $\mu = 0.4$ and to the initial conditions in the domain $x \in [0, 0.25]$, $y \in [3.96, 4]$, at the center of which rests the equilibrium point L_1. The selection of this domain is made so as to develop the part of the approximate general solution that corresponds to energy values, and hence solutions of the mass-transfer type. Indeed, as shown in Figure 6, the open neck of the 0-velocity boundary corresponding to $C = 3.96$, allows particles with energies in the range $C \in [3.96, 3.9809]$ and initial position in the range $x \in [0, 0.25]$, with $x_{L_1} \in [0, 0.25]$, to generate motions encircling both primaries and hence belonging to the mass-transfer type.

The approximation of the general solution developed for the initial conditions domain $[x, C]$ for $x : 0.02 \le x \le 0.25$, $C : 3.96 \le C \le 4$ appears in Figure 7 below. The center of this domain is occupied by L_1.

The stability and morphology of the general solution was examined by computing the eigenvalues, in effect the stability parameter, of each and all member solutions of all families as well as their paths in physical space. Figure 8 shows three such solutions.

The "**Atlas of approximate general solutions of the restricted three body problem in the vicinity of** L_1" that contains all families of periodic solutions and their properties is available on request.

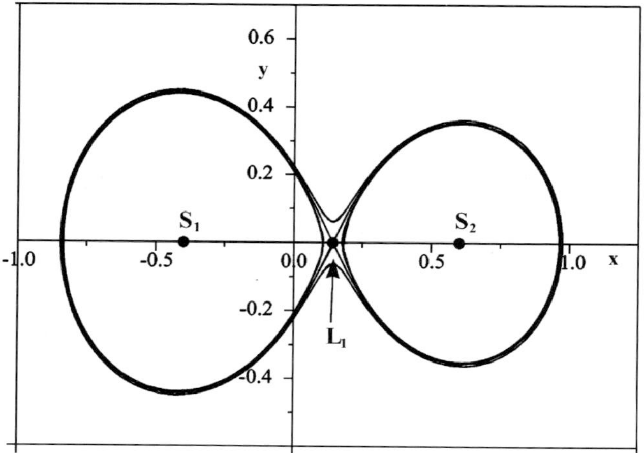

Figure 6. Open and closed 0-velocity boundaries for $C = 4$ (case of closed ovals around the primaries), $C = C_{L_1}$ (case of figure-of-eight), and $C = 3.96$ (case of possible motions around both primaries).

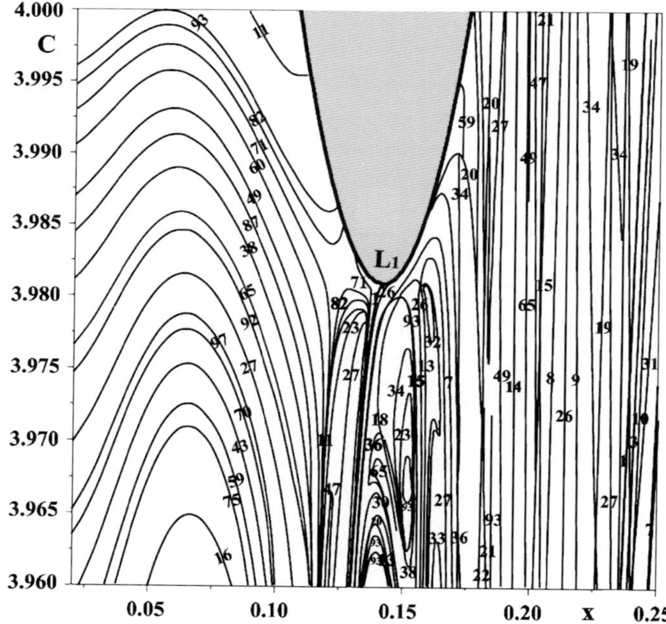

Figure 7. Approximate general solution in a domain around L_1.

5. Discussion

The approximate general solution of the Duffing and the restricted three-body problem in the neighbourhood of the equilibrium point L_1, as transpires from

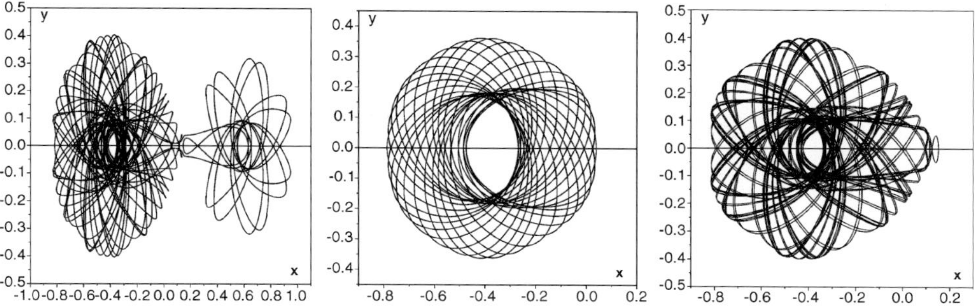

Figure 8. Three member solutions of the families of 61, 81, and 99 oscillations per period.

Figures 2 and 7, provide a satisfactory basis for the overall character of the general solution of these problems. It is clear that an improvement of the accuracy of the general solutions is easy by calculating periodic family curves within the entire space of initial conditions or parts of it. Hence, computation of a dense set of periodic family curves of density ε is clearly possible for both problems as well as for all problems of this category. However, there exists a lower limit for the accuracy of such approximations, and this is posed by the accuracy of the integrator routine and the size of the period of the families to be included in the general solution.

References

Birkhoff, G.D.: 1913, *Tran. Am. Math. Soc.* **14**, 14–22.
Birkhoff, G.D.: 1927, *Am. Math. Soc.* (rev. edn. 1966).
Goudas, C.L., Papadakis, K.E. and Valaris, E., 2003, *Ap&SS* **286**, 461–486.
Kopal, Z. 1978, *Dynamics of Close Binary Systems*, D. Reidel Publication, Dordrecht, Holland.
Poincaré, H.: 1892, Les methodes nouvelles de la mécanique Celeste, tome I, Gauthiers-Villars, Paris.
Poincaré, H.: 1912, *Rend. Cir. Mat. Palermo* **33**, 375–407.

PERIOD CHANGES OF CLOSE BINARY SYSTEMS

H. ROVITHIS-LIVANIOU

*Section of Astrophysics-Astronomy & Mechanics, Department of Physics, Athens University,
Panepistimiopolis, Zografos GR 157 84, Athens, Greece; E-mail: elivan@cc.uoa.gr*

(accepted April 2004)

Abstract. A brief review of the physical mechanisms that could produce the observed orbital period
changes in close binaries, the methods used to study them, and a general discussion is given.

Keywords: binary stars, period changes

1. Introduction

Our present knowledge of period changes in close binary systems is mainly based
on eclipsing variables, since the photometric observations of their minima can
be timed with high accuracy, far surpassing that of radial-velocity measurements
(Batten, 1973). Up to now only a test application has been carried out from non-
eclipsing data: Frasca and Lanza (2000) checked the possibility of detecting orbital period changes by means of spectroscopic orbit determination. Application
was made to AR Lac, for which the spectroscopically determined epochs of superior conjunctions could be compared with those provided by the light minima
photometry.

Concerning the minima of light, they are expected to occur when the apparent
separation of the centers of the two components becomes minimum. Several methods can be found in the literature of how to determine the time of minimum light
of an eclipsing binary. The most common one is that developed by Kwee and van
Woerden (1956).

The time at which a primary minimum is expected to occur is given by the
ephemeris of the binary: Min $I = t_0 + P_{\mathrm{orb}} \times E$; where P_{orb} is the period, t_0 the
time a minimum was originally observed, and E is the epoch.

If P_{orb} is constant, the time at which a primary minimum occurs (observed time,
symbolized by O) will coincide with the one predicted by its appropriate ephemeris
formula. That is, with the – so called – calculated minimum (C). If P_{orb} varies, the
time of an observed minimum will differ from the calculated one. In such a case
the difference $O - C$ is not equal to zero, and in this way an ($O - C$) diagram is
built up.

A period change can be determined when it causes a delay or advance in the expected time of a minimum, which must be significantly greater than the uncertainty
with which the minimum is computed. Small random variations in the intervals

Astrophysics and Space Science **296**: 91–99, 2005.

between successive minima can build up to a large, apparently systematic variation. Thus, some observed period changes might not be real; but despite of this, the statistical evidence for real period changes is very strong, and there is no doubt that they actually happen. Observations have indeed shown that in general there are small but definite changes in the orbital periods of close eclipsing binary systems. According to Yamasaki (1975), a W UMa-type binary will undergo a period change about every 17 000 cycles on average.

The orbital period variations can be real or apparent, increasing, decreasing, or have alternate sign. The smallest detectable period change, no matter if it is real or apparent, is obviously determined by the accuracy with which the period is known. Indeed, the first and very common problem in the construction of the $(O - C)$ diagram of an eclipsing binary is the quality of the observational material. It is also important to get accurate minima times. This can be achieved considering the whole light curve and using modern ways of observation and analysis. Concerning old observational data – which are often used to extend the time interval of existing observations – they might cause some other problems, besides their usually large scatter (for details see Rovithis-Livaniou, 2003).

2. Early Theoretical Considerations

From Kepler's third law, $P^2 = [4\pi^2\alpha^3]/[G(m_1 + m_2)]$, it is evident that a real change in the period must imply a change of either the major axis, α, or the total mass of the system, $(m_1 + m_2)$, or both. If the major axis is assumed to remain constant, then: $\Delta m/m = -2\Delta P/P$. This assumption is valid only if the system loses mass isotropically, which is very rarely – if at all – observed.

Both mass loss and transfer can either increase or decrease the period of a system. The effect of mass transfer depends on the direction of flow within the system; while that of mass loss depends on the direction and speed of the escaping mass (Kruszewski, 1966). Huang (1956) and Piotrowski (1964) showed that transfer of mass between the components could be more efficient in changing the period than mass loss from the system.

It was later considered that gravitational forces are not the only, or even the dominant, interaction between the star and the ejected matter. Detre (1969) suggested that ionized particles in the ejected matter can interact with the magnetic fields of the stars, and that this would be a very efficient way of changing the period.

Kopal (1959), in his book *Close Binary Systems*, represents the period variations of eclipsing binary systems caused, directly or indirectly, by mutual distortion of their components by a complicated equation, (Eqs. 7–35 on p. 88). Plavec (1960) has shown, however, that most of the terms of this equation will produce **unobservable effects** on the times of minima. From all these terms, apsidal motion and light-time effect are the two major causes of periodic variations in the orbital period of close binaries, and are referred as **apparent changes**.

3. Apparent Periodic Changes of the Orbital Period

Monotonic apparent period changes were expected because of the orbital motion of the binary stars around the center of our Galaxy, depending on their location (Kordywski, 1965). But Kreiner (1971), from the data of 137 eclipsing binaries, found that this effect is smaller than the uncertainties in his observational material. So, we will briefly refer only to the two apparent periodic period changes: apsidal motion and light-time effect.

3.1. APSIDAL MOTION – LIGHT-TIME EFFECT

The periastron line of the relative orbit of a close binary can move in space because of the tidal interaction of the two components or because of the presence of a third body accompanying the close pair. This subject, known for long, has been studied by many investigators (e.g., Slavenas, 1927; Kopal, 1959; Martynov, 1948, 1973, etc.). As a result the $(O - C)$ diagrams show a **strictly periodic** sinusoidal apparent period change (Hall, 1990).

A rotation of the periastron can also result from relativistic phenomena. The validity of the widely used classical apsidal motion formula was checked by Quataert et al. (1996), who found that it gives very accurate results, if the periods of the low-order quadrupole g, f, and p-modes are smaller than the periastron passage time by a factor of about 10 or more.

Concerning the apsidal motion rates, they are different for each of the three mentioned causes of change. Investigations of W UMa-type binaries, i.e., have shown that these coming from the tidal interaction of the two members are 100–1000 times larger than those due to relativistic phenomena, and 20–200 times larger than those caused by the presence of a third body.

The presence of a third body in a close binary, except for its effect on the periastron of the close pair orbit, also produces a periodic motion of its center of mass, as is observed from the Earth. As a result there is either a delay or an advance in the time a minimum is expected (e.g., Irwin, 1959; Martynov, 1973; Mayer, 1990; Claret and Giménez, 1993).

Both apsidal motion and light-time effect have been extensively investigated and applied to many particular systems (i.e., Company et al., 1988; Khaliullin et al., 1991; Khodykin and Vedeneyen, 1997; Smeyers et al., 1998; Ibanoğlu, 2000; Hegedüs, 2003 for the first; and, Drechsel et al., 1989; Mayer, 1997; Lorenz et al., 1998; Mayer et al.; 2001; Mayer and Wolf, 2002), for the latter effect.

4. Real Period Changes

Mass and angular momentum transfer and/or loss were the first proposed mechanisms to explain the observed period changes in a close binary. Binary components will lose or gain mass and angular momentum, to or from their companions, in a manner controlled by the Roche geometry and by the speed with which the stars react to such changes.

A star can reach its Roche lobe by expansion. This can happen – as is well known – through normal stellar evolution processes. Alternatively, the Roche lobe may shrink down onto the surface of the star as a result of orbital angular momentum loss by means of magnetic braking or gravitational wave radiation.

As the rate of mass transfer/loss from the Roche-lobe filling star is concerned, it depends on how the radius of the star changes in response to its mass changes.

4.1. MASS TRANSFER

Conservative mass transfer was the simplest case one could consider: all of the mass lost by one component is gained by its mate. So the total mass of the binary is conserved, together with the total orbital angular momentum J_{orb}. Thus, the sum $m_1 + m_2 = M_T$ will be constant, and $dm_1 = -dm_2$. In general:

$$J_{orb} = \left[Gm_1{}^2 m_2{}^2 \alpha (1 - e^2)/M_T \right]^{1/2}$$

If m_{1i}, m_{2i} and P_i denote the initial values for the two stellar masses and the orbital period, respectively, before any mass exchange process, and m_1, m_2 and P their values after mass exchange, then:

$$\frac{P}{P_i} = \left[\frac{m_{1i} m_{2i}}{m_1 m_2} \right]^3$$

From which we finally get:

$$\frac{\dot{P}}{P_i} = \frac{3\dot{m}_1 \times (m_1 - m_2)}{m_1 m_2},$$

which means that $\dot{P}$ depends on $\dot{m}_1$. The difference $(m_1 - m_2)$, in the mentioned relation can change sign as mass is transferred from one star to the other.

Mass transfer rates have been estimated for different kinds of close binaries as well as at various evolutionary stages. Average mass transfer rates, i.e. for Algols, were derived by Hall and Neff (1976) by analysing their observed period changes, while King and Watson (1987) calculated the rates for CVs.

4.2. MASS AND ANGULAR MOMENTUM LOSS

What is mostly expected to occur is non-conservative mass transfer. In this case only some fraction of the mass and angular momentum is transferred from the loser to the gainer star, while the rest is lost from the binary. This can be achieved by stellar wind or RLOF events. Although sudden catastrophic mass loss, such as from a nova or a supernova event, could also happen.

The simplest representation for mass loss due to a wind is to consider a spherically symmetric wind that does not interact with the companion star. In such case we get the simple relation:

$$\frac{\dot{P}}{P} = \frac{-2\dot{m}_1}{(m_1 + m_2)}.$$

Many other stellar wind scenarios have been suggested. Tout and Hall (1991), i.e., considered an enhanced stellar wind and investigated how the angular momentum and the period vary.

Theoretical estimates of angular momentum loss (AML) based on different assumptions for the magnetic braking law – for binaries with solar-type components – have been made by Skumanich (1972), Maceroni and van't Veer (1991), Maceroni (1992), and Stepien (1995). And according to Maceroni (1999) orbital period changes due to AML would be observable with certainty over a timespan of a century.

Various mass loss models have been proposed. According to Robertson and Eggleton (1977), i.e., a steady loss of angular momentum will drive the system towards smaller mass ratios. And evolution towards more extreme mass ratios can be found on the secondary thermal timescale, which becomes longer as the secondary transfers mass to the primary (Rahunen, 1981).

Taam (1983), on the other hand, considered the influence of angular momentum loss on the evolution of low-mass binary systems with collapsed companions and found that typical mass transfer rates, induced by the loss of angular momentum associated with magnetically coupled winds, are larger than, or approximately equal to 10^{-9} $M_\odot yr^{-1}$.

Other mass loss models have been developed by many other investigators.

4.3. MAGNETIC ACTIVITY

Nowadays, semi-regular or quasi-periodic changes of the period of rather short time scales are generally attributed to cycles of magnetic activity. Applegate and Patterson (1987) proposed that a variable quadrupole moment produced by magnetic activity in the outer convection zone of one of the components of a close binary is responsible for orbital period changes that occur on a short time scale, of the order of 10 years. And according to Applegate (1992) the magnetic field of a star could produce angular momentum transfer among its internal and external layers. This could vary its quadrupole moment and cause period changes. Rudiger et al. (2002) examined whether dynamo-degenerated fields are able to produce such quadrupole terms in the gravity potential, which can explain the observed cyclic orbital period changes of RS CVn-type active binaries. Also recently, (e.g., Ak et al., 2001;

Baptista et al., 2002, 2003) solar-type magnetic cycles in the secondary components of some CVs were reported.

5. Orbital Period Analysis Techniques

The orbital period changes of eclipsing binary systems can be found by the analysis of their $(O - C)$ diagrams. Every diagram is based on an ephemeris formula, and usually a linear one is used for its construction. Although quadratic, third or higher order approximations or even a sinusoidal description can be given, depending on the diagram's shape.

A review concerning all analysis methods of $(O - C)$ diagrams of any close eclipsing binary and a comparison between traditional and new techniques has been presented by Rovithis-Livaniou (2001).

Regarding in particular contact binaries a review concerning the problems of construction of their $(O - C)$ diagrams, their analysis and information that can be achieved has been given by Rovithis-Livaniou (2003). So, it is not necessary to repeat all of this here again. We only like to stress the superiority of the new methods, since they are not based on any assumption and are consistent with the physics of close binaries.

6. Discussion

In this work we have passed very briefly through the field of orbital period changes of close binaries. A huge and interesting subject, that has attracted the interest of both theoreticians and observers, since it is related to stellar structure and evolution problems. Mass transfer, mass and/or angular momentum loss, stellar winds, magnetic activity cycles, apsidal motion, light-time effect, or TRO cycles are the mechanisms proposed so far to produce the observed period changes in close binary systems.

Because orbital period changes are found from an $(O - C)$ diagram analysis, it is very important to pay special attention to both the way the diagrams are constructed and their analysis method. As their construction is concerned, the quality of the observational material and the problems that might arise from large E and small P values must be considered carefully to avoid any erroneous detection of abrupt period changes (Rovithis-Livaniou, 2003). Concerning the $(O - C)$ diagram analysis, the classical linear and quadratic fitting used for their analysis might not be good enough in most cases. Since they emerge as a specific application of the first continuous method (Kalimeris et al., 1994a), it is suggested to use this method, which according to Hilditch (2001), is very useful.

From an as accurate as possible description of an $(O - C)$ diagram, the way and rate of orbital period changes during the examined time interval can be immediately calculated using a pure and simple mathematical approach, as suggested by Kalimeris et al. (1994a). Moreover, the $P(E)$ function does not depend on the ephemeris formula used, as has been proven by Kalimeris et al. (1994b); so, periodic terms, if any, can be found by applying Fourier transforms to the $P(E)$ function. The results achieved so far show that in most of the examined cases the computed periodic terms correspond to the time interval covered by the observations and to half of it (Rovithis-Livaniou et al., 2001). These results rise a lot of questions with regard to the possible existence of third bodies associated with close binaries, except of course those cases, where third bodies are confirmed by other means (e.g., Hendry and Mochnacki 1998). In such cases the influence of the third body has to be removed from the $(O - C)$ diagram before its analysis, as has been done, i.e., in the case of V505 Sgr (Rovithis-Livaniou et al., 1996). If a third body really exists, its influence on the $(O - C)$ diagram of the system will appear sooner or later.

It is now believed that in solar-type stars there are two dynamos in action. One in the convection zone producing aperiodic small-scale fields, and one in the weakly turbulent overshoot layer – just below the convection zone – where the strong periodic cycle fields are generated (Hoyng, 1996). As a result, the quasi-periodic changes of the period are generally attributed to cycles of magnetic activity; although, there are some observational criteria of the Applegate mechanism, which are not always fulfilled.

Magnetic fields are related to spot activity; but the effect of cool spots does not affect RS CVn-type and contact binaries, since the shift of minima times they produce is very small. Especially for contact systems, Kalimeris et al. (2002) showed that the spots do not cause permanent slope variations like real period changes do.

The atlas of $(O - C)$ diagrams by Kreiner et al. (2001), together with old and new databases (i.e., ASAS: Pojmanski, 2002), supplied us with a wealth of data. Believing that the $(O - C)$ diagrams carry much and valuable information, it depends on us to retrieve it by analysing and translating it correctly in order to yield reliable results.

Acknowledgements

This work was partly financially supported by Athens University (grant no. 70/4/3305). The NASA ADS Abstract Service was used.

References

Ak, T., Ozkan, M.T. and Mattei, J.A.: 2001, *A&A* **369**, 882.
Applegate, J.H.: 1992, *ApJ* **385**, 621.
Applegate, A.H. and Patterson, J.: 1987, *ApJ* **322**, L99–L102.

Baptista, R., Jablonski, F., Oliveira, E., Vrielmann, S., Woudt, P.A. and Catalan, M.S.: 2002, *MNRAS* **335**, L75.

Baptista, R., Borges, B.W., Bond, H.E., Jablonski, F., Steiner, J.E. and Grauer, A.D.: 2003, *MNRAS* **345**, 889.

Batten, A.H.: 1973, Binary and Multiple Systems of Stars, Pergamon Press Oxford, New York, Toronto, Sydney, Braunschweig. Intern. Series of Monographs in *Natural Philosophy*, general Ed. D. Ter Haar, **51**.

Claret, A. and Giménez, A.: 1993, *A&A* **277**, 487.

Company, R., Portilla, M. and Gimenez, A.: 1988, *ApJ* **335**, 962.

Detre, L.: 1969, in: L. Detre (ed.), Non-Periodic Phenomena in Variable Stars D. Reidel Dordrecht, p. 3.

Drechsel, H., Lorenz, R. and Mayer, P., 1989, *A&A* **332**, 909.

Frasca, A. and Lanza, A.F.: 2000, *A&A* **356**, 267.

Hall, D.S.: 1990, in: C. Ibanoğlu (ed.), Active Close Binaries, Kluwer, Dordrecht, p. 95.

Hall, D.S. and Neff, S.G.: 1976, in: P. Eggleton, S. Mitton and J. Whelan (eds.), IAU Symp. No. 73, D. Reidel Publ. Co, Structure and Evolution of Close Binary Systems, p. 283.

Hegedüs, T.: 2003, in: O. Demircan and E., Budding (eds.), New Directions for Close Binaries Studies. The Royal Road to the Stars, p. 59.

Hendry, P.D. and Mochnacki, S.W.: 1998, *ApJ* **504**, 978.

Hilditch, R.W.: 2001, An Introduction to Close Binary Stars, Cambridge University Press.

Hoyng, P.: 1996, in: R. Pallavicini and A.K. Dupree (eds.), PASP Conf. Ser. Vol. **109**, p. 59.

Huang, S.S.: 1956, *AJ* **61**, 49.

Ibanoğlu, C.: 2000, Ibanoğlu C. (ed.), Proc. NATO ASI Ser. C: Mathematical and Physical Sciences, Vol. **544**, Variable Stars as Essential Astrophysical Tools, p. 565.

Irwin, J.B.: 1959, *AJ* **64**, 149.

Kalimeris, A., Rovithis-Livaniou, H. and Rovithis, P.: 1994a, *A&A* **282**, 775.

Kalimeris, A., Rovithis-Livaniou, H., Rovithis, P., Oprescu, G., Dumitrescu, A. and Suran, M.D.: 1994b, *A&A* **291**, 765.

Kalimeris, A., Rovithis-Livaniou, H. and Rovithis, P.: 2002, *A&A* **387**, 969.

King, A.R. and Watson, M.G.: 1987, *MNRAS* **227**, 205.

Khaliullin, Kh.F., Khuodykin, S.A. and Zakharov, A.I.: 1991, *ApJ* **375**, 314.

Khuodykin, S.A. and Vedeneyev, V.G.: 1997, *ApJ* **475**, 798.

Kopal, Z.: 1959, Close Binary Stars, Wiley.

Kordywski, H.: 1965, *Acta Astron.* **15**, 133.

Kreiner, J.M.: 1971, *Acta Astron.* **21**, 365.

Kreiner, J.M., Kim, C.-H. and Nha, T.S.: 2001, An Atlas of O-C Diagrams of Eclipsing Binaries, Cracow, Poland.

Kruszewski, A.: 1966, *Advances in Astron. & Astrophys.* **4**, 233.

Kwee, K.K. and van Woerden, H.: 1956, *Bull. Astron. Inst. Neth.* **12**, 327.

Lorenz, R., Mayer, P. and Drechsel, H.: 1998, *A&A* **332**, 909.

Martynov, D.Ya.: 1948, Izv. Engelbardt Obs. Kazan No. 25.

Martynov, D.Ya.: 1973, in: V.T. Tsesevich (ed.), Eclipsing Variable Stars, Wiley, p. 270.

Maceroni, C.: 1992, W.W. Weiss and A. Baglin (eds.), ASP Conf. Ser., Vol **40**, p. 374.

Maceroni, C. and van't Veer, F.: 1991, *A&A* **246**, 91.

Maceroni, C. and van't Veer, F.: 1993, *A&A* **277**, 515.

Massi, M., Weidhofer, J., Torricelli-Ciamponi, G. and Chiuderi-Drago, F., 1998, *A&A* **332**, 149.

Mayer, P.: 1990, *BAC* **41**, 231.

Mayer, P.: 1997, *A&A* **324**, 988.

Mayer, P. and Drechsel, H.: 1987, *A&A* **183**, 61.

Mayer, P., Lorenz, R., Drechsel, H. and Abseim, A.: 2001, *A&A* **366**, 558.

Mayer, P. and Wolf: 2002, IBVS No. 5293.

Plavec: 1960, *BAICz* **11**, 148 & 197.

Piotrowski, S.L.: 1964, *Acta Astron.* **14**, 4 & 251.

Pojmanski, G.: 2002, *Acta Astron.* **52**, 397.

Quataer, E.J., Kumar, P. and Ao, C.-O.: 1996, *ApJ* **463**, 284.

Rahunen, T.: 1981, *A&A* **102**, 81.

Robertson, J.A. and Eggleton, P.P.: 1977, *MNRAS* **179**, 359.

Rovithis-Livaniou, H.: 2001, *Odessa Astron. J.* **14**, 91.

Rovithis-Livaniou, H.: 2003, in: O. Demircan and E. Budding (eds.), New Directions for Close Binaries Studies: The Royal Road to the Stars, p. 39.

Rovithis-Livaniou, H., Fragoulopoulou, E., Sergis, N., Rovithis, P. and Kranidiotis, A.: 2001, *Ap&SS* **275**, 337.

Rovithis-Livaniou, Kranidiotis, N.A., Rovithis, P. and Kalimeris A.: in: D. Chochol, A. Skopal, and Th. Pribulla (eds.), Physical Processes in Interacting Binaries, p. 14.

Rudiger, G., Elstner, D., Lanza, A.F. and Granzer, Th.: 2002, *A&A* **392**, 605.

Skumanich, A.: 1972, *ApJ* **171**, 565.

Slavenas, P.: 1927, *Trans. Astron. Obs. Yale Univ.* **6**, 3.

Smeyers, P., Willems, B. and van Hoolst, T.: 1998, *A&A* **335**, 622.

Stepien, K.: 1995, *MNRAS* **274**, 1019.

Taam, R.E.: 1983, *ApJ* **268**, 361.

Tout, C.A. and Hall, D.S.: 1991, *MNRAS* **253**, 9.

van't Veer, F.: 1972, *A&A* **20**, 131.

van't Veer, F. and Maceroni, C.: 1992, *Bats Proc.* 237.

Yamasaki, A.: 1975, *Ap&SS* **34**, 413.

THE ECLIPSING BINARY BX ANDROMEDAE AND ITS ORBITAL PERIOD BEHAVIOUR

HELEN ROVITHIS-LIVANIOU[1], SOTIRIOS TSANTILAS[1], PETER ROVITHIS[2],
DRAHOMÍR CHOCHOL[3], AUGUSTÍN SKOPAL[3] and THEODOR PRIBULLA[3]

[1]*Section of Astrophysics, Astronomy and Mechanics, Department of Physics, University of Athens, Panepistimiopolis, Zografos, Athens, Greece; E-mail: elivan@cc.uoa.gr*
[2]*Institute of Astronomy & Astrophysics, National Observatory of Athens, P.O. Box 20048, Athens, Greece*
[3]*Astronomical Institute of the Slovak Academy of Sciences, Tatranská Lomnica, Slovakia*

(accepted April 2004)

Abstract. The orbital period variations of the eclipsing binary BX And are examined analysing its $(O–C)$ diagram 1) with the standard method, in which the minima times are fitted by the quadratic ephemeris combined with an *assumed* light-time effect, and 2) with the first continuous method. The results from the use of the two methods are, as was expected, different.

Keywords: eclipsing binaries, extrasolar planets, OGLE-TR-56

1. Introduction

The eclipsing binary BX And (HD 13078) is the brighter component of the visual binary ADS 1671. Many photoelectric observations exist for BX And (e.g. Svolopoulos, 1957; Todoran, 1965; Castelaz, 1979; Rovithis and Rovithis-Livaniou, 1984; Gulmen et al., 1988; Samec et al., 1989; Derman et al., 1989; Bell et al., 1990; Demircan et al., 1993) and only one spectroscopic study (Bell et al., 1990).

According to Chou (1959) and Ahnert (1975), a major period increase of about 0.25 s occurred around 1950, while Gulmen et al. (1988) reported an abrupt period decrease in 1981. Bell et al. (1990) noted that while the orbital period before and after these major changes does not really differ much, the period in the interval 1950–1981 was approximately 3×10^{-6} days longer. They described the $(O–C)$ diagram by a sine wave with a period of approximately 78 years and a semi-amplitude of about 0.015 days. Demircan et al. (1993) used the same approach and found a period of 71 ± 2 years and a semi-amplitude of 0.0161 ± 0.0005 days. As the explanation of the orbital period changes of BX And is still open and unclear, we decided to re-analyze the $(O–C)$ diagram of BX And, using all available data, that cover 103 years.

2. Analysis with a Traditional Method

The minima times of BX And were fitted by a simple standard best-fit method based on the assumption of a quadratic ephemeris combined with the light-time effect,

Astrophysics and Space Science **296**: 101–104, 2005.
© Springer 2005

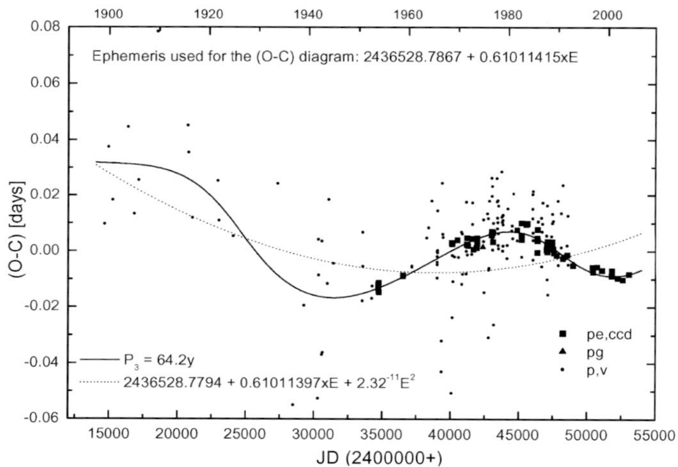

Figure 1. The (*O–C*) diagram of BX And and the best fit of minima times assuming a quadratic ephemeris and accounting for the light-time effect.

caused by the presence of a *hypothetical* third body in the system. The method was given by Chochol et al. (1998). As it is seen in Figure 1, the (*O–C*) diagram can be well described by the long-term orbital period increase (dashed line) and the light-time orbit (solid line) with a period of the third body of $P_3 = 64.2 \pm 4.4$ years. But, what will happen if another (*O–C*) diagram of BX And will be constructed, based on another ephemeris formula?

3. Analysis with the First Continuous Method

The analysis of the (*O–C*) diagram of BX And was also carried out using the first continuous method, in which both the *period* and its *rate of change* are considered to be *random functions of time* and higher-order polynomials are used to describe the (*O–C*) diagrams of any type of eclipsing binary (Kalimeris et al., 1994). Working in this way, a sixth-order polynomial was used to describe the (*O–C*) diagram of BX And, while the orbital period variation of the system and its rate of change were computed, are presented in Figure 2.

A very small long-term orbital period decrease of -1.375×10^{-11} days/E or -7.1×10^{-4} s per year is detected from Figure 2. Moreover, the wave-like shape of the $P(E) - P_e$ function, where P_e is the constant ephemeris period, may denote the action of a periodic mechanism in the system. Applying discrete Fourier transform to the $P(E) - P_e$ function, a search for periodicities was made. The result was two period terms of 103 and 51.5 years with amplitudes of 2.05×10^{-6} days and 2.17×10^{-6} days, respectively. The $P(E)$ function (after substraction of the long-term decrease) and the combined periodic term are shown in Figure 3 by the solid and dashed lines, respectively.

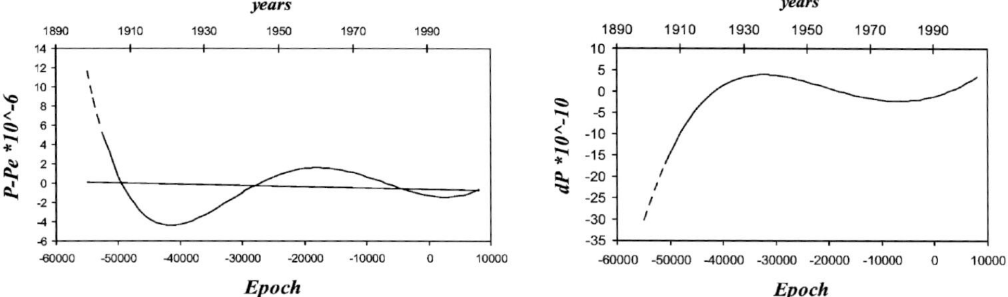

Figure 2. Period variations of BX And and their rate of change.

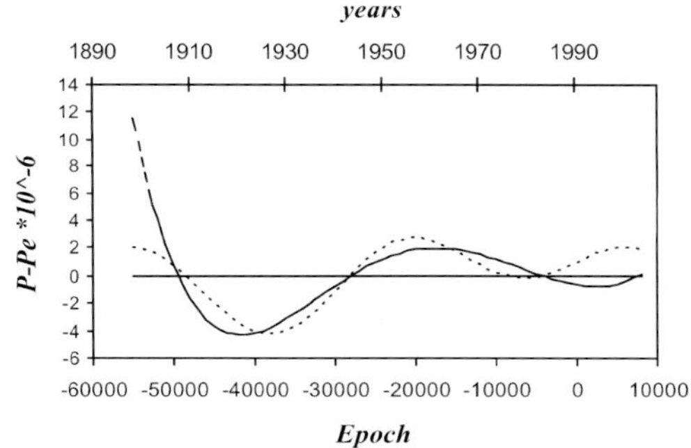

Figure 3. Comparison of the period variations of BX And and the periodic terms.

4. Discussion

Although the early data of BX And exhibit very large scatter, they were included into our analysis to cover a larger interval of minima times, as was recommended by Rovithis-Livaniou (2003).

The minima times of BX And were fitted by a standard method assuming a quadratic ephemeris modulated by the light-time effect. The fit of the data by the long-term orbital period increase combined with the 64.2 year light-time orbit is very good (see Figure 1). But, according to Rovithis-Livaniou (2001) and Kalimeris et al. (2002), the $(O-C)$ diagram *is forced* to be constructed in a *pre-described way*. Moreover, it is not known if using another ephemeris formula the description will continue to be pretty good with the *same* periodic term as in Figure 1. The application of the recent first continuous method (Kalimeris et al., 1994) to BX And revealed a very small long-term decrease of the orbital period and changing episodes of *continuously* decreasing and increasing periods. More specifically: the period was decreasing from the year 1899 on, which corresponds to the beginning of the existing observations, until about 1922; then it started to increase again. This phase lasted until about 1960, when another epoch of decreasing orbital period started and was going on until 1996. Then a new period increase began, which is still continuing to date. The periodic terms resulting from the use of the two methods are different. This was expected for the reasons explained by Rovithis-Livaniou (2001), or by Kalimeris et al. (2002). From Figure 2, where the rate of period variations of BX And is presented, it is obvious that the foregoing mentioned changes did not occur with the same rate.

Certainly, more and accurate data are needed for definite conclusions of the orbital period changes of BX And. Especially, since: a) the detected periodic terms correspond to the time interval for which observational material exist, and to half of it, and b) as is obvious from Figure 3, the wave-like shape of the $P(E) - P_e$ function is going on, while *diminishing both its length and amplitude*. This may suggest that a damping mechanism of the periodic disturbance in the system is at work.

Acknowledgements

This work was partly financially supported by Athens University (grant No. 70/4/3305) and by the VEGA grant No. 4014.

References

Ahnert, P.: 1975, *Mitt. Ver. Sterne* **6**, 189.

Bell, S.A., Rainger, P.P., Hill, G. and Hilditch, R.W.: 1990, *MNRAS* **244**, 328.

Castelaz, M.: 1979, IBVS, No. 1554.

Chochol, D., Pribulla, T., Teodorani, M., Errico, L., Vittone, A.A., Milano, L. and Barone, F.: 1998, *A&A* **340**, 415.

Chou, K.C.: 1959, *AJ* **64**, 468.

Demircan, O., Akalin, A. and Derman, E.: 1993, *A&AS* **98**, 583.

Derman, E., Akalin, A. and Demircan, O.: 1989, IBVS, No. 3325.

Gulmen, O., Gudur, N., Sezer, C. Eker, Z., Keskin, V. and Kilinc, B.: 1988, IBVS, No. 3266.

Kalimeris, A., Rovithis-Livaniou, H. and Rovithis, P.: 1994, *A&A* **282**, 775.

Kalimeris, A., Rovithis-Livaniou, H. and Rovithis, P.: 2002, *A&A* **387**, 969.

Rovithis, P. and Rovithis-Livaniou, H.: 1984, *Ap&SS* **105**, 171.

Rovithis-Livaniou, H.: 2001, *Odessa Astron. Publ.* **14**, 33.

Rovithis-Livaniou, H.: 2003, in: O. Demircan and E. Budding (eds.), New Directions for Close Binaries Studies: The Royal Road to the Stars, COMU Astrophysics Research Center Publication, Canakkale, Turkey, p. 39.

Samec, R.G., Fuller, R.E. and Kaitchuck, R.H.: 1989, *AJ* **97**, 1159.

Svolopoulos, S.N.: 1957, *AJ* **62**, 330.

Todoran, I.: 1965, *St. Cerc. Astron.* **10**, 71.

THE UNIQUE BINARY STAR α CORONAE BOREALIS

IGOR M. VOLKOV

Sternberg State Astronomical Institute, Universitetskij Prosp., Moscow, Russia;
E-mail: imv@sai.msu.ru

(accepted April 2004)

Abstract. We present a new photoelectric light curve of α Coronae Borealis. The derived rate of apsidal motion differs from the theoretical prediction. A possible solution of the problem is suggested.

Keywords: apsidal motion, double stars, rotation axis

1. Introduction

α CrB is a detached pair of stars with an eccentric orbit and a long (17^{d}) orbital period. The variability of the star was discovered by Stebbins (1928), who has observed the annular eclipses ($\Delta m = 0.1^{\mathrm{m}}$) of the bright star by its weak companion. Kron and Gordon (1953) were able to detect the skin-deep total secondary eclipse ($\Delta m = 0.017^{\mathrm{m}}$) in the near infrared region, at $\lambda = 723$ nm. Ebbighausen (1976) tried to measure the rate of apsidal motion in the system from his own spectral derivation of the longitude of periastron and from more earlier spectroscopy, but failed. Tomkin and Popper (1986) detected lines of the weak component in the $\lambda = 880$ nm region of the combined spectra of the system and derived the masses of the components.

2. Observations

α CrB was observed at the Tian-Shan High altitude mountain observatory (3000 m above sea level) of Moscow University. We used a single-channel photoelectric photometer (EMI 9863, S20 cathode) attached to the 48-cm reflector. Two interferometric filters centered at $\lambda = 460$ and 751 nm were used. The star HD 135502 served as prime standard, and HD 143761 as check star. Both stars did not show variability exceeding 0.002^{m}. Because of intrinsic variability α CrB changes its brightness between minima at a level of $0.01^{\mathrm{m}} - 0.03^{\mathrm{m}}$. So we had to bin our observations in time intervals from 5 to 20 min in order to reduce the influence of this variability. The results of our observations in minima are presented in Figure 1 along with O–C residuals from the solution presented in Table I.

Astrophysics and Space Science **296**: 105–108, 2005.
© Springer 2005

TABLE I

The absolute parameters of α CrB

Parameter	Primary	Secondary
$M/M_\odot$	2.57 ± 0.05	0.92 ± 0.04
$R/R_\odot$	2.93 ± 0.05	0.85 ± 0.03
$\log L/L_\odot$	1.79 ± 0.03	-0.24 ± 0.04
T_{eff} (K)	9460 ± 150	5520 ± 80
$\log g$ (cm s^{-2})	3.91 ± 0.03	4.54 ± 0.02
$v \sin i$ (km s^{-1})	126 ± 7	<14
M_V	$+0.43 \pm 0.01$	$+5.47 \pm 0.04$
M_{bol}	$+0.21 \pm 0.02$	$+5.29 \pm 0.04$
S_p	B9.5 IV	G7.5 V
π''	0.0436 ± 0.0008	
e	0.374 ± 0.014	
ω°	310.85 ± 0.12	
$a/R_\odot$	42.8 ± 0.8	
age (years)	$4.4 \pm 0.3 \cdot 10^8$	

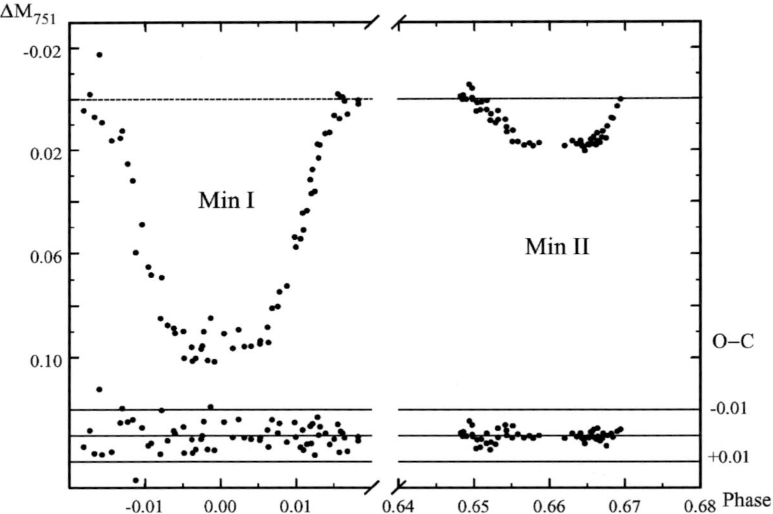

Figure 1. The light curve of α CrB close to both minima.

3. Absolute Elements and Apsidal Motion

The analysis of α CrB is based on the Ebbighausen spectroscopic observations of the primary component, on the Tomkin & Popper radial velocities of the secondary star, and on our new photoelectric narrow-band observations. We used the Hipparcos parallax, and $v \sin i$ is the mean value from all available literature sources. The age of the system was estimated by comparing the obtained parameters of the primary component with stellar models from Claret and Gimenez (1992) for stars with solar chemical composition. Because of the considerably wide separation of the system, there are no proximity effects in the light curve of α CrB. The effects of the fast rotation of the primary component (see Table I) are less than the errors of observations, and we consider the form of the stars as ideal spheres. The light curve fitting was done by means of the iterative method of differential corrections by Khaliullina and Khaliullin (1984). Volkov (1993) already presented first observations of the apsidal motion in the system. The analysis of all available observations including the Schmitt (1998) X-ray observations of the secondary minimum allowed us to find new linear ephemerides for the times of primary and secondary minima (Figure 2).

$$\text{HJD (Min I)} = 2,447,346.1174(9) + 17.3599018(12) \cdot E$$
$$\text{HJD (Min II)} = 2,447,010.3931(21) + 17.3599272(55) \cdot E.$$

The difference of these periods enables us to derive the rate of apsidal motion as $\dot{\omega}_{\text{obs}} = 0°.0104 \pm 0°.0021$ year^{-1}, which is less than the theoretical value for this star one would obtain using the data from Table I, namely $\dot{\omega}_{\text{theor}} = 0°.0186 \pm 0°.0015$ year^{-1}.

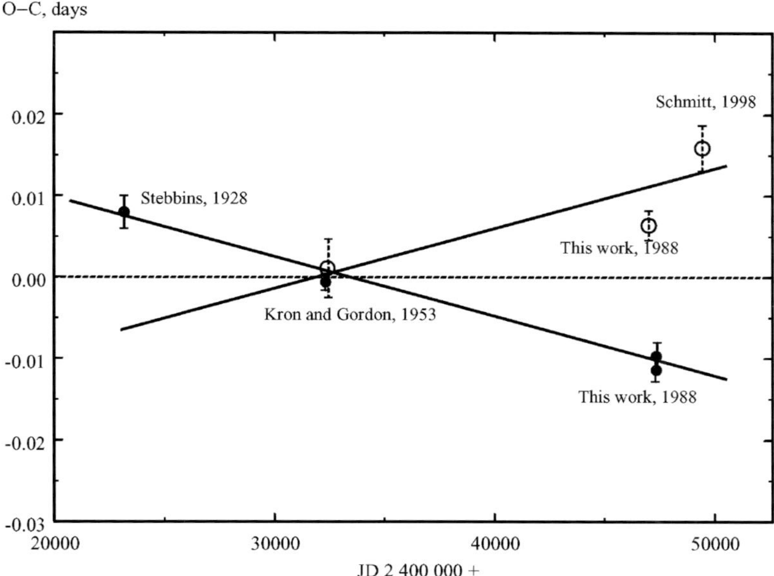

Figure 2. Linear fits to times of primary (filled dots) and secondary (circles) minima of α CrB.

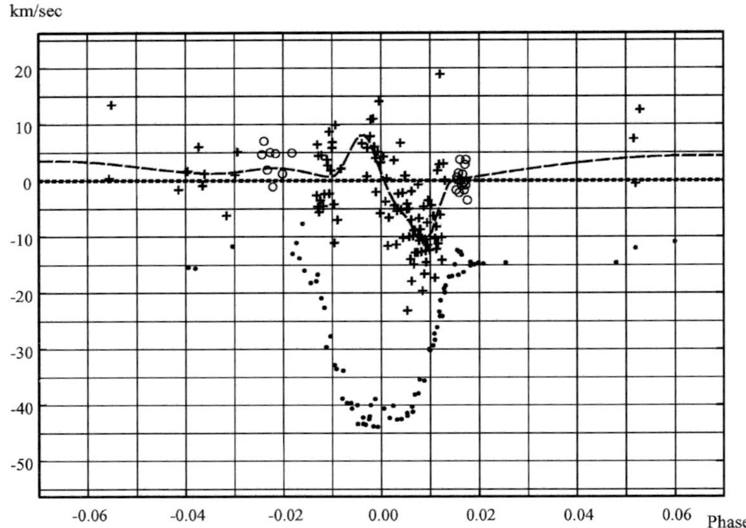

Figure 3. The radial velocities of α CrB during primary eclipse; crosses—McLaughlin (1933), circles—Ebbighausen observations; the orbital motion has been subtracted.

4. Discussion

The most probable solution to this problem is the tilt of the rotational axis of the primary to the vector of angular momentum of the double-star orbit. We see in Figure 3 that during the transit of the secondary across the disk of the primary the radial velocities change asymmetrically. A tilt of 40° of the rotational axis of the primary to the orbital plane can explain such an effect. The same tilt could also explain the observed rate of apsidal motion.

Acknowledgements

I thank Mrs. N. S. Volkova for her help in observations and Dr. P. Eggleton for valuable discussions.

References

Claret, A. and Gimenez, A.: 1992, *Astron. Astrophys. Suppl.* **96**, 255.
Ebbighausen, E.G.: 1976, *Publ. Dom. Astrophys. Obs.* **14**, 411.
Khaliullina, A.I. and Khaliullin, Kh.F.: 1984, *Soviet. Astr.* **61**, 39.
Kron, G.E. and Gordon, K.C.: 1953, *Astrophys. J.* **118**, 55.
McLaughlin, D.B.: 1933, *Publ. Michigan Obs.* **5**, 91.
Schmitt, J.H.M.M.: 1998, *Astron. Astrophys.* **333**, 199.
Stebbins, J.: 1928, *Publ. Washburn Obs.* **15**, 41.
Tomkin, J. and Popper, D.M.: 1986, *Astron. J.* **91**, 1428.
Volkov, I.M.: 1993, *IBVS* 3876.

THE ECCENTRIC ECLIPSING BINARY V889 AQUILAE

MAREK WOLF[1], ROGER DIETHELM[2] and MILOSLAV ZEJDA[3]

[1]*Astronomical Institute, Charles University, Prague, Czech Republic;*
E-mail: wolf@cesnet.cz
[2]*Astronomical Institute, University of Basel, Basel, Switzerland*
[3]*Nicolas Copernicus Observatory and Planetarium, Brno, Czech Republic*

(accepted April 2004)

Abstract. Several new times of minimum light recorded with photoelectric or CCD means have been gathered for the eccentric eclipsing binary V889 Aql ($P = 11.1$ days, $e = 0.37$). Its $O–C$ diagram is presented, and improved elements of the apsidal motion and the light-time effect (LITE) are given. We found a long apsidal motion period of about $24\,400 \pm 2400$ years and a period of the third body of about 52 ± 2 years.

Keywords: eclipsing binary, apsidal motion, light-time effect

1. Introduction

The motion of the line of apsides of a binary star is a direct consequence of the finite size of its components. The rate of motion of the apsis depends on the internal structure of each component. Determination of this rate thus provides an observational test of the theory of stellar structure and evolution. Moreover, there is a relativistic contribution for the periastron advance as observed in the orbit of the planet Mercury. On the other hand, the light-time effect (hereafter LITE) belongs to the traditional sources of our knowledge about triple and multiple stellar systems. Here we analyze the observational data and rate of apsidal motion for the well-known eclipsing system V889 Aql, which is a relatively bright northern-hemisphere object. Its orbit has been known to be eccentric and to exhibit apsidal motion.

The detached eclipsing binary V889 Aquilae (HD 181166; $V_{\mathrm{max}} = 8^{\mathrm{m}}63$; B9.5 V+A0 V) is a relatively well-studied binary with an eccentric orbit ($e \simeq 0.37$) and a rather long period of about 11.1 days. It was discovered to be a variable star by Hoffmeister (1935). The apsidal motion of V889 Aql was first discovered by Semeniuk (1968). Later, Busch (1978) and Bernardi and Scaltriti (1978) obtained further times of minima and improved the light elements. The apsidal motion rate was determined by several authors. Giménez and Scaltriti (1982) derived the period of apsidal motion $U = 24\,000 \pm 7000$ years. They also noted a large relativistic contribution in this eccentric system and its excellent agreement with their observed value. Another photometric study was presented by Khaliullin and Khaliullina (1989, hereafter KK89). From the data available at that time, they found good agreement between the observed apsidal motion rate of V889 Aql of $\dot{\omega}_{\mathrm{obs}} = 0.0155 \pm 0.0023\,^\circ$ per year and the theoretical combined relativistic and classical apsidal motion of $\dot{\omega}_{\mathrm{rel+cl}} = 0.0149\,^\circ$ per year. They derived a new value of the eccentricity $e = 0.375$ and precise relative radii $r_1 = 0.0582$ and $r_2 = 0.0524$. They predicted a third body with a contribution of third light of $L_3 = 0.185$ and a mass $M_3 \simeq 1.8\,M_\odot$ orbiting the eclipsing pair with the period $P_3 > 55$ years. Diethelm (1985, 1990, 1996) and Hegedüs et al. (1996) obtained new times of minimum light, which we incorporated in our analysis. A total of 52 times of minimum light were used in our analysis, with 26 secondary eclipses among them.

2. Apsidal Motion and LITE Analysis

The apsidal motion and the LITE in V889 Aql were studied by means of an $O\!-\!C$ diagram analysis and by the least-squares method. The computed apsidal motion elements and their internal errors (in brackets) are given in Table I. In this table P_s denotes the sidereal period, P_a the anomalistic period, e the eccentricity and $\dot{\omega}$ is the rate of periastron advance (in degrees per cycle or in degrees per year). The zero epoch is given by T_0 and the corresponding position of periastron is ω_0. The $O\!-\!C$ residuals for all times of minimum with respect to the linear part of the apsidal motion equation are shown in Figure 1. The predictions, corresponding to the fitted parameters, are plotted as continuous and dashed lines for primary and secondary eclipses, respectively. The $O\!-\!C_2$ diagram of V889 Aql after subtraction of the terms of apsidal motion is shown in Figure 2. New elements of LITE are also given in Table I, where P_3 is the orbital period of the third body, A the semi-amplitude of LITE, e_3 eccentricity of the third-body orbit, and ω_3 the length of periastron. Assuming a coplanar orbit ($i_3 = 90\,^\circ$) and a total mass of the eclipsing pair $M_1 + M_2 = 4.6\,M_\odot$ (KK89), we can obtain the value of the mass function and a lower limit for the mass of the third component $M_{3,\mathrm{min}}$. The third component could be a G5 star with a bolometric magnitude of about $+4.3$ mag. Unfortunately, such object is in contradiction with the value of third light $L_3 = 0.185$ obtained by KK89, producing only 5% of total light.

TABLE I

Apsidal motion and LITE parameters of V889 Aql

Element	Unit	Value	Element	Unit	Value
T_0	HJD	24 38241.7439 (6)	P_3	days	19 000 (660)
P_s	days	11.1207937 (25)	P_3	years	52.0 (1.8)
P_a	days	11.1208076 (25)	T_3	JD	245 0555 (25)
e		0.3745 (12)	e_3		0.00 (0.01)
$\dot{\omega}$	degree per cycle	0.00045 (5)	A	days	0.0268 (12)
$\dot{\omega}$	degree per year	0.0148 (15)	ω_3	degree	160.4 (2.5)
ω_0	degree	125.5 (0.3)	$f(m)$	$M_\odot$	0.037
U	years	24 400 (2400)	$M_{3,\mathrm{min}}$	$M_\odot$	1.06

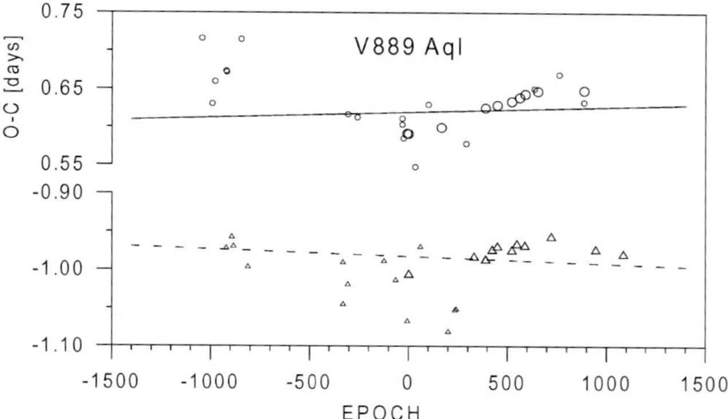

Figure 1. O–C diagram for the times of minimum of V889 Aql. The continuous and dashed lines represent predictions for the primary and secondary eclipses of the slow apsidal motion. The individual primary and secondary minima are denoted by circles and triangles, respectively. Larger symbols correspond to the photoelectric or CCD measurements, which were given higher weights in the calculations.

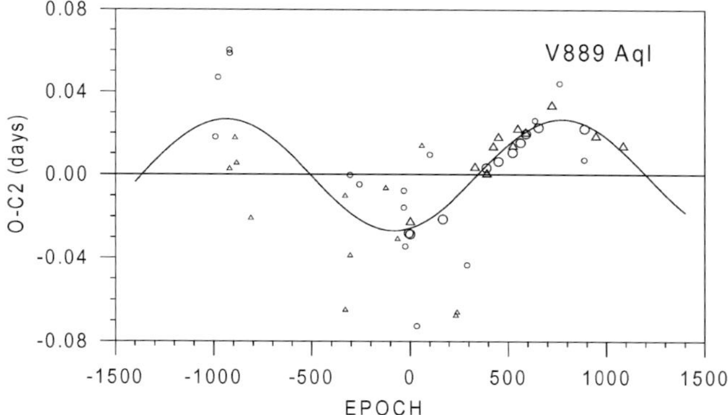

Figure 2. $O–C_2$ diagram for the times of minimum of V889 Aql after subtraction the terms of apsidal motion. The solid curve represents LITE for the third-body orbit with a period of 52 years and an amplitude of about 0.027 days.

The observed average value of the internal structure constant (ISC) $k_{2,\mathrm{obs}}$ can be derived using the well-known expression (Kopal, 1978). Taking into account the value of the eccentricity and the masses of the components, one has to subtract from $\dot{\omega}$ a relativistic correction $\dot{\omega}_{\mathrm{rel}}$. For V889 Aql, we have $\dot{\omega}_{\mathrm{rel}} = 0.00035°$ per year and $\dot{\omega}_{\mathrm{rel}}/\dot{\omega} = 78\%$. The resulting mean ISC is $\log k_{2,\mathrm{obs}} = -2.316$. The theoretical value $k_{2,\mathrm{theo}}$ according to available models for the internal stellar structure of V889 Aql computed by Claret and Giménez (1992) for a variety of masses and chemical compositions yields the value $\log k_{2,\mathrm{th}} = -2.276$. The agreement between the theoretical and observed value of ISC is relatively good.

 M. WOLF ET AL.

3. Conclusions

We improved apsidal motion and LITE parameters for the eccentric eclipsing binary V889 Aql by means of an $O-C$ diagram analysis. Our resulting parameters are in good agreement with the values previously obtained by KK89. V889 Aql is another relativistic binary system well suited to verify the theory of general relativity. Moreover, this system belongs to the important group of triple eccentric eclipsing binaries with apsidal motion (f.e. TV Cet, RU Mon, U Oph, YY Sgr, DR Vul) described recently by Wolf et al. (2001). EEBs with apsidal motion and LITE are interesting systems, which deserve a continuous photometric, spectroscopic as well as astrometric monitoring.

Acknowledgements

This investigation was supported by the Grant Agency of the Czech Republic, grant no. 205/04/2063, and has made use of NASA's Astrophysics Data System Bibliographic Services.

References

Bernardi, C. and Scaltriti, F.: 1978, *Acta Astron.* **28**, 221.
Busch, H.: 1978, Mitteilungen Bruno-H.-Bürgel-Sternwarte Hartha No. 13, 1.
Claret, A. and Giménez, A.: 1992, *A&AS* **96**, 255.
Diethelm, R.: 1985, *BBSAG Bull.* **77**, 2.
Diethelm, R.: 1990, *BBSAG Bull.* **96**, 1.
Diethelm, R.: 1996, *BBSAG Bull.* **113**, 2.
Giménez, A. and Scaltriti, F.: 1982, *A&A* **115**, 321.
Hegedüs, T., Bíro, I.B., Paragi, Z. and Borkovits, T.: 1996, IBVS No. 4340.
Hoffmeister, C.: 1935, *Astron. Nachr.* **255**, 405.
Khaliullin, Kh.F. and Khaliullina, A.I.: 1989, *AZh* **66**, 76, **KK89**.
Kopal, Z.: 1978, Dynamics of Close Binary Systems, Reidel, Dordrecht, Holland.
Semeniuk, I.: 1968, *Acta Astron.* **18**, 1.
Wolf, M., Diethelm, R. and Hornoch, K.: 2001, *A&A* **374**, 243.

TRIPLE AND MULTIPLE SYSTEMS

PAVEL MAYER

Astronomical Institute, Charles University, V Holešovičkách, Praha, Czech Republic;
E-mail: mayer@mbox.cesnet.cz

(accepted April 2004)

Abstract. The triple systems display several interesting effects; and are also important for understanding the origin and evolution of binaries. Namely mutual inclination of two orbits, changes of observed inclination, and problems with the third light will be discussed in this contribution.

Keywords: binaries: eclipsing, visual, spectroscopic, triple and multiple systems

1. Introduction

The triple and multiple systems can be studied from various points of view:

1. observability
2. dynamics
3. statistics
4. origin and evolution

I will discuss the first item only. It is the point which presents samples and examples for the others. Naturally, importance of the triple systems lies in the fourth item, which implies constraints and hints about the general problem of origin of stars.

It is well known that multiple systems can be divided into hierarchical systems and trapezium systems. However, mutual effects have been observed in the latter category only – Abt and Corbally (2000) showed that the number of "true" trapeziums is quite low (by now, 14 are known according to Abt and Corbally) and that their size is large (parsecs). Therefore, millennia would be necessary to watch the orbital motions. In what follows only hierarchical systems are discussed.

2. Orbital Inclination

It is very difficult to have complete information about a triple system: parameters of short and long orbits, including their inclinations and the mutual inclination, and parameters of all components. There are visual triples, where both orbits are known – Sterzik and Tokovinin (2002) list 22 such systems, however in 8 of them the orbital parameters are not reliable; note that in 1967, Worley was able to study only eight cases. The shortest period (of the small orbit) is 309 days, the longest (of the large orbit) 3780 years – this long period of course belongs to the less

Astrophysics and Space Science **296:** 113–119, 2005.
© Springer 2005

reliable parameters, the longest "reliable" period being 1195 years. Naturally, with the progress of interferometry the number of systems with short orbital periods will grow – the number of those with known long periods will increase mostly due to a growing time base.

One important information from these systems is, that – generally – the orbital planes are not identical, i.e., the mutual inclination is nonzero. But to find its exact value is difficult, one would also need to know both radial velocity orbits. Such an information can be obtained more probably for other types of objects than for visual triples. The various categories can be listed, e.g., as:

- visual pair and visual companion
- spectroscopic pair and visual companion
- eclipsing pair and visual companion
- spectroscopic/eclipsing pair and spectroscopic companion

(always the companion can also be a spectroscopic/eclipsing pair).

However note that Fekel (1981) found only 33% of certainly noncoplanar orbits in his sample of 20 systems – for the rest, coplanarity could not be exluded. Of course, among the noncoplanar systems, a well-known case is Algol (eclipsing plus spectroscopic/interferometric pair), where the eclipsing secondary is a radio source and its "visual" orbit was determined by radio interferometry, and the mutual orbital inclination was found close to $90°$.

The eclipsing binaries – as members of the triple or multiple systems – present effects not observable in other cases. First, light-time effect (acronym LITE or LTT, light travel time) can be detected, if of course its amplitude

$$A(\text{days}) = a \cdot m_3 \sin i (1 - e^2 \cos^2 \omega)^{1/2} (P_{\text{long}}/M)^{2/3} \tag{1}$$

is large enough ($a = 0.0057755$ if P_{long} is in years, M is the total mass of the system, and m_3 the mass of the third body, both in solar masses; e and ω are parameters of the long orbit). About 100 systems were found to display, or are supposed to display such effect. In case the nodal period of the system

$$P_{\text{node}} = b \cdot P_{\text{long}}^2 / P_{\text{short}} \tag{2}$$

(where b depends mostly on masses of components and is usually about five) is short, i.e. lasts from several tens of years to several millennia, the changing amplitude of eclipses might be observable due to precession of the orbits (causing a change of inclination). Of course, in cases where the eclipse amplitude is changing, the mutual inclination is certainly nonzero.

There are six cases where the amplitude changes due to a change of inclination: IU Aur, SV Gem, SS Las, AY Mus, RW Per, and V907 Sco. All these binaries were discovered by chance. I may say that in the case of IU Aur, the discovery was made due to bad Central-European weather – the light curve was not well covered even after several seasons. During that time, several times of minima were obtained (the light time effect was discovered), and the change of amplitude became apparent (Mayer, 1965, 1971). IU Aur is the only system for which the time dependence of the eclipsing binary inclination is known and the other angles of inclination can be estimated.

IU Aur was one of several dozen stars measured that time by me, and only this one was variable. Therefore, one would suppose that discovery of such a case should happen more often. However, nothing similar repeated in the following 40 years – the other cases were discovered using photographic data. Measuring times of minima is now common with small telescopes and CCD cameras. It should not be a problem to publish also the magnitude at minimum, and possibly also at maximum to monitor the amplitude.

Eclipsing binaries with changing amplitude of course should be discovered when old data are compared with new ones. There are several hundred of terrestrially observed light curves; new ones are often supplied by satellites and large scale surveys. Among the known six cases, the changes of amplitude have been $\geq 0.3^m$; much smaller values should also be detectable. Naturally, the detectability depends strongly on the character of the light curve, its accuracy and coverage.

For instance, I tried to compare old light curves with those obtained by the satellite HIPPARCOS (ESA, 1997). Such a comparison is not without problems. Namely in cases of sharp minima (of well detached or Algol-type binaries) the deepest parts of minima are often not covered by the satellite. In cases of low amplitudes it is difficult to judge if a change happened. And the data of large databases are usually of lower precision. Comparing about 100 light curves, no new clear case of an amplitude change was found so far.

There is another rarely studied effect connected with the changing orbital inclination: precession and nutation of the rotational axes of both close components. Formulae governing this effect were given by Kopal (see also Plavec, 1960), and recently a study of the component axes in SS Lac was published by Eggleton and Kiseleva-Eggleton (2001). Components of SS Lac are well detached, therefore the tides are minute, and the result is that the axes can move in a wide interval. Certainly a different case is IU Aur, a semidetached system; here the axes will closely follow the motion of a normal to the orbital plane, with nutation of small amplitude. Surface brightness of both components is not uniform, therefore the change of orientation of the axes should affect the light curve (level of maxima). However no such effect has been observed by now.

If the period of the third body is known the nodal period can be estimated. The nodal period of λ Tau (period of eclipses 3.95 days, of the third body only 33 days) is an order of magnitude shorter than in any other case (7–8 years, according to Söderhjelm, 1975). However, no changes of amplitude were found by Söderhjelm, and nothing can be added to his discussion. No newer measurements were published; unfortunately, HIPPARCOS data do not cover the deepest part of primary minimum, and only poorly the secondary minimum. In the other cases with relatively short nodal period, also no changes of amplitude were found: DM Per, EE Peg, FZ CMa, VV Ori. Therefore, a group of triple systems with coplanar orbits probably exists, namely among cases where the "long" period is short.

3. Difficulty with the Third Light

There are cases where the third body is a visual component, however so close, that during photometry and spectroscopy its light enters the diaphragm or the slit. The amount of the third light is usually not well known; e.g., speckle interferometry normally does not give an accurate magnitude difference. More reliable data were obtained by the satellite HIPPARCOS. Therefore, when solving the light curve, another unknown has to be included; often the correlation among third light, inclination and other parameters is strong. Only in case of total eclipses the third light has no effect on calculated inclination, radii and luminosity ratio of components (as noted by Kopal, 1959, p. 365). In spectroscopy, the third light always presents a nuisance. The added light makes the lines less deep, and, of course, line blending might be strong; the S/N ratio is always worse. A recently detected eclipsing binary, NX Vul, might be an extreme case: here the third light (due to a visual component 0.5 arcsec away) contributes nearly 80% of total light, and in the spectra, the orbital shift of lines is apparent only as slight widening and narrowing of lines, according to Gies (2004). Probably no disentangling would help. What could be learnt from such a system? The only good advice might be to leave it as it is and study other stars, at least until adaptive optics allow to separate the components at the spectrograph entrance. In less extreme cases, some more or less successful fighting with the third light might be attempted.

Table I collects those eclipsing binaries with a third light, which are also visual binaries with a separation of <2 arcsec, or spatially unresolved objects (only those stars with LITE are included, where third light was detected photometrically or spectroscopically). Certainly not all such cases are listed. Namely, the W UMa systems found to exhibit third line by Rucinski (2002) are not listed, with one exception (TU UMa, Pych and Rucinski, 2004).

It is well known that third bodies accompany close binaries quite often. Hoffleit (1996), e.g., noted 22 such cases among stars in the Bright Star Catalogue (with 220 eclipsing binaries); among 728 multiple systems contained in the Tokovinin catalogue (1997) there are 83 eclipsing binaries. However, if the third body is just a visual component without a known orbit, as it is in most cases, not much can be learnt. The ratio of orbital periods or the mutual inclination of orbits are certainly of interest, but cases with these parameters known are scarce.

Kopal requested all observers to keep an eye on the possible presence of third light in photometric data of eclipsing binaries. He advised to look for a visual component – but, of course, we know that the disturbing third light often cannot be separated even by present-state interferometry – and of possible changes of

TABLE I

Eclipsing binaries with third light

Name	Type	3.l.sp. type	Period (days)	Integral V magnitude	Separation (arcsec)	Third light total $= 1$	LITE period (year)
V1182 Aql	O8	O9?	1.622	8.2	<.5	0.25	50?
IU Aur	O9.5	B0[a]	1.811	8.2		0.20	0.81
LY Aur	O9 III	B0?	4.002	6.5	0.5	0.25	
SZ Cam	O9.5	BO?[a]	2.698	6.9	0.071	0.25	60
QZ Car	O9	O9[a]	5.998	6.2		0.49	50?
V649 Cas	B0	B0?	2.391	6.6	<.5	0.25	
δ Cir	O7 III–IV	B0.5 V	3.902	5.0		0.14	
δ Ori	O9.5	B0/1	5.732	2.1	0.3	0.22	
DI Peg	F4 IV		0.712	9.4		0.24	22?
MY Ser	O8.5	O8	3.322	7.5		0.50	
TU UMa	G?	F2?	0.377	8.8		0.55	
NX Vel	B0	O7	2.920	7.2	0.6	0.75	

[a]Third body is a binary.

systemic radial velocity of the binary. The latter request, however, involves to take more spectra and over a longer time span than necessary to determine the binary orbit.

4. Examples

Let us give some examples. A fine one is the visual triple HD 9770 (CD$-30°529$), where the periods of both orbits are known: 4.5 and 122 years. The fainter component in the short orbit is the eclipsing binary BB Scl, with a period of 0.48 days. Since integral light of the eclipsing binary is 0.8 mag fainter than the visual companion, its inclination has to be close to $90°$ in order to produce the observed depths of minima. Inclination of the short and long visual orbits are 22 and $29°$, respectively, and the noncoplanarity is only modest, however mutual inclination with the eclipsing binary is at least $70°$. The nodal period of the eclipsing binary can be estimated as 10^4 years.

The minima should display the light-time effect with a semiamplitude perhaps as large as 0.0050 day, i.e., due to the sharpness of the minima this effect might be well observable. The already obtained photometry by Drs. Cutispoto, Hearnshaw and their colleagues in the years 1993 through 1996 should be analysed from this point of view. The main result would be the mass ratio of the close visual binary.

Another four-body case is QZ Car. It is a double binary, with periods of 6.0 and 21 days, the earliest spectral type being O9. Attempt to resolve it by

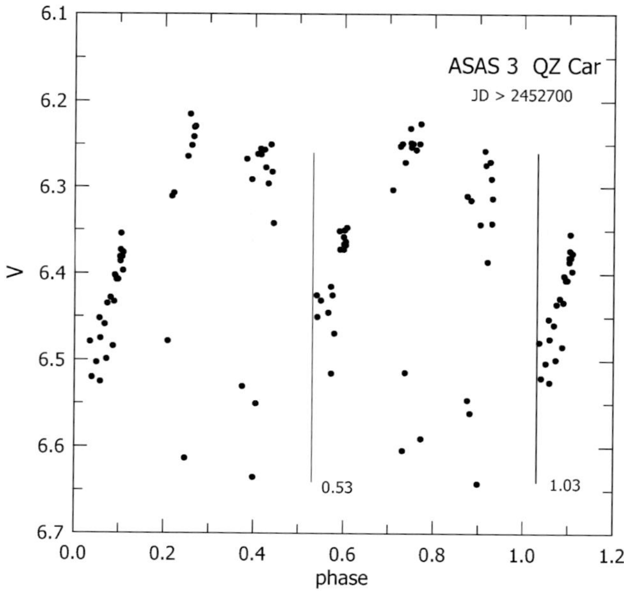

Figure 1. ASAS light curve of QZ Car from JD 2452700 to 2453050. The vertical lines are the estimated phases of the minima. Ephemeris: *Prim. Min.* = HJD 2441033.06 + 5.99857 d $\cdot E$ (used for this and next figures).

speckle interferometry was unsuccessful (Mason et al., 1998). The star was observed photometrically only rarely, nevertheless Mayer et al. (1999) attempted to construct an $O - C$ diagram. A new minimum time is strongly needed to confirm their suggestion of a period of 40–50 years for the mutual orbit of both binaries. The ASAS database is not suitable for such a bright star ($V = 6.3$ in maximum), but in spite of that, the light curve supplies some information, see Figure 1. The newer ASAS data, say taken after JD 2452700, appear to be useful, and there are 114 measurements at present. A shift of both minima from phases 0.0 and 0.5 is clearly seen; the shift can be estimated as 0.03 $\pm$ 0.01 of phase, i.e., 0.12 to 0.24 day, and this time of minimum is added to the mentioned $O - C$ diagram, see Figure 2. The change of the period is confirmed, however, no upper limit for the $O - C$ amplitude and period can yet be drawn. But the longer period means also larger angular separation, i.e., the chance for a success of interferometry certainly exists.

Note that the present period appears to be nearly exactly 6 days (5.9989 days), and in the year 2005 the minima should be observable from South America, preferably in March and December. It might be more useful to combine ingress to and egress from minima, then the best possibility for observation in 2005 is South Africa from February to June.

The inclination of the eclipsing orbit is 87.5°, inclination of the 21-day orbit – estimated using $m_1 = 20M_\odot$, $m_2 = 15M_\odot$ – is 35° (Leung et al. estimated the

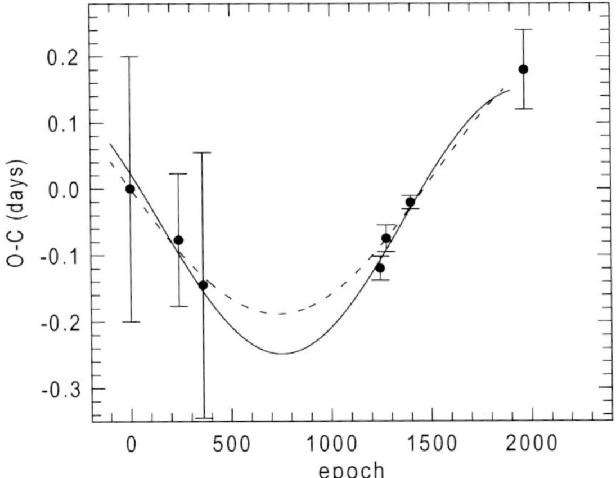

Figure 2. $O - C$ graph for QZ Car; the ASAS minimum confirms the prediction.

inclination as $60°$, assuming masses $m_1 = 40M_\odot$, $m_2 = 9M_\odot$); the inclination of the mutual orbit might be found by interferometry.

Acknowledgements

This study was supported by the research plan J13/98: 113200004 of the Czech Ministry of Education, Youth and Sports.

References

Abt, H.A. and Corbally, C.J.: 2000, *ApJ* **541**, 841.

Eggleton, P.P. and Kiseleva-Eggleton, L.: 2001, *ApJ* **562**, 1012.

ESA: 1997, Hipparcos and Tycho Catalogues, ESA SP-1200, Noordwijk, ESA.

Fekel, F.C.: 1981, *ApJ* **246**, 879.

Gies, D.R.: 2004, private communication.

Hoffleit, D.: 1996, *JAAVSO* **24**, 105.

Kopal, Z.: 1959, Close Binary Stars, Pergamon Press, London.

Leung, K.-C., Moffat, A.F.J. and Seggewiss, W.: 1979, *ApJ* **231**, 742.

Mason, B.D., Gies, D.R., Hartkopf, W.I., Bagnuolo, W.G. Jr., ten Brummelaar, T. and McAlister H.A.: 1998, *AJ* **115**, 821.

Mayer, P.: 1965, *PASP* **77**, 436.

Mayer, P.: 1971, *BAC* **22**, 168.

Mayer, P., Lorenz, R., Drechsel, H. and Abseim, A.: 1999, *A&A* **366**, 558.

Plavec, M.: 1960, *BAC* **11**, 197.

Pych, W. and Rucinski, S.M.: 2004, IBVS 5524.

Rucinski, S.M.: 2002, *AJ* **124**, 746.

Söderhjelm, S.S.: 1975, *A&A* **42**, 229.

Sterzik, M.F. and Tokovinin, A.A.: 2002, *A&A* **384**, 1030.

Tokovinin, A.A.: 1997, *A&AS* **121**, 71.

ESTIMATION OF LIGHT TIME EFFECTS FOR CLOSE BINARIES IN TRIPLE SYSTEMS

WALTER VAN HAMME[1] and R.E. WILSON[2]

[1]*Department of Physics, Florida International University, Miami, FL, U.S.A.*
E-mail: vanhamme@fiu.edu
[2]*Astronomy Department, University of Florida, Gainesville, FL, U.S.A.*

(accepted April 2004)

Abstract. We discuss implementation of light time effects in a general binary star program that solves for third body orbit parameters and binary star parameters together. The program combines radial velocities and light curves within a coherent analysis and can use data that are very unevenly distributed over time. By analyzing whole curves, the program has access to more information than only from eclipse timings. Results for λ Tau and VV Ori are shown.

1. Introduction

A significant number of eclipsing binaries are known to be part of triple systems (e.g. Chambliss, 1992). In some cases the third star is visually resolved, while in others its presence is inferred indirectly. Motion around the triple system barycenter causes a cyclic variation in the eclipsing pair's systemic velocity that is reflected in radial velocities. Photometrically, a third body may be revealed in light curves as "third light," or in eclipse timings as periodic phase excursions due to motion with respect to the ternary system barycenter (light-time effect). Parameters of that orbit can be determined well from timing residuals in cases where many minima have been observed over long baselines in time, but observed times of minima are scarce or inaccurate or both for many binaries, and velocity data are not readily utilized by the scheme. Although light and velocity curves may be spaced irregularly, with velocities at some epochs and light curves at others, the overall body of observations contains more information on light-time parameters than do times of minima. Here we discuss implementation of the light-time effect in a general binary star program and tabulate third body parameters derived from whole light and velocity curves.

2. Light Time Effects on Light and Velocity Curves

As a close binary (stars 1 and 2) orbits the barycenter of a triple system, the periodically varying light travel time produces a difference between observed or

Astrophysics and Space Science **296**: 121–126, 2005.
© Springer 2005

"barycentric" time, t_{obs}, and the eclipsing system's center of mass time, t_{sys}, which sets the orbital phase of the close pair. The difference $\Delta t = t_{\mathrm{obs}} - t_{\mathrm{sys}}$ is the light-time effect,

$$\Delta t = \frac{a' \sin i'}{c}\left(1 - \left(\frac{a}{a'}\right)^3\left(\frac{P'}{P}\right)^2\right)\frac{1 - e'^2}{1 + e' \cos f_{\mathrm{sys}}}\sin(f_{\mathrm{sys}} + \omega_{\mathrm{sys}}), \qquad (1)$$

where $a' = a_{\mathrm{sys}} + a_3$ is the semi-major axis of the outer relative orbit, i' the inclination, e' the eccentricity, ω_{sys} the argument of periastron of the close orbit's center of mass, and c the vacuum speed of light. The true anomaly of the close pair's

TABLE I

λ Tau parameters

Parameter	τ' adjusted	τ' fixed
T_0 (HJD)	2440022.4829 ± 0.0046	2440022.4829 ± 0.0045
P_0 (days)	3.9529505 ± 0.0000037	3.9529505 ± 0.0000037
i	$75°452 \pm 0°051$	$75°452 \pm 0°051$
T_1 (K)	19000	19000
T_2 (K)	8465 ± 49	8465 ± 49
a ($R_\odot$)	22.36 ± 0.27	22.36 ± 0.27
$V_{\gamma 0}$ (km/s)	14.0 ± 1.2	14.0 ± 1.2
Ω_1	3.847 ± 0.027	3.847 ± 0.067
Ω_2	2.19626	2.19626
$q = M_2/M_1$	0.2196 ± 0.0013	0.2196 ± 0.0013
a' ($R_\odot$)	95.1 ± 7.5	95.09 ± 0.80
i'	$76°0$	$76°0$
e'	0.14 ± 0.76	0.14 ± 0.14
ω_{sys}	$185° \pm 284°$	$186° \pm 10°$
P' (days)	33.21 ± 0.19	33.21 ± 0.19
τ' (HJD)	2444653 ± 30	2444663
Derived parameters		
M_1 ($M_\odot$)	7.87 ± 0.28	7.87 ± 0.28
R_1 ($R_\odot$)	6.220 ± 0.090	6.220 ± 0.090
$\log g_1$ (cm/s^2)	3.746 ± 0.029	3.746 ± 0.029
$\log L_1/L_\odot$	3.656 ± 0.016	3.656 ± 0.016
M_2 ($M_\odot$)	1.73 ± 0.06	1.73 ± 0.06
R_2 ($R_\odot$)	5.753 ± 0.079	5.753 ± 0.078
$\log g_2$ (cm/s^2)	3.156 ± 0.028	3.156 ± 0.028
$\log L_2/L_\odot$	2.184 ± 0.033	2.184 ± 0.033
M_3 ($M_\odot$)	0.86 ± 2.54	0.86 ± 0.38

center of mass, f_{sys}, is related to the mean anomaly through Kepler's equation. The mean anomaly is

$$M = \frac{2\pi}{P'}(t_{\text{sys}} - \tau'), \tag{2}$$

with P' the period and τ' the time of periastron passage. Radial velocities are affected through the eclipsing pair's systemic velocity,

$$V_\gamma = V_{\gamma 0} + \frac{2\pi}{P'} a' \sin i' \left(1 - \left(\frac{a}{a'}\right)^3 \left(\frac{P'}{P}\right)^2\right)(1 - e'^2)^{-1/2}$$
$$\times (e' \cos \omega_{\text{sys}} + \cos(f_{\text{sys}} + \omega_{\text{sys}})). \tag{3}$$

A version of the Wilson–Devinney binary star program (Wilson and Devinney, 1971; Wilson, 1979, 1990; Wilson and Van Hamme, 2003) with the above equations incorporated has been developed. This generalized program allows simultaneous fits to light and velocity curves, with third body parameters found together with those for the eclipsing pair in a coherent adjustment that also yields standard errors.

TABLE II

Parameters for synthetic λ Tau-type light curves

Parameter	Input	Final
T_0 (HJD)	2435128.73043	2435128.73051 $\pm$ 0.00014
P_0 (days)	3.952668	3.952674 $\pm$ 0.000035
i	75°.024	75°.030 $\pm$ 0°.006
T_2 (K)	8586	8583 $\pm$ 6
a ($R_\odot$)	22.5591	22.5591
$V_{\gamma 0}$ (km/s)	14.14	14.14
Ω_1	3.9116	3.9123 $\pm$ 0.0033
Ω_2	2.33328	2.33193
$q = M_2/M_1$	0.24161	0.24104 $\pm$ 0.00014
a' ($R_\odot$)	95.6945	97.1 $\pm$ 4.8
i'	76°.0	76°.0
e'	0.11	0.16 $\pm$ 0.49
ω_{sys}	168°	160° $\pm$ 28°
P' (days)	33.2	32.5 $\pm$ 2.6
τ' (HJD)	2435126.9	2435122.4 $\pm$ 4.1

TABLE III

VV Ori solution parameters

T_0 (HJD)	$2443118.58426 \pm 0.00012$
P_0 (days)	$1.48537752 \pm 0.00000021$
dP/dt (days/day)	$-3.26 \pm 0.50 \times 10^{-9}$
i	$88°721 \pm 0°041$
T_1 (K)	25000
T_2 (k)	15572 ± 31
a $(R_\odot)$	13.633 ± 0.087
$V_{\gamma 0}$ (km/s)	24.4 ± 1.8
Ω_1	3.1933 ± 0.0019
Ω_2	3.48407 ± 0.00080
$q = M_2/M_1$	0.38223 ± 0.00022
a' $(R_\odot)$	273.3 ± 1.2
i'	$89°0$
e'	0.02 ± 0.06
ω_{sys}	$183°3 \pm 7°5$
P' (days)	122.11 ± 0.12
τ' (HJD)	2443244.4 ± 3.1
Derived parameters	
M_1 $(M_\odot)$	11.14 ± 0.22
R_1 $(R_\odot)$	4.963 ± 0.032
$\log g_1$ (cm/s^2)	4.094 ± 0.014
$\log L_1/L_\odot$	3.937 ± 0.009
M_2 $(M_\odot)$	4.26 ± 0.08
R_2 $(R_\odot)$	2.412 ± 0.016
$\log g_2$ (cm/s^2)	4.303 ± 0.014
$\log L_2/L_\odot$	2.488 ± 0.016
M_3 $(M_\odot)$	2.95 ± 0.37

3. Examples

3.1. λ TAU

Simultaneous solution results based on UBV data by Grant (1959) and radial velocities by Fekel and Tomkin (1982) are in Table I. Strong correlations among outer orbit parameters lead to substantial uncertainties in a', e', ω_{sys} and τ'. To show the correlation effects, Table I lists solution parameters for a subset without τ', where significant reductions in standard errors are seen. We also generated synthetic λ Tau-like light curves from preselected sets of third body parameters, with milli-magnitude noise and with data at the same times as the real observations.

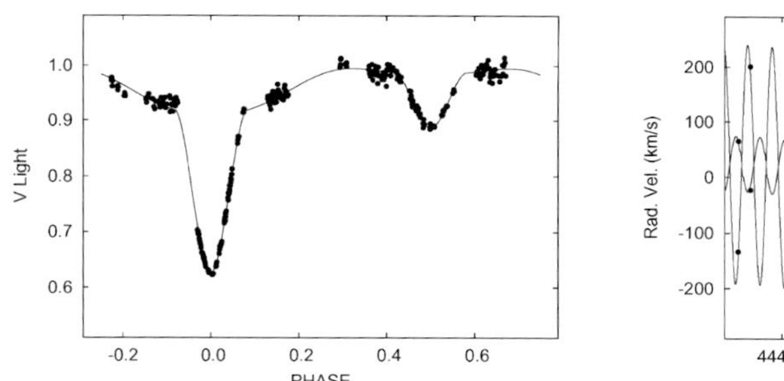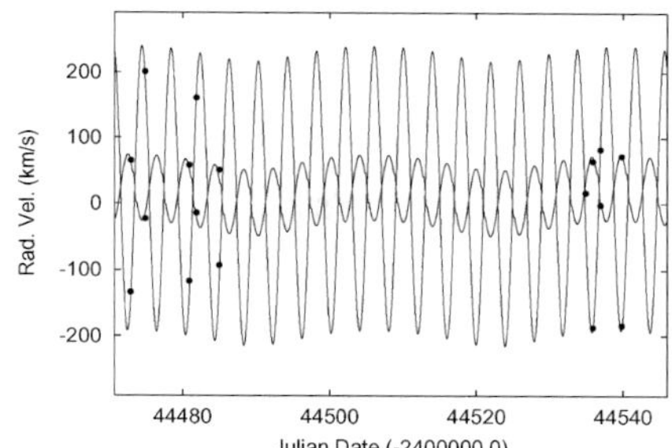

Figure 1. V light curve of λ Tau and a radial velocity curve section, with fitted curves for the simultaneous light and velocity solution including light time effects.

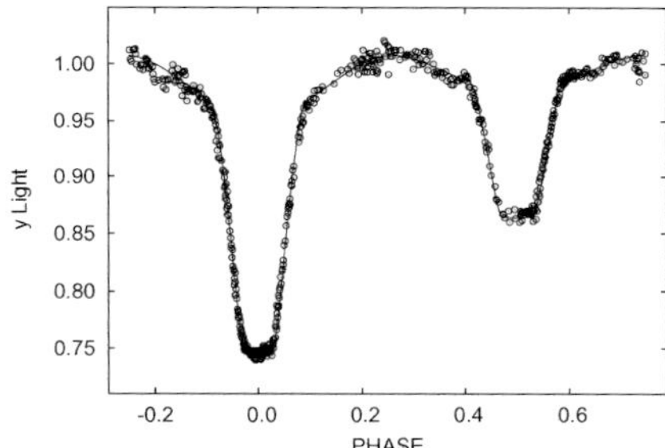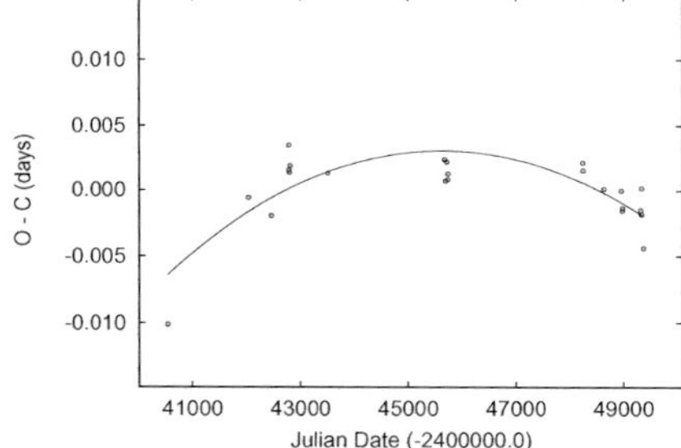

Figure 2. *Left:* VV Ori Strömgren y light curve and the solution curve from a simultaneous light and velocity solution including light time effects; *Right:* residuals from a linear ephemeris and parabolic fit.

Solutions by differential corrections recovered the known parameters within ranges appropriate to the standard errors (see Table II for an example).

3.2. VV ORI

We simultaneously solved UBV light curves and radial velocities by Duerbeck (1975), UBV and uvby observations of Chambliss and Leung (1982), and radial velocities by Popper (1993), with results in Table III. Correlations among third body parameters are smaller than for λ Tau. Figure 2 compares y data with the solution curve. Note the non-zero dP/dt which is corroborated by VV Ori's timing residuals, also in Figure 2.

Acknowledgements

This material is based upon work supported by the National Science Foundation under Grant No. 0307561.

References

Chambliss, C.R.: 1992, *PASP* **104**, 663.

Chambliss, C.R. and Leung, K.-C.: 1982, *ApJS* **49**, 531.

Duerbeck, H.W.: 1975, *A&AS* **22**, 19.

Fekel, F.C. and Tomkin, J.: 1982, *ApJ* **263**, 289.

Grant, G.: 1959, *ApJ* **129**, 78.

Popper, D.M.: 1993, *PASP* **105**, 721.

Wilson, R.E.: 1979, *ApJ* **234**, 1054.

Wilson, R.E.: 1990, *ApJ* **356**, 613.

Wilson, R.E. and Devinney, E.J.: 1971, *ApJ* **166**, 605.

Wilson, R.E. and Van Hamme, W.: 2003, *Computing Binary Star Observables*, Electronic monograph at FTP site *ftp.astro.ufl.edu*, directory *pub/wilson/lcdc2003*.

ECLIPSING BINARIES SHOWING LIGHT–TIME EFFECT

PETR ZASCHE

Astronomical Institute, Charles University, Prague, Czech Republic;
E-mail: petr.zasche@email.cz

(accepted April 2004)

Abstract. Four eclipsing binaries, which show apparent changes of period, have been studied with respect to a possible presence of the light–time effect. With a least squares method we calculated new light elements of these systems, the mass function of the predicted third body, and its minimum mass. We discuss the probability of the presence of such bodies in terms of mass function, changes in radial velocity and third light in solution of light curves.

Keywords: eclipsing binaries, period variations, O–C diagram analysis

1. Introduction

The light–time effect (hereafter LITE) in eclipsing binaries, caused by the orbital motion of the eclipsing pair around the barycenter of the triple system, produces period variations measurable by minima timings. LITE was first discussed by Irwin (1959), and the necessary criteria have been mentioned by Frieboes-Conde and Herczeg (1973). They are the following: agreement with a theoretical LITE curve, displacement of secondary minima identical to that of primary ones, a reasonable value of the mass function and corresponding variations in radial velocity are observed.

We found, that most of the necessary criteria listed above are satisfied in all systems presented below. Their O–C diagrams are also presented. The curves represent computed LITE versus epoch. The individual primary and secondary minima are denoted by dots and circles, respectively. Larger symbols correspond to the photoelectric or CCD measurements which were given higher weights in our computations. All data were taken from published papers.

2. Observations

AR Aur: (HD 34364) was discovered as a variable in 1931. For the present computations we used only accurate photoelectric and CCD minima timings. From these data we calculated new light elements:

$$\text{Min } I = \text{HJD} \ 24 \ 27887.7299 + 4\overset{d}{.}13466577 \cdot \text{E}.$$

Astrophysics and Space Science **296**: 127–130, 2005.
© Springer 2005

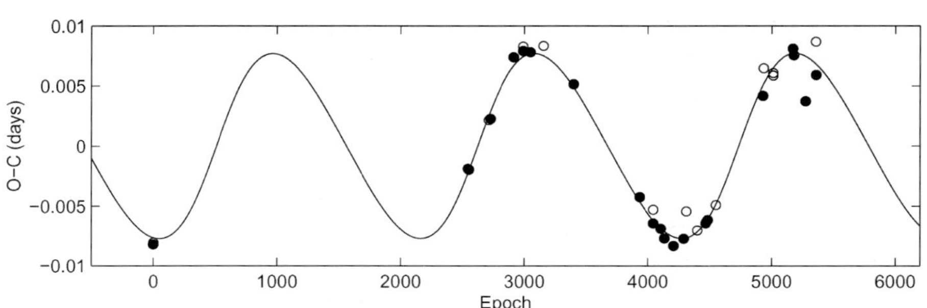

Figure 1. O–C diagram of AR Aur.

A period investigation was made by Chochol et al. (1988). With new minima timings we were able to determine the third body orbit with more plausibility (see Figure 1). Because of the relatively low mass of the unresolved component, the third light must be less than 1% of the total light. Also the variations in radial velocities, because of the low mass and errors in velocity measurements, are still inconsistent.

R CMa: It is one of the best known eclipsing binary system showing a clear LITE. Almost 300 minima times of R CMa (HD 57167) were collected. Our computations with these data led to new light elements

$$\text{Min}\,I = \text{HJD}\ 24\ 45391.2570 + 1\overset{d}{.}13594210 \cdot \text{E}.$$

We can compare our results (see Figure 2) with an analysis by Ribas et al. (2002), who calculated a shorter third body orbital period p_3, but a higher semi-amplitude, which gives a larger mass function. Variations in radial velocities and also the astrometric confirmation by Hipparcos are still not conclusive. The third body is probably a WD or an M-type dwarf.

FZ CMa: Moffat and Vogt (1983) performed a detailed spectroscopic and photometric study of FZ CMa (HD 52942), but a detailed period investigation was

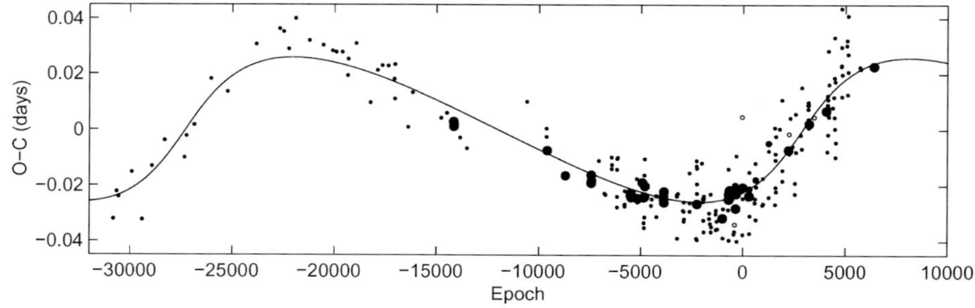

Figure 2. O–C diagram of R CMa.

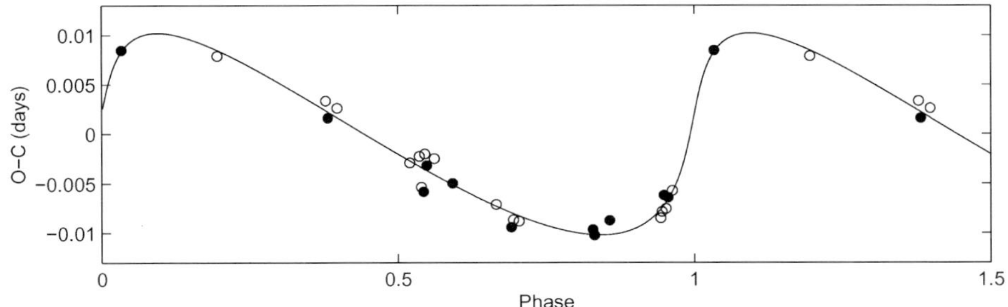

Figure 3. O–C phase diagram of FZ CMa.

realized only with minima from 1973 to 1980. New light elements are

$$\text{Min } I = \text{HJD } 24\ 41743.5915 + 1\overset{d}{.}27303611 \cdot E.$$

As we can see in Table I, the minimum mass of its third body is very high and the orbit is very eccentric (see also Figure 3). The third body should therefore be the dominant component. A possible solution is that the third body is itself a binary, what seems likely also from a spectroscopic point of view.

TX Her: (HD 156965) was found to be a variable by Miss Cannon (see Pickering, 1910). It was very often observed, what leads to a large list of visual minima times, and from the 1940s to accurate photoelectric and CCD measurements (see Figure 4), from which the following new light elements were derived:

$$\text{Min } I = \text{HJD } 24\ 40008.3686 + 2\overset{d}{.}05980975 \cdot E.$$

We can compare our results with the latest LITE study of TX Her by Ak et al. (2004), who calculated a higher value of eccentricity, a larger semi-amplitude, and a lower period p_3, so their mass function comes out nearly twice as large as ours.

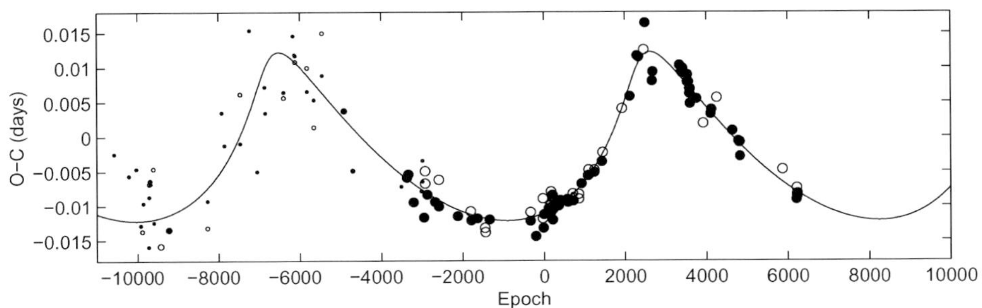

Figure 4. O–C diagram of TX Her.

TABLE I

Results for four systems showing LITE

Name of star	Spectrum	M_1 $(M_\odot)$	M_2 $(M_\odot)$	p_3 (years)	A (days)	e	ω (deg)	$f(M_3)$ $(M_\odot)$	$M_{3,\min}$ $(M_\odot)$
AR Aur	B9V + B9.5V	2.48	2.29	23.92	0.0077	0.209	2.3	0.0044	0.497
FZ CMa	B2.5IV-V + B2.5IV-V	5.40	5.30	1.511	0.0102	0.794	7.0	10.338	22.51
R CMa	F0V + K2IV	1.07	0.17	93.89	0.0258	0.498	357.8	0.0155	0.338
TX Her	A7 + F0	1.62	1.45	51.53	0.0122	0.654	48.6	0.0048	0.387

Table I contains the component masses M_i, period p_3 of the third body, semi-amplitude A of LITE (see, e.g., Mayer, 1990), eccentricity e, length of periastron ω, computed mass function $f(M_3)$, and minimum mass $M_{3,\min}$ (for $i = 90°$) of this predicted body.

3. Conclusion

We have derived new LITE parameters for four eclipsing binaries by means of an O–C analysis. However, for a final, definite decision about the presence of LITE we need further accurate timings of minima and/or appropriate confirmation by other methods (radial velocity variations, third light as result of light curve analysis, or astrometric observations).

Acknowledgements

This research has made use of the SIMBAD database, operated at CDS, Strasbourg, France, and of NASA's Astrophysics Data System Bibliographic Services. We are grateful to P. Niarchos et al. for providing us their minima measurement of FZ CMa prior to publication, This investigation was supported by the Grant Agency of the Czech Republic, grant No. 205/04/2063.

References

Ak, T., et al.: 2004, *New Astron* **9**, 265–272.
Chochol, D., et al.: 1988, *BAICz* **39**, 2.
Frieboes-Conde, H. and Herczeg, T.: 1973, *A&A S* **12**, 1.
Irwin, J.B.: 1959, *AJ* **64**, 149.
Mayer, P.: 1990, *BAICz* **41**, 231.
Moffat, A.F.J., et al.: 1983, *A&A* **120**, 278–286.
Pickering, E.C.: 1910, *Harvard Circular*, 159.
Ribas, I., et al.: 2002, *AJ* **123**, 2033–2041.

ASTROMETRIC SOLUTION OF THE MULTIPLE SYSTEM XY Leo

VOLKAN BAKIS, AHMET ERDEM, EDWIN BUDDING, OSMAN DEMIRCAN and
HICRAN BAKIS

Physics Department, Çanakkale Onsekiz Mart University, Çanakkale, Turkey;
E-mail: bakisv@physics.comu.edu.tr

(accepted April 2004)

Abstract. An astrometric solution, together with time of minimum analysis, has been made for
the multiple system XY Leonis (HIP 49136) to identify the properties of the remote companion to
the eclipsing pair (AB). From this solution, we derive the inclination of the wide orbit (AB-cd) as
$94.4° \pm 0.2°$, angle of nodes as $247.3° \pm 0.2°$, and the mass of the wide component (the dwarf binary
cd) as 0.98 ± 0.2 $M_\odot$. This study confirms that the light travel time effect can explain the sinusoidal
$O - C$ variation of the eclipsing system.

Keywords: eclipsing binaries, triple systems, astrometry, light time effect

1. Introduction

One effective way of discovering companions to close eclipsing pairs, including
long-period exosolar planets or brown dwarfs, is to use the light-time travel (LTT)
effect. A shortcoming of the LTT method is that only a few parameters of the pre-
sumed tertiary component can be determined. However, as was demonstrated in the
case of Algol (Bachmann and Hershey, 1975), LTT analysis can be complemented
with astrometry to yield the orbital inclination of the wide orbit, and thus the value
of the third mass. In the present study, we have extracted abscissal data for the
multiple system XY Leo from the Hipparcos Intermediate Astrometric Data (ESA,
1997). Available ground-based astrometric observations were also collected and all
were combined with an LTT analysis (*cf.* Yakut et al., 2003). Residuals of over 160
eclipse timings from 1944 to 2002 show a periodic (19.5 years) quasi-sinusoidal
modulation. As previously shown by Gehlich et al. (1972), Hrivnak (1985) and Pan
and Cao (1998), these variations can be suitably explained by an LTT effect caused
by the gravitational influence of a remote component. The aim of the present study
is to obtain a combined astrometric and $O - C$ solution, and specify the parameters
of this wide component to the eclipsing binary XY Leo (AB).

2. Astrometry and LTT Analysis

Hipparcos observed XY Leo between November 22, 1989 and November 9, 1992.
There are 50 one-dimensional astrometric measurements, corresponding to 26

Astrophysics and Space Science **296**: 131–134, 2005.

different epochs in the *Hipparcos Intermediate Astrometric Data* sample, as reduced by the two Hipparcos consortia FAST and NDAC. These data are available from CD-ROM 5 of the Hipparcos catalog (Perryman, 1997). Unfortunately, the time-span of these observations is smaller than the orbital period of the wide system (AB-cd), and this may result in systematic errors in the derived elements. Our database was therefore complemented by ground-based observations. Two-hundred and ninety nine times of minimum light were taken from Yakut et al. (2003) and a further 93 from Zejda (private communication). The $O - C$ data are plotted in Figure 1, using the light elements of Yakut et al. Parabolic and LTT effects were considered and, for the LTT effect, the well known equation of Irwin (1952) was applied:

$$\Delta T = \frac{a \sin i}{c} \left\{ \frac{1 - e^2}{1 + e \cos v} \sin(v + \omega) + e \sin \omega \right\}, \tag{1}$$

where c is the speed of light and a, i, e, ω and v are semimajor axis, inclination, eccentricity, argument of periastron, and true anomaly of the orbit. Here we apply this formula to the supposed two-body-like wide orbit. The line of sight and local plane of the sky standard coordinates form the reference coordinate system for the parameters of this orbit. The fit of Eq. (1) to the timing data allows estimates for a number of parameters of the wide orbit, though both the semimajor axis and mass of the companion (cd) are affected by the factor sin i, which is not determined from LTT analysis alone. The parameters i and Ω can, however, be included in a complementary fit to the astrometric data, yielding a full geometrical description of the system (*cf.* Ribas et al., 2002). Elliptic motion would produce the following effect on the standard coordinates x and y:

$$\Delta x = a \frac{1 - e^2}{1 + e \cos v} [\cos(v + \omega) \sin \Omega + \sin(v + \omega) \cos \Omega \cos i] \tag{2}$$

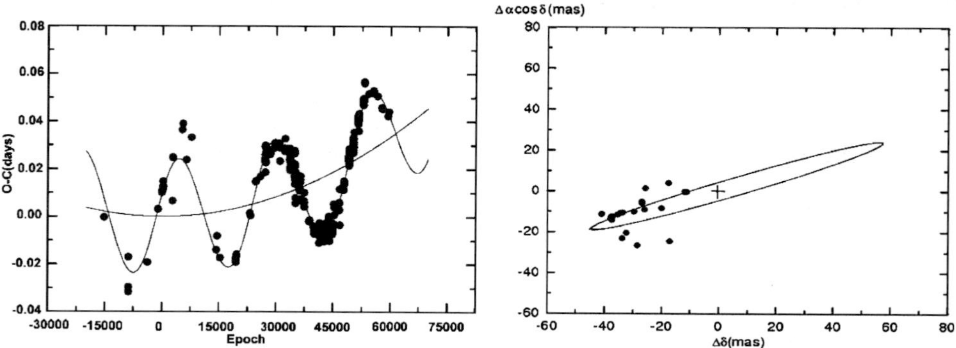

Figure 1. The best fit to the $O - C$ residuals (left) and the best fit to the astrometric data(right).

$$\Delta y = a \frac{1 - e^2}{1 + e \cos v}[\cos(v + \omega) \cos \Omega - \sin(v + \omega) \sin \Omega \cos i] \qquad (3)$$

Since Hipparcos measurements are one-dimensional, we need to express the variation of the measured abscissa Δv in terms of Δx and Δy)

$$\Delta v = \frac{\partial v}{\partial \alpha \cos \delta}(\Delta \alpha \cos \delta + \Delta x) + \frac{\partial v}{\partial \delta}(\Delta \delta + \Delta y) + \frac{\partial v}{\partial \varpi}(\Delta \varpi + \Delta z)$$
$$+ \frac{\partial v}{\partial \mu_\alpha \cos \delta}\Delta \mu_\alpha \cos \delta + \frac{\partial v}{\partial \mu_\delta}\Delta \mu_\delta, \qquad (4)$$

Here, (α, δ) are the initial equatorial coordinates, (μ_α, μ_δ) the corresponding proper motions, and ϖ is the parallax. Equation (4) allows for changes in the adopted values of these mean quantities, as well as the orbital changes. In addition to fitting the LTT and astrometric data of the wide system, corrections to the epoch and orbital period of the eclipsing pair (which could lead to a linear secular increase or decrease of the $O - C$ values) were also considered. Initial values of the period and reference epoch were adopted from Yakut et al. (2003). Finally adopted values are as follows: $a_{AB} = 5.63 \pm 0.05$, $P_{AB} = 0.28410237 \pm 4 \times 10^{-8}$ days and $T_{0AB} =$ HJD $2435484.0236 \pm 7 \times 10^{-4}$. The rate of change of period Q was found as $9.3 \pm 0.7 \times 10^{-12}$ days/cycle. Relative goodness-of-fit measures are 0.02 and 33.5 for the LTT analysis and astrometric fittings, respectively. Graphical representations of the fits to *Hipparcos Intermediate Astrometry Data* are shown in Figure 1. The resulting best-fitting parameters, together with their standard deviations, are listed in Table I.

3. Discussion and Conclusions

As well as extending the time base coverage, our study has found the inclination of the wide orbit of XY Leo by making use of available high-precision astrometry

TABLE I

Astrometric and LTT solutions for the wide orbit of XY Leo

Parameter	Value	Parameter	Value
ϖ (mas)	13.59 ± 0.08	P (year)	19.59 ± 0.06
$\mu_\alpha \cos \delta$ (mas year^{-1})	59.94	e	0.12 ± 0.02
μ_δ (ma s year^{-1})	-51.32	i (deg)	94.4 ± 0.2
a (mas)	55.6 ± 0.3	ω (deg)	247.3 ± 0.2
Ω (deg)	356 ± 14	T (HJD)	2435341 ± 284
M_{AB} ($M_\odot$)	1.32	M_{cd} ($M_\odot$)	0.98 ± 0.02

(Hipparcos). Corrected astrometric parameters are given in Table I. The parallax ϖ now corresponds to a mean distance of 73.6 pc. Using the astrometric parameters of the wide orbit given in Table I, together with the parameters derived by Barden (1987), we find the inclination of the cd orbit to be $29°.1$. The masses of the cd stars are then calculated as 0.57 and 0.41 $M_\odot$, respectively. This solution has been checked against the Fourth Catalogue of Interferometric Measurements of Binary Stars (Hartkopf et al., 2001).There is at least agreement that the AB-cd separation was less than 50 mas at the time of the interferometric measurement. The results are also compatible with the photometric study of Yakut et al. (2003).

References

Arenou, F., Mayor, M., Udry, S. and Queloz, D.: 2000, *A&A* **355**, 581.
Bachmann, P.J. and Hershey, J.L.: 1975, *AJ* **80**, 236.
Burden, S.C.: 1987, *ApJ* **317**, 333.
Gehlich, U.K., Prolss, J. and Wehmeyer, R.: 1972, *A&A* **18**, 477.
Hartkopf, W.I., Mason, B.D., Wycoff, G.L. and McAlister, H.A.: 2001, Fourth Catalog of Interferometric Measurements of Binary Stars, Available at http://www.ad.usno.navy.mil/wds/int4.html
Hrivnak, B.J.: 1985, *ApJ* **290**, 696.
Irwin, J.: 1952, *ApJ* **116**, 211.
Perryman, M.A.C.: 1997, The Hipparcos and Tycho Catalogues, ESA-SP1200.
Pan, L. and Cao, M.: 1998, *Ap&SS* **259**, 285.
Ribas, I. Arenou, F. and Guinan, E.F.: 2002, *AJ* **123**, 2033.
Yakut, K., Ibanoglu, C., Kalomeni, B. and Degirmenci, Ö.L.: 2003, *A&A* **401**, 1095.

V475 Sct (Nova Scuti 2003) – BINARY OR TRIPLE SYSTEM?

DRAHOMÍR CHOCHOL[1], THEODOR PRIBULLA[1], NATALY A. KATYSHEVA[2],
SERGEI YU. SHUGAROV[2] and IGOR M. VOLKOV[2]

[1]*Astronomical Institute, Slovak Academy of Sciences, 059 60 Tatranská Lomnica, Slovakia;*
E-mail: chochol@ta3.sk
[2]*Sternberg State Astronomical Institute, Universitetskij Prosp. 13, Moscow 119992, Russia*

(accepted April 2004)

Abstract. $UBVRI$ photometry and 4600–9000 Å spectroscopy of nova V475 Sct taken in the first 3 month after discovery is presented. The object can be classified as a Fe II type slow nova with $t_{2,V} = 48$ days, $t_{3,V} = 53$ days. The absolute magnitude of the nova at maximum, its colour excess and distance were determined. The observed 13.4 day periodicity of flares can be explained by the mass transfer bursts from the red to the white dwarf, probably caused by the periastron passage of a third body. Two sets of absorptions are seen in the P Cyg-type $H\alpha$ line profile. They arise in the expanding shell of the nova.

Keywords: stars, binaries, novae

1. Introduction

Classical novae are binary systems with orbital periods of a few hours consisting of a Roche-lobe filling red dwarf transferring matter to a white dwarf. The nova outburst is caused by a TNR of the accreted material on the surface of a white dwarf.

Classical nova V475 Sct was discovered by Nishimura (IAUC 8190) on August 28, 2003 (mag 8.5) at the coordinates $\alpha_{2000} = 18^h 49^m 37.6^s$, $\delta_{2000} = -9° 33'50.85''$ (Yamaoka, 2003). Optical spectra taken by Boeche and Munari (2003) on August 31, 1997 showed an F2 Ia absorption spectrum with weak emission of Balmer lines. The nova reached its brightness maximum $V_{\max} = 8.43$ mag, $B_{\max} = 9.33$ mag on September 1, when the F supergiant atmosphere of the outbursting white dwarf was ejected. We could not identify the precursor on POSS prints, implying an outburst amplitude of >11 mag.

2. Our Photometry and Available Spectroscopy

Our photometry consists of $UBV(RI)_C$ CCD observations with the SBIG ST10-XME camera mounted in Newton focus of the new 0.5 m reflector at the Stará Lesná Observatory; $UBVRI$ CCD observations with portable SBIG ST7, Apogee Ap7p, Pictor-416 and VersArray-1300 cameras mounted in the Cassegrain focus of the 1.25, 0.6, 0.5 and 0.38 m reflectors at Crimea Observatories at Nauchnyj (CN);

UBV photoelectric photometry with the 0.6 m reflector and the 1 m CS reflector at Simeiz (CS). Photometric data were obtained in 55 nights between August 30 and November 30, 2003. Our CCD images taken at the end of November 2003 revealed that V475 Sct is member of a visual pair. The brightness of the optical companion located at an angular distance of $3.3''$ from the nova is $B = 17.0$, $V = 16.2$, $R = 15.3$, $I = 14.6$. Our photometry of V475 Sct was corrected for the light of this component.

The 4600–9000 Å spectroscopy of V475 Sct was obtained between September 2 and October 10, 2003 by amateur astronomer Christian Buil at Castanet Tolosan (France) using the Takahashi FS-128 5-inch and CN-212 8-inch telescopes equipped with MERIS spectrograph (2.87 Å/pixel) + Audine KAF-0401E CCD camera and HIRES spectrograph (0.3805 Å/pixel) + Audine KAF-1602E CCD camera, respectively. The data are available from http://www.astrosurf.com/buil/us/nscuti.

3. Results and Discussion

The light curves (LCs) and colour indices of V475 Sct are presented in Figures 1 and 2. The V and B LCs were used to find the rates of decline $t_{2,V} = 48$ d, $t_{3,V} = 53$ d, $t_{2,B} = 50$ d, $t_{3,B} = 58$ d and to estimate the absolute magnitude at maximum $M_{V(\max)} = -7.16 \pm 0.15$, $M_{B(\max)} = -6.96 \pm 0.39$ using various maximum magnitude-rate of decline relations (Chochol and Pribulla, 1997; Downes and Duerbeck, 2000 and references therein). The latter value and formula given

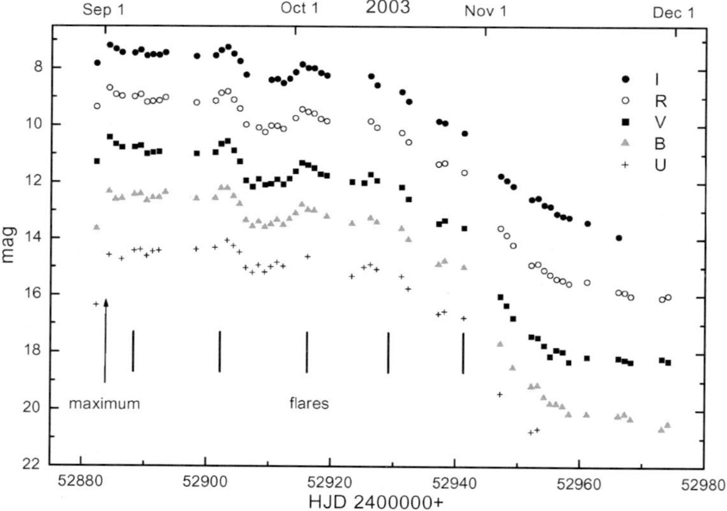

Figure 1. $UBVRI$ light curves of V475 Sct, shifted for clarity by 1 mag in R, 2 mag in V, 3 mag in B and 5 mag in U, respectively.

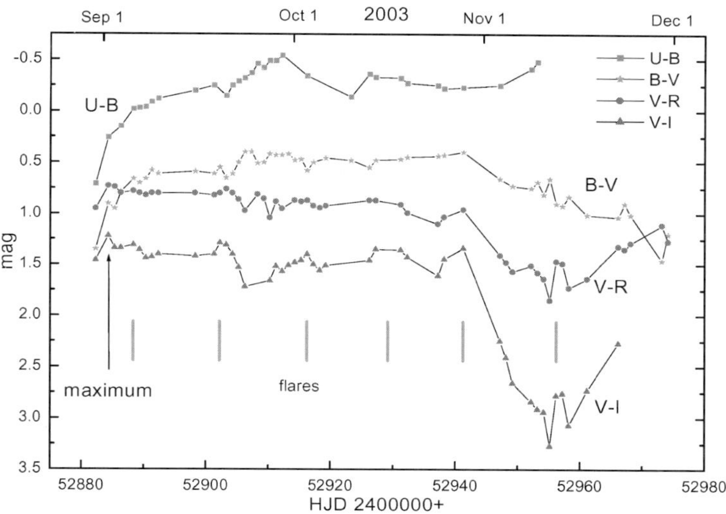

Figure 2. Colour indices of V475 Sct.

by Livio (1992) allow to determine the mass of the white dwarf in V475 Sct as $M_{wd} = 0.73 \pm 0.07\ M_\odot$. According to the classification scheme of nova LCs (Downes and Duerbeck, 2000), V475 Sct can be classified as a slow Eddington nova with standstills at maximum and dust formation at later stages. The rapid fading in the optical flux and development of an IR excess (seen in $V - R$ and $V - I$ indices), are caused by dust formation in the ejecta of the nova. Colour indices can be used to estimate the colour excess by different methods described in Chochol and Pribulla (1997). The mean value of colour excess of V475 Sct $E(B - V) = 0.69 \pm 0.05$ and the distance modulus formula provide the distance to the nova as 4.8 kpc. LCs of the nova show periodic brightness maxima (flares) clearly indicated also in the $V - I$ index. They repeat with the period 13.4 ± 0.2 days and can be explained by the brightening of the hot component due to mass transfer bursts from the red to the white dwarf probably caused by the periastron passage of a third body in the system. We observed similar behaviour in nova V723 Cas, where the flares repeated with 180-day periodicity (Chochol and Pribulla, 1998). The spectra of the nova (Figure 3) allow to classify it as Fe II class object (Williams, 1992) with an emission spectrum including also O I, Na I D, Ca II, Mg II and Balmer H I lines accompanied by P Cyg absorptions. They arise in an expanding shell ejected at maximum on September 1, 2003. Two sets of absorptions are seen in the P Cyg-type Hα line profile (Figure 4). Between September 14 and 21, 2003 they increased their RVs from -480 to -540 km s^{-1} and from -1140 to -1240 km s^{-1} (with respect to laboratory wavelength) suggesting the acceleration of the inner and outer envelope of the nova shell, where the absorptions arise, by a continuous wind.

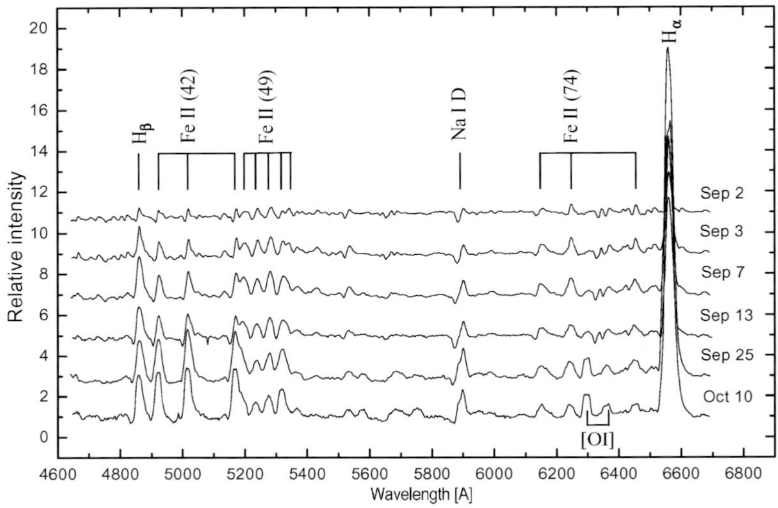

Figure 3. The spectrum of V475 Sct.

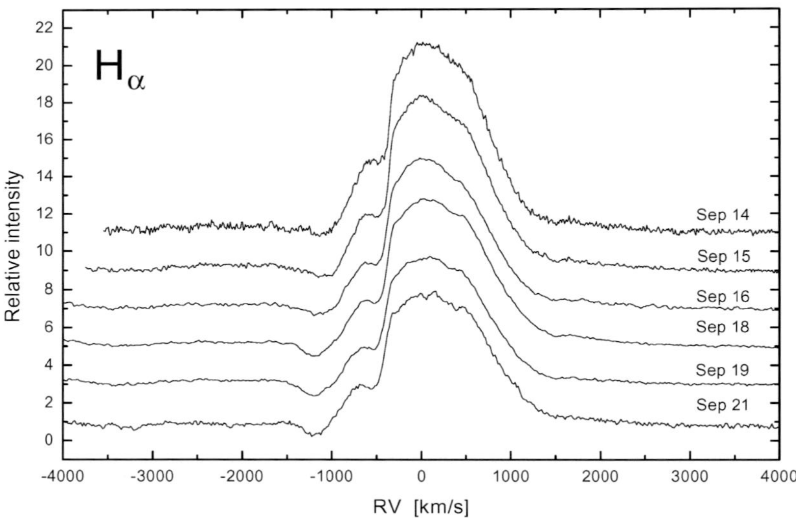

Figure 4. Hα line profile in radial velocity scale.

Acknowledgements

This work was supported by Science and Technology Assistance Agency, contract No. APVT-20-014402 and by the VEGA grant No. 2/4014/4.

References

Boeche, C. and Munari, U.: 2003, IAUC, No. 8191.
Chochol, D. and Pribulla, T.: 1997, *CoSka* **27**, 53.
Chochol, D. and Pribulla, T.: 1998, *CoSka* **28**, 121.
Downes, R.A. and Duerbeck, H.W.: 2000, *AJ* **120**, 2007.
Livio, M.: 1992, *ApJ* **393**, 516.
Williams, R.E.: 1992, *AJ* **104**, 725.
Yamaoka, H.: 2003, *IAUC* No. 8199.

PHOTOMETRIC OBSERVATIONS AND APSIDAL MOTION STUDY OF V1143 Cyg

A. DARIUSH[1], N. RIAZI[2] and A. AFROOZEH[3]

[1]*Physics Department, Arsenjan Azad University, Arsenjan, Iran*
[2]*Physics Department and Biruni Observatory, Shiraz University, Shiraz 71454, Iran,
and IPM, Farmanieh, Tehran, Iran; E-mail: riazi@physics.susc.ac.ir*
[3]*Biruni Observatory, Shiraz University, Shiraz 71454, Iran*

(accepted April 2004)

Abstract. Photometric observations of the eccentric eclipsing binary V1143 Cyg were performed during Aug.–Sep. 2000 and July 2002, in Johnson B and V bands. The analysis of both light curves was made separately using the 1998 version of Wilson's LC code. In order to find a new observed rate of apsidal motion, we followed the procedure described by Guinan and Maloney (1985). A new observed rate of apsidal motion of $3°.72/100$ yr was computed, which is close to the one reported earlier by Khaliullin (1983), Gimenez and Margrave (1985), and Burns et al. (1996).

Keywords: binaries, stars

1. Introduction

V1143 Cygni (HD 185912)is a double-lined eclipsing binary, consisting of a pair of F5 V stars with high orbital eccentricity (e = 0.540) and a relatively long period of 7.640 days. The stellar and orbital properties of this system were determined accurately by Andersen et al. (1987). It is known that V1143 Cyg is one of the best examples of eclipsing binaries with apsidal motion, in which the observed rate of apsidal motion is greater than the value predicted by general relativity and stellar evolutionary models. In the case of V1143 Cyg, the observed rate of apsidal motion is $\dot{\omega}_{\mathrm{obs}} = 3°.52/100\,\mathrm{yr} \pm 0°.72/100\,\mathrm{yr}$ (Burns at al., 1996) while Andersen at al. (1987) calculated a faster theoretical apsidal motion of $\dot{\omega}_{\mathrm{theo}} = 4°.25/100\,\mathrm{yr} \pm 0°.72/100\,\mathrm{yr}$.

2. Observations

Observations were made from July to September 2000 at Biruni Observatory of Shiraz University with a 51 cm Cassegrain telescope equipped with an uncooled RCA4509 multiplier phototube. Two stars – HD 184240 and HD 186239 – were selected as comparison and check stars, respectively. The measurements were made using B and V filters, which closely matched the Johnson UBV system specification. Data reduction and atmospheric corrections were done to obtain the complete light curves in two filters, using a computer code developed by G.P. McCook.

Astrophysics and Space Science **296**: 141–144, 2005.
© Springer 2005

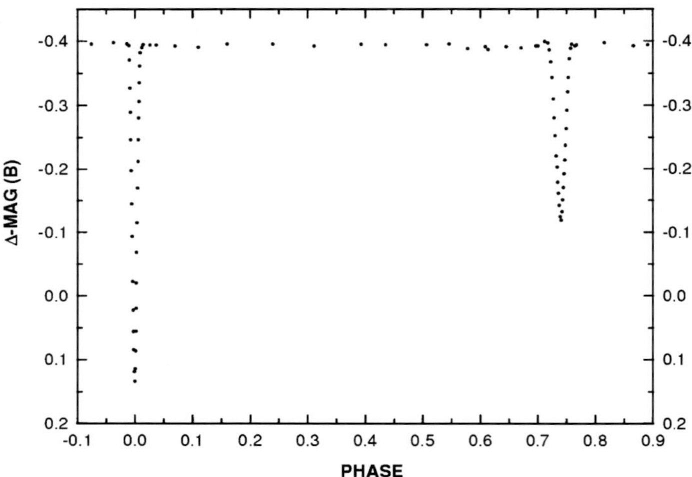

Figure 1. Observed light curve of V1143 Cygni in the *B* filter.

Figure 1 represents the observed light curve in the *B* filter. In order to enhance the accuracy of the present work , two other minima were observed on July 16 and 18, 2002. This time, the star HD 185978 (F8; $m_V = +7^m.8$) was selected as comparison star.

3. Times of Minima and Light Curve Analysis

From the observed light curves, heliocentric times of minima (two primaries and two secondaries) were computed (see Dariush et al., 2001, 2003). The *B* and *V* light curves of V1143 Cyg were analyzed separately by using the Wilson code in order to derive photometric elements of this system. Since the system V1143 Cygni is detached with both components residing well inside their respective Roche lobes (Burns et al., 1996), the solution was performed in WD mode 2. In our analysis, we assumed a value of zero for the third light ($l_3 = 0$). Before running the LC code, the position of the periastron was estimated from the apsidal motion study as discussed in Section 4. The optimized parameters are given in Table I.

4. Apsidal Motion

The rate of apsidal motion can be determined exactly by analysis of primary and secondary eclipse timings. Following the method of Guinan and Maloney (1985) the apsidal motion rate can be calculated from the change in the displacement of

TABLE I
Optimized parameters of V1143 Cygni

Parameter	Filter B	Filter V
i	87.3 ± 0.1	87.1 ± 0.1
e	0.536 ± 0.005	0.539 ± 0.005
ω	$0°.855 \pm 0.004$	$0°.855 \pm 0.004$
$q(\frac{M_2}{M_1})$	0.99 ± 0.01	0.98 ± 0.01
Ω_1	17.4 ± 0.5	18.0 ± 0.5
Ω_2	19.6 ± 0.5	20.0 ± 0.5
$\frac{L_2}{L_1}$	0.89 ± 0.05	0.82 ± 0.05

the secondary minimum from phase 0.5 according to

$$D = [(t_2 - t_1) - 0.5 \times \text{Period}]. \tag{1}$$

D in turn is related to ω, the longitude of periastron, by the formula given by Sterne (1939). The new secondary eclipse timings (Dariush et al., 2001, 2003) and those given by Snowden and Koch (1969), Koch (1977), Gimenez and Margrave (1985), and Burns et al. (1996) were used to calculate the apsidal motion by computing the slope of a line fitted to all of the these minima. We determined an observed rate of apsidal motion of $\dot{\omega}_{obs} = 3°.72/100\,\text{yr} \pm 0.37/100\,\text{yr}$ (see Figure 2).

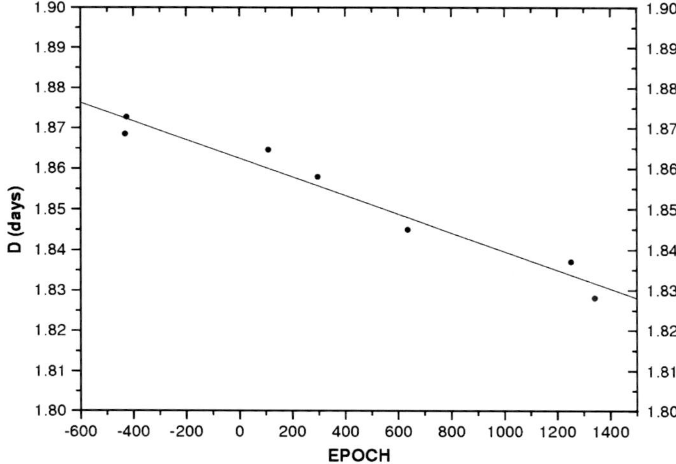

Figure 2. Plot of the displacement of secondary minima from a half period point of V1143 Cygni, in days, versus epoch.

 A. DARIUSH ET AL.

TABLE II

Determined rate of apsidal motion for V1143 Cyg

$\dot{\omega}^{\circ}_{obs}/100\,yr$	Error	U(yr)	Source
3.49	±0.38	10320	Khaliullin (1983)
3.36	±0.19	10710	Gimenez and Margrave (1985)
3.52	±0.72	10230	Burns et al. (1996)
3.72	±0.37	9680	Present study[1]

5. Results and Discussion

Table II contains independent determinations of the observed apsidal motion of V1143 Cyg together with the corresponding period of apsidal revolution U. To determine a more accurate value for $\dot{\omega}_{obs}$ we need more accurate timings of secondary minima. From Table II, it seems that expanding our observational baseline may decrease the discrepancy between $\dot{\omega}_{obs}$ and $\dot{\omega}_{theo}$. Our results for V1143 Cyg supporting $3°.72 \pm 0.37/100\,yr$ is slightly closer to the theoretical apsidal motion rate $(4°.25 \pm 0.72/100\,yr)$, computed in previous studies of this system.

References

Andersen, J., Garcia, J.M., Gimenez, A. and Nordström, B.: 1987, *Astron. Astrophys.* **174**, 107.

Burns, J.F., Guinan, E.F. and Marshall, J.J.: 1996, *IBVS* **4363**.

Dariush, A., Afroozeh, A. and Riazi, N.: 2001, *IBVS* **5136**.

Dariush, A., Zabihinpoor, S.M., Bagheri, M.R., Jafarzadeh, Sh., Mosleh, M. and Riazi, N.: 2003, *IBVS* **5456**.

Gimenez, A.: 1985, *Astrophys. J.* **297**, 405.

Gimenez, A. and Margrave, T.E.: 1985, *Astron. J.* **90**(2), 358.

Guinan, E.F. and Maloney, F.P.: 1985, *Astron. J.* **90**, 1519.

Khaliullin, Kh.F.: 1983, *Astron.Trisk* No. 1262.

Koch, R.H.: 1977, *Astron. J.* **82**, 653.

Lacy, C.H. and Fox, G.W.: 1994, *IBVS* **4009**.

Snowden, M.S. and Koch, R.H.: 1969, *Astrophys. J.* **156**.

Sterne, T.E.: 1939a, *Mon. Not. R. Astron. Soc.* **99**, 451.

Sterne, T.E.: 1939b, *Mon. Not. R. Astron. Soc.* **99**, 662.

INTRINSIC VARIABLES AS COMPONENTS OF CLOSE BINARIES

N.N. SAMUS[1,2]

[1]*Institute of Astronomy, Russian Academy of Science, 48 Pyatnitskaya Str.,
Moscow 119017, Russia; E-mail: samus@sai.msu.ru*
[2]*Sternberg Astronomical Institute, Moscow University, 13 University Avenue,
Moscow 119992, Russia*

(accepted April 2004)

Abstract. I review the current knowledge on binarity accompanying intrinsic variability of different types, according to the existing classification scheme. In many cases, binary nature not only accompanies but causes intrinsic variability.

Keywords: binary stars, variable stars

In this presentation, I will not be very particular to the word "close" in the title. Usually the term "close" refers to binaries with noticeable effects of interaction between the components, and different authors can find some systems close or not, dependent on their ideas of what a "noticeable effect" is. I will demonstrate that two other terms in the title ("intrinsic variable" and "binary") are also somewhat ambiguous, and I will not pay special attention to proof of interaction being present.

The variable star science began in 1596, with the discovery of Mira Ceti, when astronomers began to understand that some stars could vary in brightness. Variables noticed before this discovery were considered something very unusual; such "guest stars" were actually thought to be "not quite" stars. Note that of the two types of variables actually known before 1596, Novae and Supernovae, all Novae are currently believed to be binaries, with variability determined by binarity (hydrogen enrichment of the surface of a white dwarf through gas flow from a companion in a close binary, with a subsequent thermonuclear explosion). Among the first variable stars discovered within two centuries after 1596 (note that, before World War II, Novae were not even given regular variable star names!), all are now known (or suspected) binaries.

Considering the (rather complex and confused) classification system of the General Catalogue of Variable Stars (GCVS; Kholopov, 1985), we find that binarity is a quite common phenomenon for many types of variables or their larger groups and even determines processes leading to variability in some of them. We immediately find that it is not easy to define what intrinsic variability is. Eclipsing variables are not intrinsic variables. But what if they are spotted, like RS CVn stars, and spots appear, disappear, and migrate? Can we call deviations from regular orbital variations in X-ray systems intrinsic variability?

Astrophysics and Space Science **296**: 145–155, 2005.
© Springer 2005

Now let us consider the first known variable (besides Novae), Mira Ceti. It is a rather regular star for its type. However, Joy (1926) discovered its close companion, a hot star, varying with an amplitude of 2^m. This star was considered a separate variable, VZ Cet. It is now generally assumed to be a physical companion, but its period must be very large. VZ Cet is a hot star, believed to interact with Mira's wind. If so, classification of Mira as a mild symbiotic (and thus interacting binary) star (Whitelock, 1987) is justified. We immediately feel that the GCVS classification scheme does not work always well. Symbiotic stars are considered, in the GCVS, in the wide group of "cataclysmic and Nova-like" variables, and Miras are pulsating stars, and the connection between the two groups is not paid special attention in the GCVS.

However, the connection is important. Mira Ceti is comparatively close to us, so we observe VZ Cet as an individual variable. R Aqr is a much better known symbiotic Mira. Its Mira-type light curve shows serious irregularities, believed to be a result of interaction with a companion on a very eccentric orbit. Recently the pair was reported to be resolved in radio observations. Will (or should) the companion get its own GCVS name?

Symbiotics include an interesting subtype of symbiotic Novae (some of them called NC stars, very slow Novae, in the GCVS). One of the representatives of the type is RR Tel. Several years before its outburst in 1944, the star developed long-period Mira-type pulsations. Evidence for pulsations with the same period was noticed even during the peak of the outburst. Until now, the star has not reached its preoutburst brightness level. Bianchini (1990) suggests an 11-year photometric period, presumably the orbital one, for RR Tel. Thus, this binary exhibits pulsations, it underwent an outburst that, most certainly, was a result of interaction in the binary system, and it possibly shows direct manifestations of its orbital period in its light curve.

Another symbiotic Nova is PU Vul (Kuwano's object). It also shows probable long-period pulsations. After it had gradually brightened by 8 magnitudes in 1977–1979, periodic brightness variations ($P = 78^d$) were reported by Chochol et al. (1981). Now $P = 217^d$ is suggested (Chochol et al., 1998). The star's very deep minimum of 1980–1981 was originally not understood, but actually it was an eclipse (Vogel and Nussbaumer, 1992) Figure 1 (Chochol et al., 2003) is the combined many-year light curve of PU Vul. Note the second, much shallower eclipse in 1994. Again, the star shows intrinsic variability of two different kinds, one of them definitely binarity-related, and eclipsing variability.

To finish with Miras, I mention a recent result (Figure 2) where we know nothing about interaction or even about the two variables being physically related (probably not). What is interesting and instructing about this case is that V597 Sco looked like a sufficiently well-studied variable in the GCVS, with its period known. We found (Samus and Hazen, 2003) that here we had two Miras, separated by approximately 10″, and none of them had the period reported by the early investigators. Most probably, the light variations originally noticed actually referred to the blend

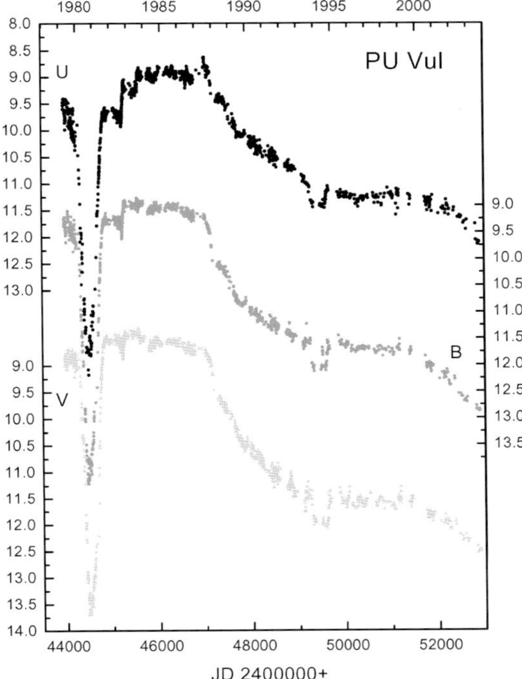

Figure 1. The light curve of PU Vul (Chochol et al., 2003). Note a deep eclipse in 1980–1981 and a shallower one in 1994.

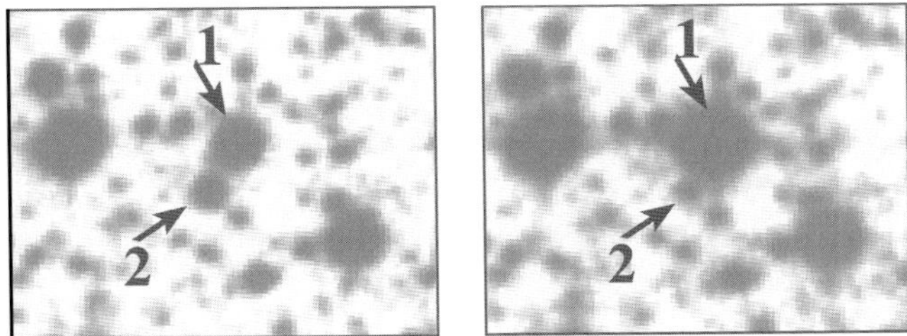

Figure 2. The two Miras at the position of V597 Sco (from red-light images of the USNO Image and Catalog Archive; Samus and Hazen, 2003).

of two variables. Our recent experience with checking information for old Harvard variables with ASAS (Pojmanski, 2002) shows that there are many cases of completely wrong types, and attention to known variable stars in ASAS is needed also for GCVS stars, not only for poorly studied stars of the NSV catalog.

Now to another type of pulsating variables, Cepheids, called "the most important stars" because of their significance for the distance scale in the Universe – and also for our understanding of stellar evolution. Classical Cepheids are Pop I pulsating supergiants. In the nineteenth and early twentieth century, many authors, among them Belopolsky (1895) interpreted Cepheids as eclipsing binaries. They already had radial velocity measurements and tried to compute orbits, which appeared quite exotic. Apparently S Sge was the first real spectroscopic binary correctly recognized among Cepheids (Aldrich, 1924). Its orbital period (682^d) is rather large, typical of binary Cepheids. Though measurements of radial velocities are not the only method to reveal binary Cepheids, it is the most straightforward and the most reliable one. In our Moscow radial velocity team, we paid considerable attention to binary Cepheids in our many-year Cepheid studies using a CORAVEL-type spectrometer designed and built by Tokovinin (1987).

A typical example of a spectroscopic-binary classical Cepheid is MW Cyg. This binary was discovered by us (Gorynya et al., 1992). It shows different levels of data points on the velocity curve for different years of observations, revealing slow orbital motion (Figure 3). The radial velocity curve can be resolved into the pulsation part and the orbital part; the pulsation part can be used, for example, for Baade–Wesselink studies, and the orbital curve permits us to restrict the mass of the companion (Figure 4). In the case of SU Cyg, the companion must be rather massive, and speculations on its unusual nature began to appear. However, the companion is now studied spectroscopically in the ultraviolet range and also appeared a binary.

The next star, TX Del, was the greatest disappointment in our Cepheid program. When, in 1989, we had become ready to announce its discovery as a spectro-scopic binary and had shown our results to L. Szabados, he produced a preprint of Harris and Welch (1989), whose authors apparently were some months ahead. The existence of several comparatively short-period binary Cepheids at high galactic

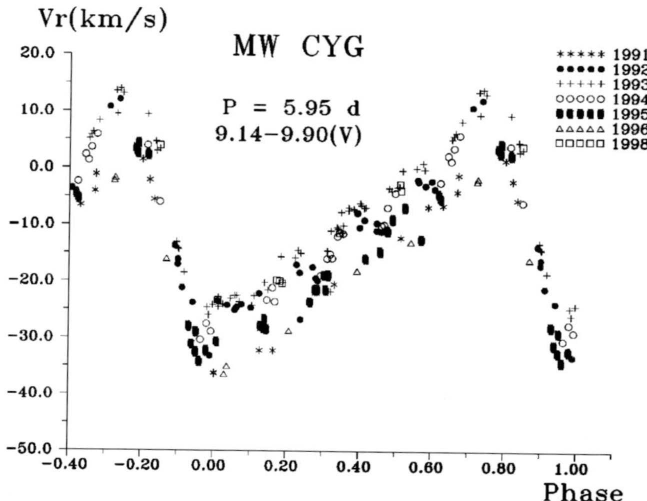

Figure 3. The radial velocity curve of MW Cyg, showing stratification of data points from different years, based on the combined data of the Moscow radial velocity team.

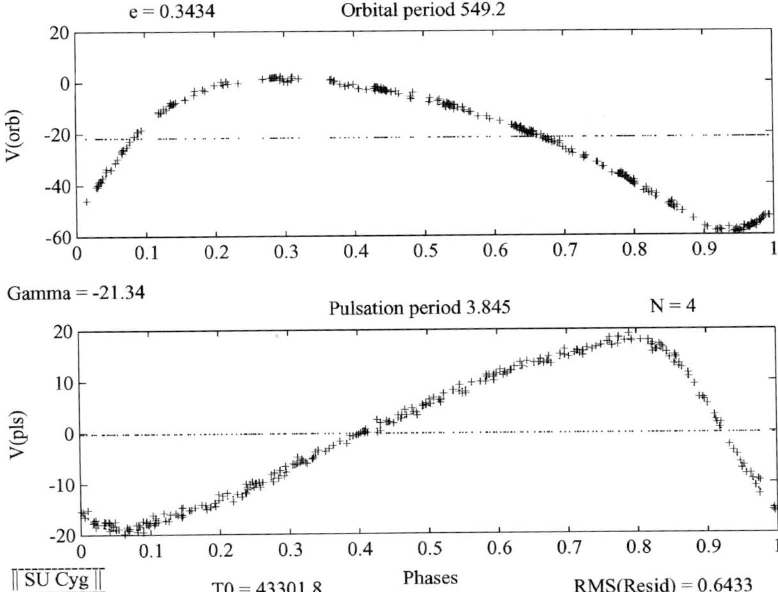

Figure 4. The radial velocity curves separately for the orbital motion and for the pulsations of SU Cyg, from Moscow data.

latitudes (TX Del, AU Peg, IX Cas) means the presence of spectroscopic binaries among Population II Cepheids. This is an important finding from the point of view of occurrence of binaries among older stars. Note, however, that the three spectroscopic-binary Population II Cepheids are not very metal poor, whereas Cepheids in globular clusters are met predominantly in very metal poor systems. Thus, spectroscopic binaries among Population II Cepheids are so far known among their metal-richer, presumably younger, representatives.

The most often met Population II pulsating variables are RR Lyrae stars. Here we do not know much about binarity. I think the only confident statement is that we are sure that binary stars do exist among RR Lyrae stars, as it was first shown by Kholopov (1971) on the base of the observations of the variable V80 in the Ursa Minor dwarf spheroidal galaxy from van Agt (1968).

On the other hand, there are quite many binaries among δ Scuti stars and some other types of pulsating variables. In the literature, there exist suggestions of δ Scuti and β Cephei pulsations being more regular and predictable in binary systems (e.g. Frolov et al., 1980). Recently, Kim et al. (2003) suggested a new variability type, oEA, for eclipsing binaries whose components show δ Scuti-like pulsations but are actually at a different, "accretion-dominated" stage of evolution. This suggestion may be quite correct from the point of view of astrophysics of these pulsating stars, but it has serious drawbacks from the point of view of practical classification,

                    N.N. SAMUS

especially for the GCVS. First, many stars will remain not classified from the day when it is shown that they are not only pulsating but also binary stars, until a very serious study shows that their evolution is really accretion dominated. Second, the abbreviation suggested is not good: imagine that we have a binary, accretion-dominated pulsating star, but its inclination is such that there are no eclipses.

Other types of pulsating variables are also often met in binaries. Among many others, I can refer to examples of an eclipsing β Cephei star EN Lac (Jerzykiewicz, 1979) and a spectroscopic-binary γ Dor star HD 19684 (Kaie and Fekel, 2003). Recently, Henry and Fekel (2003) found seven of twelve new γ Dor stars to be binaries. And the case of the pulsating sdB star in an eclipsing binary NY Vir (Kilkenny et al., 1998) is especially beautiful (Figure 5).

I will skip classical Novae (mentioned above) and proceed to dwarf Novae. The current theory of dwarf Novae connects their outbursts with thermal instability and the presence of super-outbursts, with superhumps during them, to thermal-tidal instability (cf. Osaki, 1996). We use to call outbursts of dwarf Novae intrinsic variability, despite they are completely determined by binarity, with one of the system's components being a white dwarf. The same applies to the well-known flickering of dwarf Novae and to unusual and varying shape of the eclipsing light curves for some of them. Note that superhumps are rather easily distinguished from pulsations: the amplitude of superhumps is approximately the same in B and V light, whereas for pulsations, B-band amplitudes are usually considerably larger (a notable exception are Miras).

The shapes of eclipsing light curves of dwarf Novae are unusual, and cataclysmic candidates are easily isolated among eclipsing stars just from a light curve. Light curve modeling for dwarf Novae and other cataclysmics must take into account accretion processes, existence of an accretion disk, a gas flow, and parts of the disk differing by their surface brightness. As an example, I would like to mention the interpretation of the eclipse light curves of U Gem (after Krzeminski, 1965) and

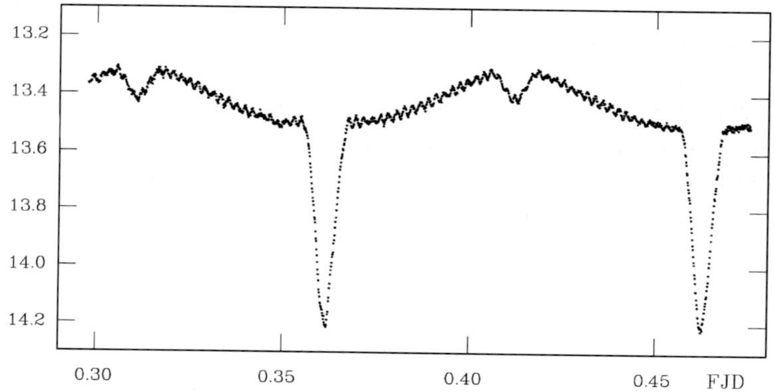

Figure 5. The light curve of NY Vir (from Kilkenny et al., 1998).

IP Peg (whose eclipse starting during an outburst was reported by Goranskii et al., 1985) in the so-called "hot-line" model (Khruzina et al., 2001, 2003), reviewed by Dr. D. Bisikalo at this conference.

The X-ray variables are also binaries, with neutron stars or black holes as compact components. HZ Her (Her X-1) was known as a variable star for decades before the detection of its X-ray nature, again confirming the necessity of studying poorly investigated GCVS stars. Already the first light curves folded with the orbital period (for example, Kurochkin, 1972) clearly revealed complex behavior, superimposed on quite regular variations due to reflection effect (the side of the normal star turned to the X-ray source, a neutron star, is much hotter). N.E. Kurochkin (1923–2003) was an excellent photographic observer, his eye estimates were very accurate, and the scatter on his light curve for HZ Her could not be explained by observation uncertainties. The newer, more accurate data (Sheffer et al., 2002) confirm that the light curve shape varies dependent on the state of the X-ray source. Again, it is not quite clear if we may call such variations intrinsic. It should also be reminded that archive photographic data show two time intervals with low-amplitude variability of HZ Her with the same period but a different light curve shape (Hudec and Wenzel, 1976). Nothing like these two episodes was observed since the advent of X-ray astronomy. On the other hand, a poorly documented episode of "on" stages of the X-ray source not coming at the expected time in 1983 was not accompanied by any significant reduction of the optical variability amplitude.

Another famous X-ray variable, V1357 Cyg (Cyg X-1), is quite different. It has no pulsar, is believed to contain a black hole, and its optical star is massive and bright, in contrast to that in the system of HZ Her. The optical variability is mainly not due to reflection effect but to a nonspherical shape of the optical star in this interacting binary. Again, the system displays various irregularities (intrinsic variations?) superimposed on the rather regular ellipsoidal variability (Karitskaya et al., 2000, 2001).

The very unusual variable V1343 Aql (SS 433) is often considered, despite of being unusual, a prototype microquasar. Its brightness variations are very irregular; in fact, variability due to disk precession is easier to find than eclipses, revealed by Cherepashchuk, 1981, only after discussion in the literature concerning their absence (Whitmire and Matese, 1981).

My last example in this group of objects is V4641 Sgr (erroneously mixed up, for some time, with its neighbor, a rather ordinary variable GM Sgr). For this star, considerable burst-like optical variability was detected first, but now its orbital brightness variations have also been established quite reliably (Goranskii et al., 2003).

Now I make a transition to young intrinsic variables through flare stars. Many flare stars are known, mainly visual, binaries. It was believed for some time that it was always the fainter component of the pair that flared. Now we know examples, including the prototype star, UV Cet, where both components flare (the orbital ephemeris of this pair, UV and BL Cet, can be found, for example, in Geyer et al.,

1988). Nevertheless, the old idea contains correct astrophysics: these absolutely faint, old stars are young by their stellar evolution stage, and the fainter the star in the pair, the lower its luminosity and mass, the "younger" it is in the evolutionary sense and hence, the higher is its activity level.

Multiplicity is also a very common phenomenon among the young intrinsic variables (classical and weak-lined T Tauri stars). Their variability, often strong, is believed to be due to effects in their environment, including accretion processes, and to rotation of stars with spotted surfaces. Some "companions" to T Tauri stars are in fact not stars but shock waves in the surrounding diffuse matter (the Herbig–Haro objects). For a number of stars, eclipses were suggested as a cause for some part of their variability. These eclipses can be caused not only by companion stars but also by nonstellar objects (sometimes referred to as comets, though I do not like the term), orbiting the primaries for a more or less long time and then possibly dissolving themselves. Again, is this binarity or is it intrinsic variability? Figure 6 shows an example of probable binarity-related variability of the prototype star, T Tau, from Ismailov and Samus (2003).

The stellar surface being spotted causes variability of stars belonging to several types. The most stable patterns of star spots are observed on the surfaces of magnetic α^2 CVn variables. Large dark spots, evidently changing their shapes and location, are the reason for variability of BY Dra stars, many of them simultaneously flare stars. Among eclipsing stars or binaries without eclipses due to orbital inclination, the well-known group of spotted stars are RS CVn variables. The prototype was long known for strange variations of its eclipse light curve, with different levels of brightness before and after the primary eclipse during different seasons. Then it was found that the distortion of the light curve was a kind of superposed wave, with a period slightly different from the orbital one. This wave is explained by spots on the surface of one of the components, with slightly asynchronous axial rotation with respect to the orbital motion. Detailed analysis of the star-spot pattern (Figure 7) reveals considerable variations from one year to another (Rodonò et al., 1995).

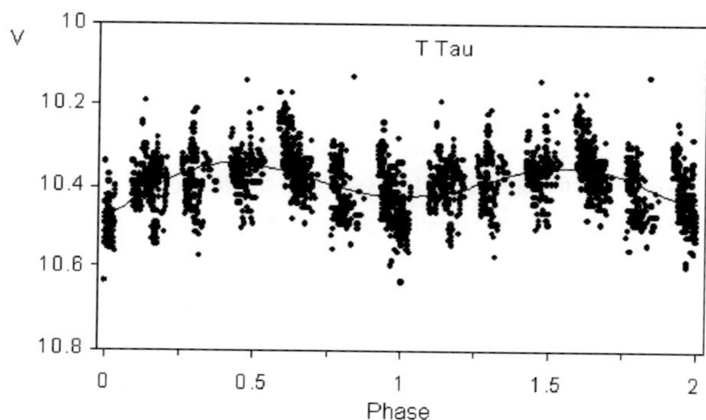

Figure 6. The light curve of T Tau folded with the period 2200^{d}.

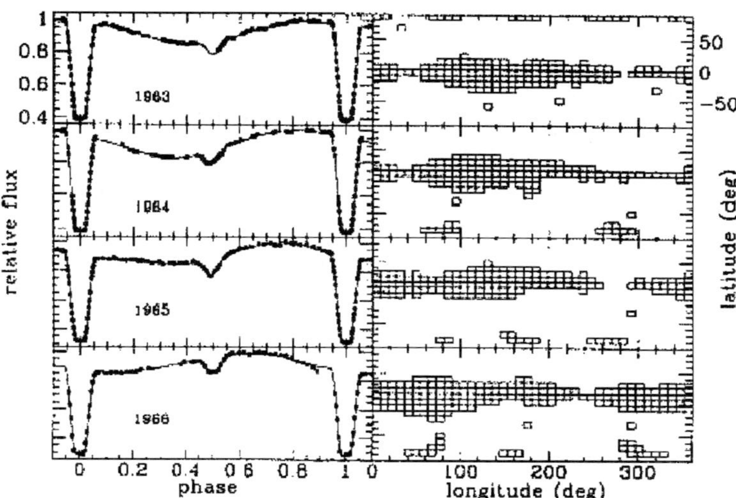

Figure 7. Several of the large number of seasonal light curves of RS CVn, with the corresponding star spot maps (Rodonò et al., 1995).

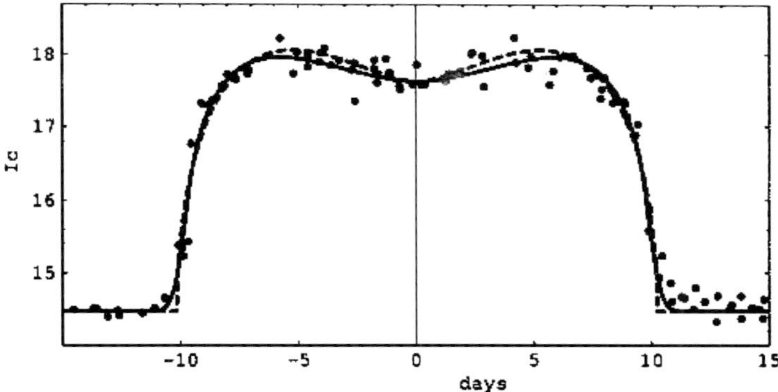

Figure 8. The light curve of the T Tau star V582 Mon (upside down, from the point of view of usual variable star conventions), represented by the model of a swarm of solid particles trapped in a giant gaseous vortex (Barge and Viton, 2003).

And to finish with, I demonstrate an eclipse light curve of the T Tauri star V582 Mon that reminded me an old variable star joke about a star eclipsed by a companion having a hole through its center. Of course the brightening in the middle of the eclipse of V582 Mon (KH15D; Figure 8, Barge and Viton, 2003) is not due to a hole through its companion. The star is eclipsed by a body of a complex geometry (in the model by Barge and Viton, 2003, it is a swarm of solid particles, trapped in a giant gaseous vortex rotating at some distance from the star), and this geometry can cause funny effects.

Acknowledgements

I would like to thank the organizers for inviting me to present this talk. It was interesting for me to collect this information, but I have found that the topic is so wide that no exhausting presentation is possible. I thank the colleagues (Drs. S. Antipin, P. Barge, D. Chochol, N. Gorynya, D. Kilkenny, M. Rodonò) who provided new figures for this paper or permitted to reproduce figures from their publications. Thanks are due to the Russian Foundation for Basic Research, to the program "Non-Stationary Processes in the Universe" of the Presidium of Russian Academy of Sciences, to the Federal Science and Technology Program "Astronomy," to the Program of Support for Leading Scientific Schools of Russia for support provided to our studies of variable stars.

References

Aldrich, J.A.: 1924, *Pop. Astron.* **32**, 218.

Barge, P. and Viton, M.: 2003, *Astrophys. J.* **593**, L117.

Belopolsky, A.: 1895, A Spectroscopic Study of the Variable Star δ Cephei (Dissertation, in Russian), Imperial Academy of Sciences, St.-Petersburg.

Bianchini, A.: 1990, *Astron. J.* **99**, 1941.

Cherepashchuk, A.M.: 1981, *Mon. Not. RAS* **194**, 761.

Chochol, D., Hric, L. and Papousek, J.: 1981, *Inform. Bull. Var. Stars* No. 2059.

Chochol, D., Pribulla T. and Tamura, S.: 1998, *Inform. Bull. Var. Stars* No. 4571.

Chochol, D., Pribulla, T., Parimicha, S. and Vanko, M.: 2003, *Baltic Astron.* **12**, 610.

Frolov, M.S., Pastukhova, E.N., Mironov, A.V. and Moshkalev, V.G.: 1980, *Inform. Bull. Var. Stars* No. 1894.

Geyer, D.W., Harrington, R.S. and Worley, C.E.: 1988, *Astron. J.* **95**, 1841.

Goranskii, V.P., Barsukova, E.A. and Burenkov, A.N.: 2003, *Astron. Rep.* **47**, 740.

Goranskii, V.P., Lyutyi, V.M. and Shugarov, S.Yu.: 1985, *Soviet Astron. Lett.* **11**, 293.

Gorynya, N.A., Irsmambetova, T.R., Rastorgouev, A.S. and Samus, N.N.: 1992, *Soviet Astron. Lett.* **18**, 316.

Harris, H.C. and Welch, D.L.: 1989, *Astron. J.* **98**, 981.

Henry, G.W. and Fekel, F.C.: 2003, *Astron. J.* **126**, 3058.

Hudec, R. and Wenzel, W.: 1976, *Bull. Astron. Inst. Czechoslovakia* **27**, 325.

Ismailov, N.Z and Samus, N.N.: 2003, *Inform. Bull. Var. Stars* No. 5382.

Jerzykiewicz, M.: 1979, *Inform. Bull. Var. Stars* No. 1552.

Joy, A.H.: 1926, *Astrophys. J.* **63**, 281.

Kaie, A.B. and Fekel, F.C.: 2003, *Inform. Bull. Var. Stars* No. 5465.

Karitskaya, E.A., Goranskii, V.P., Grankin, K.N. and Yu, Mel'nikov, S.: 2000, *Astron. Let.* **26**, 22.

Karitskaya, E.A., Voloshina, I.B., Goranskii, V.P., Grankin, K.N., Drhaniashvili, E.B., Ezhkova, O.V., Kochiasvili, N.T., Kumsiasvili, M.I., Kusakin, A.V., Lyutyi, V.M., Melnikov, S.Yu. and Metlova, N.V.: 2001, *Astron. Rep.* **45**, 350.

Kholopov, P.N.: 1971, *Perem. Zvyozdy* **18**, 117.

Kholopov, P.N. (ed.): 1985, General Catalogue of Variable Stars (4th edn., Vol. I), Nauka Publishers, Moscow.

Khruzina, T.S., Cherepashchuk, A.M., Bisikalo, D.V., Boyarchuk, A.A. and Kuznetsov, O.A.: 2001, *Astron. Rep.* **45**, 583.

Khruzina, T.S., Cherepashchuk, A.M., Bisikalo, D.V., Boyarchuk, A.A. and Kurnetsov, O.A.: 2003, *Astron. Rep.* **47**, 848.

Kilkenny, D., O'Donoghue, D., Koen, C., Lynas-Gray, A.E. and van Wyk, F.: 1998, *Mon. Not. RAS* **296**, 329.

Kim, S.-L., Lee, J.W., Kwon, S.-G., Youn, J.-H., Mkrtichian, D.E. and Kim, C.: 2003, *Astron. Astrophys.* **405**, 236.

Krzeminski, W.: 1965, *Astrophys. J.* **142**, 1051.

Kurochkin, N.E.: 1972, *Astron. Tsirk. (USSR)* No. 717.

Osaki, Y.: 1996, *Publ. Astron. Soc. Pacific* **108**, 39.

Pojmanski, G.: 2002, *Acta Astron.* **52**, 397.

Rodonò, M., Lanza, A.F. and Catalano, S.: 1995, *Astron. Astrophys.* **301**, 75.

Samus, N.N. and Hazen, M.L.: 2003, *Inform. Bull. Var. Stars* No. 5450.

Sheffer, E.K., Voloshina, I.B., Goransky, V.P. and Lyuty, V.M.: 2002, *Astron. Rep.* **46**, 981.

Tokovinin, A.A.: 1987, *Soviet Astron.* **31**, 98.

van Agt, S.L.Th.J.: 1968, *Bull. Astron. Netherl.* **19**, 275.

Vogel, M. and Nussbaumer, H.: 1992, *Astron. Astrophys.* **259**, 525.

Whitelock, P.A.: 1987, *Publ. Astron. Soc. Pacific* **99**, 573.

Whitmire, D.P. and Matese, J.J.: 1981, *Mon. Not. RAS* **194**, 293.

PRELIMINARY ANALYSIS OF *V* AND *Hp* LIGHT CURVES OF 26 CARBON MIRAS

ZDENĚK MIKULÁŠEK[1] and TOMÁŠ GRÁF[2]

[1]*Institute for Theoretical Physics and Astrophysics, Faculty of Science, Masaryk University in Brno, Kotlářská 2, 61137 Brno, Czech Republic; E-mail: mikulas@ics.muni.cz*
[2]*Observatory and Planetarium of J. Palisa, 17. listopadu 15, 70833 Ostrava-Poruba, Czech Republic*

(accepted April 2004)

Abstract. Light variations of a representative sample of 26 more or less periodically variable carbon stars were analyzed on the basis of 2220 individual observations made by the Hipparcos satellite and 33 544 visual observations listed in AFOEV and VSOLJ databases within the interval JD $= 2\,448\,000$ (1988) ± 6 cycles. We found the osculating linear ephemerides of all stars and their mean light curves, as well. We found that the light curves of the carbon Miras in our set can be satisfactorily expressed as a linear combination of only two basic light curves. The analysis was done by an own method combining robust regression and principal component analysis.

Keywords: ephemeris, carbon stars, carbon Miras

1. Introduction

Carbon stars are cool objects of the asymptotic giant branch in an advanced stage of their evolution, when products of their inner shell helium nuclear burning are mixed into outer layers. All carbon stars exhibit light variations on time scales of tens and hundreds of days, apparently caused by pulsations and propagation of shock waves through the stars. Light variations of some carbon stars are semi-regular or maybe multi-periodic (Dušek and Mikulášek, 2003), other carbon stars vary more or less regularly with one well-defined period of several hundred days. The variable stars called carbon Miras exhibit distinguished light variations of a few magnitudes in *V*; this is why they are popular targets of many visual variable star observers all over the world. Carbon Miras also show distinct spectroscopic variations, which can provide hints to the mechanism of variability of these stars. We found, e.g., that apparently all carbon Miras exhibit strongly variable Hα emission, which varies with the period of light variations (Barnbaum, 1994; Gráf, 2003).

Currently, we are preparing a review paper on the variation of a number of spectral characteristics of 30 selected carbon stars depending on their photometric phase. The spectrograms were obtained in the time interval 1988–1991. To complete the analysis we need to know the uniquely defined osculating linear light ephemeris valid in the period of spectroscopic observations. It was the motivation for the

Astrophysics and Space Science **296**: 157–160, 2005.
© Springer 2005

following analysis of the light behaviour of more than two dozen carbon Miras in the last two decades of the past century.

2. Observational Material

Our analysis was based on observations of particular carbon Miras of our list during 12 cycles of the stars centered on JD $= 2\,448\,000$, and contained in databases of Hipparcos satellite (19 stars out of 30), AFOEV (25 stars) and VSOLJ (24 stars), the latter two being associations of French and Japanese visual observers. Twenty-six out of 30 spectroscopically studied carbon stars were observed frequently enough, so that we were able to determine osculating linear ephemerides of the periodical part of their light variations. The list of all carbon stars studied is given in Table I.

In the case of Hipparcos observations, we have used only the Hp measurements, because they are much more accurate than observations made in pseudo V or B colours. Both Hipparcos and visual observations were binned to normal points with different weights corresponding to the number and quality of measurements.

3. Modeling of Light Curves

After a careful inspection of the light curves of 26 carbon Miras, we concluded that light curves in V and Hp colours can be satisfactorily described by the following model:

$$V_i(t) = \overline{V}_i + \sum_{j=1}^{p_i} b_{ij} L_j(T) + \left[1 + \sum_{k=1}^{q_i} a_{ik} L_k(T)\right] F(f)$$

$$Hp_i(t) = \overline{V}_i + \overline{(Hp - V)}_i + h_i \left\{ \sum_{j=1}^{p_i} b_{ij} L_j(T) + \left[1 + \sum_{k=1}^{q_i} a_{ik} L_k(T)\right] F(f) \right\}$$

$$f = (t - M_{0i})/P_i; \quad T = (t - 2\,448\,000)/6 P_0,$$

where $\overline{V}_i$ is the average magnitude and $\overline{(Hp - V)}_i$ is the mean colour index of the ith star, b_{ij} and a_{ik} are coefficients describing long-term changes of the mean magnitude and relative variation of the amplitude of that star, respectively. L_i is a Legendre polynomial of ith degree, $F(f, A_{im})$ is a simple representation of the periodic term of the light curve of ith star, described by a set of parameters $\{A_{im}\}$ with the meaning of an amplitude. All parameters (including P_i and M_{0i}) of this non-linear model of light curves of carbon Miras were found iteratively by the gradient method of robust regression effectively suppressing the influence of outliers (see Mikulášek et al., 2003). As an initial estimate for the ephemeris we adopted those given in SIMBAD.

TABLE I

Basic characteristics of the light curves of the studied carbon Miras

Name	Spectral type	P_{cat}	P_{oscul}	$\tilde{M}_0$	V_{mean}	A_{eff}
AU Aur	C6.3e	377	399 ± 5	$48\,071 \pm 8$	only *Hp* m.	
AZ Aur	C7.1e	414	413.3 ± 0.4	$47\,864 \pm 2$	11.76 ± 0.03	3.18 ± 0.08
S Aur	C4.4e	590	587.2 ± 1.1	$47\,926 \pm 6$	11.69 ± 0.02	1.61 ± 0.07
UV Aur	C6.2epJ	391	396.1 ± 0.7	$48\,149 \pm 4$	9.68 ± 0.03	1.63 ± 0.07
S Cam	C7.3e	326.5	325.6 ± 0.2	$47\,950 \pm 1$	9.29 ± 0.02	1.72 ± 0.05
R CMi	C7.1eJ	330.9	330.3 ± 0.4	$48\,079 \pm 2$	9.28 ± 0.02	2.46 ± 0.07
W Cas	C7.1e	399.8	402.3 ± 0.3	$47\,880 \pm 1$	10.30 ± 0.01	2.53 ± 0.04
X Cas	C5.4e	420.9	427.7 ± 0.4	$48\,214 \pm 2$	11.32 ± 0.02	1.73 ± 0.04
S Cep	C7.4e	486	485.5 ± 0.5	$48\,178 \pm 3$	9.48 ± 0.02	1.95 ± 0.04
V CrB	C6.2e	358	358.5 ± 0.3	$47\,692 \pm 2$	10.19 ± 0.03	3.25 ± 0.08
RS Cyg	C8.2e	434.1	428.0 ± 0.4	$47\,728 \pm 2$	7.93 ± 0.01	1.50 ± 0.05
U Cyg	C7.2e	460	462.7 ± 0.4	$47\,811 \pm 2$	8.81 ± 0.02	2.75 ± 0.05
V Cyg	C5.3e	417	419.7 ± 0.4	$47\,797 \pm 3$	11.52 ± 0.03	3.61 ± 0.09
WX Cyg	C8.2e	399	411.1 ± 0.4	$47\,855 \pm 2$	11.63 ± 0.02	2.23 ± 0.06
T Dra	C6.2e	422	422.6 ± 0.6	$47\,768 \pm 4$	11.35 ± 0.05	3.30 ± 0.13
R For	C4.3e	388	387.3 ± 2.0	$48\,163 \pm 9$	11.29 ± 0.10	2.50 ± 0.30
VX Gem	C7.2e	381	381.0 ± 0.5	$48\,072 \pm 3$	10.96 ± 0.03	4.15 ± 0.10
ZZ Gem	C5.3e	316	313.1 ± 0.6	$47\,854 \pm 4$	10.29 ± 0.05	2.50 ± 0.30
V Hya	C6.3e	529.2	528.3 ± 0.8	$47\,920 \pm 4$	8.78 ± 0.03	1.91 ± 0.08
R Lep	C7.6e	444	438.3 ± 0.6	$48\,190 \pm 3$	9.06 ± 0.02	2.29 ± 0.06
U Lyr	C4.5e	454	453.3 ± 0.7	$48\,034 \pm 4$	11.42 ± 0.03	2.18 ± 0.08
V Oph	C5.2e	296.1	295.3 ± 0.4	$47\,761 \pm 2$	9.02 ± 0.02	1.95 ± 0.06
RZ Peg	C9.1e	435.6	437.1 ± 0.2	$47\,857 \pm 1$	10.90 ± 0.02	4.36 ± 0.06
SY Per	C6.4	465	476.4 ± 1.5	$48\,030 \pm 7$	10.04 ± 0.08	2.11 ± 0.09
RU Vir	C8.1e	443.7	442.5 ± 0.6	$47\,785 \pm 4$	12.02 ± 0.03	2.65 ± 0.08
SS Vir	C6.3e	355	358.9 ± 0.6	$47\,903 \pm 4$	8.36 ± 0.04	1.76 ± 0.09

Star names according to GCVS catalogue; spectral classifications; catalogue period P_{cat} in days (SIMBAD), osculating period P_{oscul}, osculating JD of reference maximum $\tilde{M}_0 = M_0 - 2,440,000$, mean values of V and effective amplitudes A_{eff} of the periodic part of variations in magnitudes.

We tried to find the function $F(f, A_{im})$ in a form valid for all 26 studied carbon Miras. The function was found by means of generalized weighted principal component analysis. We have revealed that the function can be favourably expressed trough the linear combination of only two principal, mutually orthogonal normalized functions (see Figure 1) $\Phi_1(f)$ and $\Phi_2(f)$:

$$F(f, A_{1i}, A_{2i}) = A_{1i}\Phi_1(f) + A_{2i}\Phi_2(f)$$

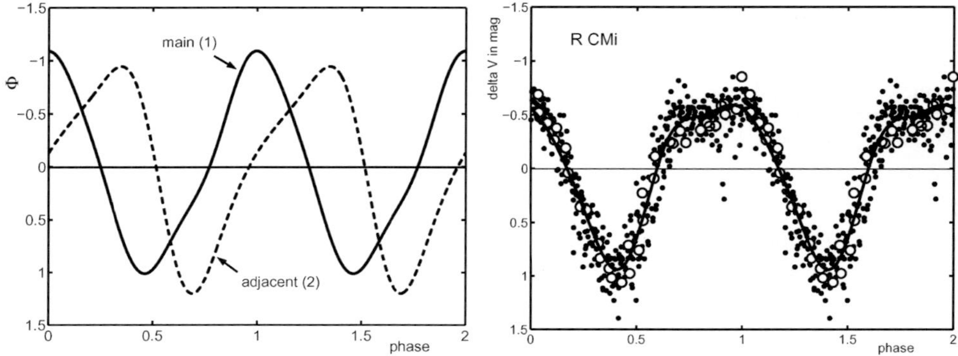

Figure 1. (a) Phase graph of the first and second principal functions $\Phi_1(f)$ and $\Phi_2(f)$. (b) Periodic component of the light curve of R CMi, one of the carbon Miras from our list. Full dots: visual measurements, circles: Hipparcos measurements transformed into V.

The first principal component reflects common properties of all carbon Miras, why we have based our new phase counting on it. We centered the zero phase to the maximum of $\Phi_1(f)$. The second, usually smaller component, expressed the observed variety of periodic parts of light curves of carbon Miras through a certain "deformation" of the main component. The extent of periodic variability of the star can be well quantified by its "effective" amplitude: $A_{\mathrm{eff}} = 2\sqrt{A_1^2 + A_2^2}$. The median of the effective amplitude in V is 2.29 mag.

A more detailed analysis of results will be published elsewhere.

Acknowledgements

The work was supported by the project 205/02/0445 of the Grant Agency of the Czech Republic.

References

Barnbaum, C.: 1994, *Astrophys. J. Suppl. Ser.* **90**, 317.
Dušek, J. and Mikulášek, Z.: 2003, *Contr. Astron. Obs. Sk. Pleso* **33**, 13.
Gráf, T.: 2003, Ph.D. Thesis, Masaryk University Brno.
Mikulášek, Z., Žižňovský, J., Zverko, J. and Polosukhina, N.S.: 2003, *Contr. Astron. Obs. Sk. Pleso* **33**, 29.

NON-RADIAL PULSATIONS IN COMPONENTS OF SYMBIOTIC STARS

N.I. BONDAR and V.V. PROKOFÉVA

Crimean Astrophysical Observatory, Nauchny, Crimea, Ukraine 98409;
E-mail: bondar@crao.crimea.ua

(accepted April 2004)

Abstract. Photometric observations of symbiotic stars in the blue and in the red spectral regions make it possible to reveal non-radial oscillations both of the cool and of the hot components. Light variations of red giants in the symbiotic systems CI Cyg and AG Peg show several periods in the 10–80^{d} range, interpreted as p-mode pulsations. These modes are excited by a bright spot produced by radiation flux from the hot component. The spot moves on the red giant's photosphere at a velocity close to the sound speed. During the active phase of the symbiotic star CH Cyg, at least 25 frequencies of oscillations in the 150–6000 s range of periods were found in the light of the white dwarf. Their features correspond to non-radial g-modes. In the frame of 2D gas dynamical non-adiabatic models, the interaction between gas flows and the accretion disk leads to formation of a system of shock waves propagating towards the compact object, which is one of possible mechanisms to excite non-radial pulsations of white dwarfs in symbiotic systems.

Keywords: stars, symbiotic – white dwarfs – stars, pulsations – stars, individual, AG Peg, CH Cyg, CI Cyg

1. Pulsations of Red Giant Components of Symbiotic Systems

Non-radial pulsations of stars are excited due to mechanisms providing energy transfer. Such mechanisms can work in complex systems like binary stars. According to Kosovichev and Skulskii (1990), oscillations in the $H\alpha$ flux of β Lyr with a period of $1^{d}.85$ can be produced by a system of quadrupole fundamental modes ($l = 2, m = 0, \pm 2$) excited in the bright component by a tidal wave. General properties of non-radial pulsations are a large number of periods, stability of phases during a long time, changes of amplitudes with time. The power spectrum shows different series of frequencies at different times.

Observational evidence for pulsations in red components of symbiotic stars comes from long-term photometry. A period of $37^{d}.2$ was found in the radiation from the symbiotic star CI Cyg in 1978–1986, whereas the periods of $50^{d}.2$ and $78^{d}.4$ were observed during 1988–1989 (Belyakina and Prokoféva, 1991; Taranova, 1987). The physical conditions in the red component of CI Cyg allow to expect the appearance of non-radial p-modes in its upper layers. The hot component's radiation produces a bright spot on the surface of the red giant, which moves due to its asynchronous rotation. The competition between mode-exciting and mode-damping processes leads to changes of periods.

Astrophysics and Space Science **296**: 161–164, 2005.
© Springer 2005

The presence of non-radial oscillations with periods of $10^{d}7$ and $9^{d}9$ in the light of the red giant was revealed in the analysis of photometry for AG Peg (Belyakina and Prokoféva, 1992). These periods remained the same during 1100 days. The source of excitation of these oscillations can be similar to that described for CI Cyg. The authors calculated that the terminator line on the giant's atmosphere moved at a velocity of 12–6 km/sec, compared to the sound speed of 9 km/sec, and the generated acoustic waves could produce non-radial p-modes in the red giant.

2. Pulsations of the White Dwarf in the Peculiar Symbiotic Binary CH Cyg

The active state of the symbiotic system CH Cyg usually begins as a strong outburst that leads to a several-fold increase of its brightness. In order to study the rapid brightness variability of the white dwarf of CH Cyg, the data obtained at the Crimean Astrophysical Observatory on four nights during June 13–August 22, 1982, when the star was in maximum brightness (Bondar' and Shakhovskaya, 2001), was analyzed. The observations were made using a 70 cm telescope with a five-channel spectral scanner; the time resolution was 20 sec. The continuum flux was registered in narrow (23 Å) bands, during 1–1.2 h in each of the four nights. A total of 646 measurements of continuum brightness at $\lambda 3737$ Å were obtained. The standard error (about $0.^{m}01$) was estimated using comparison star observations.

To search for significant frequencies, a procedure of data prewhitening with selected frequencies was used. The cleaning was made by programs providing analysis of discrete data. The power spectrum calculated for the whole observational data (Breger, 1990) shows five remarkable groups of frequencies in the range of 24–250 c/d (Figure 1a). It seems that the noise level is reached near the frequency

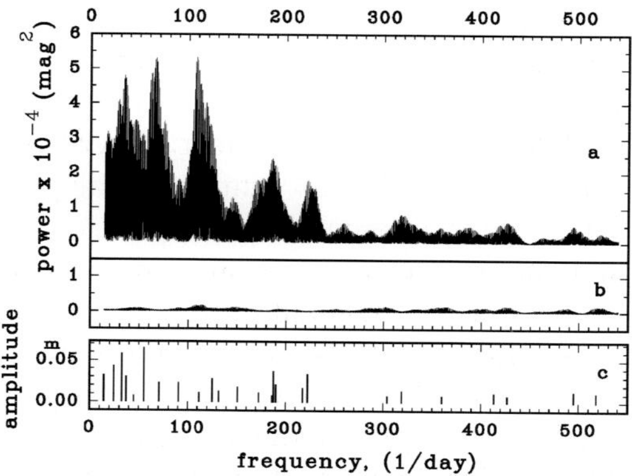

Figure 1. The power spectrum for the whole observational data set (a) and the residuals after consecutive prewhitening with the 25 determined frequencies (b); the schematic frequency distribution of the detected oscillation modes and their mean amplitudes (c).

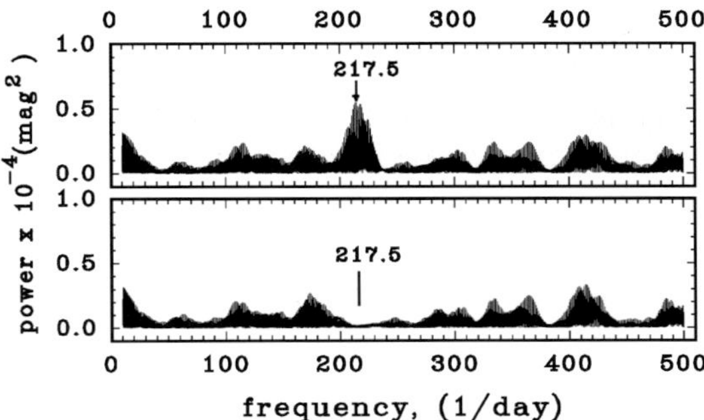

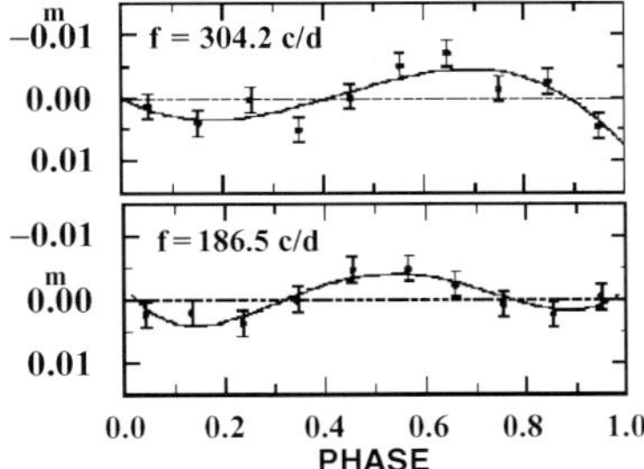

Figure 2. The power spectra of the data before (top) and after (bottom) the prewhitening with the 217.5 c/d frequency.

Figure 3. The phase diagrams for the 304.2 and 186.5 c/d periods. The vertical bars are standard errors in phase bins with widths of 0.1.

of 445 c/d, but actually it was reached only after consecutive prewhitening of data with 25 detected frequencies (Figure 1b). The distribution of these frequencies and the mean oscillation amplitudes are displayed in Figure 1c. Sample results after prewhitening with a selected frequency and phase diagrams are shown in Figures 2 and 3, respectively.

To estimate the confidence level for all of the detected frequencies, we determined the signal-to-noise ratios, $S/N = A_v/\sigma_v^2$. We found $S/N > 10$ only in four cases, and for one case, it was equal to 7. The ratio A_m/σ_m was higher than 4 for 24 frequencies, and only in one case this value was about 3.5. Here A_m is the oscillation amplitude and σ_m is the standard error.

The power spectra for individual nights show instability of amplitudes from night to night. Short-term amplitude variations are one of the properties of non-radial pulsations. For CH Cyg, we found periods of oscillations from 150 s to 6000 s. This range is wider than the 400–2000 s interval known for eigenoscillations of single white dwarfs. This indicates that the physical conditions near the hot component of this symbiotic system during its active state are more complicated compared to single white dwarfs.

3. Discussion

Typical models of gas flows in symbiotic stars make it possible to suggest a possible mechanism of excitation of eigenmodes. The system of CH Cyg contains a M6–M7 red giant and a hot star classified as a white dwarf. The cool component does not fill its Roche lobe and loses matter via stellar wind (10^{-6}–10^{-8} $M_\odot$ per year). Part of this matter is accreted onto the white dwarf, and some matter is added to the gaseous envelope. According to Mikolajewski et al. (1990), the mass of the white dwarf in CH Cyg is about 0.58 $M_\odot$, the system's luminosity is estimated as 10–70 $L_\odot$ during quiescence and about 300 $L_\odot$ in the active phase. The authors suggested that the star's active state was related to the accretion disk around the white dwarf.

The mass transfer in binary systems, where the red giant does not fill its Roche lobe, depends on parameters of the wind. In the case of a weak wind, a stable accretion disk is formed. Bisikalo et al. (1997) show how systems of shock waves appear in the disk. The group velocity of these waves is directed towards the compact object. We suppose that the shock waves reaching the white dwarf's surface can excite g-modes of non-radial pulsations. These processes can be considered a source of excitation for g-modes in hot components of symbiotic stars, in particular, of CH Cyg.

References

Belyakina, T.S. and Prokoféva, V.V.: 1991, *Sov. Astron.* **35**, 154.

Belyakina, T.S. and Prokoféva, V.V.: 1992, *Bull. Crim. Astrophys. Obs.* **86**, 42.

Bisikalo, D.V., Boyarchuk, A.A., Kuznetzov, O.A. and Chechetkin, V.M.: 1997, in: J. Mikolajewka (ed.), *Physical Processes in Symbiotic Binaries and Related Systems*, Copernicus Foundation for Polish Astronomy, Warsaw, Poland, p. 83.

Bondar', N.I. and Shakhovskaya, N.I.: 2001, *Astrophysics* **44**, 46.

Breger, M.: 1990, *Commun. Asteroseism.* **20**, 1.

Kosovichev, A.G. and Skulskii, M.Y.: *Sov. Astron. Lett.* **16**, 103.

Mikolajewski, M., Mikolajewska, J. and Khudyakova, T.N.: 1990, *Astron. Astrophys.* **235**, 219.

Taranova, O.G.: 1987, *Sov. Astron. Lett.* **13**, 173.

BE STARS IN BINARIES

PAVEL KOUBSKÝ

Astronomical Institute, Ondřejov Observatory; E-mail: koubsky@sunstel.asu.cas.cz

(accepted April 2004)

Abstract. A brief review of Be stars in binaries is presented. Attention is paid to systems, where the Be phenomenon is clearly connected to the duplicity, but is not a simple consequence of mass transfer.

Keywords: Be stars, binaries

1. Introduction

Be stars (i.e. emission-line stars of spectral class B) have been defined phenomenologically. Their defying property is the presence of Balmer emission, observed at least once during their documented history. In order to make the group more homogeneous, the term *classical* Be *stars* is sometimes introduced. The term is reserved for Be stars of luminosity classes III to V. In that sense, Be stars discussed in this contribution are classical Be stars.

A characteristic property of Be stars is their spectral, light and polarimetric variability, often unpredictable, which occurs on several distinct timescales. Especially pronounced are the long-term B $\Longleftrightarrow$ Be $\Longleftrightarrow$ Be shell transitions. There are variations on a time scale of weeks to months, which are either scaled-down events similar to long-term changes, or events related to the binary nature of particular Be stars. Analysis of line profiles and light has revealed rapid variations, which occur on time scales close to several fractions of a day. In some Be stars, changes as short as tens of minutes were described.

Most of the observed emission originates in extended enevelopes – disks, which are probably flattened towards the stellar equator and at least one order of magnitude larger that the underlying central star.

After more than 100 years of study, the formation of the Be star disk cannot be described by a single general mechanism. One of the major factors must be rapid rotation of the central star. But evidently, a number of other (secondary) factors play a role – magnetic field, pulsations, stellar wind, and duplicity (at least as a source of gravity field modification). Possible types of interaction in the Be binary stars are discussed in the next paragraphs.

2. Duplicity and the Be Phenomenon

The relevance of binarity for Be stars has been discussed by several authors. Kříž and Harmanec (1975) proposed that a large portion of Be stars (if not all) are interacting binaries undergoing mass transfer from the secondary component filling its Roche lobe. Later, when the lack of large secondaries in Be stars became evident, Harmanec (1985) tried to introduce another mode of mass transfer – originating from a rotationally unstable secondary contracting towards the helium main sequence. Gies (2001) categorized Be binaries in four groups – (1) hot Algols (i.e. interacting binaries which were the essence of the hypothesis by Kříž and Harmanec (1975), (2) Be + He stars (binary model modification by Harmanec (1985), (3) Be X-ray binaries and (4) Be + white dwarf combinations. Gies (2001) stressed that the difficulties to detect secondaries in groups (2) and (4) might influence the estimates of the fraction of binaries among the Be star population. Neglecting these problems, Porter and Rivinius (2003) found no support for a general relevance of binarity for Be stars, but admitted that binarity is an important aspect of the Be phenomenon, namely that in the past the companion might have had an influence on the Be star as a result of previous mass transfer, i.e. by spinning it up. Harmanec et al. (2002) suggested that the companion to the Be star might ease the mass loss from the primary into the disk.

3. Examples of Interaction in Be Binaries

In the spectra of Be stars emission in Balmer lines has often two peaks. The so-called violet-to-red ratio V/R is used as a parameter describing the behaviour of the envelope or disk. Phase-locked V/R variations are observed in many Be binaries. A good example of phase-locked V/R variations can be found in V360 Lac – see Figure 1. It is a binary system consisting of a B3e primary and an F9 secondary, which probably fills the Roche lobe. Orbital period is 10.08 days (Hill et al., 1997).

Examples of phase-locked V/R variations in Be binaries can be found in Gies (2000, 2001) and Harmanec (2000). In many cases these variations are evidently not caused by mass transfer. Very interesting is the case of ϕ Per, where the source of the phase-locked V/R variations is the irradiation of the disk by the hot companion (Štefl et al., 2000). Bjorkman et al. (2002) found evidence for binarity of another interesting Be star – π Aqr. They detected two systems with radial velocity changes with a period of 84 days. Their finding would lead to a system consisting of two Be stars. However, it cannot be excluded that the emission of the secondary (also a Be) star is only an artefact of V/R variations in the Hα line profile of the primary.

One has to mention another related phenomenon: the release of a new Be envelope during periastron passage in δ Sco, an eccentric binary system with a period of 10.6 years. Coté and Kerkwijk (1993) detected emission in Hα in 1990, when δ Sco was close to periastron. Ten years later, a new emission started to develop. The time history of the Hα emission is depicted in Figure 2. The spectrum taken in 2004 shows a stronger emission profile than in 2003. Thus the behaviour of the emission

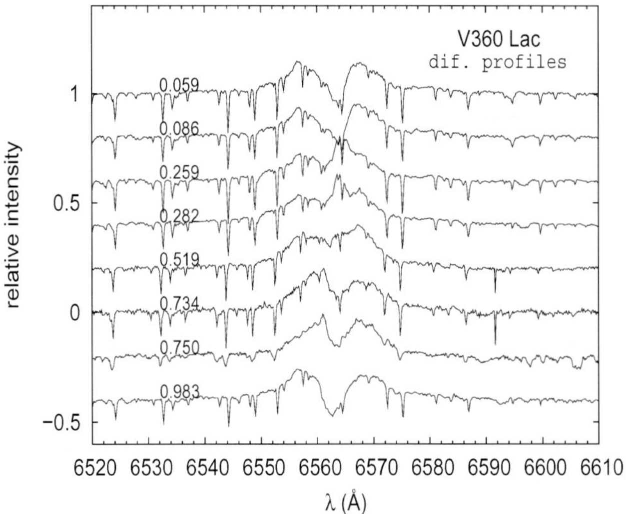

Figure 1. Orbital variations of the V/R ratio in the interacting binary V360 Lac based on observations obtained at Ondřejov and DAO Observatories. Orbital phase is given in format 0.PPP. For clarity difference profiles of Hα are displayed.

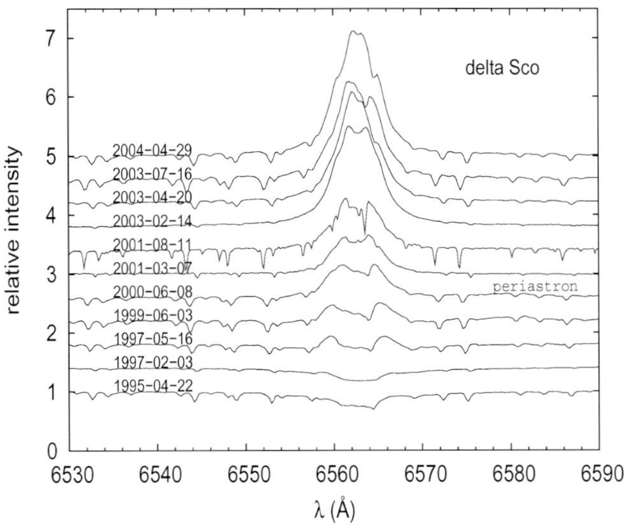

Figure 2. Hα profiles of δ Sco obtained at Ondřejov and DAO observatories. Observing dates are in the form YYYY-MM-DD. Periastron passage was in September 2000.

in Hα during the recent periastron passage (autumn 2000) is not very similar to the situation after the previous periastron passage. Without reliable hydrodynamical simulation the relationship between periastron passage and envelope release is not obvious.

                    PAVEL KOUBSKÝ

Acknowledgements

Spectra used in this study were obtained at Dominion Astrophysical Observatory in Victoria, Canada, and by several observers at Ondřejov Observatory. Their support is gratefully acknowledged. This study and my participation of the conference was supported by the grants GAČR Nos 0788 and 1000.

References

Bjorkman, K.S., Miroshnichenko, A.S. and McDavid, D.: 2002, *ApJ* **573**, 812.
Coté, J. and van Kerkwijk, M.H.: 1993, *A&A* **274**, 870.
Gies, D.R.: 2001, in: D. Vanbeveren (ed.), The Influence of Binaries on Stellar Population Studies, Kluwer, Dordrecht, The Netherlands.
Harmanec, P.: 1985, *Bull. Astron. Inst. Czech.* **36**, 327.
Harmanec, P.: 2000, *Publ. Astron, Inst. Acad. Sci.* **92**, 9.
Harmanec, P., Bisikalo, D.V. and Boyarchuk, A.A.: 2002, *A&A* **396**, 937.
Hill, G., Harmanec, P. and Pavlovski, K.: 1997, *A&A* **324**, 965.
Kříž, S. and Harmanec, P.: 1975, *Bull. Astron. Inst. Czech.* **26**, 65.
Porter, J.M. and Rivinius, Th.: 2003, *PASP* **115**, 1153.
Štefl, S., Hummel, W. and Rivinius, Th.: 2000, *A&A* **358**, 208.

ORBITAL PARAMETERS OF THE BINARY COMPANION IN o AND USING SPECTRUM DISENTANGLING

A. BUDOVIČOVÁ[1], S. ŠTEFL[1], P. HADRAVA[1], TH. RIVINIUS[2] and O. STAHL[2]

[1]*Astronomical Institute, Czech Academy of Sciences, Czech Republic*
[2]*Landessternwarte Königstuhl, Heidelberg, Germany;*
E-mail: andrea@asu.cas.cz

(accepted April 2004)

Abstract. o And is one of the most frequently observed Be stars, both in photometry and spectroscopy. It is a multiple system of at least four stars (a Be star, a close binary of spectral types B7 and B8, and an A star). For over a century, numerous observers report a highly variable spectrum, photometric changes, and a substantial range of radial velocity. The star has changed back and forth between a shell-type and a normal B-type star. The last emission phase started in 1992 and ended in 2000.

Analysis of the dynamical spectra at spectral lines Mg II 4481 Å and He I 6678 Å and radial velocity curves shows that the two binary components can be resolved. We decomposed the triple star spectra and computed orbital parameters of the binary companion using the KOREL code for spectrum disentangling.

Keywords: Be stars, binaries, o And, spectrum disentangling

1. Introduction

The bright star o And (HD 217675) is one of the most frequently observed Be/shell stars. Harmanec et al. (1987) and Hill et al. (1988, 1989) have shown the star to be a quadruple system. The total mass of the system is about 14 $M_\odot$ and the distance is 188 ± 27 pc. The principal component 'A' is the brightest star ($V = 3.8^m \pm 0.1$) in the system. o And 'A' has two resolved companions, 'a' and 'B'. Companion 'B' is a double-lined spectroscopic binary 'Ba-Bb' of spectral types B7 and B8 with a period of 33.0847 days orbiting the principal component 'A' in several decades (Sareyan et al., 1998), the possible period derived from interferometric measurements is 68.6 ± 1.7 years (Hartkopf et al., 1996). The companion 'a' is the closest star, detected by speckle interferometry, which is gravitionally bound to component 'A'. A period of 4–5 years has been proposed by McAlister et al. (1987).

2. Observations

Spectra of o And were obtained between 1992 and 2003 at various stations and with various spectrographs. The resolving power ranged from 8 500 (the coude slit spectrograph and Reticon attached to the 2-m telescope at Ondřejov Observatory)

to 20 000 with the HEROS spectrograph (Kaufer, 1998; Rivinius et al., 1998) linked to the telescopes at the Calar Alto, Wendelstein, and Ondřejov Observatories.

Customized versions of the MIDAS echelle context (Stahl et al., 1995) were used for reducing and calibrating the spectra from the HEROS spectrograph. Spectra taken with the Ondřejov coudé slit spectrograph in 1992–2000 were reduced using the data-reduction code SPEFO (Škoda, 1996).

3. Radial Velocities of the Binary Companion

Dynamical spectra for Hill's period of 33.0847 days can help to identify the lines in which both components of o And 'B' can be separated. The two components are best visible in the lines of Mg II 4481 Å and He I 6678 Å. The resulting phase diagram (see Figure 1) was computed for Mg II 4481 Å. We used the same line for further analysis of radial velocities. We estimated a preliminary orbital period, which was used as a starting parameter in spectral disentangling.

4. Spectrum Disentangling

The method developed by Hadrava (1995, 1997; computer code KOREL) enables the determination of orbital elements and the decomposition of composite spectra of multiple systems into individual components. The parameters taken from Hill et al. (1988), interferometry (Hartkopf et al., 1996), and parameters computed from radial velocities using Time Series Analysis implemented in MIDAS were used as input values for the procedure. We used seasons 2000, 2001 and 2002 when no emission activity was present. We decomposed the spectral regions around Mg II 4481 Å, He I 4471 Å, He I 6678 Å and O I 7773 Å. An example of a decomposed spectrum is given in Figure 2.

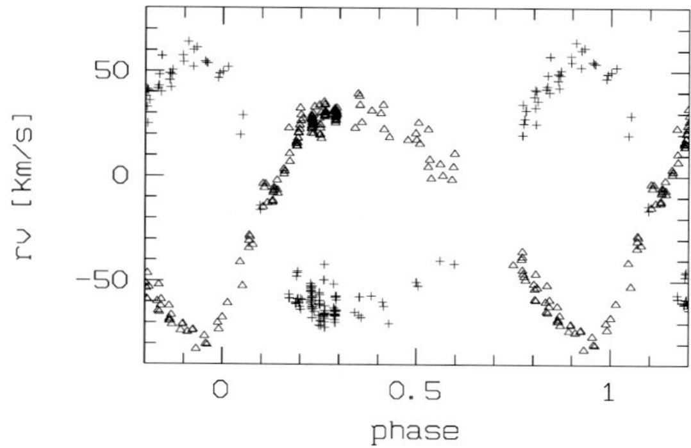

Figure 1. The radial velocity phase diagram of binary components of o And 'B' in the line of Mg II 4481 Å.

TABLE I

Orbital parameters of the close binary in o And

	KOREL	Hill et al. (1988)
Period (days)	33.0877 ± 0.0003	33.01 ± 0.02
$T_{\mathrm{per}}(24\,00000+)$	51920.69 ± 0.07	46925.3 ± 0.02
e	0.28 ± 0.01	0.24 ± 0.01
ω (degrees)	234.4 ± 0.2	226.2 ± 2.3
K_1 (km s^{-1})	52.1 ± 0.5	54.8 ± 0.8
K_2 (km s^{-1})	69.7 ± 2.0	71.6 ± 0.8
M_2/M_1	0.75 ± 0.01	0.74

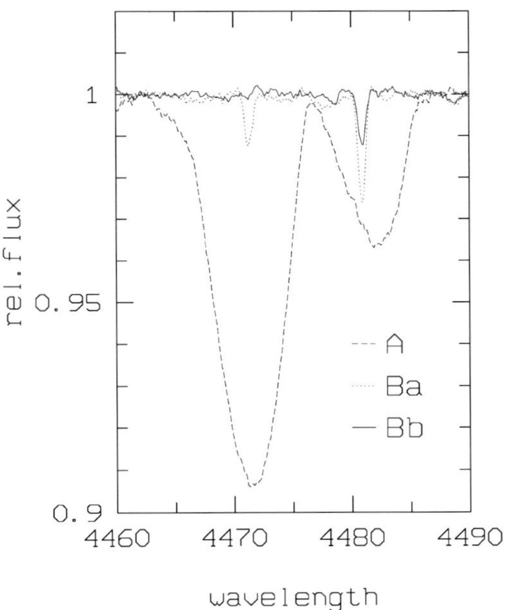

Figure 2. The decomposed spectrum of o And in the lines of He I and Mg II.

The resulting orbital parameters for the binary companion Ba-Bb of o And 'B' are given in Table I.

5. Summary

Spectrum disentangling is a very powerful method for determining the orbital parameters in multiple stellar systems. It allows the decomposition of triple spectra of o And 'A' and 'B' into individual components, which can be used for spectral classification. A good solution of the orbital parameters of the binary companion *o* And 'B' has been found.

The question about the number of components and their orbital parameters in this system is still open. Hill et al. (1988) have shown that there must be a minimum of four stars in the system of *o* And. More interferometric measurements and longer monitoring of these stars are needed to constrain the system parameters.

References

Hadrava, P.: 1995, *A&AS* **114**, 393.
Hadrava, P.: 1997, *A&AS* **122**, 581.
Harmanec, P., Oláh, K., Božič, H. et al.: 1987, in: A. Slettebak and T.P. Snow, (eds.), *IAU Coll. 92, Physics of Be Stars*, Cambridge University Press, Cambridge, p. 456.
Hill, G.M., Walker, G.A.H., Dinshaw, N., Yang, S. and Harmanec, P.: 1988, *PASP* **100**, 243.
Hill, G.M., Walker, G.A.H. and Harmanec, P.: 1989, *PASP* **100**, 258.
Hartkopf, W.I., Mason, B.D. and McAlister, H.A.: 1996, *ApJ* **111**, 370.
Kaufer, A.: 1998, in: E. Schielicke, (ed.), Variable Circumstellar Structure of Luminous Hot Star: The Impact of Spectroscopic Long-term Campaigns, *Review in Modern Astronomy*, No. 11.
McAlister, H.A., Hartkopf, W.I. and Hutter, D.J.: 1987, *ApJ* **93**, 688.
Rivinius, Th. Baade, D. and Štefl, S.: 1998, *A&A* **333**, 125.
Sareyan, J.P., Gonzalez-Bedola, S., Guerrero, G., Chauville, J., Huang, L., Hao, J.X., Guo, Z.H., Adelman, S.J., Briot, D. and Alvarez, M.: 1998, *A&A* **332**, 155.
Stahl, O., Kaufer, A., Wolf, B. et al.: 1995, *J. Astronom. Data* **1**, 3 (on CD–ROM).
Škoda, P.: 1996, in: G.H. Jacoby and J. Barnes (eds.), *Astronomical Data Analysis Software and Systems V, ASP Conference Series* Vol. 101, p. 187.

SPECTRUM DISENTANGLING AND ORBITAL SOLUTION FOR κ Dra

S.M. SAAD[1,2], J. KUBÁT[1], P. HADRAVA[1], P. HARMANEC[1,3], P. KOUBSKÝ[1],
P. ŠKODA[1], M. ŠLECHTA[1], D. KORČÁKOVÁ[1] and S. YANG[4]

[1]*Astronomický ústav AV ČR, CZ-251 65 Ondřejov, Czech Republic;*
E-mail: Somaya@Sunstel.asu.cas.cz
[2]*National Research Institute of Astronomy and Geophysics, 11421 Helwan, Cairo, Egypt*
[3]*Astronomický ústav UK, V Holešovičkách 2, CZ-180 00 Praha 8, Czech Republic*
[4]*Department of Physics and Astronomy, University of Victoria, Victoria, BC V8W 3P6, Canada*

(accepted April 2004)

Abstract. Based on a new set of electronic spectra in a relatively wide spectral range (3500–8300 Å) and using the methods of spectrum disentangling (code *KOREL*) and solution of RV curves (code *FOTEL*), we determined new orbital elements of the binary star κ Dra. The solution of the radial velocity curves for Balmer and some other strong metallic lines suggested a circular orbit and led to the following orbital elements: period $P = 61.555 \pm 0.029$ days, epoch of periastron passage $T_{\mathrm{periast}} = 49980.22 \pm 0.59$, RV semi-amplitude $K_1 = 6.81 \pm 0.24\,\mathrm{km\,s}^{-1}$, and a mass function of $f(m) = 0.002\,M_\odot$. Lines of the secondary were not detected. In addition, moving absorption bumps in the violet peaks of Hα and Hβ lines were found to be phase-locked with the orbital period. Their presence suggests some kind of interaction between the binary components.

1. Introduction

κ Dra (5 Dra, HD 109387, HR 4787, BD + 70 703, MWC 222, HIP 61281) is a bright variable ($\bar{m}_V$) = 3.75–3.95), single lined spectroscopic binary Be star. Its duplicity was discovered by Juza et al. (1991). A detailed study of the long term variability of this star has been published recently (Saad et al., 2004). Here we describe an improved orbital solution of the system using the spectrum disentangling method.

2. Observations and Measurements

2.1. OBSERVATIONS

Our spectra contain data obtained in both coudé and Cassegrain foci of the Ondřejov 2m telescope and at the Dominion Astrophysical Observatory. Most of the spectrograms covering the spectral range 6300–6700 Å were obtained with the Reticon detector in the coudé focus of the Ondřejov 2m telescope between 1993 and 2000. They were supplemented by recent observations using the CCD detector. 30 spectrograms covering the whole visible and near infrared region were obtained in

Astrophysics and Space Science **296**: 173–177, 2005.
© Springer 2005

the Cassegrain focus of the 2m telescope using the *HEROS* echelle spectrograph between 2000 and 2003. 12 recent CCD spectra were secured at the DAO observatory and are covering the red region 6200–6750 Å.

2.2. RADIAL VELOCITY ANALYSIS

Radial velocity (hereafter RV) measurements were performed interactively using the computer program *SPEFO*. RV measurements obtained in the red region were wavelength-calibrated using measurements of telluric absorption lines. The RV analysis was based mainly on the Hα line and on strong metallic lines in its vicinity (He I 6678 Å and Si II 6347, 6371 Å). In addition, for *HEROS* (extended) spectra, RVs for Balmer lines up to Hε and for some other metallic lines (Mg II 4481 Å and He I 4471 Å) were measured. In all cases, RVs for these lines were measured based on the outermost parts of absorption wings, neglecting the details in the core.

3. Solutions for Orbital Elements

We first used *KOREL* (Hadrava, 1995, 1997) to find a one-component orbital solution. Using this program we obtained preliminary orbital elements together with RVs from different regions of the stellar spectrum. A complementary set of the orbital solutions for the binary system was then obtained with the *FOTEL* code (Hadrava, 1990) using our direct RV measurements. We fixed the period at the value obtained from the previous *KOREL* solution ($P = 61^{\rm d}.555$). The final solution is based on all available data sets (7 data sets) – Si II 6347 Å, Si II 6371 Å, He I 6678 Å, He I 4471 Å, Mg II 4481 Å, O I 7772-5 Å, and Hε. Fixing the values $e = 0$ and $\omega = 0$ we obtained a solution with the ephemeris $T_{\rm max.RV} = {\rm HJD}\,2449980.22 \pm 0.59 + 61^{\rm d}.555 \pm 0^{\rm d}.029 \times E$, $f(m) = 0.002\,M_{\odot}$, and $K_1 = 6.81 \pm 0.24\,{\rm km\,s^{-1}}$, with an rms per observation of $4.75\,{\rm km\,s^{-1}}$. The radial velocity solution is plotted in Figure 1.

3.1. SECONDARY COMPONENT

An attempt to find lines of the secondary component using the *KOREL* disentangling technique failed. No reasonable traces of lines of the secondary were found in the disentangled spectrum, as can be seen from Figure 2. Taking into account the spectroscopic mass of κ Dra ($M_1 = 4.8\,M_{\odot}$) determined by Saad et al. (2004) and results of the *FOTEL* solution, we calculated several possible values of the secondary mass M_2 for different inclination angles i (see Table I). If the rotation axis of the primary and the orbital axis are parallel, then for the stellar radius $R = 6.4\,R_{\odot}$ and rotation velocity $170\,{\rm km\,s^{-1}}$ (Saad et al., 2004) the minimum inclination angle $i \approx 27°$. Thus if κ Dra rotates near its break-up velocity, the most

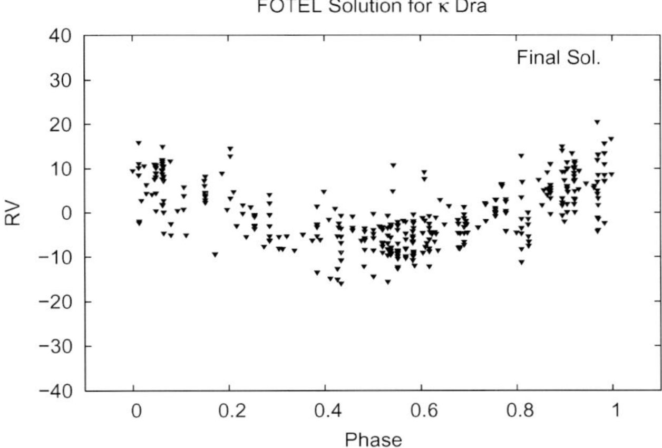

Figure 1. Radial velocity curve based on the *FOTEL* solution.

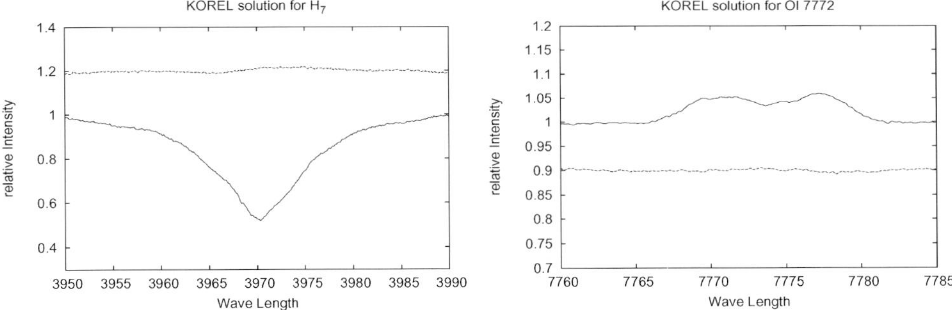

Figure 2. KOREL two-component solution. The primary spectrum is plotted using the full line, the result for the secondary spectrum is plotted using the dotted line. Note that no discernible trace of a line is present in the secondary spectrum.

appropriate inclination angle is about $30°$. A corresponding mass of the secondary is then $M_2 \approx 0.8 \, M_\odot$.

Moving absorption bumps (MAB) were found simultaneously in the violet peaks of both Hα and Hβ lines over their observing run. They are moving across the violet peak redwards and their radial velocity varies from -130 to $-80 \, \mathrm{km \, s^{-1}}$ for Hα

TABLE I

Mass of the secondary at different inclination angles

Inclination	$20°$	$30°$	$40°$	$50°$	$60°$	$70°$	$80°$
M_2 ($M_\odot$)	0.43	0.53	0.61	0.68	0.72	0.76	0.785
M_2/M_1	0.09	0.11	0.127	0.141	0.15	0.158	0.163

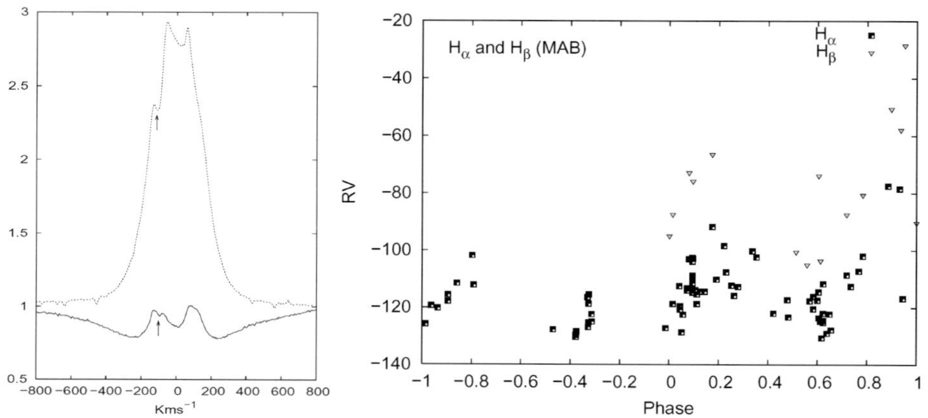

Figure 3. Moving absorption bumps (MAB – left panel) and their radial velocity changes (right panel).

and from -100 to $-20\,\mathrm{km\,s^{-1}}$ for Hβ. Their strength varies from a deep absorption to a shallow one. The feature disappears completely in some particular phases. It does not show long-term temporal variations comparable to that found in light variation, line strength and line intensity. However, its variability is related to the orbital period. It is not clear whether this bump comes from the interaction of the invisible secondary with a circumstellar disk of a primary or with a more extended circumbinary envelope. Similar absorptions with much more pronounced velocity variations have already been found for another Be binary, 4 Her, by Koubský et al. (1997).

4. Conclusions

The binary nature of κ Dra was studied using a new homogeneous set of electronic spectra. Using spectrum disentangling and the *FOTEL* code, we determined a new orbital solution of κ Dra. It confirms the earlier result of Juza et al. (1991). However, the nature of the secondary still remains unclear, since using *KOREL* for spectrum disentangling we did not find any clear evidence for lines of the secondary component in the whole visible and near infrared regions. There is a possibility of an interaction of the invisible secondary with the extended envelope of the primary.

Acknowledgements

This work was supported by grants GA ČR 205/02/0445 and 205/03/0788. The Astronomical Institute Ondřejov is supported by projects K2043105 and Z1003909.

References

Hadrava, P.: 1990, *Contrib. Astron. Obs. Skalnaté Pleso* **20**, 23.
Hadrava, P.: 1995, *A&AS* **114**, 393.
Hadrava, P.: 1997, *A&AS* **122**, 581.
Juza, K., Harmanec, P. and Hill, G.M.: 1991, *Bull. Astron. Inst. Czech.* **42**, 39.
Koubský, P., Harmanec, P. and Kubát, J.: 1997, *A&A* **328**, 551.
Saad, S.M., Kubát, J. and Koubský, P.: 2004, *A&A* **419**, 607.

AN OUTBURST DETECTED IN THE SPECTRUM OF HD 6226

MIROSLAV ŠLECHTA and PETR ŠKODA

Astronomický ústav AV ČR, CZ-251 65 Ondřejov, Czech Republic;
E-mail: slechta@sunstel.asu.cas.cz

(accepted April 2004)

Abstract. HD 6226 is a bright binary Be star at visual magnitude 6.81 (Hipparcos database). The emission and absorption phases occur in cycles, which are probably not periodic. The suspected period of about 630 days (derived from photometric measurements) is not confirmed by our spectroscopic survey. The latest emission phase developed in the beginning of 2003, then the emission strength systematically decreased and disappeared between July 21, 2003 and August 4, 2003. The last (absorption) spectrum was exposed on August 25, 2003. Unexpected very strong emission appeared in a spectrum exposed on October 28, 2003. A short-term photometric brightening followed this "outburst".

Long-term spectroscopic RV studies revealed a 2.615 d period modulated by a 29.7 d period (in the He I 6678 line), which perhaps may be interpreted as orbital period of a binary. Nevertheless, the physical nature of the dominant short 2.615 d period is not yet clear.

The last detected emission episode has changed considerably our view of the interesting object HD 6226. We hope this study will reveal more details of the physical properties of the Be phenomenon.

1. Physical Nature of HD 6226

HD 6226 is a star of spectral type B2 IV–B3 III; the photometry and the comparison of spectroscopic measurements with synthetic spectra allow to estimate some physical properties of HD 6226, supposing it is a single star:

$$T_{\mathrm{eff}} = 17000\,\mathrm{K}; \quad \log g = 3.0; \quad v \sin i = 70\ \mathrm{km\ s}^{-1}; \quad V_0 = 6\overset{\mathrm{m}}{.}402;$$

$$R = 11\,\mathrm{R}_{\odot}; \quad M = 5\,\mathrm{M}_{\odot}.$$

Nevertheless, the parameters may considerably change in case the star will be recognized as a binary system—see, e.g. Božič et. al. (2004).

HD 6226 is the first star where the spectroscopic variations were predicted from photometric measurements (Božič and Harmanec, 1998). They supposed that the long-term spectroscopic variability should be correlated with the photometric one, having a 481.3 d period (in case it is really a periodic process). The first spectroscopic studies (McCollum et al., 2000) really confirmed the emission in the Hα line.

The detailed spectroscopic investigation led to the recognition of four different periods in radial velocities measured in the absorption core of He I 6678: 2.615 d,

Astrophysics and Space Science **296**: 179–182, 2005.
© Springer 2005

2.719 d, 24.92 d and 29.686 d (Božič et. al., 2004). The 29.7 d period is usually interpreted as an orbital period in a possible binary system. Nevertheless, the physical nature of the two shorter periods is unclear. See also Castelaz and McCollum (2003).

2. Last Spectroscopic Observations

We have observed the star HD 6226 during the year 2003. At the beginning of the year the emission stage continued. The strength of emission decreased irregularly up to August 11, 2003, when the last (faint) emission was detected. A later spectrum, exposed on August 25, 2003, showed typical absoption features without any emission remnants. It was the last spectrum of this "cycle".

Unexpected strong emission in the $H\alpha$ line appeared on October 28, 2003. The strength of the emission varied irregularly, but decreased on a long-term scale. The

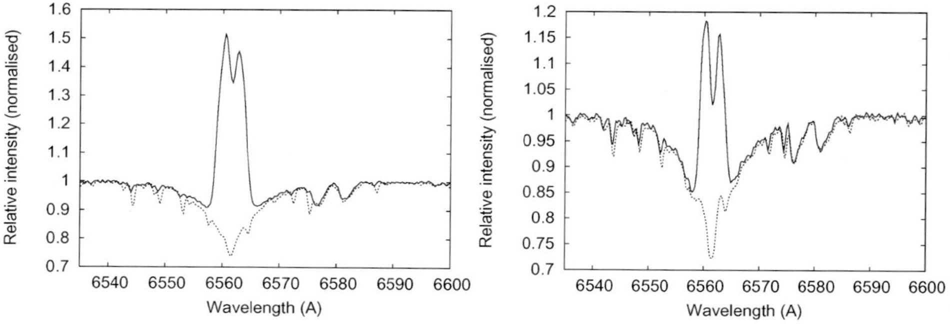

Figure 1. Unexpected behaviour of HD 6226 in the $H\alpha$ region. Left: the first occurrence of emission between August 25, 2003 (dashed line) and October 28, 2003 (solid line). Right: the last detected emission on January 14, 2004 (solid line), and an absorption spectrum secured on March 22, 2003 (dashed line).

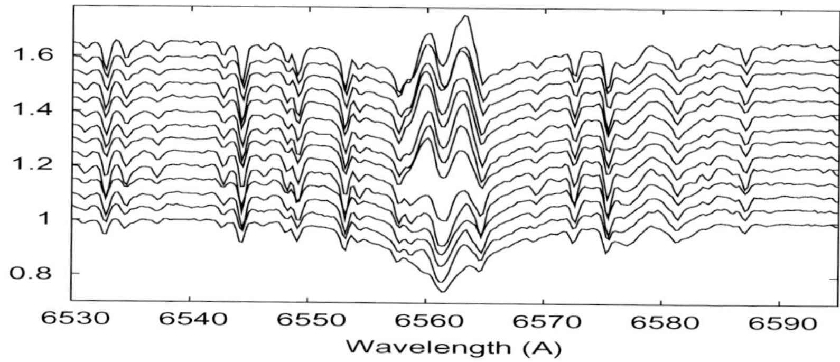

Figure 2. Systematic decrease of $H\alpha$ emission between January (top) and August (bottom) 2003.

last emission spectrum was acquired on January 14, 2004. Then, unfortunately, bad weather conditions did not allow to prolong the observing run, and thus the following spectrum was only secured on March 22, 2004. Surprisingly, no emission was found. A spectrum taken on March 30, 2004 confirmed the absorption phase of HD 6226.

3. Phase Coverage

For this preliminary announcement, we have decided to include data exposed with the Ondřejov observatory 2 m telescope only. HD 6226 is a rarely studied object despite of its brightness.

First, two spectra of HD 6226 were secured using the fiber-fed echelle spectrograph HEROS (Heidelberg Extended Range Optical Spectrograph), which has been installed in the Cassegrain focus of the Ondřejov 2 m telescope. Nevertheless, this device was originally designed to be connected to smaller telescopes; therefore its optical efficiency was limited in Ondřejov, and the signal-to-noise ratio allowed only for a rough analysis of the spectrum (which covers 3650–5650 Å with the blue camera and 5830–8350 Å with the red one).

Considering the limitation of HEROS, we have decided to exclude the echelle data from this study. Later, we have exposed all data using the single-order slit coude spectrograph in the Hα region covering 6255–6770 Å.

Altogether, 35 spectra were exposed between July 11, 2003, and March 22, 2004. Weather conditions forced some "gaps" in our data sampling. Nevertheless, a qualitative summary is possible even with these data.

4. Preliminary Results and Conclusions

We have studied five spectral lines: Si 6347, Si 6371, Ne 6402, Hα and He I 6678.

The He I 6678 line is very convenient for the study of the principal nature of HD 6226. A well defined absorption core enables us to derive the radial velocity curve using a limited number of points; the 2.615 d period is evident from Figure 3.

We have plotted there all data, i.e. both before and after the outburst mentioned above. As expected, the radial velocities follow the same curve. It is not possible to draw the same conclusion from other studied spectral lines.

The Hα line can be separated into five parts: broad absorption wings, absorption core (during the absoption phase), steep emission peak, V and R emission peaks and a central core between them (the last three components can be measured only if present, of course). One has to point out that the measurement of broad absorption wings is very sensitive to the normalization procedure (i.e. to the continuum determination) and is possible only in excellent spectra.

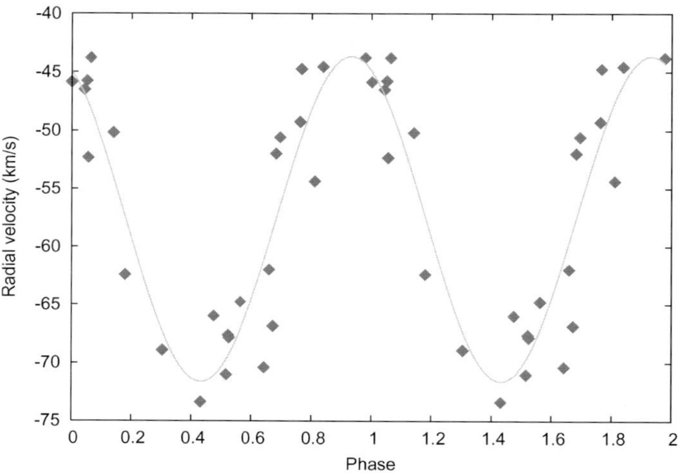

Figure 3. Radial velocity curve of HD 6226 based on He I 6678 line measurements. The curve is phased with the 2.615 d period.

The behaviour of the Hα profile seemed to be almost chaotic. None of the assumed periods could be reliably confirmed. Tentatively, we suggest two possible explanations. Maybe many physical effects contribute to the Hα shape. It is also possible that multiple periods are interfering, and the sampling of our data is not good enough to resolve them.

The line profiles of the Si doublet and of Ne 6402 are quite complicated. Sometimes it is almost impossible to identify the core of the line. The radial velocity curve and periods cannot be confirmed using these lines. Some indication of the 2.6 d period can be seen in radial velocities of the Ne 6402 line, but rms is too large.

In conclusion, we can say that HD 6226 is a very interesting object. It is advisable to continue the spectroscopic and photometric monitoring. The spectroscopic behaviour is probably driven by some process of creation and dissolution of the radiative envelope. The next emission phase is impatiently expected. We believe that new data will help to recognize the processes at work and help to explain some open questions connected with the Be phenomenon.

References

Božič H. and Harmanec P.: 1998, *A&A* **330**, 222.

Božič, H., Harmanec, P., Yang, S., Žižňovský, J., Percy, J.R., Ruždjak, D., Sudar, D., Šlechta, M., Škoda P., Krpata, J. and Buil, C.: 2004, *A&A* **416**, 669.

Castelaz, M.W. and McCollum, B.: 2003, *BAAS* **202**, 3911C.

McCollum, B., Castelaz, M.W. and Caton, D.B.: 2000, *BAAS* **33**, 713.

Škoda, P., Šlechta, M. and Honsa, J.: 2002, *Publ. Astron. Inst. Acad. Sci. Czech Republic* **90**.

BIMODAL STRUCTURE OF GAS STREAMS IN INTERACTING BINARIES

ZDISLAV ŠÍMA

Astronomical Institute, Prague; E-mail: sima@ig.cas.cz

(accepted April 2004)

Abstract. There are two modes of gas streams in close binary systems: geometrically thick for low mass transfer rate or geometrically thin for higher mass transfer. Geometrical thickness of the streams is not proportional to the amount of transferred mass. The limit between the two possibilities is discussed.

Keywords: interacting binaries, mass transfer, gaseous streams, WZ Sge, Algol-type

1. Introduction

Long time ago two types or "modes" of gas rings around the primary component of an interacting binary were described by Smak (1985). They can be either geometrically thick in a hot case, or geometrically thin in a cold case. The difference between both is due to thermal properties of the flowing gas which influence its geometrical structure.

A similar dichotomy can be valid even in a more pronounced form for the streams originating from the L_1-point in binaries with Roche-lobe overflow. However, the physical reason for such a behaviour of gaseous matter in this case is partly different from the situation in disks.

Šíma and Hadrava (1987) compared results of four different methods for calculation of stream thickness in z-direction. Two of the models are physically consistent: the adiabatically expanding stream (cool case, geometrically thin, higher density) and the stream in a state of radiative equilibrium (hot case, geometrically thick, lower density).

2. Scenario

Let us suppose a typical interacting binary with a gas stream from L_1 towards the primary star. The physical background of the dichotomy in question is as follows. When the mass exchange rate $\dot{M}$ is small, the flowing gas can be heated up by the radiation of the stars. This means that the stream will be hot and consequently it will be geometrically thick, with low density. When $\dot{M}$ increases, the radiative field in the binary (mainly from the hotter primary star) is no longer able to heat up the streaming gas, which is adiabatically cooled by the expansion along the stream

Astrophysics and Space Science **296**: 183–184, 2005.
© Springer 2005

trajectory. The temperature in the stream will fall dramatically, and so the stream will be cold and dense. As a consequence, its cross-section will be smaller (size of about 30% of the preceding case). Despite the fact that cooling by expansion and radiative heating take always place simultaneously, the transition between these two extreme cases needs not to be smooth.

If there are two modes of gaseous streams there must also be a limiting case between them. The critical mass transfer rate $\dot{M}_s$ between these two modes is a function of the radiation field in the binary, i.e. (1) of the stars' temperature; and (2) of the geometry of the system. It can be supposed that in real cases an hysteresis effect takes place: with increasing mass transfer rate in the case of radiative equilibrium the stream thickness is large, and consequently the stream absorbs more energy radiated by the component stars. It will thus remain in the radiative equilibrium mode even if the mass transfer rate $\dot{M}$ surpasses the limit $\dot{M}_s$, and the transition of the stream into the adiabatic expansion regime is delayed. On the other hand when $\dot{M}$ is high – i.e. in the case of adiabatic expansion – and slowly decreases, the adiabatic expansion can persist even if $\dot{M} < \dot{M}_s$, because the narrow stream cannot absorb enough energy to switch over into the radiative equilibrium mode. The models of gas rings (Bisikalo, 2004) show a complicated structure of the gas around primary stars. The interaction of the stream with the ring is a rather complex process. However, the above mentioned dichotomy of the streams can happen even in catalysmic variables with complicated accretion disks and streams (e.g., WZ Sge, etc.).

A much simpler case of streams is expected for binaries with larger (geometrically more extended) primaries, i.e. in classical Algol-type binaries. The simplest situation can be expected in cases where the stream is directly falling onto the surface of the primary star without forming a ring around it. It can be supposed that the observational evidence of the effect will also be more straightforward in these cases than for binaries with a compact primary. Two modes of gas streams could exist in combination with two modes of accretion disks giving rise to several possible forms of 'hot spots' or better 'hot lines,' two of which are of greatest importance: the case of a "central punch" and that of "pincers." From here several possibilities for forming the S-wave follow, and these may be spectroscopically distinguished from each other.

Acknowledgements

The author wants to thank Dr. Petr Hadrava for his very kind help and inspiring comments.

References

Bisikalo, D.V.: 2004, this volume.
Smak, J.: 1985, *Acta Astron* **35**, 351.
Šíma, Z. and Hadrava, P.: 1987, *Ap&SS* **130**, 151.

ENERGY TRANSFER IN W UMa SYSTEMS

SZILÁRD CSIZMADIA

*Konkoly Observatory of the Hungarian Academy of Sciences, H-1525 Budapest, P. O. Box 67,
Hungary; E-mail: csizmadia@konkoly.hu*

(accepted April 2004)

Abstract. W UMa stars consist of two main sequence stars touching each other. Mass and energy is
transferred from the primary star to the secondary one throughout the neck between the stars. The rate
of the energy transfer is studied. The energy transfer rate depends on the mass and luminosity ratio.
A new group of W UMa-type stars is introduced: it is called H-subtype systems meaning high mass
ratio systems. They have a mass ratio larger than 0.72 and a different energy transfer rate relation than
other types of W UMa systems.

Keywords: W UMa-type stars, contact binaries

1. Introduction

W Ursae Maioris-type is one of the main types of eclipsing binary stars with short
($\sim$8 h) orbital periods and continuous light variation during a cycle. Binnendijk
(1965) has shown that the mass ratio-luminosity ratio relation of these binaries is
like $L_2/L_1 \approx M_2/M_1$ and this is in contradiction with their spectral type: since
they consist of two dwarf stars of F, G, or K spectral class (sometimes earlier
ones, but no M-type contact binary was observed by now), they should follow the
main-sequence mass-luminosity ratio and hence their mass ratio-luminosity ratio
relation should scale like $L_2/L_1 = (M_2/M_1)^4$. He also found that the mass ratio-
radius ratio relation is $R_2/R_1 \approx (M_2/M_1)^{1/2}$. The exponent in this expression is
highly different from the case of two detached main-sequence stars.

In order to explain simultaneously the light curve shape and peculiar mass
ratio-luminosity ratio relation of W UMa-type stars, Lucy (1968a,b) has constructed the contact model. In this model the two stars are in geometrical as
well as in thermal contact, and luminosity is transferred from the more massive
component to the less massive secondary one. For various reasons this model was
improved (a recent overview on the structural and evolutionary theories of contact
binaries can be found in Webbink, 2003), but the main characteristics of the model
remained unchanged. The assumption that the components fill out or overfill
their own Roche-lobe successfully explained the peculiar mass ratio-radius ratio
relation as well and allowed the opportunity of mass and luminosity transfer. Note
that W UMa-type stars represent the third class of *Zdenék Kopal*'s classification
scheme: he divided the binary stars into detached, semi-detached and contact ones
according to their Roche-lobe filling rate. (A fourth one, the over-contact type – in

Astrophysics and Space Science **296**: 185–188, 2005.

which both components highly overfill their Roche-lobes – was introduced later). A review on the evolution of terminology of the Roche-lobe filling types can be found in Wilson (2001) together with many historical points.

The predicted mass ratio-luminosity ratio relation for W UMa-type stars is given in Lucy (1968a) and has the form $L_2/L_1 \approx (M_2/M_1)^{0.92}$. Kalimeris and Rovithis-Livaniou (2001) investigated the energy transfer in W UMa systems and found that the transferred energy rate depends on the secondary's luminosity. We also examined the validity of this theoretical relation and found slightly different results than Kalimeris and Rovithis-Livaniou (2001), because it depends not only on the luminosity ratio, but on the type of contact binary, too. Therefore a catalogue of light curve solutions of W UMa-type stars has been compiled containing information about the results of light curve solutions of 159 systems. For the sake of homogenity, we collected only solutions produced by the Wilson-Devinney code (see Wilson, 1998 and references therein). The following data have been included in the catalogue: RA and D of the system for 2000.0; epoch and period; photometric and spectroscopic mass ratios; fill-out factor; inclination; surface temperatures of the components; fractional radii of the components; dimensionless surface potential of the components (the so-called Ω parameter); the fractional luminosity of the primary star in different colours; third light in different colours; results of spot modeling; values of reflection coefficients and gravity darkening exponents used in the modeling.

The catalogue is available on the Konkoly Observatory's homepage (www. konkoly.hu) with its references. Any comments and additional information on new or already known light curve solutions are welcome to the e-mail address: csizmadia@konkoly.hu.

2. Energy Transfer in W UMa Stars

V luminosity ratios determined by light curve modeling (which were carried out by different investigators whose list can be found in the catalogue mentioned above) were transformed into bolometric ones applying Flower's (1996) tables. The mass ratio-luminosity ratio diagram can be seen in Figure 1. Note that contact binaries – except B and H subtype ones[1] – seem to follow Lucy's (1968a) relation, but the scatter is remarkable.

We empirically studied this diversity and defined the transfer parameter β as follows:

$$\beta = \frac{L_{1,\mathrm{obs}}}{L_{1,\mathrm{ZAMS}}} = \frac{1 + q^{4.6}}{1 + \lambda} \tag{1}$$

where λ is the bolometric luminosity ratio of the components.

---

[1]Subtypes of W UMa stars are defined as follows: W: the smaller star is the hotter; A: the larger star is the hotter; B: the temperature difference between the components is larger than 1000 K; H: A, B or W subtype systems with high mass ratio ($q > 0.72$).

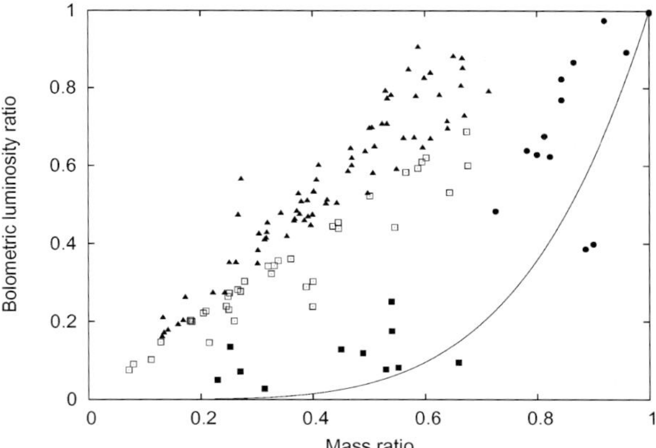

Figure 1. Mass ratio-luminosity ratio diagram of contact binary stars. Open squares: A, filled squares: B, filled circles: H, filled triangles: W subtype systems. For further explanation see text.

We found that in W, B and A subtype systems there is a rigorous relationship between β, the bolometric luminosity ratio, and the mass ratio:

$$\beta = 1.003(\pm 0.005) - 1.02(\pm 0.04)\lambda + 0.92(\pm 0.09)\lambda^2 \tag{2}$$
$$- (0.40 \pm 0.06)\lambda^3 + 0.42(\pm 0.02)q^{4.6\pm 0.3} \tag{3}$$

H subtype systems obey a different relation:

$$\beta = 1.67(\pm 0.21) - 1.02(\pm 0.04)\lambda + 0.92(\pm 0.09)\lambda^2 \tag{4}$$
$$- (0.40 \pm 0.06)\lambda^3 - 1.06(\pm 0.25)q \tag{5}$$

This difference exists not only between H and other subtype systems in the energy transfer rate. There are A and W systems, which produce the same luminosity ratio as H-type systems, although their mass ratios are quite different. Note that W and some A systems between approximately $q = 0.55$ and 0.72 produce the same luminosity ratio as H-type systems with $q > 0.72$. It is very interesting that the same luminosity ratio can be observed in systems with very different mass ratios and transfer rates. It is evident that H-type has a lower energy transfer rate, because they have mass ratios close to unity, and hence less luminosity should be transferred in order to equalize the surface temperatures. Therefore they have a higher transfer parameter, but at the same luminosity ratio, where we observe H-type systems, we find A as well as W type systems with a lower transfer parameter (see Figure 2). It should be emphasized that all systems, which are far from the envelope in Figure 2, have $q > 0.72$ without any exception. Therefore we conclude that these systems form a new class of contact binary stars.

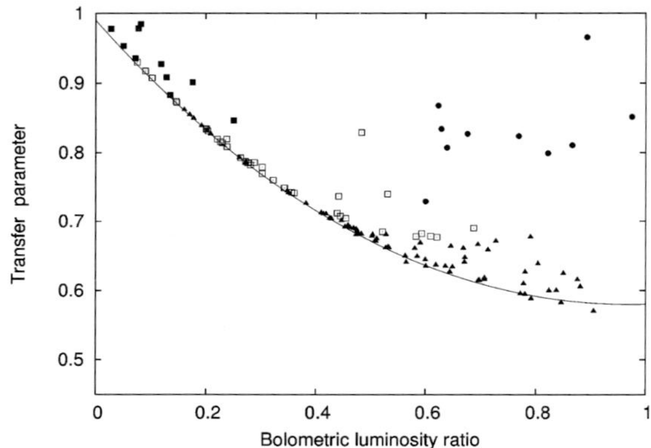

Figure 2. Transfer parameter β vs. luminosity ratio λ. Symbols are the same as in Figure 1. Note that all H systems are on the right side of the diagram above the solid line. The solid line describes the rate of transferred luminosity in A and W type systems in first approximation.

Acknowledgements

This work was supported by the Hungarian Science Fund under number OTKA T 034 551.

References

Binnendijk, L.: 1965, *Kleine Veröff. der Remeis-Sternwarte Bamberg* **40**, 36.
Flower, P.J.: 1996, *ApJ* **469**, 355.
Kalimeris, A. and Rovithis-Livaniou, H.: 2001, *Odessa Astron. Publ.* **14**, 33.
Lucy, L.B.: 1968a, *ApJ* **151**, 877.
Lucy, L.B.: 1968b, *ApJ* **153**, 1123.
Webbink, R.F.: 2003, *ASP Conf. Ser.* **293**, 76.
Wilson, R.E.: 2001, IBVS 5076.

IN BETWEEN β LYRAE AND ALGOL: THE CASE OF V356 Sgr

D. DOMINIS[1], P. MIMICA[2], K. PAVLOVSKI[3] and E. TAMAJO[3]

[1]*Institut für Astrophsik, Universität Potsdam, Germany;*
E-mail: dijana@astro.physik.uni-potsdam.de
[2]*Max-Planck Institut für Astrophysik, Garching, Germany*
[3]*Department of Physics, University of Zagreb, Croatia*

(accepted April 2004)

Abstract. The eclipsing binary system V356 Sgr is of considerable interest, since it is probably at the very end of its mass transfer phase, i.e. between β Lyrae and Algol. Hence, the binary provides an opportunity to directly examine the exposed core of a star for signatures of nuclear burning, and to test stellar evolution models. The system is composed of an early B star accreting matter from a Roche-lobe filling A2 II star. Recently, with progress in the UV spectral region, significant revision of previous values for absolute parameters has been made. Therefore, we find it justified and important to present a new photometric solution. Our model is compared to an early disk model, and is discussed in the framework of mass transfer processes in this binary system.

Keywords: binary stars, accretion disk, V356 Sgr

1. Introduction

V356 Sgr (HD 173787, $V_{\max} = 6.8$ mag, $P = 8.93$ d) is an unusual binary system. This was recognized by Popper (1955) in an admirable study in which he had analyzed both spectrographic and photometric observations. Lines of both components are visible in the spectra. During primary minimum he found the spectrum to be of type A2 II. Because lines of the hotter component of spectral type B2 or B3 are considerably broadened by rotation, its RV curve was less precise than the RV curve of the sharp-lined A component. There are a number of anomalies in the light curves, so that Popper was not able to fit a model with methods of the 50s.

A major improvement in the model of the V356 Sgr system has been achieved by Wilson and Caldwell (1978), who have calculated synthetic light curves with a model in which the hotter and more massive component is hidden by an opaque and thick circumstellar disk.

While the effect of the disk eclipse by the secondary will be slight, since the disk emits only a small fraction of total light, the eclipse of the secondary by the disk should be appreciable. Thus the secondary eclipse should be wider than the primary eclipse, and it should also be deeper than it would be without an eclipse by the disk. With this disk model some of the anomalies in the light curves were explained.

Astrophysics and Space Science **296**: 189–192, 2005.
© Springer 2005

Our work on the new photometric solution has been motivated by: (i) Important revision of some fundamental quantities for the components of V356 Sgr made by Polidan and co-workers (c.f. Polidan 1989, Roby et al. (1996), (ii) Brief note by Wilson and Woodward (1995) who challenged the earlier opaque disk model by Wilson and Caldwell in favour of an only rotationally distorted primary component in a double contact configuration, and (iii) Importance of abundance studies for interacting binaries, which need reliable fundamental stellar data for understanding the chemical evolution of close binary systems, and in particular, mass-transfer processes in these binaries (c.f. Tomkin and Lambert 1994).

2. Modeling and Optimization

The models of the binary system considered in the present calculations are: (1) a 'disk model', in which a cool component, mass-losing star, is filling its critical equipotential surface. A mass-gaining component is detached from its Roche lobe and is surrounded by an optically thick accretion disk, and (2) a 'rotation model', in which a hot component is rotationally distorted. Its surface is defined by the equipotential surface of a single star rotating as a solid body. Two parameters define its shape: the polar radius and flattening coefficient.

The optimal solution was reached by simultaneously adjusting a set of parameters to obtain the best agreement between theoretical model, and observed data (Popper's 1995 *UBV* observations). Light curves are calculated with the code described by Pavlovski and Kříž (1985), while an optimization algorithm based on the laws of natural selection has been used after Charbonneau (1995), who coded a general purpose optimization subroutine PIKAIA.

3. Results and Conclusion

The 'disk model' solution was found to fit the measurements better than the 'rotational model', both in detail and globally with smaller residuals (Figure 1). In general, our new photometric solution with an accretion disk, which surrounds the hotter component, corroborates recent results of on-going work by Polidan and co-workers, in particular a lower mass ratio $q \sim 0.27$, effective temperature of the hotter, partially hidden, component $T_{B,eff} \sim 23\,000$ K, and its high rotational velocity close to the centrifugal break-up limit. Recently, from IUE high-resolution UV spectra of V356 Sgr, Fraser et al. (1997) revised Popper's mass ratio to $q = 0.25$. Also, from the same set of observations, Roby et al. (1996) derived effective temperatures for both components, which are, in particular for the hotter component, in substantial discordance with the temperature obtained earlier by Wilson and Caldwell in their photometric study.

Parameters derived in the present study (Table I) position the star on the borderline where a stable accretion disk could be formed around the mass-gaining star

TABLE I

Absolute parameters for the components of V356 Sgr
derived from optimization with the 'disk model'

Quantity/Component	B	A
Mass M [M$_\odot$]	10.4	2.8
Radius R [R$_\odot$]	5.2	11.7
log g [cgs]	4.0	2.7
$T_{\rm eff}$ [K]	23 000	9 000

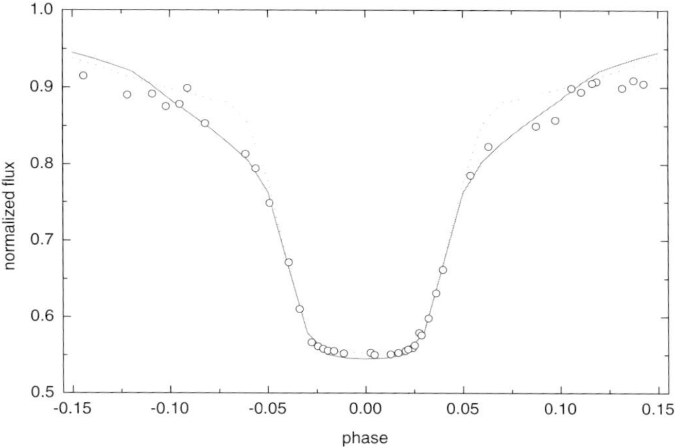

Figure 1. Optimal solution for 'disk' (solid line) and 'rotation' (dotted line) models compared to Popper's (1955) V band light curve of V356 Sgr.

(Figure 2). High rotation velocity is expected in such a mass-transfer binary, which is also confirmed by the present solution.

Additional support for lower q comes from precise $v \sin i$ measurements. On our request Dr. Jocelyn Tomkin kindly estimated the projected rotational velocity of the A component from high-resolution high S/N CCD spectra obtained at McDonald Observatory for an abundance analysis (Tomkin and Lambert 1994). He has measured the widths of two clean N I lines at 7442 and 7468 Å, respectively, and estimates $v \sin i = 66 \pm 7$ km s^{-1} based on these measurements. This result, on the basis of high quality spectra, again strongly supports a mass ratio $q \sim 0.27$, in excellent agreement with (i) the average of the scattered published values, (ii) the direct determination by Fraser et al. (1996) from RV curves measured in UV spectra, and (iii) our new photometric solution with the disk model.

Our new photometric solution represents a good initial model for more detailed modeling. Such a new step in modeling will require a new set of precise multi-colour

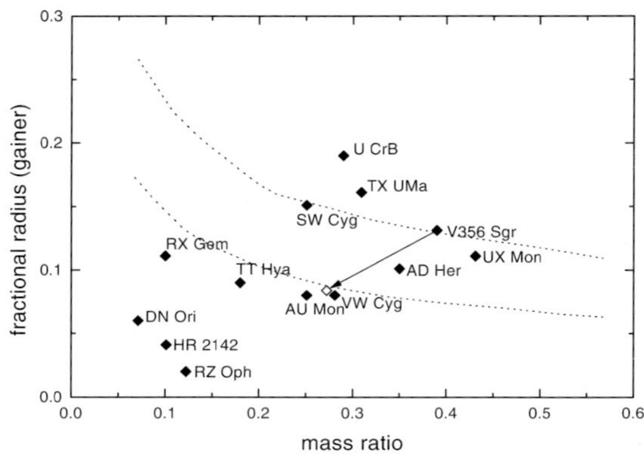

Figure 2. Optimal solutions for the 'disk' (solid line) and 'rotation' (dotted line) models compared to Popper's (1955) V band light curve of V356 Sgr. Position of V356 Sgr among Algols according to the parameters derived in the present study.

photometric measurements. This should not be a major challenge for one of the brightest binaries in the sky, and we appeal to southern observers to observe light curves of this astrophysically important and bright binary star.

References

Charbonneau, P.: 1995, *ApJS* **101**, 309.
Fraser, K.L., Roby, S.W. and Polidan, R.S.: 1997, *AAS* **191**, 4513.
Pavlovski, K. and Kříž, S.: 1985, *Bull. Astron. Inst. Czech.* **35**, 120.
Popper, D.M.: 1955, *ApJ* **222**, 155.
Roby, S.W., Polidan, R.S., Fraser, K., Hassett, J.M. and Mauser, K.E.: 1996, *AAS* **189**, 7706.
Tomkin, J. and Lambert, D.L.: 1978, *PASP* **106**, 365.
Wilson, R.E. and Caldwell, C.N.: 1978, *ApJ* **221**, 917.
Wilson, R.E. and Woodward, E.J.: 1995, *PASP* **107**, 132.

III: MATHEMATICAL PHYSICS AND NUMERICAL MODELING

III.1: Solution of Eclipse Light Curves

EB LIGHT CURVE MODELS – WHAT'S NEXT?

R.E. WILSON

Astronomy Department, University of Florida, Gainsville, FL; E-mail: wilson@astro.ufl.edu

(accepted April 2004)

Abstract. Enhanced unity in binary system observables models is discussed, including development strategy and connections between morphology and solutions. Distance estimation can be made direct (one step) via simultaneous light/velocity solutions that incorporate a simple flux-scaling procedure. Potential importance of time-variable polarization work is emphasized, especially for Algols, with mention of needed improvements in polarimetric analysis and reporting of data. Recent developments in general all-data (light, velocity, etc.) ephemerides are reviewed, with projections for further generalization, and ideas and formalisms for analysis of period change *events* (perhaps discontinuous) are introduced. Advantages and practical problems of unified [light, velocity, pulse arrival time] analyses for X-ray binaries, possibly constrained by measured X-ray eclipse durations, are discussed.

Keywords: binary stars, distances, polarization, ephemerides, X-ray binaries

1. Strategic Overview

1.1. MODEL DEVELOPMENT CONSIDERATIONS

Interests of the author and collaborators have led over the years to the present WD binary star model, where the primary references to the public version are Wilson and Devinney (1971), Wilson (1979), and Wilson (1990). In addition to the public version, numerous special purpose versions have been developed to handle, for example, photospheric and circumstellar polarization (Wilson and Liou, 1993), X-ray pulse arrival times (Wilson and Terrell, 1998), and wind/chromosphere fluorescence (Wilson, 1998; Chochol and Wilson, 2001; Van Hamme and Wilson, 2002). Criteria for inclusion of phenomena in the distributed version concern whether there is sufficient user interest and whether enough observations exist to justify the very time consuming tests required for a public program. The philosophy has been that the model should be as general as is reasonably consistent with real usage, with external simplicity balanced against overall generality. Measures to attain effective simplicity in ordinary use of a very general computer model can include input control integers to turn features on and off, other control integers that tailor program operation to various astrophysical situations, and internally computed quantities that sense situations and direct operations accordingly. Few of us like to read manuals, so the idea is to have only one (rather general) program, with simple and astrophysically intuitive controls, that can figure out what needs to be done or at least needs only minimal instruction. However we also can have the familiar problem of "the

Astrophysics and Space Science **296**: 197–207, 2005.

machine having a mind of its own", meaning that it is programmed to do things without consulting the user, who may be taken by surprise. Still, in most realistic situations the user should face only those decisions that involve real astrophysical options, while those that are set by mainstream astrophysical theory are decided by the program. Striking a balance between "the user sees only those things that he/she has to worry about" vs. "the machine shouldn't have a mind of its own" is an evolving process. In another evolutionary area, capabilities may be developed in special program versions, used only a few times, and then lie dormant for lack of data or lack of widespread interest. Some capabilities described here are of that type, while others are anticipated for the future. Growing generality naturally begets a growing parameter list, with 34 adjustable parameters now in WD and perhaps 20 to be added over the next several years. Curiously, some persons are disturbed by long parameter lists, although having many adjustable parameters obviously is not a problem if they are used responsibly. Each parameter represents an option that need not be exercised, so even with 100 parameters built into a model, only a few may be utilized in typical solutions.

1.2. EB SOLUTIONS: BASIC AND ADVANCED

Conceptual eclipsing binary (EB) light curve analysis is reviewed in Wilson (1994) and Wilson (2000). Advanced issues are discussed in Wilson and Wyithe (2003), including automation via the field of Expert Systems in dealing with large datasets, and effective follow-up with large optics of promising EB's that are found in survey data. The first topic covers the broad rationale of what an expert machine program or person can do, which ideally would be the same except for the great speed and freedom from mistakes of the machine. The second topic covers the relative usefulness of well detached, semi-detached (SD), and overcontact (OC) binaries for distance estimation in regard to numbers of free parameters, solution constraints, and efficient use of observing time.

2. Direct Solutions for Distance

Light curves establish relative EB dimensions – not only mean radii but, with help from a little physics, star figures as well. Then the same physics (potential theory, now bolstered by radiative and convective energy transfer theory) allows inference of surface brightness distributions from the light curves. So we have a picture with a relative intensity scale but so far without a length scale. The relative intensities can be made absolute if we know effective temperatures T_1^{ref} and T_2^{ref} at reference points on the two stars and bring in a stellar atmosphere model, while the length scale comes from radial velocities. Naturally the rather complete geometric and radiative model thus derived allows one to compute observable flux in a given photometric band for an arbitrary location, assuming that interstellar extinction can be reliably estimated (or better, is negligible). Comparison of fluxes thus computed with directly observed fluxes produces distance estimates. As emphasized previously (Wilson and Wyithe, 2003; Wilson, 2004), EB's are not standard candles, but are *individually* measurable distance indicators. *If there were only one* EB, *we could, in principle, find its distance.* The absolute character of EB distance measurement eliminates need for calibration via nearby examples and so eliminates calibration error, thus enhancing accuracy.

All of this was known long ago, but only recently have the required stellar atmosphere models begun to approach fully satisfactory levels. Many papers over the past five years have found distances in this way, with primary attention paid to EB's in the Magellanic Clouds (e.g. Guinan et al., 1998; Clausen et al., 2003; Ostrov and Lapasset, 2003). Most of the authors strongly preferred well detached EB's (stars nearly spherical), for which the full power of modern EB models is not utilized. Wyithe and Wilson (2002) ran statistical simulations to see whether EB's of the SD type might have significant advantages over well detached EB's for distance estimation and concluded that they do. Arguments can be made in favor of distance-estimation preference for SD's over well detached EB's on conceptual grounds and in favor of OC's as even better (Wilson, 2004), although most OC's are too faint to be presently useful at Magellanic Cloud distances.

Direct distance estimation, as the term is used here, means that distance, d, is evaluated by a "one step" fitting process. Solutions are iterative as usual but no side calculations are needed, except for an estimate of interstellar extinction, and light curves are processed simultaneously with velocity curves. The input includes parameter d which, like any solution parameter, iterates to a value with a standard error and a place in the correlation matrix. The input also will include interstellar extinction, although it will be completely correlated with d if attempts are made to solve for both. Direct distance estimation improves on accuracy and on the realism of standard errors in several ways. It treats the light and radial velocity measures as a unified data set and thereby takes their joint uncertainties into proper account, including correlations, and it combines mass information from light and velocity curves in a logical way. It also accounts automatically for proximity effects, as well as any other phenomena such as surface features that may be built into the model, so one need not deal with ill-defined mean radii. Of course the reduction in human work is important and will become more so as data from *GAIA* and other large scale surveys come in.

A basic requirement is to begin from a model that keeps track of physics and geometry without arbitrary normalization so as to produce absolute light curves. While it may not be widely realized, not only does the **WD** model generate absolute light, but even the earliest versions did so. The synthesized output light (flux) may appear to be on an arbitrary scale, as the numbers are of order unity in ordinary applications, but that is because the input luminosities, L_1 and L_2, are usually set to achieve that result, thus producing a good match between observed and synthesized light. **WD** effectively gives absolute fluxes, as **WD** flux can easily be converted to ordinary physical units. To see this point, note that bandpass-integrated flux (directly observable) is in units of energy/(time · area · Δ wavelength) while bandpass-integrated luminosity (inferred, not directly observable) is in energy/(time·Δ wavelength). Also note that program light or flux, l, scales with luminosity when other parameters (p_i) and the phase (ϕ) are fixed [i.e. $l_1 = f_1(p_1, p_2, \ldots, p_n, \phi)L_1$ and $l_2 = f_2(p_1, p_2, \ldots, p_n, \phi)L_2$], with WD computing the factors f_1 and f_2. Therefore all units except those of area cancel in the ratio of flux to luminosity and the area units give no trouble if we adopt a standard distance for a hypothetical observer. The area at issue is that of the observer's light collector, but we can imagine that collector to cover a steradian at distance a (i.e. one semi-major axis length from the binary) and then re-scale to the actual observer distance. The units of (output) flux are chosen by the user, who sets them by the choice of (input) luminosity. For example, if luminosities are entered in erg/(sec · micron), fluxes will be in erg/(sec · micron) and will refer to 1 steradian of collecting area. So there is a hidden area unit for fluxes that happens to be unity because of the WD model convention of 1 steradian of collecting area at distance a. WD's output fluxes were designed from the earliest versions to correspond to that 1 steradian configuration – of course without finite distance perspective effects, which become negligible at realistic distances.

The scaling relation of Wilson (2004) produces theoretical flux in physical units, F_d^{abs}, with input being absolute radiative intensities from a stellar atmosphere model ($I_1^{\mathrm{abs}}, I_2^{\mathrm{abs}}$) and from the WD program ($I_1^{\mathrm{prog}}, I_2^{\mathrm{prog}}$) for reference points on the two stars, program fluxes $F_{a,1}^{\mathrm{prog}}$ and $F_{a,2}^{\mathrm{prog}}$ for distance $d = a$, and interstellar extinction in stellar magnitudes, A. The reference points are the $+z$ ("north") poles. The scaling relation is repeated here for completeness:

$$F_d^{\mathrm{abs}} = 10^{-.4A}\left[F_{a,1}^{\mathrm{prog}}\left(\frac{I_1^{\mathrm{abs}}}{I_1^{\mathrm{prog}}}\right) + F_{a,2}^{\mathrm{prog}}\left(\frac{I_2^{\mathrm{abs}}}{I_2^{\mathrm{prog}}}\right)\right]\left[\frac{a}{d}\right]^2. \tag{1}$$

As a matter of convenience, input to the analysis program can include the conversion factor between relative and absolute flux, so that the time-varying flux can be entered on a relative scale.

3. Polarization

Astronomical polarization work faces a problem similar to that encountered by high definition television (HDTV). Television stations have been slow to bear the expense of HDTV programming, because there were few HDTV receivers, while few viewers have been willing to buy expensive HDTV's for lack of HDTV programs, so start-up has been slow. Synthesis of theoretical time-dependent polarimetry for comparison with observations of binaries is in the same situation, with little work on models for lack of data and little incentive to observe for lack of model and analysis algorithms. Yes, there are polarimetric surveys, but there is very little in the way of timewise polarization curves for Algols and other potentially interesting binaries. Algols are inviting, as many have hot primaries with high photospheric electron densities that produce polarization due to Thomson scattering. They also have streams of low density transferring gas, and some Algols also have circumstellar scattering disks, where former stream material resides before accreting onto its star. Algol itself does have useful polarimetric curves (Kemp et al., 1983) that have been matched reasonably well in terms of disk and photospheric scattering (Wilson and Liou, 1993). If extended to other Algols, the model could estimate densities and dimensions of circumstellar disks, provide statistics of episodic flows, test the theory of photospheric ("limb") polarization, probe stellar atmospheres, determine orbital orientations (nodal position angle) and sense of motion, and measure rotation via polar flattening. Of course the measurements can serve as checks on binary star evolution and/or stellar statistics.

3.1. BROAD BRUSH THOUGHTS ON EB POLARIZATION

Useful binary system polarimetry tutorials may seem absent from the literature, but are indeed to be found in most papers on polarimetry of particular binaries – if one just does the opposite of whatever their authors recommend. For example, scientific methodology supposes to recognize a clear distinction between theory and observation. Although we compare one with the other, theory can be changed, while observations are definite except for corrections that are considered part of the observing process. Yet stellar polarimetry authors who are interested in one effect may consider other *astrophysical* polarization sources as nuisances. Accordingly they often pre-calculate the nuisance effects and *remove* them from the data, perhaps as simply as by subtraction or division, intending to make subsequent analysis for the "interesting" effect relatively easy. Now it may seem immaterial whether an effect is removed from the data or included in the model, but several problems afflict removal schemes. First, most "nuisance" effects can be pre-computed only with knowledge of parameters that are to come from subsequent analysis, so one either must accept an otherwise unneccesary level of iteration or guess the needed values. Usually the remedy is to guess. Second, the appropriate removal scheme may not be as simple as subtraction or division (is the phenomenon additive, multiplicative, or more complicated?). Third, matters can become intricate when more than one effect is to be removed. Notice that all three difficulties disappear, if we return to the simple scientific maxim *"put the theory into the model rather than correct the data!"*

Much of the required theory is the same as needed for light curves, so we are almost there if we begin from a good EB light curve model. Given that several such models are readily available, there is no point in using a simple geometrical model. We also want to fit the data directly, free of artificial intermediaries such as Fourier series that may be fitted to the direct polarimetry curves, and are common in the polarimetric literature. Also unnecessary is the approximation whereby polarimetric eclipse signatures are computed for a thin ring at the limb rather than for the entire visible face of a star. Wilson and Liou (1993) show that the thin ring approximation gives only a crude representation of polarimetric eclipse curves. As a last point for modelers/analysts, let us avoid the commonly seen graphical parameter estimations and carry out solutions impersonally with generation of standard errors. As a last point for observers, be sure to publish *times* of observations. Occasional papers list only phases in a 0–1 range, ignoring the fact that circumstellar polarization is episodic rather than periodic. Of course one should publish times of observations for essentially all data, but it is crucial for polarimetry.

3.1.1. *Polarimetry from Space*

EB photospheric polarization curves measured from space can exploit the important increase of photospheric polarization amplitudes at short wavelengths that are unobservable from the ground. The reason is in terms of three factors – we want strongly *asymmetric* radiation fields at given wavelength in order to have large polarization amplitudes, as net linear polarization cancels for symmetric fields. Strong directional dependence of monochromatic intensity requires not only a strong temperature gradient (factor 1), which is natural for a stellar photosphere, but also a strong dependence of intensity on temperature (factor 2). Factor 1 is unrelated to wavelength while factor 2 is important on the short wavelength side of the spectral energy peak but not on the long wavelength (Rayleigh-Jeans) side, as can be seen in a simple way by comparing blackbody curves stepped in temperature. Factor 3 is obvious – increasing temperature leads to increasing numbers of the free electrons that account for Thomson scattering. So now consider what happens in visible bands (say U, B, V) as we go to hotter stars to have more Thomson scattering: Yes, there will be more Thomson scattering, but we are pushed ever further into the Rayleigh-Jeans region, thus rendering intensities more symmetric and taking away from gains by Thomson scattering. We want to be at short wavelengths such as 0.20μ or 0.15μ where factors 2 and 3 are both important, giving strongly asymmetrical radiation fields with much Thomson scattering.

4. Generalized All-Data Ephemerides

This section is not about a refinement in analyzing ephemerides, but about an alternative that need *not* involve traditional timing diagrams. The idea is to utilize all or much of existing data (not just times of eclipse minima). We shall relinquish use of eclipse symmetry in defining T_0 and therefore need an alternative definition. Logical is *time of conjunction*. Begin by asking what is the best light curve template. Obviously it is a theoretical light curve for a specific EB, and of course likewise for velocity curves and any other kinds of synthesized curves. We will have approximations to these idealized templates that become better as we approach our final solution. Observed curves of various types, or a chosen subset, are to be fitted simultaneously. The central idea is to build ephemeris parameters in at ground level and find their values by an inversion algorithm, perhaps the often used Differential Corrections (DC). So we put "all" parameters in, not only reference epoch and period (P), but also dP/dt, parameters associated with episodic changes in P, third body light time parameters, etc., and adjust the ephemeris parameters along with the other ("astrophysical") parameters. In brief we will have Multiple-Origin Unified Curve Ephemerides (new acronym *MOUCE*, pronounced *mouse*).

4.1. RECENT *MOUCE* HISTORY

As with many EB modeling developments, *MOUCE*'s beginnings were stimulated by a particular binary. The [K giant, Be giant] binary AX Monocerotis lacks eclipses, with radial velocities from two epochs being the only reliable means to an ephemeris. Since tick marks similar to those of light curve eclipses are absent in radial velocity curves, a good algorithm was needed for treating whole curves, and the idea for AX Mon was to work within a full binary star observables program (in this case, WD), rigorously compute phases from time, and let WD take care of all sophistications of the synthesized velocity curves. As long as whole curves are involved, they might as well be of general type – any kind that the program can produce, so a natural extension to light curves and mixed types of curves soon was applied to other binaries. Actually the output need not even be a curve or curves, but can be any time-dependent observables.

The time to phase conversion implemented for AX Mon is basic to all subsequent work in this area, and includes dP/dt as well as T_0 and P. Phase intervals for constant dP/dt are given by

$$\Delta\phi = \frac{\ln\left[1 + \frac{(t-T_0)dP/dt}{P_0}\right]}{dP/dt}. \tag{2}$$

A series expansion to avoid a 0/0 form for the case of $dP/dt \longrightarrow 0$ is

$$\Delta\phi = \frac{t - T_0}{P_0}\left[1 - \frac{(t - T_0)dP/dt}{2P_0} + \frac{(t - T_0)^2(dP/dt)^2}{3P_0}\right. \\ \left. - \frac{(t - T_0)^3(dP/dt)^3}{4P_0} + \dots\right]. \tag{3}$$

Going the other way (phase to time) at constant dP/dt,

$$t - T_0 = (e^{\Delta\phi dP/dt} - 1)\left(\frac{P_0}{dP/dt}\right), \tag{4}$$

and again to avoid a 0/0 form when dP/dt approaches zero, use a series approximation

$$t - T_0 = P\Delta\phi\left(1 + \frac{\Delta\phi dP/dt}{2} + \frac{(\Delta\phi dP/dt)^2}{6} + \frac{(\Delta\phi dP/dt)^3}{24} + \dots\right). \quad (5)$$

The scheme thereby allows one to work with any kind of curve that the observables program can synthesize, although the AX Mon ephemeris was based only on radial velocities. Unfortunately I did not think to explain even the basics of the scheme in the resulting AX Mon paper (Elias et al., 1997), with the section on the ephemeris giving only results (which did find a dP/dt at the 3.6σ level). Papers with W. Van Hamme extended *MOUCE* to find apsidal motion via parameter $d\omega/dt$ (Van Hamme and Wilson, 1998) and periodic light time effects via third body orbit parameters (Van Hamme et al., 2002; Van Hamme and Wilson, this meeting). The $d\omega/dt$ parameter has been in the public version of WD for about 6 years, while the six light time parameters should be added reasonably soon. *MOUCE* was applied by Wilson and Van Hamme (1999) to β Lyr ($[T_0, P]$ from velocities), by Van Hamme et al. (2001) to CN And ($[T_0, P, dP/dt]$ simultaneously from light and velocities), by Chochol and Wilson (2001) to V1329 Cyg ($[T_0, P, dP/dt]$ from light curves), by Van Hamme and Wilson (2002) to AG Dra ($[T_0, P, dP/dt]$ simultaneously from light and velocities), and by Terrell et al. (2003) to TU Mus ($[T_0, P, dP/dt]$ simultaneously from light and velocities). Good agreement was found in cases where *MOUCE* and eclipse timing ephemerides could be compared.

4.2. PERIOD CHANGE *events* (PERHAPS DISCONTINUOUS)

We now have constant dP/dt as a parameter, but this is a rather simple case as we cannot expect dP/dt to be the same at all times. Examples of nearly discontinuous period changes can be seen in timing diagrams of binaries such as SW Lac, ER Ori, and W UMa in the atlas by Kreiner et al. (2000) or on the website//www.as.ap.krakow.pl/o-c/index.php3 by the same authors. We could introduce second, third, etc. period derivatives, but more useful in the light of experience with real binaries is to consider episodic changes with individually constant dP/dt's. That is, we consider idealized multiple period change events. We think of each event as due to a (usually un-identified) physical cause that lasts a time δt. The events might be of mass loss or mass transfer and can overlap. Discontinuous period changes can be limiting cases where $dP/dt \longrightarrow \infty$ and $\delta t \longrightarrow 0$. Solutions can approximate that ideal with very large dP/dt's and very small δt's. Usually one would fix δt at a value below the effective time resolution and solve for dP/dt, or realistically $\log(dP/dt)$, and publish $dP/dt \times \delta t$ as the period jump δP. The episodic period change parameters should be subscripted, as there may be several period change episodes, so we have dP/dt_k, Δt_k. One can ask about the value of all this if we do not identify a physical cause for a given episode. However, armed with *impersonal quantitative* episodic data, we can draw *statistical* inferences about causes. Implementation is now in progress.

What about traditional times of minima? Of course timing residual graphs will continue to be needed for illustration, and times of minima have three advantages – (1) they need only eclipse data, (2) they need only simple analysis programs, and (3) large historical databases already exist. The obvious strategy is to encompass traditional times of minima within *MOUCE*, thereby treating times of minima together with whole curves. WD is ready to handle times of minima, as it routinely computes conjunction times. A minor issue concerns redundancy if one uses a time of mimimum derived from data that also is entered in light curve form, but the remedy is trivial – enter the curve or the time of minimum but not both.

5. X-ray Pulse Arrival Times

The astonishingly accurate and specific information derived from pulse arrival times is a modern astrophysical legend, acting like a caliper applied to orbital dimensions and orientations. However most information about the *normal stars* in [X-ray pulsar]/[normal star] binaries comes from more traditional data such as light and velocity curves. These distinct data types are usually analyzed separately for arrival time, light curve, and radial velocity parameters, with relatively complete pictures of the binaries then formed by considering the collection of all such parameters. However accuracy is to be gained by solving for all parameters simultaneously, as some parameters are common to the separate analysis problems. Pulse arrival time is an analytic function of the parameters and is analytically differentiable (the derivatives are in Wilson and Terrell, 1999), so hardly any computation time is needed for the derivatives in DC solutions. Also helpful is that no new parameters are involved except for the obvious ones of pulse period and initial pulse epoch – pulse arrival time just helps in finding the 'old' parameters.

For X-ray binaries we sometimes know the duration of X-ray eclipse, perhaps with a transition interval, which is controlled by surface potential, mass ratio, rotation, orbital eccentricity, inclination, and argument of periastron. Solutions constrained by X-ray eclipse duration have been in WD for decades, with the mathematics of duration computation for the general eccentric, non-synchronous case in Wilson (1979). Pulse arrival times tell the X-ray star's instantaneous location along the line of sight relative to the system center of mass. How about handling radial velocities and pulse arrivals (and perhaps optical light curves) in one program, conditioned by observed X-ray eclipse duration? Algorithms to carry out such simultaneous solutions were outlined in Wilson (1979), tried by Wilson and Terrell (1994), and further developed in Wilson and Terrell (1999) for the example GP Vel. The independent variable in this algorithm is *pulse arrival time*, as opposed to most X-ray pulse analysis work where it is *pulse delay*. While the distinction may seem immaterial, adoption of *arrival time* as independent variable has practical advantages that are discussed in Wilson and Terrell (1999). The algorithms have not been placed in the public WD program, but could be so placed if greater quantities of suitable observations accumulate. Primary incentives for constrained simultaneous solutions are the incorporation of reliably known geometry via the eclipse duration constraint and correct balancing of the influences of the multiple data types, with parameter correlations taken properly into account. Mass estimates derive from a confluence of the various kinds of data and are important for the neutron star mass limit and mass function, and for understanding the companion's evolutionary state.

As with any simultaneous solutions, weighting is important in velocity/pulse or velocity/light/pulse solutions and needs to be thought through. The essential considerations are in Wilson (1979) and Kallrath and Milone (1999). Then there are the well known idiosyncracies of X-ray binaries. For example, the optical stars often have continual light curve transients that may be due to dynamical tides in eccentric orbits, especially in high mass X-ray binaries (HMXB's). X-ray irradiation pressure fluctuations and instabilities in low mass X-ray binaries (LMXB's) also may produce transients, although many LMXB's are not candidates for simultaneous solutions for lack of detectable pulses. Other problems are that wild variations in pulse period (mainly of HMXB's) require allowance for nearly discontinuous behavior, and that partially understood or not-understood problems beset individual binaries.

6. A Unified Model

One can conceive of a light/velocity/polarization/pulse/whatever unified model with the physical realism of modern "light curve" models plus capabilities of direct distance and all-data ephemeris analysis that are discussed above, and indeed other features whose discussion would exceed available space. The main impediment to realization of such a computer model is the large amount of testing needed to ensure proper operation in a *distributed* program, although feedback from users can help eliminate bugs. Of course, star surfaces are to be equipotentials (many X-ray publications are for spheres), orbits can be eccentric without resort to some common approximations, and stars can rotate asynchronously. Lobe filling at periastron can be specified, information in X-ray eclipse durations can be utilized, and data of mixed types can be fitted impersonally and simultaneously. The usual convergence enhancing tricks for DC solutions naturally apply. Such a general model will come, and eventually even more general models. Existence of data and interest in applications will drive their development.

Acknowledgements

This material is based upon work supported by the National Science Foundation under Grant No. 0307561.

References

Chochol, D. and Wilson, R.E.: 2001, *Mon. Not. R. Astron. Soc.* **326**, 437–452.

Elias, N.M., Wilson, R.E., Olson, E.C., Aufdenberg, J.P., Guinan, E.F., Güdel, M., Van Hamme, W. and Stevens, H.: 1997, *ApJ* **484**, 394–411.

Guinan, E.F., Fitzpatrick, E.L., DeWarf, L.E., Maloney, F.P., Maurone, P.A., Ribas, I., Pritchard, J.D., Bradstreet, D.H. and Gimenez, A.: 1998, *ApJ* **509**, L21–L24.

Kallrath, J. and Milone, E.F.: 1999, Eclipsing Binary Stars: Modeling and Analysis, Springer, Berlin.

Kemp, J.C., Henson, G.D., Barbour, M.S., Kraus, D.J. and Collins, G.W.: 1983, *ApJ* **273**, L85–L88.

Kreiner, J.M., Kim, C. and Nha, I.: 2000, An Atlas of O-C Diagrams of Eclipsing Binary Stars, Wydawnictwo Naukowe, Krakow, Poland.

Ostrov, P.G. and Lapasset, E.: 2003, *Mon. Not. R. Astron. Soc.* **338**, 141–146.

Terrell, D., Munari, U., Zwitter, T. and Nelson, R.H.: 2003, *ApJ* **126**, 2988–2996.

Van Hamme, W., Samec, R.G., Gothard, N.W., Wilson, R.E., Faulkner, D.R. and Branly, R.M.: 2001, *ApJ* **122**, 3436–3446.

Van Hamme, W. and Wilson, R.E.: 1998, *Bull. A.A.S.* **30**, 1402.

Van Hamme, W. and Wilson, R.E.: 2002, *ASP Conf. Ser.* **279**, 161–166.

Van Hamme, W., Wilson, R.E. and Branly, R.M.: 2002, *ASP Conf. Ser.* **279**, 155–160.

Wilson, R.E.: 1979, *ApJ* **234**, 1054–1066.

Wilson, R.E.: 1990, *ApJ* **356**, 613–622.

Wilson, R.E.: 1994, *IAPPP* **55**, 1–20.

Wilson, R.E.: 1998, A Fluorescence and Scattering Model for Binaries, in: R. Dvorak, H.F. Haupt and K. Wodnar (eds.), Modern Astrometry and Astrodynamics, Austrian Academy of Science, Vienna, Austria, pp. 219–232.

Wilson, R.E.: 2000, Eclipsing Binary Stars, Encyclopedia of Astronomy and Astrophysics, Institute of Physics Publishing and Nature Publishing Group, London, pp. 706–712, http://www.ency-astro.com.

Wilson, R.E.: 2004, *New Astr. Rev.*, (http://dx.doi.org/10.1016/j.newar.2004.03.015), in press.

Wilson, R.E. and Devinney, E.J.: 1971, *ApJ* **166**, 605–619.

Wilson, R.E. and Liou, J.C.: 1993, *ApJ* **413**, 670–679.

Wilson, R.E. and Terrell, D.: 1994. *AIP Conf. Proc.* **308**, 483–486.

Wilson, R.E. and Terrell, D.: 1999, *Mon. Not. R. Astron. Soc.* **296**, 33–43.

Wilson, R.E. and Van Hamme, W.: 1999, *Mon. Not. R. Astron. Soc.* **303**, 736–754.

Wilson, R.E. and Wyithe, S.B.: 2003, *ASP Conf. Ser.* **298**, 313–322.

Wyithe, S.B. and Wilson, R.E.: 2002, *ApJ* **571**, 293–319.

BASIC FUNCTIONS OF THE LIGHT CURVE ANALYSIS OF ECLIPSING VARIABLES IN THE FREQUENCY DOMAIN

OSMAN DEMIRCAN

Physics Department, Faculty of Science and Arts, Çanakkale Onsekiz Mart University, Çanakkale, Turkey; E-mail: demircan@comu.edu.tr

(accepted April 2004)

Abstract. An analytically tractable method of transforming the problem of light curve analysis of eclipsing binaries from the time domain into the frequency domain was introduced by Kopal (1975, 1979, 1990). This method uses a new general formulation of eclipse functions α, the so-called moments A_{2m}, and their combinations as $g_{2m} = A_{2m+2}/(A_{2m}A_{2m+4})$ functions for the basic spherical model. In this paper, I will review the use of these functions in the light curve analysis of eclipsing binaries.

Keywords: eclipsing variables, eclipsing binaries, light curve analysis, eclipse functions, Fourier analysis

1. Introduction

Kopal's (1975a,b, 1979, 1990) Fourier analysis of the light curves of close binary systems is based on the integral transforms called "moments" of the light curves. The physical and geometrical factors forming the light curves could be written separately in terms of the moments as

$$
\begin{aligned}
\bar{A}_{2m} &= \int_0^{\pi/2} [l(\pi/2) - l(\theta)]d\sin^{2m}\theta \\
&= \sum_{j=0}^{n} c_j \int_0^{\pi/2} \cos^j\theta\, d\sin^{2m}\theta + L_1 \sum_{n=0}^{N} c^n \int_0^{\pi/2} \left[\alpha_n^o + f^{(n)}\right] d\sin^{2m}\theta \\
&= B_{2m} + A_{2m} + P_{2m}
\end{aligned}
\tag{1}
$$

where the terms on the right side stand for the proximity effect, the spherical eclipse effect, and the photometric perturbations (caused by the non-sphericity of the components), respectively. The proximity effect B_{2m} represents the reflection and the ellipticity effects together and is deduced directly from the observations (for details see Niarchos, 2004). The whole method is based on the Roche model. The first term in the expansion contains the spherical α-functions, and they form the A_{2m} moments in turn after integration with respect to $\sin^{2m}\theta$. The remaining terms of the expansion of the Roche model forms the perturbation term P_{2m}, which is written explicitly in terms of the physical and geometrical parameters of the binary system (see Chapter VI.3 and 4 of Kopal, 1975, 1990, and Livaniou, 1977). The $\bar{A}_{2m}$ term on the left side is the observational moment of index $2m$ (by the way m

Astrophysics and Space Science **296**: 209–220, 2005.
© Springer 2005

is not necessarily integer, but can be any real number). The A_{2m} is the area under the function of measured light l, plotted against $\sin^{2m}\theta$, and thus an empirically determined quantity. Thus, as many equations as we need for the solution of n parameters (such as fractional radii, inclination, fractional luminosities, limb darkening coefficients, gravity darkening coefficients, etc.) can be formed for different real values of index m of Eq. (1), as

$$\bar{A}_{2m} = (A_{2m})_{\text{obs}} = (A_{2m})_{\text{theo}} = B_{2m} + A_{2m} + P_{2m} \tag{2}$$

and can be used for the solution of n parameters, by using known numerical techniques. Kopal, however, followed a kind of "classical rectification" technique, which applies subsequent subtractions of B_{2m} and P_{2m} terms from the observational moments $(A_{2m})_{\text{obs}}$ in order to obtain the cleaned spherical moments A_{2m}. The spherical moments A_{2m} are valid for the spherical model and they are used for the solution of four basic elements ($r_{1,2}$, i, and L_1) of the close binary system.

In this paper I will review the geometrical behaviour and the use of the spherical moments A_{2m} and related f and g-functions in the solution of four basic eclipse elements.

2. The Eclipse Function α

In Kopal's Fourier analysis method, the projection of the close binary system on the celestial sphere is considered. The loss of light during the eclipses in this case is defined as a cross-correlation between two arbitrarily brightened (e.g. limb and gravity darkening) figures representing the projection of two stellar components and is considered in two terms: first, the spherical part representing the loss of light due to two spherical stars and second, perturbation terms as the residuals between the eclipse effect of actual equipotential figures and two spheres. The spherical part of the loss of light due to eclipses comes out to be the Hankel transforms of zero order of the products of two Bessel functions J and can be expanded into fast convergent series in terms of shifted Jacobi polynomials R as

$$\alpha_n^0 = 2^\nu T(\nu) b \int_0^\infty (ay)^{-\nu} J_\nu(ay) J_1(by) J_0(cy)\, dy$$

$$= b^2(1-c^2)^{(\nu+1)} T(\nu) \sum_{n=0}^\infty \frac{n!(\nu+2n+2)}{(n+1)\Gamma(\nu+n+1)} \left[R_n^{(1,\nu)}(a)\right]^2 R_n^{(\nu+1,0)}(c^2)$$

where $\nu = (n+2)/2$, $a = r_1/(r_1+r_2)$, $b = r_2/(r_1+r_2) = 1-a$, $c = \delta/(r_1+r_2)$, $\delta^2 = 1 - \cos^2\theta \sin^2 i$ such that for any type of eclipse $0 < a, b, c < 1$, and the Jacobi polynomials and the coefficients of the expansion can be generated recursively (for details see Kopal, 1990, Chapter III).

3. The Spherical Moments A_{2m}

The spherical moments A_{2m} of the light curves are defined as

$$A_{2m} = \int_0^{\theta_1} (1 - l)\, d\,(\sin^{2m}\theta),$$
$$= L_1 \int_0^{\theta} \alpha d(\sin^{2m}\theta), \tag{3}$$

where

$$1 - l = L_1\alpha = L_1 \sum_{n=0}^{\Lambda} C^{(n)}\alpha_n^0$$

Substitution of α from Eq. (3) and term by term integration reveals that

$$A_{2m} = L_1 b^2 \Gamma(m+1)\sin^{2m}\theta_1 \sum_{l=0}^{\Lambda} C^{(l)}\Gamma(v)\bigl(1 - c_0^2\bigr)^{v+1}$$
$$\sum_{n=0}^{\infty} \frac{n!(n+v+1)(2n+v+2)}{(n+1)\Gamma(m+n+v+2)}\bigl[R_n^{(v,1)}(a)^2\bigr]R_n^{(m+v+1,-m)}(c)_0^2 \tag{4}$$

where $C^{(0)} = \frac{1-u_1-u_2-\ldots}{1-\sum_{n=1}^{\Lambda}\frac{nu_n}{n+2}}$ and $C^{(n)} = \frac{u_n}{1-\sum_{n=1}^{\Lambda}\frac{nu_n}{n+2}}$ in terms of the limb darkening coefficients $u_1, u_2, \ldots, u_n$ of nth degree. It is interesting to note that Eq. (5) for
A_{2m} reduces to the loss of light α for $m = 0$ (see Eq. (3)). For alternative forms of
expansion 5 see Kopal (1978, 1990, Chapter V) and Demircan (1978a,b).

4. The Empirical Moments $(A_{2m})_{\mathrm{obs}}$

The empirical moments are obtained from the observations by using the definition
either by polarimetry or by numerical integration. They represent the areas between
the lines $l = 1$, $\sin^{2m}\theta = 0$ and $l \cdot \sin^{2m}\theta$, and thus should be more accurate than
the single or normal points of the observed light curves. Their accuracy is limited
by the accuracy of the unit of light U, and can be given as

$$\Delta A_{2m} \simeq \Delta U \sin^{2m}\theta \tag{5}$$

To give an impression I present in Table I some numerical values of the moments
A_0, A_2, A_4 and A_6 for different light curves of well known close binary systems.
The empirical values of the $g_2 = A_2^2/(A_0 A_4)$ and $g_4 = A_4^2/(A_0 A_6)$ functions are also
listed in Table I. It is seen in Table I that the numerical values of A_{2m} decrease very

TABLE I

The empirical moments A_0, A_2, A_4 and A_6 obtained from different light curves of well known binaries, where $g_2 = A_2^2/(A_0 A_4)$ and $g_4 = A_4^2/(A_2 A_6)$

		A_0	A_2	A_4	A_6	g_2	g_4
Algol occ.	λ4350	.702	.04335	.00429	.00052	.624	.816
	V (narrov)	.687	.04830	.00391	.00046	.621	.814
	V	.683	.04114	.00401	.00048	.618	.814
YZ Cas occ.	λ4500	.0622	.001274	.00003434	.000001099	.760	.842
	λ6700	.1022	.002048	.00005357	.000001701	.766	.824
YZ Cas tra.	λ4500	.3070	.005243	.0001283	.000003809	.698	.824
	λ6700	.2670	.004895	.0001225	.000003652	.733	.839
VV Ori	λ3320	.081	.0107775	.0015919	.000280	.901	.840
	λ3320	.081	.010751	.0015885	.000280	.898	.838
	λ2460	.063	.009541	.001505	.000280	.960	.848
	λ2980	.078	.0101162	.0014196	.000280	.922	.711
RW Tau	U	.9896	.059128	.0054860	.000631	.644	.807
	B	.9814	.058633	.005355	.000603	.654	.811
	V	.9597	.057342	.005334	.000613	.642	.809
U Sge	U	.9762	.081931	.010743	.001740	.640	.810
	B	.8555	.080190	.010527	.001713	.639	.807
	V	.9050	.075996	.009939	.001606	.642	.809
SZ Cam	U	.134	.01691	.00286	.00065	.746	.744
	B	.146	.01883	.00351	.00074	.694	.885

rapidly with increasing value of the index m. This effect is stronger in well detached systems. As an example, the m dependent A_{2m} values in the case of the well detached system V401 Lac are plotted in Figure 1, where according to Eq. (6) above, the S/N ratio becomes less than 10 for $m > 1.5$, if $S/N = 30$ for the unit of light U.

5. The Moments A_{2m} for Total Eclipses

It was shown by Kopal (1979) that the theoretical expressions for the moments A_{2m} for total eclipses are very much simpler than the general expressions given by Eq. (5). For $m = 0$, 1, 2, and 3 for total eclipses the moments A_{2m} become

$$A_0 = L_1 \tag{6}$$

$$A_2 = L_1 \bar{C}_3 \tag{7}$$

$$A_4 = L_1\left(\bar{C}_3^2 + \bar{C}_2^2\right) \tag{8}$$

$$A_6 = L_1\left(\bar{C}_3^3 + 3\bar{C}_2^2\bar{C}_3 + \bar{C}_1\bar{C}_2^2\right) \tag{9}$$

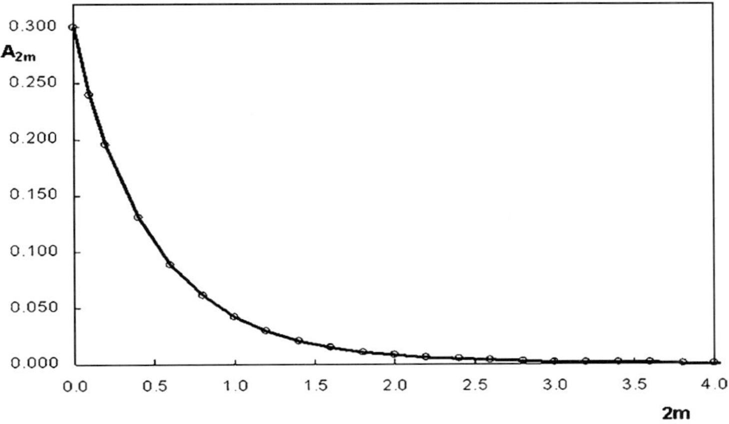

Figure 1. The *m* dependence of the empirical moments A_{2m} for the well detached system V401 Lac.

where

$$\bar{C}_3 = r_2^2(\csc^2 i - \cot^2 i) \tag{10}$$

$$\bar{C}_2^2 = r_1^2 r_2^2 \csc^4 i \sum_{n=0}^{\Lambda} \frac{2!C^n}{\nu(\nu+1)} \tag{11}$$

$$\bar{C}_1\bar{C}_2^2 = r_1^2 r_2^4 \csc^6 i \sum_{n=0}^{\Lambda} \frac{3!C^n}{\nu(\nu+1)(\nu+2)} \tag{12}$$

For any degree Λ of the law of limb-darkening. The summations on the right side of Eqs. (12) and (13) become

$$\frac{15 - 7u_1}{5(3 - u_1)}, \quad \frac{3(15 - 7u_1)}{5(3 - u_1)} \quad \text{and} \quad \frac{3(35 - 19u_1)}{35(3 - u_1)}, \tag{13}$$

respectively, for the linear law of limb darkening; and

$$\frac{2(15 - 7u_1 - 10u_2)}{5(6 - 2u_1 - 3u_2)}, \quad \frac{6(15 - 7u_1 - 10u_2)}{5(6 - 2u_1 - 3u_2)} \quad \text{and} \quad \frac{3(140 - 76u_1 - 105u_2)}{70(6 - 2u_1 - 3u_2)} \tag{14}$$

for the quadratic law of limb-darkening. Equations (7)–(10) further reduce to a simpler form as given below for a uniform brightness distribution ($u = 0$) over the component stars:

$$A_0 = L_1 \tag{15}$$

$$A_2 = L_1 C_3 \tag{16}$$

$$A_4 = L_1\left(C_3^2 + C_2^2\right) \tag{17}$$

$$A_6 = L_1\left(C_3^2 + 3C_2^2 C_3 + C_1 C_2^2\right) \tag{18}$$

where

$$C_1 = r_1^2 \csc^2 i \tag{19}$$

$$C_2^2 = r_1 r_2 \csc^2 i \tag{20}$$

$$C_3 = r_2^2 \csc^2 i - \cot^2 i \tag{21}$$

In practice, first the $C_{1,2,3}$ or $\bar{C}_{1,2,3}$ values are inverted from Eqs. (16)–(19) (or Eqs. (7)–(10)) by using empirical moments $A_{0,2,4,6}$, and once this has been done, an inversion of Eqs. (20)–(22) (or Eqs. (11)–(13)) yield the desired elements of the system in the form

$$r_{1,2}^2 = \frac{C_{1,2}^2}{(1 - C_3 C_1) + C_2^2} \tag{22}$$

and

$$\sin^2 i = \frac{C_1}{(1 - C_3 C_1) + C_2^2} \tag{23}$$

6. The f- and g-Functions

The general expression given by Eq. (5) for the moments A_{2m} can be rewritten as

$$A_{2m} = L_1 b^2 \sin^{2m} \theta_1 f_{2m}(a, c_0) \tag{24}$$

By using the ratio of two moments as

$$\frac{A_{2m}}{A_{2\mu}} = \sin^{2m-2\mu} \theta_1 \frac{f_{2m}(a, c_0)}{f_{2\mu}(a, c_0)} \tag{25}$$

we can eliminate one basic parameter L_1 and are left with three unknowns θ_1, a and c_0 on the right hand side.

Forming a quadratic ratio as

$$g(a, c_0) = \frac{A_{2m} A_{2\mu'}}{A_{2\mu} A_{2\mu'}} = \sin^{2m+2m'-2\mu-2\mu'} \theta_1 \frac{f_{2m} f_{2m'}}{f_{2\mu} f_{2\mu'}} \tag{26}$$

by considering that $m + m' = \mu + \mu'$, we can further eliminate the first term on the right hand side containing θ_1.

Such quadratic ratios of the moments are called g-functions, which depend only on a and c_0 parameters. In particular if $m = m' = 1$ and $\mu = 2$, $\mu' = 0$

$$g_2(a, c_0) = \frac{(A_2)^2}{A_0 A_4} = \frac{f_2^{f\,2}}{f_0 f_6} \tag{27}$$

while if $m = m' = 2$ and $\mu = 3$, $\mu' = 1$

$$g_4(a, c_0) = \frac{(A_4)^2}{A_2 A_6} = \frac{(A_2)^2}{A_0 A_4} = \frac{f_4^{f\,2}}{f_2 f_6} \tag{28}$$

If the moments $A_{0,2,4,6}$ are obtained from the observations and substituted in Eqs. (28) and (29), then the functions $g_{2,4}\,(a, c_0)$ constitute two independent relations between the unknown parameters a and c_0; and can be solved for them by using Eq. (5). The basic elements then follow as

$$L_1 = D/f_0 \tag{29}$$
$$r_1^2 = \frac{a_2}{f_0 A_0 + \left(\frac{c_0}{a}\right)^2 A_2} \tag{30}$$
$$\cos i = c_0(r_1 + r_2) \tag{31}$$

where D stands for the depth of the respective eclipse minimum. The necessary and sufficient condition for the solution of a and c_0 is the requirement that the functions $g_{2,4}$ do not simulate too closely functional dependence. In order to see the functional behaviour of these functions, four digit numerical values of f_{2m}'s and g_{2m}'s have been calculated (Demircan, 1978) for different m's between zero and three, under the assumption of plane parallel atmospheres. It was also found that the limb darkening effect on the f- and g-functions increases by decreasing m, and is for example less than 5% in the g_2-function.

In order to understand the determinacy of the parameters a and c_0, the functional behaviour of the functions $f_0 = \alpha$, g_2, g_4 and $g_{1/4} = A_{1/4}^2/A_0 A_{1/2}$ is shown in Figures 2–6.

7. Discussion

As it was noted in the introduction, the frequency domain analysis of the light curves of eclipsing binaries applies a classical "rectification" method, where the solution for the eclipse elements of distorted systems is reduced to a basic spherical model. To do so, the proximity effects containing reflection and ellipticity effects are obtained directly from observations and subtracted from the distorted moments together with theoretical perturbations (caused by the non-sphericity of the component stars) (see, e.g., Kopal, 1990, p. 126, and Niarchos, 2004). In the solution of

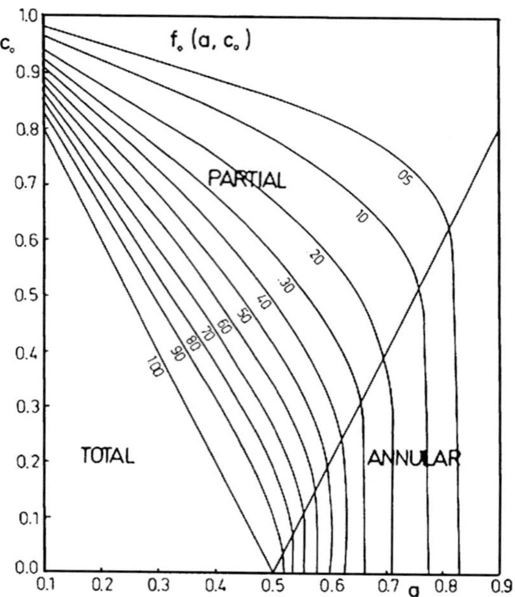

Figure 2. Functional behaviour of f_0 as defined by Eq. (25), in the (a, c_0) plane (after Demircan, 1978d).

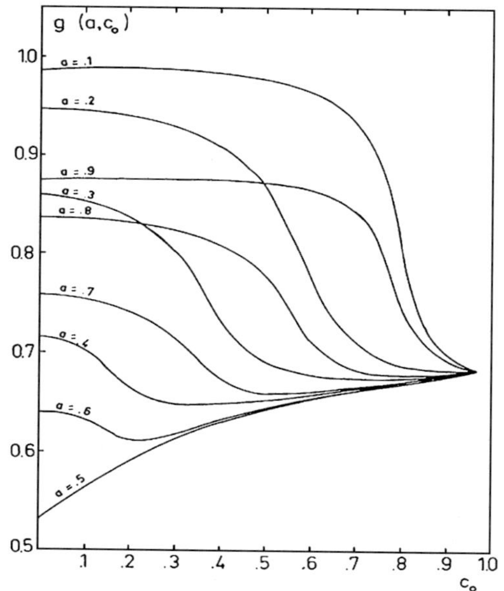

Figure 3. A plot of the function g_2 versus c_0 for fixed a (after Kopal and Demircan, 1978). Note that for any system $(g_2)_{occ} \geq (g_2)_{tra}$.

the eclipse parameters the low indexed combinations as g-functions of the moments A_{2m} of the deduced spherical model are used. However, in the previous sections the formulation of all α, A (or f) and g-functions and their use in the solution of eclipse parameter functions is given only for the primary eclipse minimum. In order to render a solution more determinate, consideration should be given to combined photometric evidence; a simultaneous treatment of two photometric eclipse minima, or if present, of more photometric light curves obtained at different epochs or at different wavelengths will obviously increase the determinacy of the eclipse elements. Moreover, in order to further increase the determinacy, additional evidences as spectroscopic radial velocity curves should be included in the analysis, as is done in later versions of the Wilson-Devinney (1971) code.

It was noted by Kopal (1990, p. 113) that the estimation of the first contact point θ_1 of the eclipse in the photometric observations can be used as a great advantage. The iterative solution for θ_1, a and c_0 can be commenced by using for example Eq. (26), where the denominator can be either from the same or the alternate minimum. Once this has been accomplished for a judicious choice of m and μ, the geometrical elements $r_{1,2}$ and i then follow from

$$r_1^2 = \frac{a^2 \sin^2 \theta_1}{1 - c_0^2 \cos^2 \theta_1} \tag{32}$$

$$r_2^2 = \frac{b^2 \sin^2 \theta_1}{1 - c_0^2 \cos^2 \theta_1} \tag{33}$$

and

$$\sin^2 i = \frac{a^2 c_0^2}{1 - c_0^2 \cos^2 \theta_1} \tag{34}$$

replacing Eqs. (23) and (24), in which the same elements were expressed in terms of the auxiliary constants $C_{1,2,3}$, and

$$\sin^2 \theta_1 = (r_1 + r_2)^2 \csc^2 i - \cot^2 i = C_1 + 2C_2 + C_3. \tag{35}$$

In Figures 4–6 the geometrical behaviour of the g-functions looks similiar, although they are formed by different indexed moments A_{2m}. The vertical lines at $a = 0.5$ in Figures 4 and 5 separate occultation and transit-type eclipses, while the geometrical behaviour of the g-functions also appears similiar on both sides of the $a = 0.5$ vertical line. In order to use, e.g., Eqs. (28) and (29) to locate the a and c_0 values satisfying given g-values, it is necessary to plot say g_2 and g_4 in the same diagram. We will then notice that two different g-functions will always intersect at low angles in such diagrams. This means that the determinacy of the parameters a and

O. DEMIRCAN

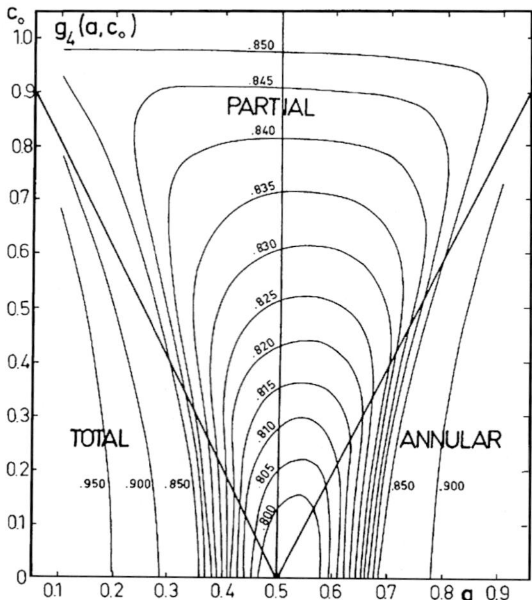

Figure 4. A diagrammatic representation of the function g_4 in the (a, c_0) plane (after Kopal and Demircan, 1978).

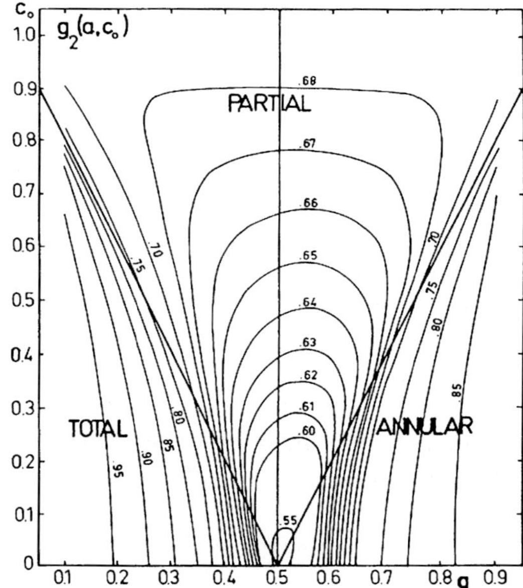

Figure 5. A diagrammatic representation of the function g_2 in the (a, c_0) plane (after Kopal and Demircan, 1978).

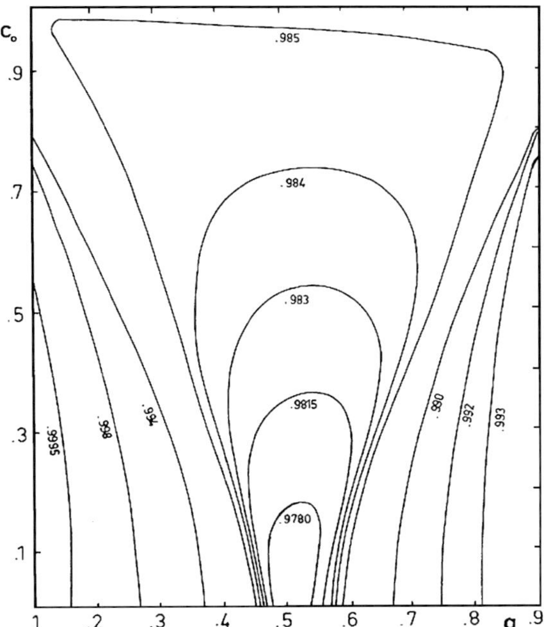

Figure 6. A diagrammatic representation of the function $g_{1/4}$ in the (a, c_0) plane (after Demircan, 1978b).

c_0 derived from g-functions is generally weak, i.e. the Jacobian

$$\frac{\partial(g_m, g_{m'})}{\partial(a, c_0)} \neq 0 \tag{36}$$

is not far from zero. However, this condition enables us to optimize our solution by maximizing the Jacobian for a judicious choice of the indices m and m' (which need not necessarily be integers). It was shown by means of trial solutions (see Figures 4–6 and Demircan, 1978a,b) that the larger the separation between the m's for employed moments the better is the achievable determinacy in four basic eclipse elements. On the other hand, it was well known from the theory of approximation of functions by Fourier series that the higher the value of m the more will an empirical determination of the moments A_{2m} be exposed to high frequency noise arising from the dispersion of observations. For this reason, the use of low indexed moments seems advisable for any application.

References

Demircan, O.: 1978a, *Ap&SS* **56**, 453.
Demircan, O.: 1978b, *Ap&SS* **59**, 313.
Kopal, Z.: 1975a, *Ap&SS* **34**, 431.

                                    O. DEMIRCAN

Kopal, Z.: 1975b, *Ap&SS* **38**, 191.
Kopal, Z.: 1979, *Language of the Stars*, D. Reidel, Dordrecht, The Netherlands.
Kopal, Z.: 1990, *Mathematical Theory of Stellar Eclipses*, Kluwer Academic, Norwell, MA.
Kopal, Z. and Demircan, O.: 1978, Kluwer Academic, Norwell, MA **55**, 241.
Livaniou, H.: 1977, Kluwer Academic Press **51**, 77.
Niarchos, P.: 2005, *Ap&SS* **296**, 359.

PHOTOMETRIC MASS RATIOS OF ECLIPSING BINARY STARS

DIRK TERRELL[1] and R.E. WILSON[2]

[1]*Department of Space Studies, Southwest Research Institute, Boulder, CO, USA;*
E-mail: terrell@boulder.swri.edu
[2]*Department of Astronomy, University of Florida, Gainesville, FL, USA*

(accepted April 2004)

Abstract. Following a brief history of measurement of eclipsing binary mass ratios from light curves, we show that photometric mass ratios for overcontact and semi-detached binaries are reliable because the relative stellar radii, R/a, are accurately measured and not, as commonly claimed, because of information in the light variation outside eclipse. We explore the accuracy of photometric mass ratios by solving synthetic data of typical precisions for a semi-detached and an overcontact binary for orbital inclinations from $89°$ down into the partial eclipse range.

Keywords: mass ratio, binaries:eclipsing, binaries:overcontact, binaries:semi-detached

1. Introduction

Photometric mass ratios (q_{ptm}) have an interesting history, although for the most part we give only citations to that history for lack of space. Historical essentials of the overall eclipsing binary (EB) field and of q_{ptm}'s are in Wilson (1994), where the conceptual basics of q_{ptm} also are explained. Concepts preceded applications for semi-detached (SD) cases, with the main early SD references to q_{ptm} being Wood (1946), Russell (1948), and Kopal (1954, 1959), whereas applications came first for overcontact[1] (OC) binaries in papers by Mochnacki and Doughty (1972a,b), Wilson and Devinney (1973), and Lucy (1973). Some recent OC applications are Yang and Liu (2003), Xinag and Zhou (2003), and Hiller et al. (2004). Basically one cannot avoid finding an accurate q_{ptm} for an OC binary when q is adjusted by means of a correct physical EB light curve model, and one need not understand the situation to succeed. The earliest explanation known to us of why good mass ratios are found from (completely eclipsing) OC light curves, where a connected logical flow from light curve features to star radii to mass ratio is made, is in Wilson (1978). Although that connection may now seem obvious, occasional published comments entirely miss the point, with authors under the incorrect impression that q_{ptm} derives from ellipsoidal variation (i.e. tides).

Success is not quite so inevitable for SD's, where not only must the light curve model be physically correct but also the solution must be carried out under a constraint of accurate lobe filling, as in the WD model (Wilson and Devinney,

---

[1]viz. Wilson (2001) for the meaning of *overcontact*.

Astrophysics and Space Science **296:** 221–230, 2005.
© Springer 2005

1971; Wilson, 1979, 1990) since we learn the mass ratio specifically from the lobe size. Wilson and Wyithe (2002) cast these issues in terms of the number of measured radii that contribute to q_{ptm} – two for OC's, one for SD's, and none for detached binaries. The q_{ptm} concept easily works best in examples with complete eclipses, as those have by far the best determined relative radii (R/a), and Maceroni (1986) has demonstrated that point quite well for V523 Cas. Experience and intuition agree that the q_{ptm} concept essentially does not apply to detached binaries.

2. When and Why q_{ptm} Works

Accurate photometric mass ratios can be obtained when certain constraints are applied to the sizes of one or both stars, with the OC and SD cases differing in some important respects – it is not just the same story. Particularly straightforward is the case of OC binaries of the W UMa type with complete eclipses, where there is a steep relation between light curve *amplitude* and mass ratio (Wilson, 1978). The relation allows one to assess q *at a glance* from the star's light curve. It is all very simple in practice – equal masses $\longrightarrow$ big amplitude, unequal masses $\longrightarrow$ small amplitude, with classic examples of the unequal mass case such as AW UMa and FG Hya showing their extreme q's without any formal analysis. The relation is a consequence of two conditions, the hydrostatic condition that relates q to dimensions for OC's, and the stellar structure condition that equalizes surface brightnesses for W UMa components. Since the surface brightnesses of the two stars are nearly the same, eclipse depths are determined by the fractional area[2] covered in annular eclipse, which then gives the ratio of the radii in one algebraic step. With the ratio of the radii known, the Roche geometry (enforced by the constraint that the surface potentials be equal) strongly limits the range of possible q's. Of course eclipse depth translates into light curve amplitude, so we have an amplitude versus q relation. The q_{ptm} situation is nearly as favorable for an OC that has significantly unequal surface brightnesses, although the simple amplitude versus q relation is replaced by more intricate behavior requiring formal analysis.

Figure 1 shows the relationship between the ratio of the radii and the mass ratio for OC binaries. The steepness of the function illustrates the fact that if the ratio of the radii (and to a lesser extent, the degree of overcontact) is known, then the mass ratio is well determined. Actually we have further valuable radius information, as $r_{1,2} = R_{1,2}/a$ are measured individually and each one is closely tied to q. Their sum is nearly independent of mass ratio and depends almost entirely on level of overcontact for OC's, allowing overcontact to be measured from $r_1 + r_2$, which is to say from eclipse width. The accuracy of the mass ratio thus hinges on the circumstances under which the relative radii are best determined. For partial

---

[2]Fractional area means eclipsed fraction of the total projected surface area of the two stars.

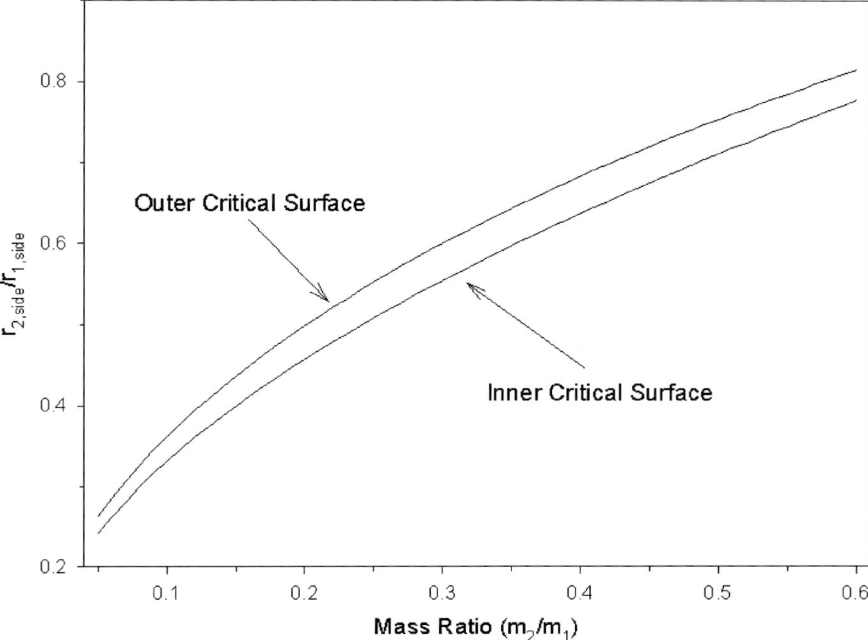

Figure 1. Ratio of radii versus mass ratio for overcontact binaries. The relationship is shown for the extremes of the degree of overcontact.

eclipses a change in eclipse depths due to a change in the ratio of the radii can nearly be compensated by a change in the inclination. However *complete* eclipse depths, acting together with the duration of totality, greatly restrict the ranges of possible relative radii and inclinations, thus breaking the degeneracy between the radii and inclination (viz. Wyithe and Wilson, 2001). The result is that we expect to find accurate q_{ptm}'s for completely eclipsing overcontact binaries.

We also have a size constraint for semi-detached binaries, namely that one star accurately fills its Roche lobe. Figure 2 shows that there is a unique value for the mass ratio if the relative radius of the lobe-filling star is known. Since complete eclipses lead to accurate radii, we expect to find accurate mass ratios for completely eclipsing SD systems. It is sometimes claimed that ellipsoidal variation allows SD q_{ptm}'s to be determined, but the fact that accurate q_{ptm}'s for Algols are found argues against that idea. There is very little ellipsoidal variation in Algols, because the low luminosity secondary is the distorted star, yet accurate values of q_{ptm} are found (e.g. Van Hamme and Wilson, 1993).

There is observational evidence that the nature of the eclipses determines the accuracy of q_{ptm}. Rucinski and colleagues (Pych et al., 2004 and references therein) have measured radial velocities for a number of OC binaries that have q_{ptm}'s from earlier papers. Figure 3 shows that for a sample of systems from the

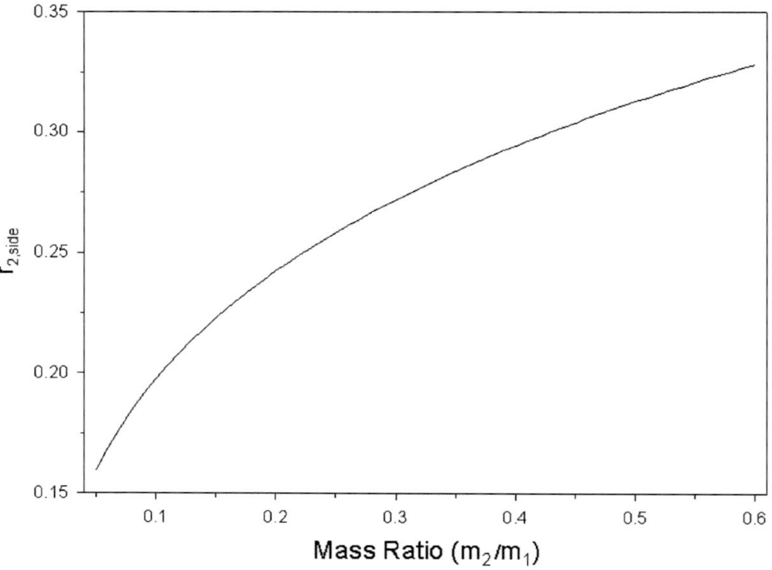

Figure 2. The relative side radius versus mass ratio for the less-massive, lobe-filling star in semi-detached binaries.

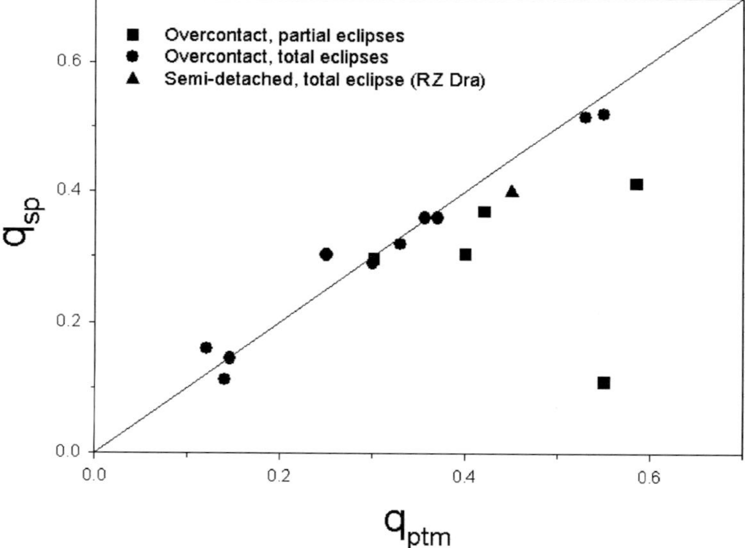

Figure 3. Spectroscopic versus photometric mass ratios for a sample of OC binaries. The line shows $q_{sp} = q_{ptm}$. The agreement for complete eclipses is good, while that for partial eclipses is significantly worse. One SD binary with complete eclipses, RZ Dra, is also shown to have good agreement between q_{sp} and q_{ptm}.

Rucinski papers with mass ratios measured by both methods, OC's with complete eclipses show good agreement between q_{ptm} and q_{sp}, while those with partial eclipses show much poorer agreement. Since the Rucinski velocities are from spectra of good resolution reduced via the broadening function approach, the q_{sp}'s are likely to be reliable. We conclude that the degradation in agreement between q_{sp} and q_{ptm} for partially eclipsing systems arises from loss of radius information.

3. Numerical Simulations

3.1. METHODOLOGY

To explore the effect of various parameters on the accuracy of derived photometric mass ratios, we synthesized and solved light curves of various precisions with the 2003 version of the WD program. We thereby can gauge the accuracy of the derived q_{ptm} because the correct value is known. Our procedure was as follows.

- Choose parameters to represent an SD and an OC binary.
- Use WD's light curve (LC) program to generate V-band light curves of 1000 points with Gaussian scatter over a range of orbital inclinations covering both complete and partial eclipses.
- Use WD's differential corrections (DC) program to solve each light curve, starting from 20 randomly selected places in parameter space. Initial values of the adjusted parameters were usually $\pm 10\%$ of their true values, although in some of the simulations larger ranges were used.

We performed simulations for the OC and SD binaries with the parameters given in Table I. We explored the effect of observational scatter by solving light curves of various precisions. To simulate typical data from surveys we used a 5% standard

TABLE I

Parameters of the OC and SD binaries used
in the numerical simulations

Parameter	OC	SD
q	0.6	0.2
Ω_1	3.0	6.0
Ω_2	3.0	2.233
T_1	5500 K	12,000 K
T_2	5350 K	5000 K
$L_1/(L_1 + L_2)$	0.644	0.916

deviation. We also analyzed light curves of 1% and 0.2% standard deviations that are more representative of targeted observing programs.

The DC solutions were automated and required no user input, thus eliminating bias. The adjusted parameters for both the OC and SD cases consisted of the mass ratio (q), the modified surface potential of star 1 (Ω_1), the mean effective temperature of star 2 (T_2), the orbital inclination (i), and the bandpass luminosity of star 1 (L_1). Three outcomes for the DC runs are possible:

- Convergence was achieved. We define convergence as corrections for all parameters being at least 10 times smaller than the standard errors.
- The maximum number of iterations (50) was reached and the iterations were wandering around a small volume of parameter space, but convergence by the foregoing definition was not achieved.
- The corrections diverged and resulted in unphysical parameter values.

We found that for strongly determined solutions (i.e., complete eclipses), the same (correct) minimum was found for all starting points, but weaker cases did not fully converge, and the q_{ptm}'s of the figures are those at the stopping point (50-iteration limit).

3.2. SIMULATIONS OF AN OVERCONTACT BINARY

Our experiments were designed to test whether eclipse character (complete versus partial) or ellipsoidal variation is the main consideration in explaining q_{ptm}. If the former, then the accuracy of q_{ptm} should quickly degrade as the eclipses go from complete to partial, while the latter predicts that the accuracy should degrade gradually as orbital inclination[3] decreases. In our first set of simulations we generated light curves for the OC binary at inclinations from $89°$ to $61°$ in $1°$ increments. In Figure 4 we plot the recovered mass ratio versus the true inclination for simulated light curves with 1% scatter for formally converged and "wandering" solutions. It shows very clearly that the accuracy of q_{ptm} drops dramatically as the eclipses go from complete to partial. Figure 5 shows a similar result for data with a 0.2% scatter.

We then did simulations of higher resolution by stepping the inclination in $0.1°$ increments in the region where the complete to partial transition takes place. This finer grid confirmed the critical nature of the eclipse circumstances in determining the mass ratio. Figure 6 shows the recovered mass ratio versus true inclination for simulated data with 1% scatter. A few incorrect solutions for q_{ptm} are seen on the "complete eclipse" side of the plot and all are cases, where the DC solution found a partial eclipse configuration. Visual inspection of these light curves reveals subtle but noticeable departures from good fits in the bottoms of eclipses.

---

[3] Actually it is as $\sin i$ decreases, since $i > 90°$ can be meaningful if there are polarimetric data.

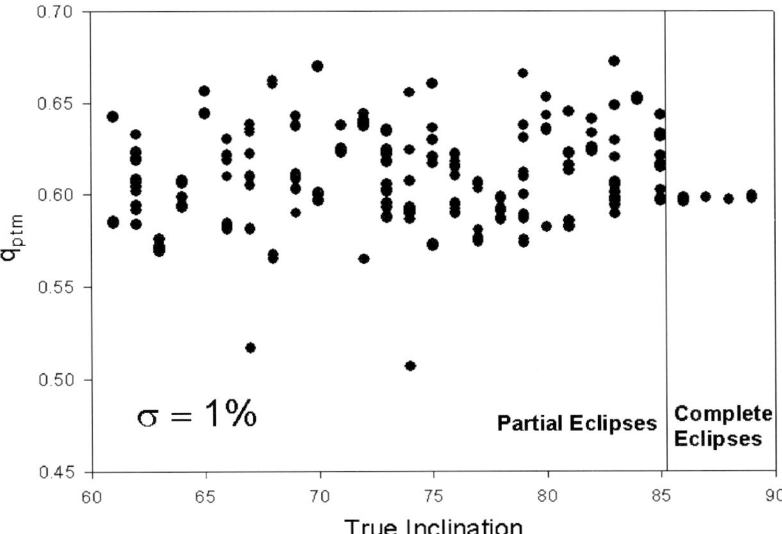

Figure 4. Value of q_{ptm} versus true inclination in $1°$ increments from solutions to simulated data of an OC binary with a Gaussian scatter of 1%. The true value of q_{ptm} is 0.6 and that value is reliably recovered, when the eclipses are complete. Even when the eclipses are only very slightly partial, the accuracy of q_{ptm} drops dramatically.

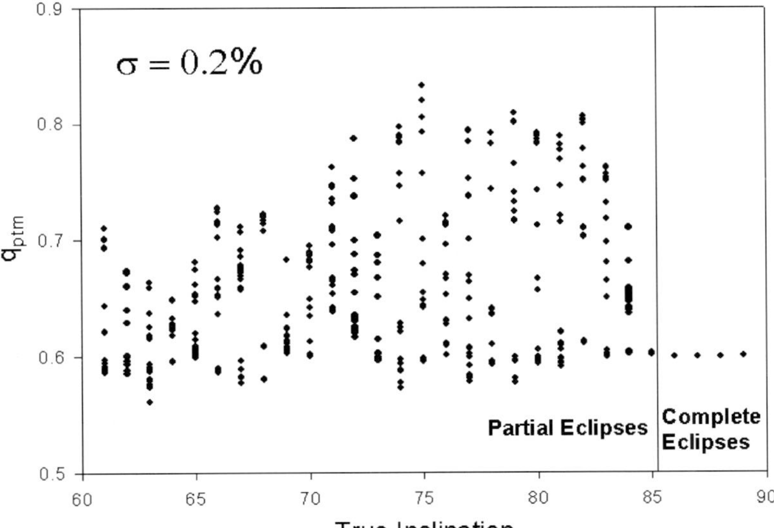

Figure 5. Value of q_{ptm} versus true inclination in $1°$ increments from solutions to simulated data of an OC binary with a Gaussian scatter of 0.2%.

    D. TERRELL AND R.E. WILSON

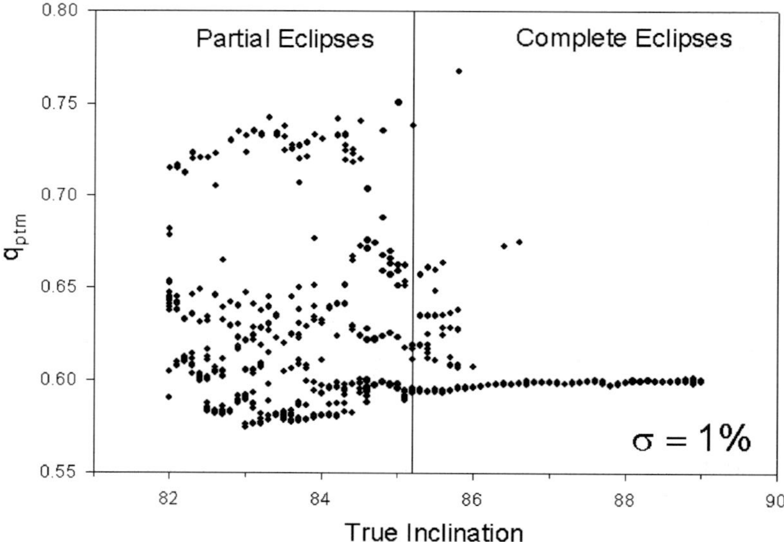

Figure 6. Value of q_{ptm} versus true inclination in $0.1°$ increments from solutions to simulated data of an OC binary with a Gaussian scatter of 1%. The true value of q_{ptm} is 0.6.

3.3. SIMULATIONS OF A SEMI-DETACHED BINARY

We also did simulations of an SD binary to see if reliable values of q_{ptm} could be measured. The approach was the same as for the OC case except that we allowed the starting mass ratio in the DC solutions to range from 0.2 to 0.6 and i extended only down to $74°$, corresponding to the reduced i range of SD eclipses. Figure 7 shows that the accuracy of q_{ptm} is very good for SD systems with complete eclipses and noticeably poorer for those with partial eclipses. Because of the larger range of initial mass ratio values, the number of simulations that resulted in divergence was much larger than for the OC cases, resulting in fewer points in Figure 7 as compared to Figure 4.

4. Conclusions and Future Work

Our simulations show unambiguously that eclipse circumstance (complete versus partial) rather than ellipsoidal variation governs photometric mass ratios for over-contact and semi-detached binaries. Complete eclipses lead to accurate radii and these, coupled with size constraints based on the Roche geometry, can lead to accurate mass ratios. Future work will involve simulations on additional OC and SD systems, exploration of the effect of third light, and quantitative assessment of the reliability of mass ratio error estimates from light curve solutions.

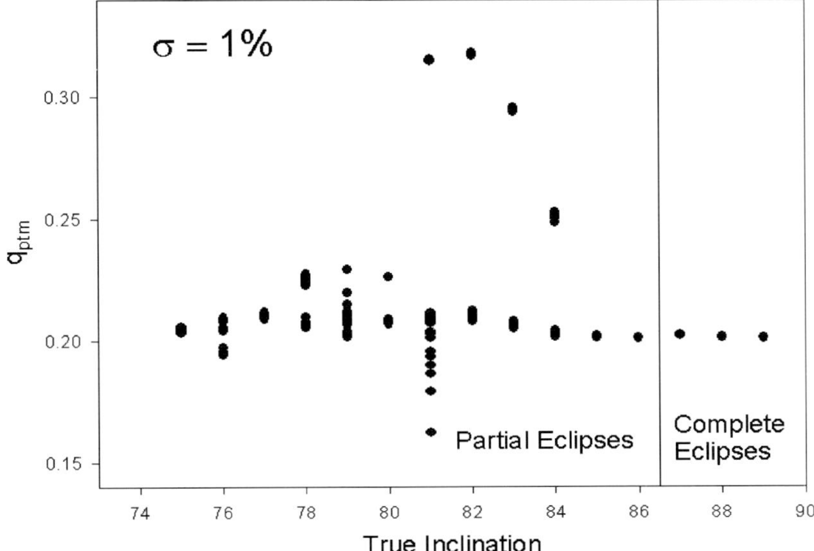

Figure 7. Value of q_{ptm} versus true inclination in 1° increments from solutions to simulated data of an SD binary with a Gaussian scatter of 1%. The true value of q_{ptm} is 0.2.

Acknowledgements

This material is based upon work supported by the National Science Foundation under Grant No. 0307561. DT thanks Alan Stern for financial support to attend the meeting.

References

Hiller, M.E., Osborn, W. and Terrell, D.: 2004, *PASP* **116**, 337.
Kopal, Z.: 1954, *Jodrell Bank Ann.* **1**, 37.
Kopal, Z.: 1959, *Close Binary Systems*, John Wiley Publishing Company, pp. 490–496.
Lucy, L.B.: 1973, *Ap&SS* **22**, 381.
Maceroni, C.: 1986, *A&A* **170**, 43.
Mochnacki, S.W. and Doughty, N.A.: 1972a, *MNRAS* **156**, 51.
Mochnacki, S.W. and Doughty, N.A.: 1972b, *MNRAS* **156**, 243.
Pych, W., Rucinski, S.M., DeBond, H., Thomson, J.R., Capobianco, C.C., Blake, R.M., Ogloza, W., Stachowski, G., Rogoziecki, P., Ligeza, P. and Gazeas, K.: 2004, *AJ* **127**, 1712.
Russell, H.N.: 1948, *Harvard Obs. Monograph No. 7*, p. 181.
Van Hamme, W. and Wilson, R.E.: 1993, *MNRAS* **262**, 220.
Wilson, R.E.: 1978, *ApJ* **224**, 885.
Wilson, R.E.: 1979, *ApJ* **234**, 1054.
Wilson, R.E.: 1990, *ApJ* **356**, 613.
Wilson, R.E.: 1994, *PASP* **106**, 921.

Wilson, R.E.: 2001, *Inf. Bull. Var. Stars* **5076**.
Wilson, R.E. and Devinney, E.J.: 1971, *ApJ* **166**, 605.
Wilson, R.E. and Devinney, E.J.: 1973, *ApJ* **182**, 539.
Wilson, R.E. and Wyithe, S.B.: 2002, *ASP Conf. Ser.* **298**, 313.
Wood, F.B.: 1946, *Princeton Obs. Contr.* No. 21.
Wyithe, J.S.B. and Wilson, R.E.: 2001, *ApJ* **559**, 260.
Yang, Y. and Liu, Q.: 2003, *AJ* **126**, 1960.
Xiang, F.Y. and Zhou, Y.C.: 2003, *New Astron* **9**, 273.

THE OGLE-TR-56 STAR–PLANET SYSTEM

TODD VACCARO[1] and WALTER VAN HAMME[2]

[1]*Department of Physics & Astronomy, Louisiana State University, 202 Nicholson Hall, Baton Rouge, LO, USA; E-mail: vaccaro@rouge.phys.lsu.edu*
[2]*Florida International University Miami, FL, USA*

(accepted April 2004)

Abstract. We simultaneously fitted light and velocity data for the star–planet system OGLE-TR-56 with the Wilson–Devinney (WD) binary star program. We solved for orbital and planet parameters, along with the ephemeris using all currently available observational data. Parameters for the star (OGLE-TR-56a) were kept fixed at values derived from spectral characteristics and stellar evolutionary tracks. Our results are in good agreement with parameters obtained by other authors and have slightly smaller errors. We found no significant change in orbital period that may be due to orbital decay.

Keywords: eclipsing binaries, extrasolar planets, OGLE-TR-56

1. Introduction

The Optical Gravitational Lensing Experiment (OGLE) has provided a wealth of light curve data. The 2001–2002 campaign (OGLE III) has revealed 137 stars with low-amplitude transits that may be caused by low-mass companions, several of which possibly being extrasolar planets (Udalski et al., 2003). Konacki et al. (2003) found OGLE-TR-56 to be a solar type star ($T = 5900\,\mathrm{K}$ and solar metal abundance) with radial velocities indicative of a planetary companion. Their original announcement was based on only three velocities that appeared to agree phase-wise with the light data. Follow-up high-resolution spectra (Torres et al., 2004) yielded an additional eight velocities and confirmed prior results. The OGLE database now contains over 1100 photometric observations (http://bulge.astro.princeton.edu/~ogle/ogle3/transits/ogle56.html) of OGLE-TR-56, including 13 transits, and 11 radial velocities.

Extrasolar planets that transit their stars provide tight constraints on the orbital inclination and, when coupled with radial velocity data, allow the determination of planetary masses and sizes. Parameters of the parent star still need to be determined independently. The temperature can be estimated from the star's spectral characteristics, while the mass and size can be derived from stellar evolutionary models (Cody and Sasselov, 2002; Sasselov, 2003). We demonstrate the use of a general binary star program to model light and velocities of stars with transiting extrasolar planets. The model is that of Wilson (1979) including the latest updates (Wilson and Van Hamme, 2003). We simultaneously solve the available light and velocity data of OGLE-TR-56 and compare our results with other solutions in the literature.

Astrophysics and Space Science **296**: 231–234, 2005.
© Springer 2005

2. Analysis and Conclusions

A star–planet (or other low-luminosity object) system, with transits and radial velocities for the star only, is analogous to a single-lined spectroscopic and eclipsing detached binary. In the WD program, adjustable radial velocity related parameters

TABLE I

OGLE-TR-56 parameters

Parameter	This paper	Torres et al. (2004)
T_0 (HJD)	$2452444.73945 \pm 0.00097$	2452075.1046 ± 0.0017
P_0 (days)	1.2119198 ± 0.0000041	1.2119189 ± 0.0000059
a (AU)	0.0225 ± 0.0028	0.0225 ± 0.0004
V_γ (km s^{-1})	-48.317 ± 0.026	-48.317 ± 0.045
$i(°)$	80.94 ± 0.40	81.0 ± 2.2
g_1, g_2	$0.32, 0.32$	
F_1, F_2	$0.06, 1.0$	
T_1 (K)	5900	
T_2 (K)	1900	
A_1, A_2	$0.5, 0.5$	
Ω_1	4.419 ± 0.083	
Ω_2	1.5492 ± 0.0012	
$q = M_2/M_1$	0.0013289	0.00132
x_1, y_1 (I)	$0.596, 0.256^{a}$	0.56 ± 0.06^{b}
$L_1/(L_1 + L_2)$ (I)	0.99998 ± 0.00016	
Derived parameters		
$\langle r_1 \rangle$	0.2264 ± 0.0043	
$\langle r_1 \rangle / \langle r_1 \rangle_{\mathrm{lobe}}$	0.3450 ± 0.0054	
$\langle r_2 \rangle$	0.0268 ± 0.0007	
$\langle r_2 \rangle / \langle r_2 \rangle_{\mathrm{lobe}}$	0.506 ± 0.013	
$M_1(M_\odot)$	1.04 ± 0.39	1.04 ± 0.05
$R_1(R_\odot)$	1.10 ± 0.14	1.10 ± 0.10
$M_{\mathrm{bol}\,1}$	4.45 ± 0.28	
$\log(L_1/L_\odot)$	0.12 ± 0.11	
$\log g_1$ (cm s^{-2})	4.37 ± 0.28	
$M_2(M_{\mathrm{JUP}})$	1.45 ± 0.55	1.45 ± 0.23
$R_2(R_{\mathrm{JUP}})$	1.26 ± 0.16	1.23 ± 0.16
$\log g_2$ (cm s^{-2})	3.35 ± 0.28	
Density (g cm^{-3})	0.89 ± 0.48	1.0 ± 0.3

[a]Logarithmic limb darkening law.
[b]Linear limb darkening law.

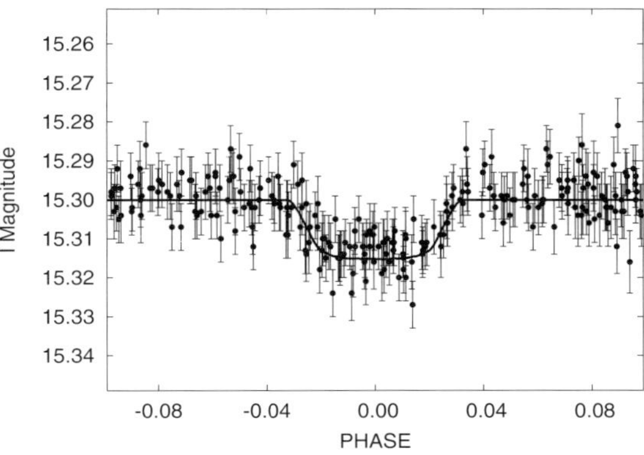

Figure 1. Phased *I* band observations of OGLE-TR-56 and computed curve.

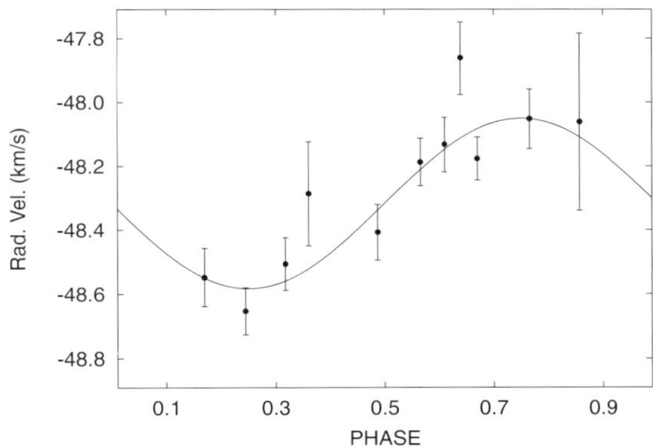

Figure 2. OGLE-TR-56 radial velocities (Torres et al., 2004) and solution curve.

are the semi-major axis (a), systemic velocity (V_γ) and mass ratio (q). In a single-lined binary, q cannot be determined from the velocity curve. A value for the mass ratio needs to be adopted, and the mass of the star can then be computed from Kepler's Third Law after a and P have been obtained from the fitting process. We selected q ($q = M_{\mathrm{planet}}/M_{\mathrm{star}}$) so that Kepler's Third Law yielded a stellar mass consistent with that derived from evolutionary models (1.04 $M_\odot$, see Sasselov, 2003). Time instead of phase was the independent variable, and ephemeris parameters (reference epoch T_0, period P, and possibly dP/dt) were fitted together with the other parameters. We adopted a logarithmic limb darkening law with coefficients x, y from Van Hamme (1993). The star's rotation rate (F_1) was set to 0.06 corresponding to a rotational period of 20 days (Sasselov, 2003). Data were assigned individual weights inversely proportional to the square of their listed mean errors. Table I lists solution parameters, including standard errors of parameters that were adjusted. Parameters obtained by Torres et al. (2004) are listed for comparison. The agreement is very good. Figure 1 shows the observed light curve near the transit together with our solution curve. Radial velocities are shown in Figure 2. We checked for tidally induced orbit decay as discussed by Sasselov (2003). Including dP/dt as an adjustable parameter did not indicate a value significantly different from zero.

The WD program lends itself very well to modeling star–planet transit curves and radial velocities. We also note that the stellar radius, found by fitting the light curve data, agrees very well with the value found by other authors who fit evolutionary tracks. The program's ability to fit ephemeris parameters, including dP/dt, will provide a useful tool for checking orbital decay of extrasolar planet systems.

Acknowledgements

This work was supported in part by NSF grant AST-0097895 to Louisiana State University.

References

Cody, A.M. and Sasselov, D.: 2002, *ApJ* **569**, 451.

Konacki, M., Torres, G., Jha, S., and Sasselov, D.: 2003, *Nature* **421**, 507.

Sasselov, D.: 2003, *ApJ* **596**, 1327.

Torres, G., Konacki, M., Sasselov, D. and Jha, S.: 2004, *ApJ*, preprint, doi:10.1086/421286.

Udalski, A., Pietrzynski, G., Szymanski, M., Kubiak, M., Zebrun, K., Soszynski, I., Szewczyk, O. and Wyrzykowski, L.: 2003, *Acta Astron.* **53**, 133.

Van Hamme, W.: 1993, *AJ* **106**, 2096.

Wilson, R.E.: 1979, *ApJ* **234**, 1054.

Wilson, R.E. and Van Hamme, W.: 2003, Computing Binary Star Observables, Monograph available at FTP site *ftp.astro.ufl.edu*, directory *pub/wilson/lcdc2003*.

LIGHT CURVE ANALYSIS OF EARLY-TYPE CLOSE LMC BINARIES

STEFAN NESSLINGER

Dr. Remeis Observatory Bamberg, Astronomical Institute of the University Erlangen-Nürnberg, Germany; E-mail: ai38@sternwarte.uni-erlangen.de

(accepted April 2004)

Abstract. Results are presented of an analysis of eclipsing binaries in the Large Magellanic Cloud. The sample of close OB-type stars was taken from the MACHO microlensing survey. The present study was restricted to systems with orbital periods shorter than 2 days and V and R light curves with large eclipse amplitudes, high S/N and homogeneous and dense phase coverage. Problems encountered during the analysis are discussed, especially with respect to the degeneracy of photometric mass ratios and other parameter correlations.

Keywords: binaries, light curves, LMC, OB-type stars

1. Introduction

A vast amount of eclipsing binary light curves has become available as a by-product of the OGLE, EROS and MACHO microlensing surveys conducted during recent years. While EROS discovered 79 new eclipsing binaries in the LMC (Grison et al., 1995), MACHO produced more than 600 light curves of EBs in the LMC (Alcock et al., 1997), whereas OGLE yielded about 4000 EB light curves both in the LMC and SMC (Udalski et al., 1998; Wyrzykowski et al., 2003).

2. Data Selection

This work is based on MACHO data, which mostly provide a good S/N ratio and excellent phase coverage: typically, MACHO light curves comprise between 500 and 1500 measurements per V and R filter. The newly detected EB's naturally comprise numerous luminous early-type systems, which deserve special attention due to their generally small abundance.

Among more than 600 light curves listed by Alcock et al. (1995), the most suitable ones were selected for closer analysis. Since emphasis was put on close early-type binaries, colour index and periods were restricted to $V - R \leq -0.10$ and $P \leq 2$d. Further selection criteria were small observational scatter (high S/N), deep eclipse minima and accurately determined orbital periods.

The selection process resulted in 16 eclipsing binaries remaining for analysis.

Astrophysics and Space Science **296**: 235–238, 2005.
© Springer 2005

3. Data Preparation

Besides V and R light curves and orbital periods, no further information on the program stars was initially available. The determination of some additional system parameters was necessary for the numerical solution.

First the effective temperature of the primary components had to be estimated. The MACHO catalogue provides a $V - R$ colour index; however, a reliable temperature calibration of hot stars is not possible based on this index alone. So a comparison was made with EROS objects, which are simultaneously contained in the MACHO sample. As EROS colours are given in $B - V$ by Grison et al. (1995), an empirical relation between $V - R$ and $B - V$ could be derived. The corresponding temperatures were then taken from Schaifer et al. (1995).

Additionally, interstellar extinction had to be taken into account. For this purpose an extinction map for the LMC region published by Schwering and Israel (1991) was used, which has a spacial resolution of 48′.

As the primary minimum epochs given by Alcock et al. (1995) are not always sufficiently precise, they were directly derived from the light curves by fitting a parabolic function to the inner parts of the minima. Where, e.g., flat minima required a different approach, the method described by Kwee and van Woerden (1956) as implemented in a programme by Harmanec (2003) was applied.

In accordance with Lucy (1969) and Ruciński (1976), albedos and gravity darkening coefficients were set to 1, while the limb darkening coefficients were individually taken from Wade and Ruciński (1985) for each system.

4. Data Analysis

For the light curve analysis the software package MORO was used, which is based on the Wilson–Devinney-code (Wilson and Devinney, 1971), and has been subsequently developed at Remeis Observatory Bamberg (Drechsel et al., 1995). It employs the Simplex algorithm as parameter optimization process, ensuring fast and reliable convergence of the solutions. Seventeen parameters are adjustable, five of which depend on wavelength.

Adjustable parameters in this study were inclination (i), effective temperature of the secondary component ($T_{\text{eff},2}$), surface potentials (Ω_1, Ω_2), mass ratio (q), and luminosity of the primary component (L_1). Radiation pressure δ parameters and third light were only adjusted if required, i.e. if their inclusion led to definitely improved solutions.

5. Problems and Results

As had been anticipated, determination of the mass ratio q from photometric data alone turned out to be problematic. For most of the analyzed stars, solutions of

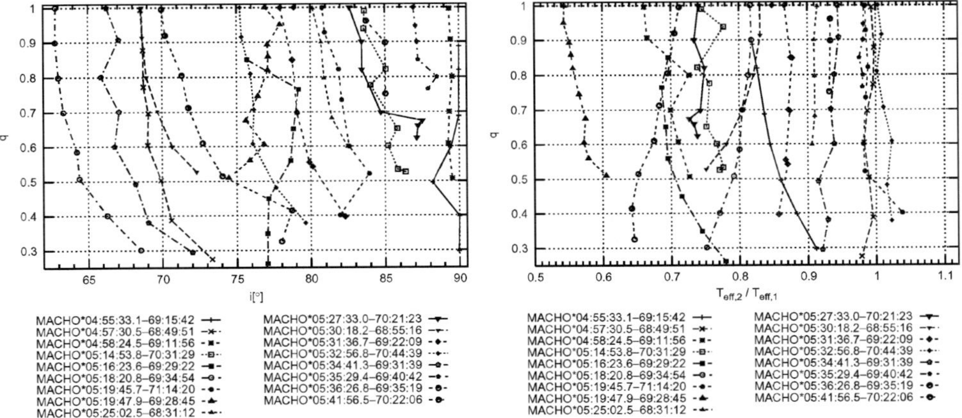

Figure 1. Correlation of mass ratio with inclination (left) and temperature ratio (right).

nearly identical numerical quality could be produced in a broad range of the mass ratio.

This is due to correlations of q with other parameters, most notably the luminosity ratios and the surface potentials of the stars. Therefore, as pointed out by Wyithe and Wilson (2002), the problem is less severe in contact and semi-contact configurations, because one or both surface potentials are fixed by the inner Lagrangean point.

Effective temperatures of the secondary components and inclinations were quite well defined. Only slight deviations of these parameters occurred with variation of q. For a fixed mass ratio, all parameters could be derived consistently (see Figure 1).

There are several ways to confine the mass ratio. The most precise one is employing spectroscopic data and deriving radial velocities. If not available, other approaches must be considered. For unevolved systems one can imply the mass-luminosity relation. Schönberner and Harmanec (1995) provide an empirical correlation between mass ratio $q = M_2/M_1$ and temperature ratio $T_{\mathrm{eff},2}/T_{\mathrm{eff},1}$ derived from carefully analyzed hot binaries. It could be used to dissolve the degeneracy of solutions with respect to q (Figure 1; right), if the validity of such relation is justified. However, deviations have to be envisaged, especially in the case of interacting or evolved binaries.

Considering only the numerically best solutions mostly resulted in getting closer to the empirical relation, however – as anticipated – not in all cases. Attaining spectroscopic mass ratios therefore seems inevitable to derive reliable and consistent system parameters.

References

Alcock, C., et al.: 1997, *AJ* **114**, 326.
Drechsel, H., Haas, H., Lorenz, R. and Gayler, S.: 1995, *A&A* **294**, 723.

Grison, P., et al.: 1995, *A&A Supl. Ser.* **109**, 447.

Harmanec, P.: 2003, Private communication.

Kwee, K.K. and van Woerden, H.: 1956, *Bull. Astron. Inst. Netherlands* **12**, 327.

Lucy, L.B.: 1976, *AJ* **205**, 208.

Ruciński, S.M.: 1969, *Acta Astronomica* **19**, 245.

Schaifers, K., et al.: 1982, In: K. Schaifers, H.H. Voigt, H. Landolt, R. Börnstein, and K.H. Hellwege (eds.), Astronomy and Astrophysics. B: Stars and Star Clusters, Landolt-Börnstein: Numerical Data and Functional Relationships in Science and Technology. New Series, Springer, Berlin.

Schönberner, D. and Harmanec, P.: 1995, *A&A* **294**, 509.

Schwering, P.B.W. and Israel, F.P.: 1991, *A&A* **246**, 231.

Udalski, A., et al.: 1998, *Acta Astronomica* **48**, 563.

Wade, R.A. and Ruciński, S.M.: 1985, *A&ASS* **60**, 471.

Wilson, R.E. and Devinney, E.J.: 1971, *AJ* **166**, 605.

Wyithe, J.S.B. and Wilson, R.E.: 2002, *AJ* **571**, 293.

Wyrzykowski, L., et al.: 2003, *Acta Astronomica* **53**, 1.

NEW TECHNIQUES AND LIMITATIONS OF LIGHT CURVE ANALYSIS

PETR HADRAVA

Astronomical Institute, Academy of Sciences, CZ 251 65 Ondřejov, Czech Republic;
E-mail: had@sunstel.asu.cas.cz

(accepted April 2004)

Abstract. The solution of light curves of eclipsing binaries provides the simplest way of measuring the basic physical parameters of stars. The development of computers and numerical methods enables us to perform an analysis of large amount of photometric data by comparing it to more and more sophisticated physical models which are also suitable for stellar systems with different peculiarities. Nevertheless, some parameters of stellar systems are poorly determined by light curves alone, and simultaneous fitting together with other data is preferable. The progress of instrumentation not only improves the precision and resolution of data, but also overcomes the classical gaps between stellar systems observable as visual, photometric and spectroscopic binaries. This opens new problems in the development of techniques of data analysis, which can be treated as a generalization of light curve modeling. Some of these directions are outlined here. In particular, the disentangling of spectra with line-profile variability due to proximity effects is discussed.

Keywords: close binaries, light curves, spectroscopy

1. Introduction – History and Aim of Light Curve Analysis

According to Proclus Diadochus (411–485 AD), Euclid of Alexandria (325–265 BC) rejected the demand of his king Ptolemy to introduce him by a shortcut to his art saying that there is no royal road to geometry. Following Henry Norris Russell (1877–1957) and Zdeněk Kopal (1914–1993) there are important exceptions to this Euclidian rule; at least the observation of eclipsing binaries is an advantageous shortcut not only to the geometry, but also to the physics of stars, which is a good starting point to investigating the many other secrets of Nature. (However, to advocate Ptolemy, let us note that this way would not be easy to follow for an ordinary king.)

The interest in investigating fixed stars became reasonable at an early modern time on the basis of the achievements by scientists in the 16th and 17th centuries like N. Copernicus, T. Digges, T. Brahe, T. Hájek, J. Kepler, G. Bruno, G. Galilei, R. Descartes, and others. In particular, the discovery of the variability of stars by D. Fabricius for Mira in 1596 and by G. Montanari for Algol in 1667 was important for our present topic. However, 'the immense distance of stars' did not allow to estimate the scales of their sizes and masses at that time.

The discovery of periodicity of Algol's fadings and the ingenious explanation as a result of eclipses by an orbiting companion made by J. Goodricke in 1782 can be

Astrophysics and Space Science **296**: 239–249, 2005.
© Springer 2005

treated as the beginning of light curve analysis and, at the same time, the first step to the determining of basic physical parameters of stars, especially their relative dimensions (as it was done by E.C. Pickering for Algol in 1880 only).

It is important to note the insufficiency of the light curve analysis method to yield absolute values of stellar radii unless the size of the orbit is known. The latter can be determined spectroscopically for close binaries, for which eclipses are also most probable and radial velocity changes most pronounced. Historically this method dates back to Ch. Doppler's effect (1842), its application to shifts of spectral lines by H. Fizeau (1848), the first measurement of radial velocity of Sirius by W. Huggins (1868) and RV variations of (again!) Algol by H. Vogel (1888).

In the meantime, the efforts made by W. Herschel to measure the annual parallax of stars (started in 1774) led to the discovery of visual binaries in 1802. Together with the parallax finally found independently by F.W. Bessel, T. Henderson and F.G.W. Struve in 1838, the dimensions of orbits of (wide, i.e. long-period) binaries and consequently also their masses could be determined. It is thus an independent and a parallel way of obtaining at least some of the basic stellar parameters applicable, however, to a set of stellar systems, which was practically disjunct with the set of spectroscopic and eclipsing binaries until recently.

Nowadays, the steadily increasing sensitivity of observational techniques, especially interferometry, enables us to use various alternative methods for the same objects and thus to test their mutual consistency and reliability. Moreover, once the nature and physics of stars became understood, additional exact as well as statistical methods of determining the basic stellar parameters were opened. Nonetheless, the classical methods are still dominant and needed to tune the details of our understanding of stars or to reveal discrepancies in some objects deserving special attention.

Generally it is taken that the binarity of stars yields an admirable opportunity to get more knowledge about the component stars, which can be then applied to other stars as well. However, the mutual interaction of binary components can also produce many peculiarities in their behaviour. The theoretical study of this problem has to agree with observational results. This direction also has a long-lasting tradition in which remarkable results were obtained, e.g. by E.A. Roche (model of tidal interaction in 1849–50) or about one century later by Z. Kopal.

In the present contribution we shall not deal with the details of the most advanced techniques of light curve analysis presently used;[1] instead, let us try to extrapolate the historical development of the subject and to generalize the present experience for suggesting the directions to be followed in near future.

---

[1] For this subject we can refer the reader to, e.g. the review by Kallrath and Milone (1999) or to the literature and documentation on the probably most commonly used Wilson–Devinney code (cf. Wilson and Devinney, 1971; Wilson, 1979, 1990; Kallrath et al., 1998).

Figure 1. The allegory of astronomy in context of other sciences.

2. General Links Between Theory and Observations

The procedure by which light curve analysis contributes to the advance of science consists of several steps elaborating the source information. This chain of procedures can be represented by the allegory of sciences (cf. Figure 1) on the ceiling fresco, painted after 1740, in the New Mathematical Hall of the Jesuit college Clementinum (C) in Prague. The stars[2] (S) in space emit light, which is detected by telescopes (T) or some other instruments. This is the only one-way part of the flow of information in the whole procedure, because unlike the situation in experimental sciences $(E_{1,2,3})$ we cannot influence the objects of our study in stellar astrophysics, which is an observational science by nature. The rough data observed by the observer (O) have to be processed and fitted by an interpreter (I) to a theoretical model (M) using its free parameters. This model must be constructed by a theoretical astrophysicist (A) on the basis of theoretical physics (P) and mathematics, which is a common background shared with other branches of physics $(E_{1,2,3})$, technical sciences (involved already in the instrument design D)

---

[2]Note that they are clearly drawn with extrasolar planetary systems around and shared cometary objects; this follows the tradition of the then prohibited Copernicus' heliocentric opinion generalized by Giordano Bruno and René Descartes.

and applied sciences, which are together with arts depicted in the left part of the
fresco skipped in our reproduction, but which obviously give meaning to the whole
above described procedure.

The purpose why these well known facts are repeated here again is to point
out that (i) the strength of the whole chain is limited by its weakest link, and (ii)
that the light curve analysis as a part of the complex cognitive process should
reflect both demands and offers of related fields. The first item implies that it is
of little use to improve significantly the computational technique of the solution
if the model itself is physically oversimplified. Concerning the second item, we
have already mentioned the complementarity of the information yielded by light
curves and by other methods. It can thus be foreseen that the future development
of light curve analysis should lead to its merging with the other methods. On
the other hand, applications to new observations in deep space (enabled by the
advance of instrumental techniques) promise a revival of interest in this already
classical method, which can contribute seriously to scaling the ladder of cosmic
distances.

3. Uncertainties of Physical Assumptions

Practically all present codes for finding solutions of light curves of eclipsing binaries
use models based on the Roche model. It means that the shape of the component stars
is approximated by an equipotential surface of the effective potential consisting of
two potentials of point-like massive centers of the stars and the centrifugal potential
due to the orbital motion (to which the rotation of components should be bound).
We know, however, that even in the framework of Newtonian theory the gravity
potential of a non-spherical body is more complex due to higher order multipoles of
density distribution. It can be argued that the high increase of mass density towards
the centre of most stars justifies to neglect the higher multipoles. This is mostly
acceptable in comparison with the following much more crude simplifications. On
the other hand, the higher multipoles influence the dynamics of binaries and they
are observable by the secular changes of orbital parameters, and therefore they
deserve attention as well.

A more serious problem of applicability of the Roche model to real binaries
consists in its basic assumption that the stars must be in hydrostatic equilibrium, in
which case they must be completely homogeneous on equipotentials (as it follows
from the equation of hydrostatic equilibrium). It is obvious that the component
stars cannot be in hydrostatic equilibrium, if they have a non-synchronous rotation or if their orbital motion is eccentric, which are both very frequent cases.
The Roche potential can then be modified to describe the particle motion in the
frame corotating with the star (Plavec, 1958). However, it cannot save the hydrostatic equilibrium and thus justify the commonly used approximation of the star's
shape by the instantaneous equipotential surface. An additional problem of the

proper choice of the potential value on the surface arises in this approximation, because even if the star would follow a set of instantaneous hydrostatic equilibria, its structure and consequently also its volume will change with time (cf. Hadrava, 2004).

The light curves of eclipsing binaries are determined not only by the geometry of their tidally distorted components, but also by the distribution of intensity on their surfaces, which is influenced by the standard limb-darkening being present also in spherically symmetric stars. Its theoretical values predicted by model stellar atmospheres can be, in principle, tested by observations of eclipsing binaries. However, two important proximity effects can significantly change the simple picture based on plane-parallel or spherical atmospheres. These are the reflection effect and gravity darkening. The latter effect should be taken into account for single rotating stars as well. It is usually estimated using the von Zeipel theorem, according to which the local radiative flux at the surface of a non-spherical star should be (following the diffusion approximation of radiative transfer) proportional to the sub-photospheric temperature gradient, i.e. (assuming again homogeneity at the equipotentials) proportional to the local gravitational acceleration. The reprocessing of the radiative flux by which the companion star illuminates the atmosphere from outside can be estimated using the same degree of approximation, e.g. by adding the incident flux to the proper local (gravitationally darkened) flux of the star and calculating the radiated surface intensity as that of the model atmosphere with corresponding effective temperature and gravity acceleration. However, the outcome of such a procedure is in contradiction with its underlying assumption of homogeneity on equipotentials, and thus also with the hydrostatic equilibrium of the subsurface layers of the nonspherical star. Exactly speaking, the diffusion approximation itself is inconsistent with radiative equilibrium in nonspherical stars because of an unequal dilution of vertical lines across the stellar surface.

To get a physically self-consistent model of a tidally distorted (or illuminated) stellar atmosphere one should solve the 3-dimensional radiative transfer equation or actually the problem of radiative hydrodynamics. Such a model (involving also an analogue of meridional circulations induced by the asymmetry) will generally deflect from the Roche geometry. In fact, it should also replace the approximation of a sharp stellar surface by a model of a semitransparent atmosphere in which the fading of the eclipsed companion is smooth and frequency dependent. The common use of precise Roche geometry is thus the consequence of the fact that it is known in analytical form, however, its higher order terms do not need to correspond to real binaries. This is even more pronounced for interacting binaries in which the circumstellar matter changes the geometry of stellar disks completely (cf., e.g., the results of 3D-hydrodynamic models of circumstellar matter by Bisikalo et al., 1998).

Additional effects which may significantly influence the light curves of eclipsing binaries (as well as of single or non-eclipsed stars) are free or forced pulsations or surface spots. Different methods of mapping such phenomena have been developed. Quite often, these effects can mimic each other, so that the observational data can be interpreted by different models with comparable success. The availability of different kinds of parallel observations is extremely welcome in such cases.

4. Errors of Parameter Values

The problem of uniqueness and thus also of the correctness of a solution does not only concern the involved effects, but also the numerical values of the free parameters of a particular model. Because of observational errors and simplifications of the model, any fit to the data by a model has some residual $O - C$, which is minimized (usually by a least squares technique) to get the best solution. However, the problem usually has many local minima,[3] between which it is not easy to find the absolute minimum. Moreover, the observational errors can mimic a shift of the minimum to slightly (or completely) different values of the parameters. The data thus only define the probability distribution of each combination of free parameters. The light curve solution or the result of an analysis of other type of data should thus be given not only by the parameter values of the best fit, but also by (at least) the corresponding covariance matrix, i.e. the errors and correlations of the parameters.

Let us demonstrate it by an example of simulated data of a light curve and radial-velocity (RV) curve of a binary with an eccentric orbit to which random noise (of the typical order of 0.01 mag and 1 km s^{-1}, respectively) is added – cf. Table I. The solutions of both the light curve and the RV curve have small errors of the time of periastron passage (T), eccentricity (e) as well as longitude of periastron (ω), if the other parameters are fixed. However, while e is well defined by the shape of the RV curve, the changes of e and ω influence mainly the distance between the primary and the secondary minima in the light curve and they can well compensate each other. There is thus a strong correlation between these parameters in the light curve solution (the correlation between T and ω is strong for both solutions owing to the small value of e). Only the light curve can yield information about the inclination or component radii. It is thus advantageous to solve both curves simultaneously to eliminate the shortcomings of individual types of data. (Judging the compatibility of solutions of different data, it is insufficient to simply check for an overlap of error boxes of each parameter independently, but the correlations of parameters should also be taken into account).

---

[3]The number of local minima depends on precision and phase distribution of the input data, but also on the number of free parameters. According to the author's experience with his code FOTEL (cf. Hadrava, 1991–2003), too many free parameters of second-order importance (allowed by some users in an effort to maximize the gain from their data) makes the model much 'softer' and capable to accommodate at some shallow local minimum.

TABLE I

Resulting values of parameters, their errors and correlations for solution of a simulated light curve (with minima depths of about 0.8 and 0.2 mag), radial-velocity curve (with K velocities of 20 and 40 km s^{-1}) and their simultaneous solution, which gives mean values of correlations

Element	Correct value	Light curve	RV- curve	phot. + RV
T	0.0	0.00099	$-.00125$	$-.00099$
		$\pm.00056$	$\pm.00130$	$\pm.00081$
e	0.2	0.20109	0.19875	0.19858
		$\pm.00148$	$\pm.00163$	$\pm.00143$
ω	1.0 rad	1.00854	0.99208	0.99381
		$\pm.00383$	$\pm.00872$	$\pm.00600$
c_{Te}		0.7892	0.0133	0.5707
$c_{T\omega}$		0.9578	0.9804	0.9726
$c_{e\omega}$		0.6029	0.0044	0.4458

5. Parallel Observational Information

The above given example illustrates the obvious fact that it is always advantageous to combine the information contained in different type of data – best of all by a simultaneous solution. In this sense the light curve analysis is inseparable from the analysis of the other data. Naturally, the reliability of these data and the applicability of the adopted model for their fitting has to be critically judged, and this task depends a lot on the experience and responsibility of the interpreter.

First of all, the photometric light curves should be solved in all spectral bands simultaneously. The data are quite often very numerous, and it is often tempting to analyze the mean phase curves instead of the individual original magnitudes as a function of time. It can lead to different systematic errors, e.g. due to incorrect period, due to periastron advance and other secular changes, or simply due to incompatibilities of data obtained by different instruments. The use of data covering a long time base can help to find a more precise value of the period; on the other hand they can be more misleading by secular changes. The common practice of analysis of times of minima instead of the complete light curves is commented on in Appendix B.

The already discussed combination of photometry of eclipsing binaries with their spectroscopy (cf., e.g., Wilson, 1979 for simultaneous solution of light and RV curves) is the most common one. The complete spectroscopic data are often replaced by measurements of radial velocities only. This is dangerous if the spectral lines are blended. In these cases the method of spectra disentangling discussed in more detail in Appendix A can give more reliable orbital parameters and, in addition, it gives the spectra of the individual components. Generalizations of the method (e.g. the line-photometry, Hadrava, 1997) can contribute to the determination of the geometry of the binary components. On the other hand, spectra with high S/N ratio, which are preferable for this method, usually do not cover a long enough time basis such as the photometric light curves or older RV measurements do. It is thus self-suggesting to merge the methods for obtaining light curve and RV curve solutions with the disentangling into one procedure. The spectra are a much richer source of information in any case – for instance, the parameters of the component atmospheres can be obtained by a detailed comparison with model atmospheres. Special techniques (e.g. the Doppler imaging) can contribute to our understanding of objects with significant line profile variations, which can be caused by pulsations, by rotation of spotted stellar disks or other effects.

Until recently, the set of eclipsing and spectroscopic binaries was naturally disjunct with the set of visual binaries. However, the use of high-resolution interferometry changes the situation rapidly and can significantly supplement the results of the above discussed methods. The analysis of the original observed visibility functions is quite a demanding procedure by itself. However, the fitting of angular distances and position angles between the binary components would be a useful and quite straightforward enrichment of the light and radial velocity curve solution.

Another technique contributing to the investigation of close binaries is polarimetry.

Appendix A: Disentangling with 'Light Curve Models' of Line Profiles

Assuming that the spectrum $I(x, t)$ of a multiple stellar system (e.g. binary, $n = 2$) is a superposition of Doppler shifted component spectra $I_j(x, t)|_{j=1}^{n}$, one can write

$$I(x, t) = \sum_{j=1}^{n} I_j(x) * \Delta_j(x, t, p), \tag{1}$$

where $x = c \ln \lambda$ is the logarithmic wavelength scale. If the spectra of the components are time-independent, the broadening functions Δ_j are given by delta-functions

$$\Delta_j(x, t, p) = \delta(x - v_j(t, p)) \tag{2}$$

shifted by the instantaneous radial velocities v_j of the components. The principle of spectra disentangling consists of fitting the observed spectra simultaneously by I_j and either v_j or directly the orbital parameters p (cf. Simon and Sturm, 1994; Hadrava, 1995).

However, due to the proximity effects (like tidal distortion or reflection effect) and due to the rotational effect in the case of eclipsing binaries with fast rotating components, the line profile variations can shift the effective centre of the spectral line and thus cause a distortion of the radial-velocity curve (cf. Wilson, 1994 and references therein). To treat properly the line profile variations in the procedure of disentangling, it is necessary to replace the summation of spectra of the components by integrating over their visible surface, s

$$I(x, t) = \sum_{j=1}^{n} \int_{s} \mu I_j(x, s, \mu, t) * \delta(x - v_j(s, t)) \, d^2 s, \tag{3}$$

where $\mu = \mu(s, t) \equiv \cos \vartheta$ is the directional cosine of the escaping ray with respect to the stellar surface. It can be shown (Hadrava, 2001; Hadrava and Kubát, 2003) that in a reasonably good approximation the local intensity can be separated in the form of a few terms

$$I_j(x, s, \mu, t) = \sum_{k} f_j^k(s, \mu, t) I_j^k(x), \tag{4}$$

so that the expression (1) takes the form

$$I(x, t) = \sum_{j,k} I_j^k(x) * \Delta_j^k(x, t, p) \tag{5}$$

of superposition of several spectral functions I_j^k corresponding to different limb darkening laws for the same component j, each one convolved with its own broadening function

$$\Delta_j^k(x, t, p) = \int_{s} \mu f_j^k(s, \mu, t) \delta(x - v_j(s, t)) \, d^2 s. \tag{6}$$

In comparison with (2) these functions now also express, in addition to the Doppler shifts, the broadening of spectral lines, e.g. due to rotation or other effects. (Note, that the often used approximation of a single convolution of the overall spectral flux in the rest frame of the stellar atmosphere with the rotational broadening neglects the change of limb-darkening across the line profile.)

Equation (5) can be used to disentangle the observed spectra, i.e. to fit them again simultaneously by the unknown spectral functions I_j^k (yielding now also some information about the structure of the atmosphere) and the parameters p of the broadening (orbital parameters but possibly also tidal distortion, rotation, etc.). If the Fourier transform $\tilde{\Delta}_j^k$ of broadenings Δ_j^k is calculated either in an analytical form – e.g. $\tilde{\Delta}_j(y, t, p) = \exp(iyv_j(t, p))$ for (2) – or numerically from a 'light curve model' of the surface intensity distribution, it can be easily performed in the

Fourier domain, where the equations for individual Fourier modes (i.e. for chosen values of y) are independent,

$$\tilde{I}(y, t) = \sum_{j,k} \tilde{I}_j^k(y) \tilde{\Delta}_j^k(y, t, p). \tag{7}$$

Appendix B: Comment on The Method of Light Curve Minima $O - C$

Determining the times of light curve minima is a traditional approach among observational studies of eclipsing binaries. It is advantageous especially for amateur astronomers, because it can be performed using quite simple and cheap technical means and because it can reward the observer with a particular result every night, thus stimulating his next effort. The study of $O - C$ of a sufficiently long series of light curve minima can reveal period changes and consequently can help in selecting objects that are worth enough to be observed intensively with more powerful instruments.

During the Litomyšl conference a discussion came up on the reliability of this method. An objection with respect to its use could be based on the fact that in the case of asymmetry of the eclipse, which can arise for manifold reasons, the determination of the minimum (or in fact of some effective centre of eclipse, because the precise minimum can be hidden in observational noise) is spurious and its meaning for orbital and other binary parameters is questionable. Moreover, the desirable information about the profile of the minimum is missing in this method of observation. For this reason the fitting of the observed magnitudes by light curve models with free parameters corresponding to the period changes is preferable.

On the contrary one could argue that an aprioristic light curve model does not need to be capable of properly describing the irregular period changes often observed in different binary systems. It is thus better to see the unbiased time dependence of $O - C$ first and to choose the proper model only after.[4] The reduction of the complex profile of the minimum to a single value of its centre can eliminate the problems with proper calibration of the photometric data and it thus extracts the most reliable information from the observations of quality insufficient for light curve fitting.

In fact the difference between the method of $O - C$ and light curve fitting is very similar to the difference between the radial-velocity measurements and spectra disentangling. The more sophisticated method can usually yield more information, provided its data are of sufficient quality and the model is properly chosen. However, simpler methods can also contribute a great deal to the information on the basis of cheaper but often more extensive data covering a longer interval of time. The

---

[4]See the contribution by R.E. Wilson (2004) in this volume, where the problem of light curve solution in the case of period changes is discussed.

information yielded by each from the couple of photometric methods (as well as by each from the couple of spectroscopic methods) is complementary, and it would thus be desirable to join all of them into one simultaneous solution. In particular, the codes for light curve fitting should be able to make use of an enriched input data set, which combines the magnitude dependence on time with $O - C$ data (i.e. with times of primary or secondary minima).

Acknowledgements

This work has been supported by grants GA ČR 202/02/0735 and 205/03/H144 and by projects K1048102 and Z1003909. The author thanks A. Kawka, H. Rovithis-Livaniou, and R.E. Wilson for helpful discussions and comments.

References

Bisikalo, D., Boyarchuk, A.A., Kuznetsov, O.A. and Chechetkin, V.M.: 1998, *Astron. Rep.* **42**, 621.
Hadrava, P.: 1991, *FOTEL – User's guide*, http://www.asu.cas.cz/~had/fotel.html (version 2003).
Hadrava, P.: 1995, *A&AS* **114**, 393.
Hadrava, P.: 1997, *A&AS* **122**, 581.
Hadrava, P.: 2001, *DSc. thesis*, Astronomical Institute, Ondřejov.
Hadrava, P.: 2004 in press, *Roche lobe in eccentric orbits*, in: 'Limites de Roche, Lobes de Roche', ed. J.-M. Faidit, Société Astronomique de France and Vuibert Editions.
Hadrava, P. and Kubát, J.: 2003, *ASP Conf. Ser.* **288**, 149.
Kallrath, J. and Milone, E.F.: 1999, *Eclipsing Binary Stars: Modeling and Analysis*, Springer, New York, Berlin.
Kallrath, J., Milone, E.F., Terrell, D. and Young, A.T.: 1998, *ApJ* **508**, 308.
Kopal, Z.: 1989, *The Roche Problem*, Kluwer Academic Publishers, Dordrecht.
Plavec, M.: 1958, *Mém. Soc. Roy. Sci. Liège* **20**, 411.
Simon, K.P. and Sturm, E.: 1994, *A&A* **281**, 286.
Wilson, R.E.: 1979, *ApJ* **234**, 1054.
Wilson, R.E.: 1990, *ApJ* **356**, 613.
Wilson, R.E.: 1994, *PASP* **106**, 921.
Wilson, R.E.: 2004, *ASTR* **xx**, xx.
Wilson, R.E. and Devinney, E.J.: *ApJ* **166**, 605.

A NEW PACKAGE OF COMPUTER CODES FOR ANALYZING LIGHT CURVES OF ECLIPSING PRE-CATACLYSMIC BINARIES

V.-V. PUSTYNSKI[1] and I.B. PUSTYLNIK[2]

[1]*Tallinn Technical University, Tartu Observatory, Estonia*
[2]*Tartu Observatory, Estonia; E-mail: izold@aai.ee*

(accepted April 2004)

Abstract. Using the new package of computer codes for analyzing light curves of the two eclipsing pre-cataclysmic binary systems (PCBs) UU Sge and V471 Lyr we find updated values of the physical parameters and discuss the evolutionary state of these PCBs.

Keywords: close binaries, pre-cataclysmic variables, evolution

1. Introduction

Pre-cataclysmic binaries (PCBs) compose a small group of detached close binary systems. The primary star in the pair normally is a hot white dwarf or a subdwarf, while the secondary usually is an unevolved low-mass cool companion with $M_2 < 1 M_\odot$. The substantial proximity of the components and high temperatures of the hot primary lead to an extraordinarily large irradiation effect observed in these systems. The number of known PCB systems is small and eclipsing binaries among them are rare exceptions. PCBs are mostly noticed just due to the pronounced irradiation effect.

During the last several years we have been dealing with reevaluation of the extant methods of analysis applied to PCBs (Pustylnik and Pustynski, 2001; Pustynski et al., 2001). Our current studies are based on the following considerations: (i) Conventional treatment of the reflection effect may not be adequate for binaries with separations of about ~ 1 to $5\, R_\odot$ and primary temperatures of about $T_1 \sim 3 \times 10^4 - 10^5$ K. UV radiation and even soft X-rays may be a reason for instability of various types. Thus, significant departures from the standard mass–radius relation valid for main sequence stars may be expected. (ii) Modeling physical conditions in cool atmospheres irradiated by strong Lyman continuum flux (L_c) from subdwarf companions should give an independent method for estimation of temperatures of subdwarfs.

We have compiled an original two-layer model of the secondary atmosphere and incorporated it in a computer code, which calculates model atmospheres at different regions of the illuminated secondary surface and finally constructs continuum light curves for a selected wavelength. Both occultation and transit-type eclipses are taken into account.

Astrophysics and Space Science **296**: 251–254, 2005.

2. Underlying Model

We have chosen a two-layer model of the secondary atmosphere with the following simplifications: (i) The whole L_c flux is absorbed in the upper atmosphere, which is supposed to be transparent for optical photons. The validity of such simplification was checked post-factum and proved to be well founded; (ii) The free path of a photon is considered to be negligible compared to the layer thickness; (iii) The L_c flux energy absorbed in the upper layer is considered to be spent on recombination radiation; (iv) The standard Eddington diffusion approximation is used for the deeper atmospheric layer. Upper and deeper layer models are coordinated *via* common boundary variables. We find the impinging primary flux using the method described by Napier (1968): the secondary disk is divided into circular zones symmetrical around the sub-stellar point, the flux is calculated taking into account the possible occultation of the primary disk by the local horizon, for each zone an independent model is built and emergent radiation intensity is found. Thereafter, intensities from different zones are integrated to yield the total luminosity of the system, and inputs of possibly eclipsed portions of both stellar disks are consequently subtracted.

The initial parameters of the model are as follows: stellar masses, radii, system semi-major axis, effective temperatures, electron pressure at the top of the atmosphere and also hydrogen, helium and preselected metal abundances, as well as the method of mean absorption coefficient calculation.

3. Computational Results and Discussion

The basic results of computations are that UV radiation penetrates into the upper atmosphere until a certain depth, while the ionization degree remains close to unity. Thereafter, almost all of the flux is absorbed in a very thin layer of practically neutral hydrogen. (i) The temperature declines steeply at the top of the atmosphere. Then the temperature gradient remains rather low until one reaches the regions, where the ionization changes dramatically. (ii) The uppermost layers of the irradiated atmosphere are severely overheated. The temperature inversion in the uppermost regions of the irradiated atmosphere is caused by the specific dependence of the adopted cooling function on the temperature and the density of the atmospheric layers. The behaviour of the cooling coefficient indicates presence of thermal instability. (iii) In deep layers where the optical depth in the L_c is 5–10, roughly only 25–50% of the incoming flux is used for ionization of H with subsequent recombinations. The remaining portion is expended on heating the electronic gas. (iv) The thermal time scale for the layer, where the L_c flux is effectively absorbed, is comparable to the orbital period of a typical PCB (for $dm \simeq 10^{-3}$ g cm^{-2} and an effective thickness of the layer of 10^7 cm). For typical values of $T_e \sim 10^5$ K and $N_e \sim 10^{12}$ cm^{-3} we find a cooling time scale $t_{th} \simeq 10^4$ s. The results of our model calculations suggest that transient effects on a time scale comparable to the orbital period of PCBs are anticipated (for more details see Pustynski and Pustylnik, 2004).

4. Application of the Model to UU Sge and V477 Lyr

We have applied the computer code based on our model to two PCBs, UU Sge and V477 Lyr.

UU Sge is the central star of the planetary nebula Abell 63, an eclipsing binary with a period of 0.465 days. Using V observations provided by Polacco and Bell (1993), we found two sets of parameters that fit well the observational light curve (Pustynski and Pustylnik, 2004). They claim for a primary temperature ranging from about $T_1 = 80\,000$ to $85\,000$ K, which constrains the T_1 estimates of other authors ($T_1 \simeq 35\,000$–$117\,000$ K). The secondary temperature estimate of about $T_2 = 5500$ K turns out to be lower than what Pollaco and Bell propose ($T_2 = 7300$ K).

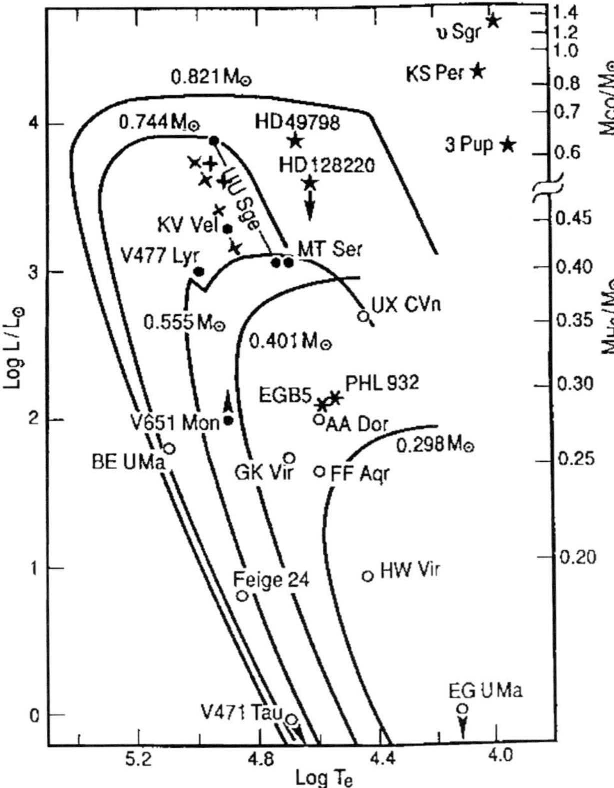

Figure 1. The cooling tracks for white dwarfs according to Iben and Tutukov (1993).

V477 Lyr is the eclipsing central star of Abell 46. Based on observations by Polacco and Bell (1994), we found four sets of parameters satisfying the observational light curve. The range of the primary temperature in our fits remains extensive: $T_1 = 74\,000\text{--}102\,000$ K.

Using our new determination of T_1, R_1 for UU Sge and V477 Lyr, we inserted the data in the plot of cooling tracks for white dwarfs, constructed by Iben and Tutukov (1993). Our updated positions of the hot subdwarfs in the H–R diagram are designated by crosses for UU Sge and by plus signs for V477 Lyr (Figure 1). Please, note a considerable change in the location for V477 Lyr. As one can see from the plot, it is impossible to discriminate between a degenerate He or CO core white dwarf in the case of V477 Lyr.

Acknowledgement

Supporting Grant 5760 from Estonian Science Foundation is gratefully acknowledged.

References

Napier, W.McD.: 1968, *Ap&SS* **2**, 61.

Polacco, D.L. and Bell, S.A.: 1993, *Monthly Notices RAS* **262**, 377.

Polacco, D.L. and Bell, S.A.: 1994, *Monthly Notices RAS* **267**, 452.

Pustylnik, I.B. and Pustynski, V.-V.: 2001, in: *12th European Workshop on White Dwarfs, PASP, Conference Series* **226**, 253.

Pustynski, V.V. and Pustylnik, I.B.: 2004, *Baltic Astron.* **13**, 122.

Pustylnik, I.B., Pustynski, V.-V. and Kubat, J.: 2001, *Odessa Astron. Publ.* **14**, 87.

Iben, I.Jr. and Tutukov, A.V.: 1993, *Astrophys. J.* **418**, 343.

LIGHT CURVE ANALYSIS OF EARLY-TYPE OVERCONTACT SYSTEMS

MICHAEL BAUER

Dr Remeis-Sternwarte Bamberg, Astronomisches Institut der Universität Erlangen-Nürnberg, Sternwartstraße 7, D-96049 Bamberg, Germany; E-mail: bauer@sternwarte.uni-erlangen.de

(accepted April 2004)

Abstract. A method is presented to include irradiation effects in the modeling of over-contact binaries. The Roche potential is numerically modified in a way to account for mechanical effects of the mutual irradiation of the two binary components. The efficiency of radiative interaction is parametrized by means of the fraction of radiative over gravitational forces for each component of the binary. The modified Roche geometry is implemented as binary model in an eclipse light curve solution code, which is based on the general Wilson–Devinney scheme. As an example the method is applied to the early-type over-contact system V606 Cen.

Keywords: eclipsing binary stars, light curve, radiation pressure, numerical analysis, V606 Cen

1. Introduction

Even though radiation pressure effects for over-contact eclipsing binaries of, e.g., W UMa type are of less importance due to the low temperatures of their components, some systems were discovered, where radiation pressure might indeed have a noticeable effect on their light curves. These systems consist of early-type components with temperatures above 20,000 K. Drechsel et al. (1995) have developed a method to account for radiation pressure effects in close early-type systems with detached or semi-detached configuration. In this work we report on the extension of the numerical approach to over-contact binaries of early spectral type.

2. Parametrization of Radiation Pressure and the Modified Roche Potential

Radiation pressure can be parametrized in this method by a scalar value, defined by the fraction of radiative force over gravitational force:

$$\delta = \frac{F_{\mathrm{rad}}}{F_{\mathrm{grav}}} \tag{1}$$

This parameter gives the effectiveness of radiation pressure compared to gravitational force. In the following a distinction will be made between so-called inner and outer radiation pressure. Inner radiation pressure, which will be denoted by δ, is the pressure on the star's surface due to radiation from inside the star. This pressure is

Astrophysics and Space Science **296**: 255–260, 2005.
© Springer 2005

essentially constant over the surface, as gravitation and radiation forces can both be treated as central forces in this approximation. On the other hand, outer radiation pressure, denoted by δ^*, is the pressure on the star's surface due to radiation from the other component. This is of course geometry-dependent, as one has to account for shadowing effects, the distance between the radiating and the irradiated surface elements, and their mutual inclination. Therefore $0 \leq \delta^* = f(x, y, z) \leq \delta$.

The parameters for inner radiation pressure $\delta_{1/2}$ (one for each component of the binary) are input parameters of the light curve analysis program. The outer radiation pressure parameters $\delta_{1/2}^*$, as they are dependent on the position on the surface, have to be calculated numerically for each surface element.

Now an effective gravitational force can be defined:

$$\vec{F}_{grav}^{eff} = \vec{F}_{grav} - \vec{F}_{rad} = (1 - \delta)\vec{F}_{grav} \tag{2}$$

The Roche potential can be modified to account for radiation pressure effects:

$$\Omega = \frac{1 - \delta_1 - \delta_1^*(x, y, z)}{r_1} + q\frac{1 - \delta_2 - \delta_2^*(x, y, z)}{r_2} + \frac{q + 1}{2} \times (x^2 + y^2) - qx \tag{3}$$

with the mass ratio $q = m_2/m_1$ and the distances to the corresponding mass centers $r_1 = \sqrt{x^2+y^2+z^2}$ and $r_2 = \sqrt{(d-x)^2+y^2+z^2}$; d is the distance between the two mass centers.

3. Modeling of the Surface

The first step when modeling the surface is to determine the position of the inner Lagrangian point L_1. This has to be done as δ_1 and δ_2 are dependent on the position of the plane which separates the two components in the binary. This position will be identified with the position of L_1, which will be denoted by x_{L_1}. The dependency of δ_1 and δ_2 on the separation plane can be illustrated as follows. It does not make physical sense to speak of inner radiation pressure inside the secondary component originating from the primary component and vice versa, so δ_1 has values different from zero only when $x < x_{L_1}$. Likewise δ_2 is different form zero only for $x > x_{L_1}$.

The process for determining x_{L_1} is iterative, as the position of L_1 is affected by δ_1 and δ_2 through the potential, but on the other hand δ_1 and δ_2 are affected by x_{L_1}.

This process sets a limitation on the radiation pressure parameters. For too large parameters the position of L_1 is no longer well defined, as more than one extremum can appear in $\Omega(x, y = 0, z = 0)$ with $x_{m_1} < x < x_{m_2}$.

The next step is to generate an approximated surface, which can later be used to calculate the outer radiation pressure parameters. This approximated surface is calculated from the modified Roche potential (Eq. 3) by accounting only for inner radiation pressure effects, so $\delta_1 \geq 0$, $\delta_2 \geq 0$, but $\delta_1^* = 0$, $\delta_2^* = 0$.

After having constructed the entire approximated surface of the binary, the outer radiation pressure parameter for each surface element is computed. This is done with the definition of the radiation pressure parameter

$$\delta_{1/2}^* = \frac{F_{\mathrm{rad}}}{F_{\mathrm{grav}}}. \tag{4}$$

As it is not possible in general to calculate the gravitational force, especially when neither absolute masses nor the distance are given, F_{grav} can be eliminated by using the equation for inner radiation pressure of the other component

$$\delta_{2/1} = \frac{\hat{F}_{\mathrm{rad}}}{F_{\mathrm{grav}}}. \tag{5}$$

$\hat{F}_{\mathrm{rad}}$ means the virtual radiative force acting in the absence of any shadowing effects. Only then the fraction of forces is equal to the inner radiation pressure parameter.

As the radiative force is proportional to the flux one can write

$$\delta_{1/2}^* = \delta_{2/1} \frac{S}{\hat{S}}. \tag{6}$$

The fluxes S and $\hat{S}$ can be calculated by integrating over the flux contributions of all visible surface elements on the radiating component with respect to distance to the irradiated surface element, limb darkening and their tilt relative to the line of sight.

To determine the final surface of the binary the position of each surface element has to be recalculated with Eq. 3, as $\delta_{1/2}$ and $\delta_{1/2}^*$ now are known values.

The above model was implemented into an eclipse light curve solution code otherwise based on the general Wilson–Devinney model (Wilson and Devinney, 1971).

4. Simulated Configurations and Corresponding Light Curves

To show the effects on the surface due to radiation pressure three configurations with equal components were simulated. The radiation pressure parameters were chosen to be 0.00, 0.05 and 0.10. To ensure comparable surfaces, where no radiation pressure effects apply, the surface potentials were adjusted to keep the backside radii

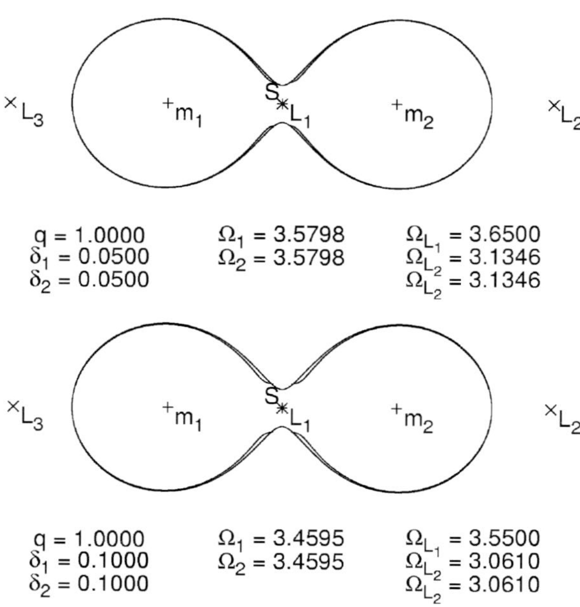

$q = 1.0000$
$\delta_1 = 0.0500$
$\delta_2 = 0.0500$

$\Omega_1 = 3.5798$
$\Omega_2 = 3.5798$

$\Omega_{L_1} = 3.6500$
$\Omega_{L_2}^1 = 3.1346$
$\Omega_{L_2}^2 = 3.1346$

$q = 1.0000$
$\delta_1 = 0.1000$
$\delta_2 = 0.1000$

$\Omega_1 = 3.4595$
$\Omega_2 = 3.4595$

$\Omega_{L_1} = 3.5500$
$\Omega_{L_2}^1 = 3.0610$
$\Omega_{L_2}^2 = 3.0610$

Figure 1. Simulation of two configurations: $\delta_{1/2} = 0.05$ and $\delta_{1/2} = 0.10$. The outer line in the plots is a reference surface calculated with $\delta_{1/2} = 0.00$.

of the two components at the same value. Meridional cuts of the three configurations are shown in Figure 1. The configuration with $\delta_{1/2} = 0.00$ is overplotted in both configurations (outer line) to give a reference surface. In Figure 2 the corresponding light curves are shown for comparison.

5. Application to V606 Cen

As an example the method was applied to V606 Cen, which has already been analyzed by Lorenz et al. (1999) without accounting for radiation pressure effects. Their solution tended towards an over-contact configuration. Now, as it is possible to account for radiation pressure also in over-contact systems, a new analysis was performed (for the fitted light curves see Figure 3).

The new solution now shows radiation pressure parameters of $\delta_1 = 0.99\% \pm 0.05\%$ and $\delta_2 = 0.51\% \pm 0.15\%$. Except for some deviations of the limb darkening parameters the rest of the parameters agree quite well with the solution found by Lorenz et al. The quality of fit (standard deviation in units of normalized light) improves from 0.0051 to 0.0037.

This result agrees well with the prediction of Lorenz et al. that, based on detached systems, δ_1 could be expected to be $\sim 1\%$ for this system.

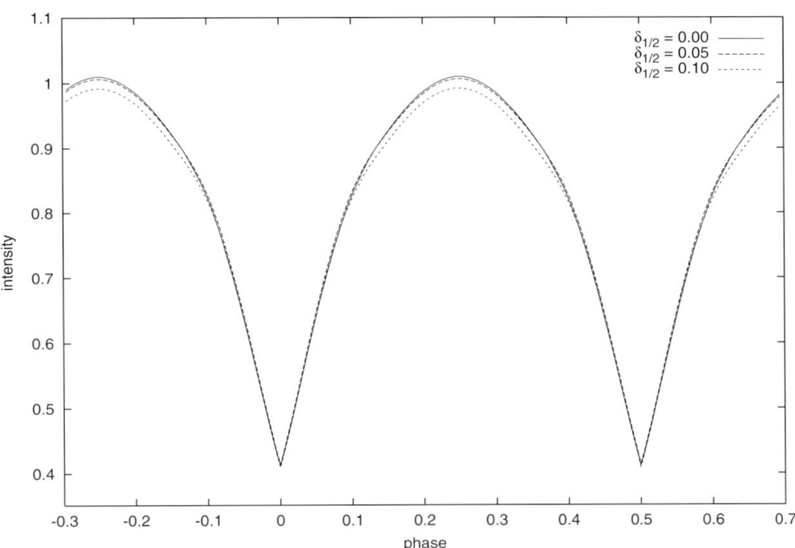

Figure 2. Light curves corresponding to the simulated configurations in Figure 1. The remaining parameters were chosen to be: $i = 90°$, $g_1 = g_2 = 1.0$, $T_1 = T_2 = 30,000\,K$, $A_1 = A_2 = 1.0$, $L_1 = L_2 = 0.5$, $x_1 = x_2 = 0.44$, $l_3 = 0.00$.

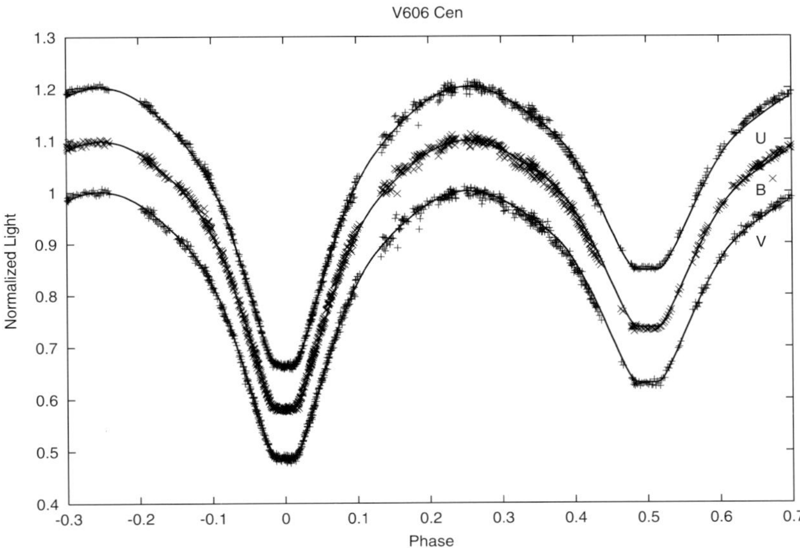

Figure 3. Light curves of V606 Cen in U,B and V bands. Plottet are measured data points and the synthetic light curves (solid lines).

6. Conclusion

With this work it is now possible to include radiation pressure effects also in the modeling and light curve analysis of over-contact binaries. The results correspond fairly well to the predictions based on detached systems, as shown for V606 Cen.

References

Drechsel, H., Haas, S., Lorenz, R. and Gayler, S.: 1995, *A&A* **294**, 723.
Lorenz, R., Mayer, P. and Drechsel, H.: 1999, *A&A* **345**, 531.
Wilson, R.E. and Devinney, E.J.: 1971, *ApJ* **166**, 605.

A PHOTOMETRIC STUDY OF THE ACTIVITY OF SW Lac

G. DJURAŠEVIĆ[1], B. ALBAYRAK[2], S. ERKAPIĆ[1] and T. TANRIVERDI[2]

[1]*Astronomical Observatory, Volgina 7, 11160 Belgrade, Serbia and Montenegro;*
E-mail: gdjuvasevic@aob.bg.ac.yu
[2]*Department of Astronomy and Space Sciences, Faculty of Science, Ankara University, 06100,*
Tandoğan, Ankara, Turkey

(accepted April 2004)

Abstract. New BV photoelectric observations of the W UMa-type system SW Lac obtained at the Ankara University Observatory in 2001, 2002, and 2003 are used to study the activity of the system.

Keywords: binaries, eclipsing binaries, close individual, SW Lac

1. Introduction

In this paper we have analyzed our own B and V seasonal light curves of the short-period ($P \sim 0^d.32$) eclipsing binary SW Lac ($V_{\max} = 8^m.91$, HD 216598, BD+37° 4717, HIP113052, G3 V+G3 V) with the aim to study the activity of SW Lac and to estimate the orbital and physical parameters of the system. The observations were collected during the years 2001 (on two nights), 2002 (one night), and 2003 (two nights) at the Ankara University Observatory using an uncooled SSP-5A photometer attached to a 30-cm Maksutov telescope.

2. Results and Discussion

We have analyzed the SW Lac light curves using the synthetic light curve programme by Djurašević (1992a) modified for contact configurations (Djurašević et al., 1998), which is based on the Roche model. The light curve analysis has made use of the inverse-problem method developed by Djurašević (1992b). For more details on the analysis procedure for this particular system see Albayrak et al. (2004).

Since the stars in the system have external convective envelopes, which can exhibit magnetic activity, we have assumed cool spot regions of the same nature as solar magnetic spots to model the asymmetry and deformation of SW Lac seasonal light curves. By testing based on the simultaneous analysis of B and V light curves we have chosen the Roche model with dark spotted areas on the more massive (cooler) component as the optimum model to fit the observations. The cooler component has a deeper convective layer, suggesting that more intense active regions can be expected.

Astrophysics and Space Science **296**: 261–264, 2005.
© Springer 2005

As an immaculate configuration (spotfree, the reference light level) the maximum light level of the 2002 seasonal light curve at phase 0.25 has been chosen. The simultaneous analysis of the light curves of 2002 provided us with estimates of basic system parameters and of parameters for the active cool spot region, which deforms the light curve. This set of basic system parameters has been used as the initial one in the inverse problem to fit the light curves for three epochs of observations.

The analysis proved (see Table I and Figure 1) that the 2001 and 2002 seasonal light curves can be successfully simulated with the Roche model including one cool spot on the more massive (cooler) secondary. A conspicuous depression of the light level in the 2003 curve forced us to employ a more complex Roche model with two cool spot regions, which provided a good fit.

Table I comprises the results of our simultaneous fitting of the available light curves in B and V photometric bands for three epochs of observations with the same set of basic system parameters. The solutions for each seasonal light curve are graphically presented in Figure 1.

The relative size of the circular spotted area is calculated assuming a spherical shape of the star. It can be used as an indicator of the system activity. The 2001 solution yielded a smaller spotted area ($\sigma_{S1} \sim 7\%$) located relatively close to the stellar polar region. The 2002 solution suggests a larger spot covering a significant part of the stellar surface ($\sigma_{S1} \sim 14\%$) on the same hemisphere, which gives rise to the conspicuous asymmetry in the light curve. It appears that the system activity has increased in 2002. The 2003 light curve is less asymmetric, but exhibits a large depression of the light level. With two cool spots on the same component, the first one larger ($\sigma_{S1} \sim 16\%$) and approximately at the same location, and the additional second one smaller ($\sigma_{S2} \sim 5\%$) at almost opposite longitude and moderate latitude, we obtain a good fit. These two spots cover approximately 21% of the secondary surface, pointing again to increased activity.

3. Conclusions

Our results derived from the analysis of SW Lac light curves in 2001, 2002 and 2003 show that the Roche model with spot areas on the cooler component provides a satisfying fit of the observations. During that period, the main variations in the light curves can be explained by the development and migration of spot regions on the cooler component. The solutions obtained by solving the inverse problem do not show changes of the system basic parameters among different seasonal light curves. The solutions presented here show that SW Lac is in an overcontact configuration with a relatively high degree of overcontact $f_{over} \sim 31\%$, and with a small temperature difference between the components ($\Delta T \sim 280$ K) indicating a thermal contact between the components.

TABLE I

Results of the analysis of the SW Lac seasonal light curves

Quantity	2001	2002	2003
n	470	362	810
$\Sigma(O-C)^2$	0.1624	0.1342	0.1976
σ	0.0186	0.0193	0.0156
$q = m_\mathrm{c}/m_\mathrm{h}$	1.255	1.255	1.255
$f_{\mathrm{h,c}}$	1.0	1.0	1.0
$\beta_{\mathrm{h,c}}$	0.08	0.08	0.08
$A_{\mathrm{h,c}}$	0.5	0.5	0.5
T_h	5630	5630	5630
$A_{\mathrm{S1}} = T_{\mathrm{S1}}/T_\mathrm{c}$	0.84 ± 0.01	0.85 ± 0.01	0.87 ± 0.01
θ_{S1}	31.8 ± 0.5	44.4 ± 0.4	46.9 ± 0.4
λ_{S1}	278.9 ± 1.6	294.3 ± 1.1	263.6 ± 1.3
φ_{S1}	48.3 ± 1.6	45.4 ± 1.1	61.9 ± 1.2
$\sigma_{\mathrm{S1}}\,[\%]$	~ 7	~ 14	~ 16
$A_{\mathrm{S2}} = T_{\mathrm{S2}}/T_\mathrm{c}$			0.89 ± 0.01
θ_{S2}			26.4 ± 0.4
λ_{S2}			87.8 ± 1.6
φ_{S2}			30.5 ± 2.7
$\sigma_{\mathrm{S2}}\,[\%]$			~ 5
T_c	5347 ± 12	5348 ± 11	5345 ± 13
F_h	1.059 ± 0.001	1.060 ± 0.001	1.060 ± 0.001
$i\,[^\circ]$	79.85 ± 0.11	79.85 ± 0.14	79.85 ± 0.09
$\Omega_{\mathrm{h,c}}$	3.9801	3.9796	3.9795
$f_{\mathrm{over}}\,[\%]$	30.86	30.95	30.96
$R_\mathrm{h}, R_\mathrm{c}[D=1]$	0.357, 0.395	0.357, 0.395	0.357, 0.395
$L_\mathrm{h}/(L_\mathrm{h}+L_\mathrm{c})(B;V)$	0.538; 0.530	0.548; 0.540	0.557; 0.548

Note. n – number of observations, $\Sigma(O-C)^2$ – final sum of squares of residuals between observed (LCO) and synthetic (LCC) light curves, σ – standard deviation of observations, $q = m_\mathrm{c}/m_\mathrm{h}$ – mass ratio of components, $f_{\mathrm{h,c}}$, $\beta_{\mathrm{h,c}}$, $A_{\mathrm{h,c}}$ – nonsynchronous rotation coefficients, gravity-darkening exponents, and albedos of components, $T_{\mathrm{h,c}}$ – temperature of hotter primary and cooler secondary, $A_{\mathrm{S1,2}}$, $\theta_{\mathrm{S1,2}}$, $\lambda_{\mathrm{S1,2}}$ and $\varphi_{\mathrm{S1,2}}$ – spot temperature coefficient, angular dimension, longitude and latitude (in arc degrees), $\sigma_{\mathrm{S1,S2}}[\%]$ – spot area in percent of stellar surface, $F_{\mathrm{h,c}}$ – filling factors for critical Roche lobe of hotter (less massive) and cooler (more massive) star, $i\,[^\circ]$ – orbital inclination (in arc degrees), $\Omega_{\mathrm{h,c}}$ – dimensionless surface potentials of primary and secondary, $f_{\mathrm{over}}[\%] = 100 \cdot (\Omega_{\mathrm{h,c}} - \Omega_{\mathrm{in}})/(\Omega_{\mathrm{out}} - \Omega_{\mathrm{in}})$ – degree of overcontact, $R_{\mathrm{h,c}}$ – polar radii of components in units of distance between component centers, and $L_\mathrm{h}/(L_\mathrm{h}+L_\mathrm{c})$ – BV luminosities of hotter star (including spot on the cooler one).

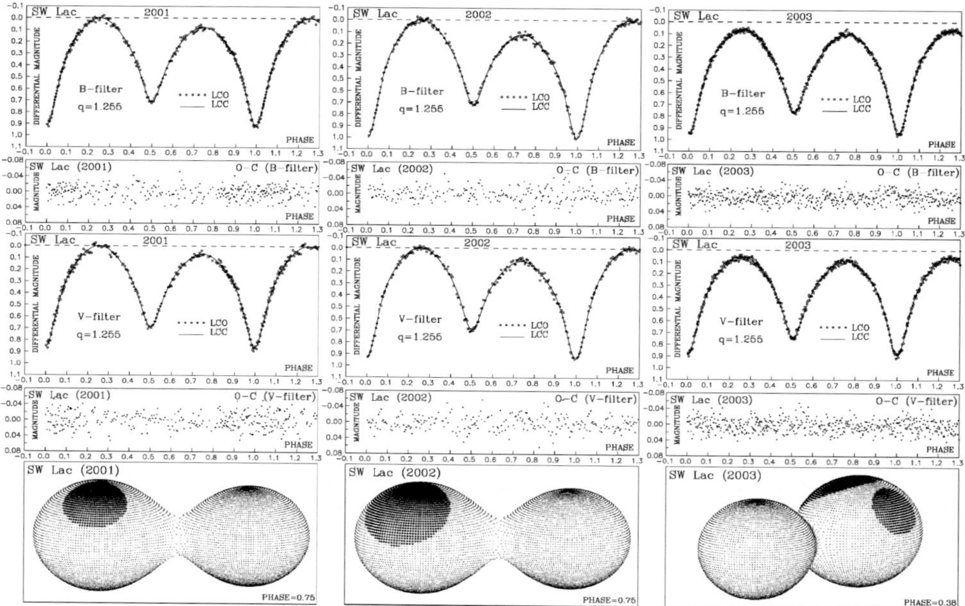

Figure 1. Observed (LCO) and final synthetic (LCC) light curves of SW Lac with final $O-C$ residuals obtained by analyzing seasonal (left panel for 2001, middle panel for 2002, right panel for 2003) B and V observations, and view of the system at orbital phase 0.75, corresponding to the derived parameters.

The spot areas were continously increasing during the whole examined period. So, future monitoring (for at least 10 years) of this active system will be of great interest.

References

Albayrak, B., Djurašević, G., Erkpić, S. and Tanrıverdi, T.: 2004, *Astron. Astrophys.* **420**, 1039.
Djurašević, G.: 1992a, *Astrophys. Space Sci.* **196**, 241.
Djurašević, G.: 1992b, *Astrophys. Space Sci.* **197**, 17.
Djurašević, G., Zakirov, M., Hojaev, A. and Arzumanyants, G.: 1998, *Astron. Astrophys. Suppl. Ser.* **131**, 17.

CCD PHOTOMETRY AND MODELING OF THE OVERCONTACT BINARY SYSTEMS NN VIR AND YY CrB

K.D. GAZEAS and P.G. NIARCHOS

Department of Astrophysics, Astronomy and Mechanics, National and Kapodistrian University of Athens, GR 157 84 Zografos, Athens, Greece; E-mail: kgaze@skiathos.physics.auth.gr

(accepted April 2004)

Abstract. New ground-based BV RI CCD observations of the eclipsing binary systems NN Vir and YY CrB are analyzed and illustrated. New times of minima are given and new ephemerides are proposed. The light curves are analyzed with the Wilson-Devinney light curve synthesis code and new geometric and photometric elements are derived. These elements are used together with the available spectroscopic data to compute absolute elements. The evolutionary status of each system is studied by means of mass-radius diagrams. The systems NN Vir and YY CrB are found to be A-type and W-type W UMa systems, respectively.

Keywords: stars, binaries, eclipsing-stars, fundamental parameters

1. Introduction

The EW-type eclipsing binary systems YY CrB and NN Vir were discovered by Hipparcos satellite mission (ESA, 1997) and were listed in the 74th special name list of variable stars (Kazarovets et al., 1999). The light curves of YY CrB have no significant asymmetries, as it was shown from ground-based observations made by Sipahi et al. (2000) and Demircan et al. (2003). This system was assumed to be an A-type W UMa system, as it was not clear which minimum was the deeper one. The mass ratio was found spectroscopically to be $q = 0.243$ and the spectral type is F8 V (Rucinski et al., 2000). According to Woitas (1997), NN Vir is a RR Lyrae variable but, after verification by several observers (Gomez-Forrellad and Garcia-Melendo, 1997), it was found that NN Vir is a W UMa type eclipsing binary with a period of 0.48 days. Rucinski and Lu (1999) observed this system spectroscopically and found a mass ratio of $q = 0.491$ and a spectral type of F0 V/F1 V.

2. Observations

The ground-based observations of YY CrB were made on three consecutive nights on July 15, 16 and 17, 2002. The instruments were the 1.22 m Cassegrain reflector at the Kryonerion Astronomical Station, Greece, and a Photometrics CH250 CCD camera. A total of 1867 frames were obtained in B, V, R and I filters. The ground-based observations of NN Vir were made on March 26, 28, April 2, 8,

9, May 7 and June 1, 2, 4, 2003. The instruments were the 0.40 m Cassegrain reflector at the University of Athens Observatory, Greece, and a ST-8 CCD camera. A total of 2724 frames were obtained in B, V, R and I filters. The images were processed with the AIP4WIN program of Berry and Burnell (2000). The standard deviations of our measurements were 0.002–0.010 mag in the four bands.

3. Light Curve Analysis

New times of minima were computed by the method of Kwee and van Woerden (1956) and new ephemerides for both systems were calculated. For NN Vir we used our 9 times of minima, while for YY CrB we used 21 times of minima from 1998 until 2003, plus two more from our observations. Our computed times of minima are in Table I. The ephemerides below were used for the subsequent analysis:

$$\text{NN Vir: Min } I_{\text{HJD}} = 2452795.4456 + 0^{\text{d}}.480690 \times E \tag{1}$$
$$\text{YY CrB: Min } I_{\text{HJD}} = 2452472.3516 + 0^{\text{d}}.376564 \times E \tag{2}$$

Our light curves were analyzed with the WD (Wilson and Devinney, 1971; Wilson, 1979) DC program, running in Mode 3. The usual values from the new tables of Claret (2000) and Claret et al. (1995) were used for the gravity exponents and bolometric albedos, i.e. $g_1 = g_2 = 0.32$ and $A_1 = A_2 = 0.50$. The third light was assumed equal to zero. The results are in Table II, where estimated uncertainties are standard errors.

TABLE I

New times of minima for YY CrB and NN Vir

	HJD	σ	Epoch	Type	O–C	Filter
YY CrB	2452472.3516	0.0003	0	I	−0.0625	B, V, R, I
	2452473.4815	0.0005	3	I	−0.0623	B, V, R, I
NN Vir	2452725.5075	0.0003	−145.5	II	0.0023	B, V, R, I
	2452727.4294	0.0003	−145.5	II	0.0014	B, V, R
	2452732.4766	0.0003	−131	I	0.0014	B, V, R, I
	2452738.4843	0.0004	−118.5	II	0.0005	B, V, R, I
	2452739.4465	0.0006	−116.5	II	0.0013	B, V, R, I
	2452767.3272	0.0004	−58.5	II	0.0020	B, V, R, I
	2452793.2847	0.0007	−4.5	II	0.0022	B, V, R, I
	2452793.5231	0.0003	−4	I	0.0003	V, R
	2452795.4456	0.0001	0	I	0.0000	B, V, R

TABLE II

Light curve solutions for YY CrB and NN Vir

Parameter	YY CrB	NN Vir
ϕ	0.0007 ± 0.0003	-0.0011 ± 0.0002
i (deg)	80.40 ± 0.53	66.20 ± 0.04
T_1 (K)	6200^*	7200^*
T_2 (K)	6138 ± 17	7142 ± 13
$\Omega_1 = \Omega_2$	2.2688 ± 0.0055	2.6991 ± 0.0005
$q = m_2/m_1$	4.115^*	0.491^*
degree of overcontact	41%	54%
$L_1/(L_1 + L_2)$ (B)	0.2988 ± 0.0042	0.6180 ± 0.0036
$L_1/(L_1 + L_2)$ (V)	0.2841 ± 0.0035	0.6149 ± 0.0043
$L_1/(L_1 + L_2)$ (R)	0.2811 ± 0.0029	0.6132 ± 0.0025
$L_1/(L_1 + L_2)$ (I)	0.2752 ± 0.0024	0.6107 ± 0.0018
r_1 (vol)	0.2895 ± 0.0015	0.4840 ± 0.0013
r_2 (vol)	0.5284 ± 0.0016	0.3621 ± 0.0015
$\Sigma w(\mathrm{res})^2$	0.0298	0.0279

*Assumed.

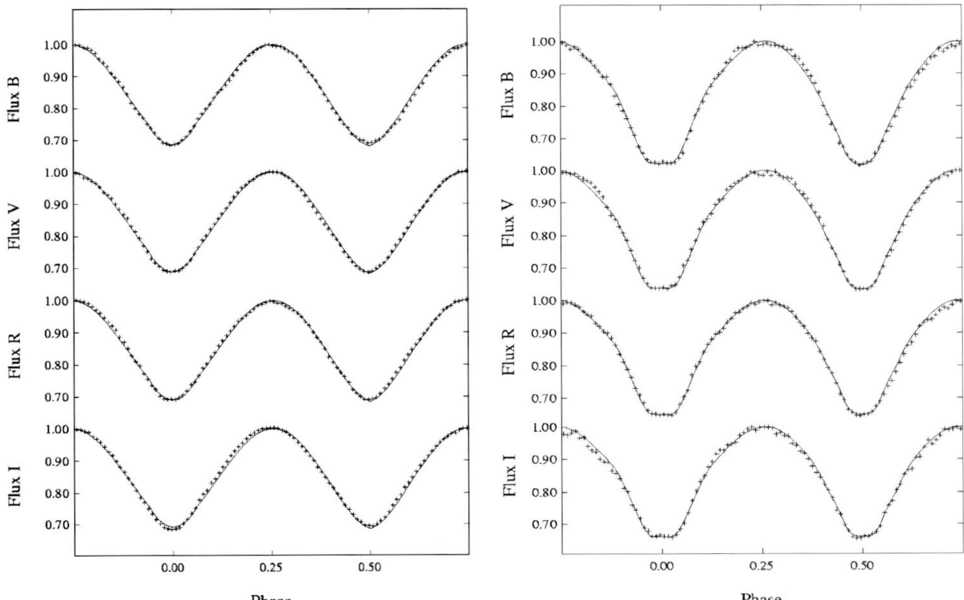

Figure 1. Observed and theoretical light curves of YY CrB (left) and NN Vir (right).

The derived parameters were used to construct theoretical light curves, which are shown along with the observed light curves in Figure 1. The absolute physical parameters (radii, masses and luminosities in solar units) of YY CrB are: $R_1 = 0.760 \pm 0.048$, $R_2 = 1.387 \pm 0.087$, $M_1 = 0.336 \pm 0.240$, $M_2 = 1.382 \pm 0.987$, $L_1 = 0.751 \pm 0.010$, $L_2 = 2.404 \pm 0.043$, $M_{(bol)1} = 5.059$, and $M_{(bol)2} = 3.796$, and for NN Vir: $R_1 = 1.698 \pm 0.107$, $R_2 = 1.271 \pm 0.080$, $M_1 = 1.689 \pm 0.251$, $M_2 = 0.830 \pm 0.167$, $L_1 = 6.994 \pm 0.046$, $L_2 = 3.856 \pm 0.046$, $M_{(bol)1} = 2.636$ and $M_{(bol)2} = 3.283$.

4. Summary and Conclusions

The observed light curves of YY CrB and NN Vir are rather symmetrical with the same height at phases 0.25 and 0.75, an indication that there is little magnetic spot activity. The analysis shows a highly overcontact configuration with fill-out factors 41% for YY CrB and 54% for NN Vir. The results of the photometric solutions were used to calculate absolute parameters. In both systems the primary components lie between the ZAMS and TAMS lines in the mass-radius diagram. The secondary components are well above the TAMS line, indicating that both stars are evolved or at least oversized for their Main Sequence masses. NN Vir and YY CrB are A-type and W-type W UMa systems, respectively.

References

Berry, R. and Burnell, J.: 2000, The Handbook of Astronomical Image Processing. Willmann-Bell, Richmond, VA.

Claret, A., Diaz-Cordoves, J. and Gimenez, A.: 1995, *A&AS* **114**, 247.

Demircan, O.: 2003, IBVS No. 5364.

ESA: 1997, The Hipparcos and Tycho Catalogues. ESA SP-1200.

Gomez-Forrellad, J.M. and Garcia-Melendo, E.: 1997 IBVS, No. 4469.

Kazarovets, A.V., Samus, N.N., Durlevich, O.V., Frolov, M.S., Antipin, S.V., Kireeva, N.N. and Pastukhova, E.N.: 1999, IBVS, No. 4659.

Kwee, K.K. and van Woerden, H.: 1956, *Bull. Astron. Inst. Neth.* **12**, 327.

Rucinski, S.M. and Lu, W.: 1999, *ApJ* **118**, 2451.

Rucinski, S.M., Lu, W. and Mochnacki, S.W.: 2000, *ApJ* **120**, 1133.

Sipahi, E., Keskin, V. and Yasarsoy, B.: 2000, IBVS No. 4859.

Wilson, R.E.: 1979, *ApJ* **234**, 1054.

Wilson, R.E. and Devinney, E.J.: 1971, *ApJ* **166**, 605.

Woitas, J.: 1997, IBVS, No. 4444.

RADIAL VELOCITY AND LIGHT CURVE ANALYSIS
OF THE ECLIPSING BINARY NN Vir

R. PAZHOUHESH[1] and E.G. MELENDO[2]

[1]*Physics Department, Faculty of Sciences, Birjand University, Birjand, Iran;*
E-mail: pazhouhesh@fastmail.ca
[2]*Esteve Duran Observatory, El Montanya, Seva 08553 Seva, Barcelona, Spain;*
E-mail: duranobs@astrogea.org

(accepted April 2004)

Abstract. The eclipsing binary NN Vir is a short period system showing an EW-type light curve. Photometric observations of NN Vir were done by Gomez–Ferrellad and Garcia–Melendo (1997) at Esteve Duran Observatory. The first spectroscopic observations of this system were obtained by Rucinski and Lu (1999). The radial velocity and light curves analysis was made with the latest version of the Wilson program (1998), and the geometric and physical elements of the system are derived. From the simultaneous solutions of the system, we determined the masses and radii of the components: 1.89 $M_\odot$ and 1.65 $R_\odot$ for the primary component; 0.93 $M_\odot$ and 1.23 $R_\odot$ for the secondary component. We estimated effective temperatures of 7030 K for the primary and 6977 K for the secondary component.

1. Introduction

NN Vir ($=$ HD 125488 $=$ BD $+ 06°2869 =$ GSC 323.930) is one of the stars, whose variability was detected by the Hipparcos satellite. An analysis of the photometric data from the Tycho Mean Photometric Catalogue and the Tycho Photometric Observations Catalogue performed by Woitas (1997) yielded a list of 43 new bright variables. According to Woitas, NN Vir is an RR Lyr variable with a 0.20 day period.

Photometric observations of NN Vir were made by Gomez–Ferrellad and Garcia-Melendo (1997) at Esteve Duran Observatory. Their observations showed that HD 125488 is not an RR Lyr star, but an eclipsing binary system with a period of 0.48 days, and the following ephemeris was computed:

$$Min I = HJD\ 2450525.6434 + 0^d.48069 \times E \qquad (1)$$

The first spectroscopic observations of NN Vir were done by Rucinski and Lu (1999). They found that this system has a spectral classification F0-F1, systemic velocity $V_\gamma = -6.24 \pm 0.65$ km/s, and a relatively large mass ratio $q_{sp} = 0.491 \pm 0.011$, which is rather uncommon among A-type contact binary systems.

Astrophysics and Space Science **296:** 269–272, 2005.

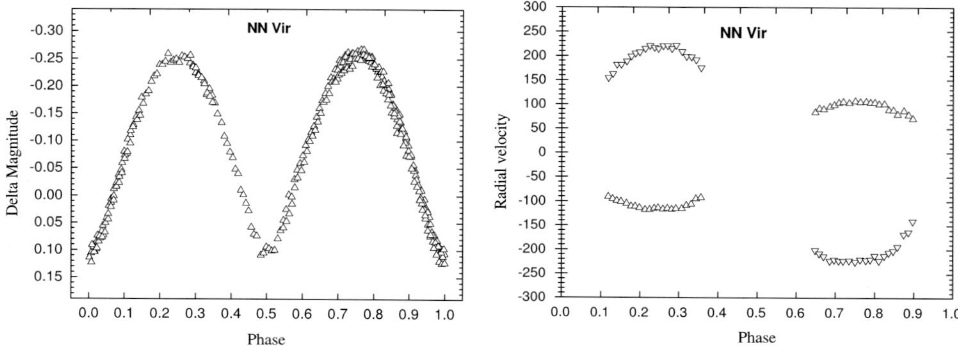

Figure 1. Light and radial velocity curves of NN Vir plotted vs. orbital phase as obtained at Esteve Duran and David Dunlap Observatories, respectively.

As CCD data in the *V* colour bandpass (329 points) as well as radial velocity data (46 points) of this system proved to be highly precise and reliable, we utilized these data for analysis. Radial velocity and light curves, phased with ephemeris (1), are shown in Figure 1.

2. Radial Velocity and Light Curve Analysis

Photometric and spectroscopic solutions of NN Vir were obtained by means of the latest version of the Wilson program (1998).

Based on photometric data of Gomez–Ferrellad and Garcia–Melendo (1997), magnitude differences were converted to intensities and used as photometric input data. For the radial velocity analysis we used all 46 points data given by Rucinski and Lu (1999). Two radial velocity and one light curve were employed simultaneously for determination of the geometric and physical elements of the system.

For our detailed analysis mode 3 of the Wilson program (1998), suitable for contact binaries with differences in the components' surface brightnesses, was used with constraints for gravity darkening exponents $g_1 = g_2$, bolometric albedos $A_1 = A_2$, modified surface potentials of two components $\Omega_1 = \Omega_2$, limb darkening coefficients $x_1 = x_2$, $y_1 = y_2$, and the luminosity of the secondary component coupled to its temperature. The gravity darkening coefficients were adopted to be $g_1 = g_2 = 0.32$ (Lucy, 1967), and bolometric albedo to be $A_1 = A_2 = 0.50$ (Rucinski, 1969), in accordance with the stellar convective envelope. Stellar rotation was assumed to be synchronized for both components.

For both components, we used the bolometric logarithmic limb darkening law of Klinglesmith and Sobieski (1970); the *x* and *y* parameters of both components were fixed to their theoretical values, interpolated with the Vhlimb program of Van Hamme (1993). According to the spectral type of the system, F0–F1, the temperature of the primary component was adopted to be $T_1 = 7030$ K. We also assumed that the stars had no spots and no third light, $l_3 = 0.0$. The following parameters were adjusted: the orbital inclination i, the mass ratio $q = m_2/m_1$, the mean surface temperature of secondary component T_2, the non-dimensional surface potential of primary component Ω_1 $(= \Omega_2)$, the monochromatic luminosity of primary component L_1, the semi-major axis a, the systemic velocity V_γ, the orbital period p, the zero point of the orbital ephemeris HJD_0, and the first time derivative of orbital period dp/dt. These parameters were varied until the solution converged. The adopted solution for NN Vir is summarized in Table I and Table II, and the theoretical radial velocity and light curves are shown in Figure 2.

TABLE I

Elements of the binary system NN Vir (most quantities are defined in the text)

Element	Value	Element	Value
i	66.270 ± 0.834	q	0.491 ± 0.005
$a(R_\odot)$	2.982 ± 0.005	V_γ	6.110 ± 0.004
$\Omega_1 = \Omega_2$	2.801 ± 0.027		
$g_1 = g_2$	0.320^*	$A_1 = A_2$	0.500^*
x_1	0.639^*	x_2	0.641^*
y_1	0.255^*	y_2	0.250^*
$T_1(K)$	7030^*	$T_2(K)$	6977 ± 174
$L_{1V}/(L_1 + L_2)$	0.650 ± 0.004	$L_{2V}/(L_1 + L_2)$	0.350 ± 0.004
$r_{1(pole)}$	0.426 ± 0.005	$r_{2(pole)}$	0.309 ± 0.005
$r_{1(side)}$	0.405 ± 0.006	$r_{2(side)}$	0.328 ± 0.006
$r_{1(back)}$	0.480 ± 0.008	$r_{2(back)}$	0.372 ± 0.011
period	$0^d.4806547 \pm 0.0000053$	dp/dt	0.133×10^{-5}
HJD_0	$2450595.34564 \pm 0.00124$	$(\sum(\text{res})^2)$	0.0522

*Assumed.

TABLE II

Physical parameters of NN Vir

Parameter	Value	Parameter	Value
R_1 $(R_\odot)$	1.65	R_2 $(R_\odot)$	1.23
M_1 $(M_\odot)$	1.89	M_2 $(M_\odot)$	0.93
L_1 $(L_\odot)$	6.00	L_2 $(L_\odot)$	3.22
T_1 (K)	7030	T_2 (K)	6977
$(M_{bol})_1$	2.84	$(M_{bol})_2$	3.52

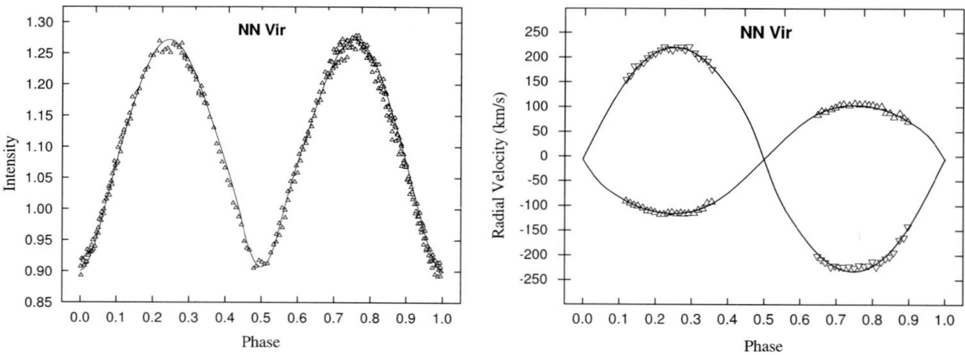

Figure 2. Triangles show individual observations of NN Vir, the model fits are shown as continuous lines.

3. Summary and Discussion

In the present work, the results from a radial velocity and light curve analysis of NN Vir, based on photoelectric data of Gomez-Ferrellad and Garcia-Melendo (1997) and the first radial velocity data of Rucinski and Lu (1999) are presented and discussed. We used the latest version of the Wilson program (1998) for eclipsing binary stars. Two radial velocity and one light curve were employed simultaneously for the determination of the geometric and physical elements of the system. The values of the system parameters are given in Tables I and II.

References

Ferrellad, G. and Melendo, E.G.: 1997, IBVS, No. **4469**.
Klinglesmith, D.A. and Sobieski, S.: 1970, *AJ* **75**, 175.
Lucy, L.B.: 1967, *Z. Astrophys.* **65**, 89.
Rucinski, S.M.: 1969, *Acta Astron.* **19**, 245.
Rucinski, S.M. and Lu, W.: 1999, *AJ* **118**, 2459.
Van Hamme, W.: 1993, *AJ* **106**, 2096.
Wilson, R.E.: 1990, *ApJ* **356**, 613.
Wilson, R.E.: 2002, *The Program for Eclipsing Binary Stars*. Available online at: ftp.ufl.astro.pub/wilson/lcdcprog.
Woitas, J.: 1997, IBVS, No. **4444**.

ON THE OVERCONTACT BINARY V2388 Oph AND AN IMPROVED M–R RELATION FOR LTCBs

KADRI YAKUT

*Department of Astronomy and Space Sciences, Faculty of Science, University of EGE,
Bornova 35100, İzmir, Turkey; E-mail: yakut@astronomy.sci.ege.edu.tr*

(accepted April 2004)

Abstract. The evolutionary stage of the low-temperature contact binary (LTCB) V2388 Oph has been investigated. V2388 Oph was previously classified as an A-type W UMa star, and is the brighter member of the visual binary Fin 381. When compared to other well-known LTCBs it is evident that the primary component has evolved to the TAMS, and the companion also seems to be more evolved than a ZAMS star. Mochnacki proposed a new subgroup of W UMa stars, namely of OO Aql type, distinct from A and W types. V2388 Oph is suggested to be a member of this new group.

Keywords: stars, binaries, eclipsing-stars, binaries, visual-stars, fundamental parameters, stars, individual, V2388 Oph-techniques, photometry

1. Introduction

The components of contact binaries are in physical contact. There are two kinds of contact binaries: one group, known as W UMa systems, which are extensively studied low-temperature contact binaries (LTCBs), whose components share a common convective envelope, while another group consists of high-temperature contact binaries (ETCBs) with common radiative envelopes.

LTCBs have been divided into two subgroups of A- and W-type by Binnendijk (1970). If the more massive component is eclipsed during the deeper (primary) minimum, we are dealing with an A-type system, and if the less massive star is the more luminous one, the system is called W-type. The spectral types of A-type and W-type systems range from A to G and F to K, respectively. Although about 500 contact binary systems are known, only part of them has been studied both photometrically and spectroscopically. Besides many other researchers, especially Rucinski and colleagues performed spectral studies, which helped to define the physical parameters of these systems.

2. V2388 Oph

A total of 25 observing nights were spent on V2388 Oph. The light curve obtained in the year 2000 season was published by Yakut and İbanoğlu (2000). New light curves with additional observational data were published by Yakut et al. (2004)

TABLE I

Absolute parameters of V2388 Oph

Parameter	Primary	Secondary
M ($M_\odot$)	1.80 (2)	0.34 (1)
R ($R_\odot$)	2.60 (2)	1.30 (1)
T (K)	6900	6349 (23)
L ($L_\odot$)	13.5 (1.9)	2.43 (7)

(YKİ). The photometric observations were obtained at Ege University Observatory using the 48 cm Cassegrain and 30 cm Schmidt–Cassegrain telescopes. Detailed information about the observations can be found in the paper by YKİ.

V2388 Oph is the brighter member of the visual binary Fin 381. The parameters of the visual binary were obtained by numerous authors. Recently Söderhjelm (1999) determined the system's orbital parameters as $P = 8.9$ years, $a = 0''.088$, $e = 0.32$, $i = 157°$, $\omega = 233°$ and $\Omega = 169°$. Independently the variability of V2388 Oph was reported by Rodriguez et al. (1999) and Hipparcos (ESA, 1997). Rucinski et al. (2002) obtained the radial velocity curve of the system and determined the amplitudes as $K_1 = 44.62$ km s^{-1}, $K_2 = 240.22$ km s^{-1}, and hence a mass ratio of 0.186.

The light curve of the system was solved by YKİ with the Wilson–Deviney (WD) code (Wilson and Devinney, 1971; Wilson, 1994; Wilson and van Hamme, 2003). In the present study, the same light curves were solved with the 2003 version of the WD code. The latest version has some additions to the contact mode 3 compared to previous versions. Now the A_1, A_2, x_1, x_2, g_1, and g_2 parameters can be treated as free parameters. Our light curve solution is nevertheless similar to the previous one (Figure 1a), and the physical parameters of the system are consistent with those given in Table I.

3. Discussion and Conclusions

LTCB and near-contact systems have been studied extensively in the past. Yakut and Eggleton (2004) have recently listed those which were subject to simultaneous photometric and spectroscopic studies, and for which physical parameters are well determined. Figures (1c–1e) show the $M - L$, $M - R$, and $T - M_{\mathrm{bol}}$ relations for LTCBs with most reliable physical parameters. As is apparent from the figure, the primary component of V2388 Oph seems to be highly evolved. Also its companion is more evolved than those of other systems like, e.g., OO Aql, AH Aur, or EF Dra. Mochnacki (1981) proposed that W UMa systems may be reclassified by adding a third class besides A- and W-type, namely OO Aql (sometimes also called B-type). V2388 Oph may be a member of this new class, besides AH Aur and EF Dra. A

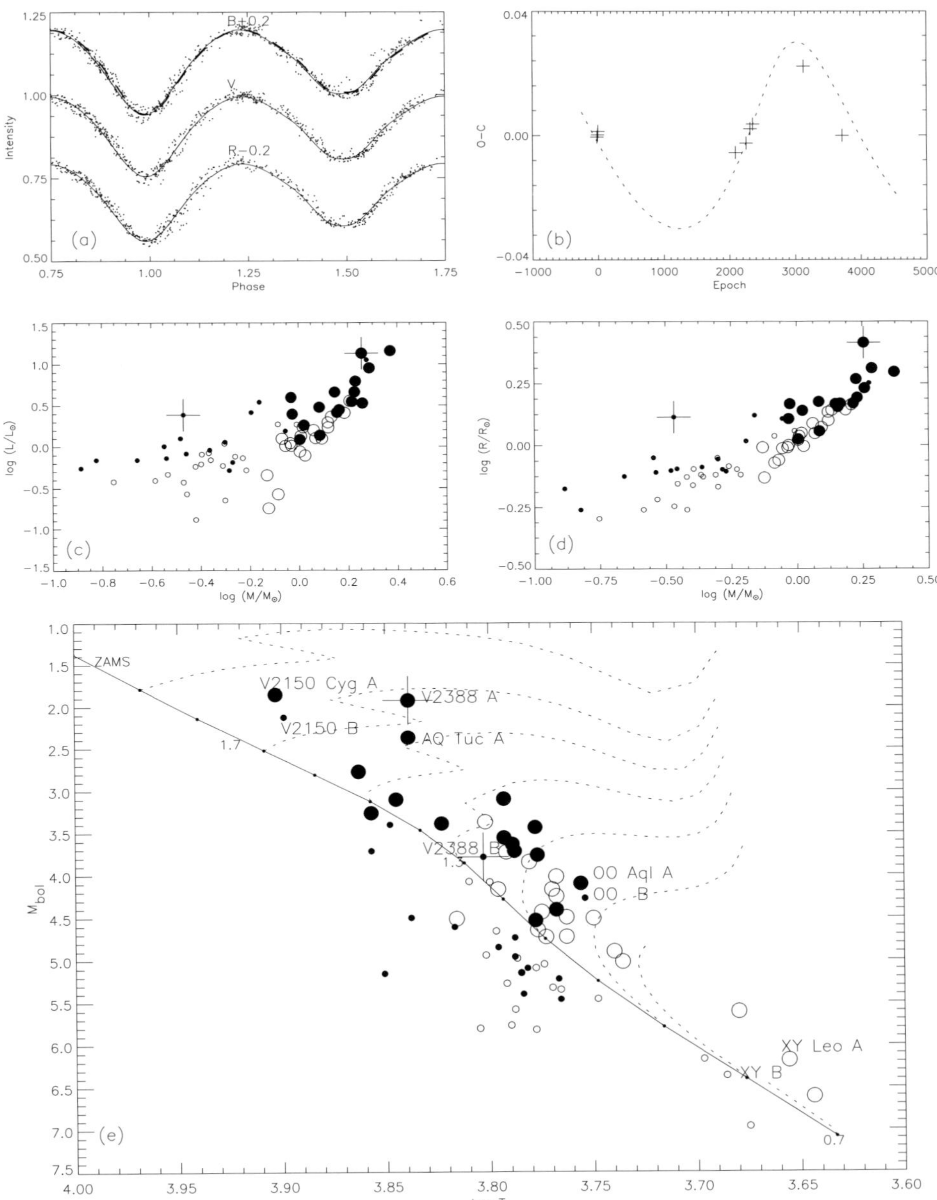

Figure 1. a) Comparison of observed and calculated *B V R* light curves; solid line corresponds to WD 1998 version, bold-dashed line to WD 2003 version; b) O–C diagram; the dashed line was derived from the parameters of the visual binary orbit; c) mass-luminosity diagram for LTCBs; filled circles are for A-type, open circles for W-type contact binaries, smaller symbols are for secondary companions; d) mass-radius diagram for LTCBs; e) LTCBs in the HR diagram; V2388 Oph components are marked by "+" symbols.

276 K. YAKUT

least-squares fit of W-type LTCB systems done by Maceroni et al. (1985) yielded a near-linear mass-radius relation with 15% standard deviation (see their Eq. 3). On the basis of new data this relation could be improved to a standard deviation of 6% as follows:

$$\log R = 0.814(0.079)\log M + 0.039(0.012). \tag{1}$$

Acknowledgements

I am very grateful to Drs Drechsel, İbanoğlu and Pekünlü for carefully reading this paper, and TÜBİTAK-BAYG for their support. This study has been partly supported by Ege University Research Project.

References

Binnendijk, L.: 1970, vistas in *Astronomy* **12** 217.
ESA: 1997, The Hipparcos and Tycho Catalogues, SP-1200.
Maceroni, C., Milano, L. and Russo, G.: 1985, *MNRAS* **217** 843.
Mochnacki, S.W.: 1981, *ApJ* **245** 650.
Rodriguez, E., Claret, A., Garcia, J.M., et al.: 1998, *A&A* **336** 920.
Rucinski, S.M., Lu, W., Capobianco C.C., Mochnacki, S.W., Blake, R.M., Thomson, J.R., Ogloza, W. and Stachowski, G.: 2002, *AJ* **124** 1738.
Rahunen, T.: 1981, *A&A* **102** 81.
Söderhjelm, S.: 1999, *A&A* **341** 121.
Wilson, R.E.: 1992, Document of Eclipsing Binary Com. Model, University of Florida, Florida.
Wilson R.E.: 1994, *PASP* **106** 921.
Wilson, R.E. and Devinney, E.J.: 1971, *ApJ* **166** 605.
Wilson, R.E. and Van Hamme, W.: 2003, Document of Eclipsing Binary Com. Model, University of Florida, Florida.
Yakut, K. and İbanoğlu, C.: 2000, IBVS No. 5002.
Yakut, K., Kalomeni, B. and İbanoğlu, C.: 2004, *A&A* **417** 725.
Yakut, K. and Ulaş, B.: 2004, submitted.

AQ Psc – ANALYSIS OF NEW LIGHT CURVES

ATSUMA YAMASAKI

Department of Earth & Ocean Sciences, National Defense Academy, Yokosuka, Japan;
E-mail: yamasaki@nda.ac.jp

(accepted April 2004)

Abstract. New BV light curves of the A-type W UMa star AQ Psc ($P = 0.476$d) have been observed and are described. A few times of minimum light are obtained and the ephemeris is improved. The light curves are analyzed for the binary parameters with a light curve synthesis method. Combining the results with Lu and Rucinski's spectroscopic mass ratio we determined the masses and radii of the components: $M_1 = 1.69M_\odot$, $M_2 = 0.38M_\odot$, $R_1 = 1.77R_\odot$, and $R_2 = 0.89R_\odot$.

Keywords: contact binaries, CCD photometry, light curve, AQ Psc

1. Introduction

While making observations of the RS CVn star, HD 8357, Sarma and Radhakrishnan (1982) noticed light variations of a check star, HD 8152 ($V = 8.67$). Subsequently they made photometry of HD 8152 and found that the star shows nearly equal depths of primary and secondary minima and continuous light variation outside the eclipses, characteristic of W UMa binaries. They determined the ephemeris of this new W UMa star, AQ Psc ($=$ HD 8152 $=$ BD $+$ 06 203 $=$ HIP 6307) to be Min I $=$ HJD 2444562.4691 $+$ 0.47564 $\cdot$ E, and estimated the spectral type to be F8 from the colour.

Lu and Rucinski (1999) observed the radial velocities of AQ Psc and determined the spectroscopic orbit: A-type, $K_1 = 59.5$ km s^{-1}, $K_2 = 263.3$ km s^{-1}, $q_{sp} = M_2/M_1 = 0.226$.

However, no extensive photometric study of AQ Psc has been made since then – only a few times of minima have been reported.

2. CCD Photometry

Photometry of AQ Psc was made at National Defense Academy (Yokosuka) between October 2003–February 2004 with a 20-cm Schmidt-Cassegrain telescope and a CCD, SBIG ST-8E, through BV filters. HD 7997 ($=$ BD $+$ 06 197, $V = 8.42$, $B = 9.33$) was chosen as the comparison star. V observations were made on six nights (October 23, 24, 27, 29, 30, and February 4; 769 points), and B observations on eight nights (November 14, 17, 21, 26, December 2, 22, 25, and February 5; 822 points).

Astrophysics and Space Science **296:** 277–280, 2005.
© Springer 2005

TABLE I

Times of minima

Time[a]	Ecl	$O - C$	Ref.[b]	Time[a]	Ecl	$O - C$	Ref.[b]
44562.4691	I	−0.0096	1	52193.3573	II	0.0622	6
49283.3250	I	−0.0207	2	52189.5488	I	0.0546	7
49326.3696	II	−0.0159	2	52194.5459	II	0.0614	8
49327.3203	II	−0.0169	2	52936.9761	I	0.0821	9
49596.4629	II	−0.1224	3	52942.2081	I	0.0828	9
49596.5183	II	−0.0059	4	52943.1611	I	0.0866	9
50333.4720	I	0.0000	5	53039.9469	II	0.0867	9

[a]Time denotes minimum time − 2400000.

[b]References − 1: Sarma and Radhakrishnan (1982) UBV; 2: Müyesseroğlu et al. (1996) UBV; 3: Demircan et al. (1994) B; 4: Demircan et al. (1994) V; 5: Lu and Rucinski (1999) Sp; 6: Derman and Kalci (2003) UBV; 7: Müyesseroğlu et al. (2003) BVR II?; 8: Müyesseroğlu et al. (2003) BV I?; 9: present V.

Four times of minima have been obtained and are given in Table I, where earlier times of minima from the literature are also listed. (Note that the minima of Müyesseroğlu et al. (2003) seem to be designated in the opposite sense.) $O - C$ residuals (Table I) are computed with Lu and Rucinski's (1999) ephemeris:

$$\text{Min I} = \text{HJD}\,2449283.3292 + 0.4756056 \cdot E. \tag{1}$$

Figure 1 suggests that a period change occurred around HJD 2450000, and the period stays constant since then. We improved Lu and Rucinski's ephemeris (1) to be

$$\text{Min I} = \text{HJD}\,2450333.47192 + 0.47561287 \cdot E. \tag{2}$$
$$\pm 8 \qquad\qquad \pm 25$$

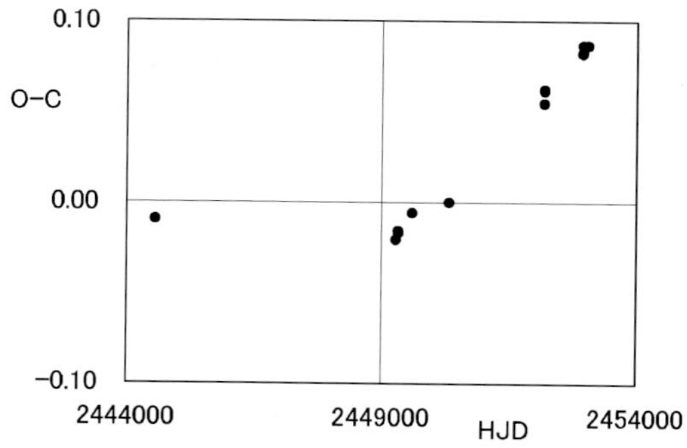

Figure 1. Residuals of times of minima of AQ Psc using Lu and Rucinski's (1999) ephemeris.

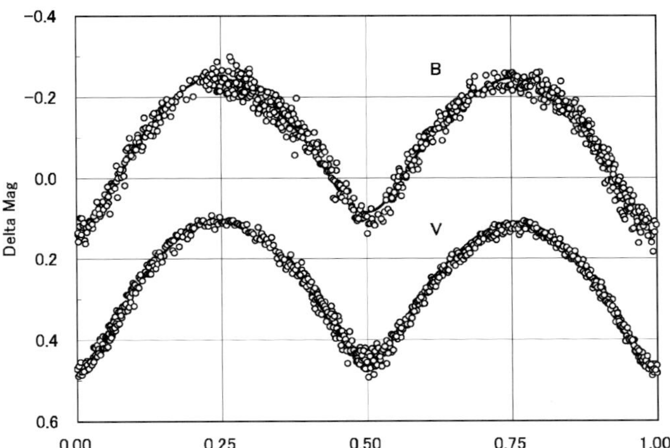

Figure 2. Light curves of AQ Psc. Delta mag (var–comp) versus orbital phase. *B* (top); *V* (bottom). Solid lines denote theoretical curves.

3. Light Curves

Light variations were plotted against orbital phase computed with the ephemeris (2). Figure 2 shows BV light curves of AQ Psc. The B light curve shows rather large scatter due to the small telescope used and the low quantum efficiency of the CCD in B. Both light curves show nearly equal depths of the primary and secondary minima and continuous light variations outside eclipses, confirming the finding of Sarma and Radhakrishnan (1982). The height of Max II (phase 0.75) is a little fainter than that of Max I (phase 0.25).

4. Light-Curve Analysis

BV light curves were analyzed with a light-curve synthesis method based on the Roche model (Yamasaki, 1975, unpublished; I further improved the method during 1977–1979 (Yamasaki, 1981), while I stayed at Manchester University – I wish to express my deep appreciation to late Professor Zdeněk Kopal). In the analysis, the mass ratio was fixed to $q = 0.226$ (Lu and Rucinski, 1999), and the limb darkening coefficients were adopted from Van Hamme (1993) with $T_{\mathrm{eff1}} = 6250$ K for F8 V (Sarma and Radhakrishnan, 1982). Convective atmospheres were assumed for both components.

The fitted models are

$$V : i = 68.3 \pm 0.1, \ \Omega_1 = \Omega_2 = 2.2248 \pm 0.0056,$$

$$r_1 = 0.542 \pm 0.001, r_2 = 0.273 \pm 0.001, L_1/(L_1 + L_2) = 0.770 \pm 0.002,$$

$$\Sigma \, (O - C)^2 = 0.0962, \sigma \, (\text{fit error}) = 0.011;$$

$$B: i = 69.3 \pm 0.1, \Omega_1 = \Omega_2 = 2.2309 \pm 0.0060,$$
$$r_1 = 0.539 \pm 0.001, r_2 = 0.271 \pm 0.001, L_1/(L_1 + L_2) = 0.794 \pm 0.003,$$
$$\Sigma (O - C)^2 = 0.2380, \sigma(\text{fit error}) = 0.017;$$

with $q = 0.226$, $\Omega_{in} = 2.2962$, $\Omega_{out} = 2.1529$ being fixed. Solid curves in Figure 2 show the respective theoretical models. Apparently the theoretical curve in V fits the observations better than that in B. Therefore, the V-solution is adopted as the system solution.

By combining the photometric solution with the spectroscopic elements given by Lu and Rucinski (1999), the following absolute parameters are obtained.

$$M_1 = 1.69 \pm 0.04 M_\odot, M_2 = 0.38 \pm 0.01 M_\odot,$$
$$R_1 = 1.77 \pm 0.02 R_\odot, R_2 = 0.89 \pm 0.01 R_\odot,$$
$$T_{\text{eff1}} = 6250 \text{ K (assumed)}, T_{\text{eff2}} = 6180 \pm 60 \text{ K}.$$

The distance to AQ Psc is estimated to be 125 pc from the HIPPARCOS Catalogue (ESA, 1997). This leads to an absolute magnitude of AQ Psc of $M_v \simeq 3.0$ at light maximum.

Acknowledgements

I thank M. Hirahara and S. Inada for their help with the observations.

References

Demircan, O. et al.: 1994, *IBVS* 4126.
Derman, E. and Kalci, R.: 2003, *IBVS* 5439.
ESA: 1997, *HIPPARCOS TYCHO Catalogues* ESA SP-1200.
Lu, W. and Rucinski, S.M.: 1999, *Astron. J.* **118**, 515–526.
Müyesseroğlu, Z., Gürol, B. and Selam S.O.: 1996, *IBVS* 4380.
Müyesseroğlu, Z., Törün, E., Özdemir, T., Gürol, B., Özavci, I., Tunç, T. and Kaya, F.: 2003, *IBVS* 5463.
Sarma, M.B.K. and Radhakrishnan, K.R.: 1982, *IBVS* 2073.
Van Hamme, W.: 1993, *Astron. J.* **106**, 2096–2117.
Yamasaki, A.: 1981, *Astrophys. Space Sci.* **77**, 75–109.

CCD PHOTOMETRY OF THE NEGLECTED CONTACT BINARIES V344 Lac AND V1191 Cyg

THEODOR PRIBULLA[1], MARTIN VAŇKO[1], DRAHOMÍR CHOCHOL[1],
ŠTEFAN PARIMUCHA[2] and DANIEL BALUĎANSKÝ[3]

[1]*Astronomical Institute, Slovak Academy of Sciences, Tatranská Lomnica, Slovakia;*
E-mail: pribulla@ta3.sk
[2]*Faculty of Theoretical Physics and Astrophysics, University of P.J. Šafárik, Moyzesova, Košice,*
Slovakia
[3]*Observatory Roztoky, Roztoky, Slovakia*

(accepted April 2004)

Abstract. New photoelectric and CCD observations of the eclipsing contact binary systems V344 Lac and V1191 Cyg are presented and analyzed. All available times of minimum light were used to study period changes of the systems and determine up-to-date ephemerides. The orbital period of V1191 Cyg is found to be increasing at a very fast rate. The photometric elements were determined using the new light curve, radial-velocity curve and broadening function fitting code ROCHE.

Keywords: stars, eclipsing binaries

1. Introduction

Many bright ($V < 13$ mag) contact systems have unknown photometric and absolute elements. This fact was the main motivation for the investigation of these objects. The main aim is the analysis of the light curves (hereafter LCs), period changes and evolution of neglected contact binaries.

V344 Lac (GSC 3619-847, $V_{max} = 12.3$, $V_{min} = 13.0$, sp. type A3:) was discovered to be a variable by Miller and Wachmann (1973). First photoelectric minima were published by Hoffmann (1983). Since then the system has been observed only visually until 1996, when more regular CCD observations by BBSAG members started. GCVS 4 ephemeris (Min I $= 2\,445\,222.5635 + 0.39222768 \times E$) does not predict recent CCD minima well.

V1191 Cyg (GSC 3159-1512, $V_{max} = 10.82$, $V_{min} = 11.15$) was found to be a variable star by Mayer (1965). GCVS 4 ephemeris (Min I $= 2\,438\,634.5471 + 0.313377 \times E$) predicts minima half an hour earlier than observed at present.

For both systems neither CCD nor photoelectric LCs have been published and analyzed yet. The systems were not observed spectroscopically.

Astrophysics and Space Science **296**: 281–284, 2005.
© Springer 2005

2. CCD Photometry

The first part of CCD photometry was obtained using the 50 cm Newton tele-
scope of the Stará Lesná Observatory (SL) of the Astronomical Institute of the
Slovak Academy of Sciences. The telescope is equipped with the SBIG ST-10
MXE camera. The observations of both systems were obtained in $BV(RI)_c$ fil-
ters. For a detailed description of the instrument see Pribulla and Chochol (2003).
V344 Lac was observed on September 13, 18, and 20, 2003; V1191 Cyg on
September 19 and 22, 2003.

Another part of CCD photometry was obtained at the Roztoky Observatory
(RO) ($\lambda = 21°28'54''$ E, $\varphi = 49°33'57''$ N). The 40-cm Cassegrain telescope
is equipped with the SBIG ST-8 CCD camera. The observations of V1191 Cyg
were obtained in $V(RI)_c$ filters. V344 Lac was observed in the I passband only.
V344 Lac was observed on Nov. 11–13, 2003; V344 Lac on May 18, June 19 and 30,
2002.

The CCD frames were reduced in the usual way (bias and dark subtraction,
flat-field correction) with the MIDAS reduction package using procedures written
by the first author. At RO the frames were reduced using the MuniPack package
(http://www.ian.cz/munipack/).

Our observations enabled us to determine seven times of minima of V344 Lac
and seven times of minima of V1191 Cyg. The times of minima were determined separately for all filters using the Kwee-van Woerden method. The minima of V1191 Cyg up to September 2002 were already published (Pribulla et al.,
2002).

3. Orbital Period Analysis

For **V344 Lac** there are more than 60 photographic, visual, CCD and photoelectric
(pe) times of minima available. The visual minima times (1989–1998) are hard to
interpret. The only reliable part of the data are CCD observations (1996–2003).
Excluding one deviating observation at 2 452 134.3950 of Agerer and Hübscher
(2003) we get Min I = 2 452 901.4175(6) $+ 0.39224203(22) \times E$. This ephemeris
describes our CCD observations well, and the above Min I coincides with the deeper
minimum in our BVR LCs.

For **V1191 Cyg** there are only 29 CCD and pe times of minimum light.
The period study of the system is complicated due to a large gap in the data –
the first data after the discovery observations of Mayer (1965) appear only in
1993. This causes ambiguity in number of cycles elapsed in between. The last,
continuous sequence of minima displays an unusually fast period increase: Min I =
2 452 548.5128(6)$+0.31338505(12) \times E + 5.67(3) 10^{-10} \times E^2$. The average rate of
period increase is $\Delta P/P = 4.216(19) 10^{-6}$ per year. Other contact binaries, which
show a fast continuous period change, are AP Aur ($\Delta P/P = 2.12 10^{-6}$ year^{-1}) and

EZ Hya ($\Delta P/P = -1.46\ 10^{-6}$ year^{-1}) (Pribulla et al., 2003). This period change could be interpreted as the result of mass transfer from the less to the more massive component.

4. Light-Curve Analysis

For the LC analysis of both systems the new LC, radial–velocity curve and broadening function fitting code ROCHE (Pribulla, in press) was applied. Since the mass ratio of both systems is unknown we performed grid searches in q and in the $q - i$ plane. For both systems the model atmosphere option (assuming solar metalicity) and a logarithmic limb darkening law was used.

For **V344 Lac** we adopted a polar temperature of the primary component of $T_1 = 8560$ K according to the observed colour index. For the computation of the photometric elements we took observations obtained at SL (taken in the course of a week). The times of the observations were not phased, so the ephemeris could also be optimized.

Since the estimated spectral type is close to the margin between convective and radiative envelopes ($T_{\rm eff} \approx 7500$ K), we calculated two sets of solutions. The convective solution was found to be much better. Since there is no spectroscopic mass ratio, we performed a grid search for the best solution in the $q - i$ plane. χ^2 varies only weakly for mass ratios from $0.24 < q < 0.36$ because of the fact that the eclipses are partial. Hence the estimated mass ratio needs to be verified spectroscopically. The fit has been obtained under the assumption that the primary (deeper) minimum is the transit of the smaller (less massive) component. This corresponds to an A-subtype classification of the system. We tested the other possibility, but ended up with a χ^2 about 30% higher. The resulting geometric elements of the system are $i = 77.5(2)°$, $q = 0.24$, $f = 0.45(1)$.

For **V1191 Cygni** the estimate of the spectral type has not been published yet. According to $B - V = 0.62$, obtained from our observations, we adopted $T_1 = 5860$ K. For the computation of photometric elements we used only CCD observations from SL. Since the time interval of the observations is very short, only the phase shift was optimized. Coefficients of bolometric albedo and gravity darkening were fixed appropriate to the convective case of energy transfer. Similarly as in V344 Lac, we performed a grid search in the interval of mass ratios $q = 0.07 - 0.15$. The minimum of χ^2 is relatively well defined due to the fact that the eclipse is total. The system is definitely of W-subtype, i.e. the more massive, primary component is the cooler one. The resulting geometric elements of the system are $i = 80.4(4)°$, $q = 0.094$, $f = 0.46(2)$. The best fits to the data for both systems are shown in Figure 1. In the case of V1191 Cyg, a systematic deviation of the fit in the R passband is visible around primary minimum. A possible explanation is the presence of third light.

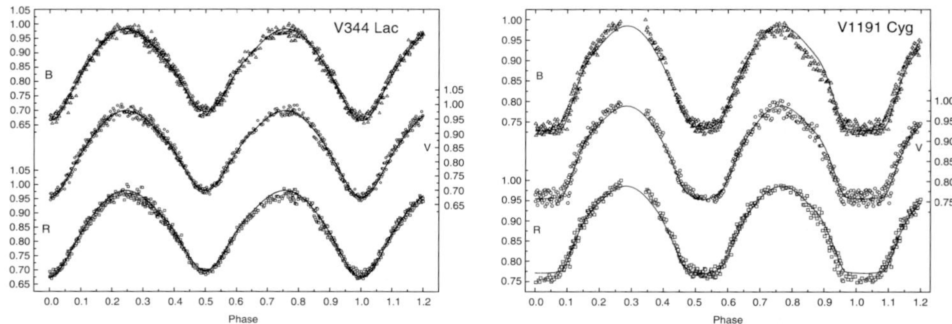

Figure 1. Best fits to the light curves of V1191 Cyg and V344 Lac.

A detailed discussion of the evolutionary status of both systems is in preparation. Since V1191 Cyg is rather bright, it deserves a thorough spectroscopic study.

Acknowledgements

This work was supported by the Science and Technology Assistance Agency, contract No. APVT-20-014402, and by VEGA grant No. 2/4014/4.

References

Agerer, F. and Hübscher, J.: 2003, *IBVS* 5484.
Hoffmann, M.: 1983, *IBVS* 2344.
Mayer, P.: 1965, *Bull. Astron. Obs. Czechosl.* **16**, 255.
Miller, W.J. and Wachmann, A.A.: 1973, *Ricerche Astron.* **8**, 367.
Pribulla, T.: in press, in: R.W. Hilditch, et al. (eds.), Spectroscopically and Spatially Resolving the Components of Close Binary Stars, ASP Conference Series.
Pribulla, T. and Chochol, D.: 2003, *Baltic Astron.* **12**, 555.
Pribulla, T., Kreiner, J.M. and Tremko, J.: 2003, *CoSka* **33**, 38.
Pribulla, T., Vaňko, M., Parimucha, Š. and Chochol, D.: 2002, *IBVS* 5341.

CCD PHOTOMETRY OF NEGLECTED ECLIPSING BINARIES – KZ Dra, LR Cam AND IM Vul

ONDŘEJ PEJCHA[1], DAVID MOTL[1], MILOSLAV ZEJDA[1], KAREL KOSS[1],
JITKA KUDRNÁČOVÁ[1], PETR HÁJEK[1], LADISLAV ŠMELCER[2],
PAVOL A. DUBOVSKÝ[3], EVA ŠAFÁŘOVÁ[3] and MAREK WOLF[4]

[1]*Nicholas Copernicus Observatory and Planetarium, Brno, Czech Republic;*
E-mail: pejcha@astro.sci.muni.cz
[2]*Valašské Meziříčí Observatory, Valašské Meziříčí, Czech Republic*
[3]*BRNO – VSS of the Czech Astronomical Society, Czech Republic*
[4]*Astronomical Institute, Charles University Prague, Czech Republic*

(accepted April 2004)

Abstract. We present ephemerides and solutions of one Algol-type (KZ Dra) and two overcontact systems (LR Cam and IM Vul) based on $V(RI)_C$ CCD observations obtained in the project *Prosper* (network of amateur observers).

Keywords: photometry, eclipsing binary, model, KZ Dra, IM Vul, LR Cam

1. Introduction

BRNO – Variable Star Section of the Czech Astronomical Society has been supporting amateur astronomers interested in the field of variable stars since its establishment in the early twentieth century. First members of the VSS, including Zdeněk Kopal, observed semiregular and irregular stars and LPVs. In the sixties, Oto Obůrka restored the VSS along with a new observational programme – short period variable stars and especially eclipsing binaries.

Nowadays, the observational programme of BRNO includes more than 1500 eclipsing binaries with known ephemerides (sometimes inaccurate or incorrect, though). Recently, several most dedicated observers were looking for larger scientific impact and established a new project – *Prosper* – dedicated to neglected eclipsing binaries. Inspired by GEOS, a selection of about 20 suspected eclipsing binaries was made and an extensive visual and CCD monitoring has started.

This paper contains several results of our 2-year effort. Observations presented here were acquired using three 30–40-cm class telescopes spread over the Czech Republic. SBIG ST-7 CCD cameras equipped with $V(RI)_C$ filters were used for imaging and standard aperture photometry was performed using *Munipack* (Hroch et al., 2001; based on DAOPHOT routines).

Astrophysics and Space Science **296**: 285–288, 2005.
© Springer 2005

2. Analysis of Observations

For each star, new accurate ephemerides were determined and used for the phasing of their light curves. Normal points with a binning size of $0.01 \times P$ were computed and used as input for the solution with the 2003 version of the Wilson–Devinney code (Wilson, 1993) and with the *Nightfall* program (Wichmann, 2003). Results of both programmes agree very well, and the actual values of computed parameters are given in Table I.

The overcontact systems discussed in this paper (IM Vul and LR Cam) have low inclinations and exhibit only partial eclipses. Therefore, their photometric mass ratios should be taken with care. The given errors can only be regarded as formal ones (cf. Terrell and Wilson, 2004).

3. Notes on Individual Stars

3.1. KZ Dra

The preliminary ephemerides of KZ Dra ($=$ TmzV131 $=$ GSC 4446 1025) were determined by a visual observer (PAD), and CCD multi-colour photometry followed. The system might be slightly eccentric. Light curves along with the fit are shown in Figure 1.

3.2. LR Cam

For a detailed review of the history of LR Cam ($=$ NSV 2544 $=$ GSC 4344 0123) see Pejcha et al. (2001). The fit was computed using recent observations only as there

TABLE I

Computed parameters of studied systems

Parameter	KZ Dra	LR Cam	IM Vul
M_0	2452794.466 (1)	2453047.4938 (3)	2452907.295 (1)
P (d)	2.233664 (10)	0.43413875 (31)	0.4542670 (77)
i (deg)	85.6 (4)	76.3 (3)	66.5 (5)
q	0.15 (1)	0.39 (1)	0.34 (1)
T_1 (K)	12000 (1000)	5500 (200)	6200 (1000)
T_2 (K)	5100 (300)	5200 (200)	6000 (1000)
$r_{1_{\text{pole}}}$	0.24 (1)	0.49 (1)	0.47 (1)
$r_{2_{\text{pole}}}$	0.25 (1)	0.32 (1)	0.29 (1)
Ω_1	4.54 (9)	2.65 (2)	2.48 (1)
Ω_2	2.11[a]	2.65 (2)	2.48 (1)

Note. Errors in the last digits are given in parentheses.

[a] Ω_2 was kept fixed at its critical value during fitting.

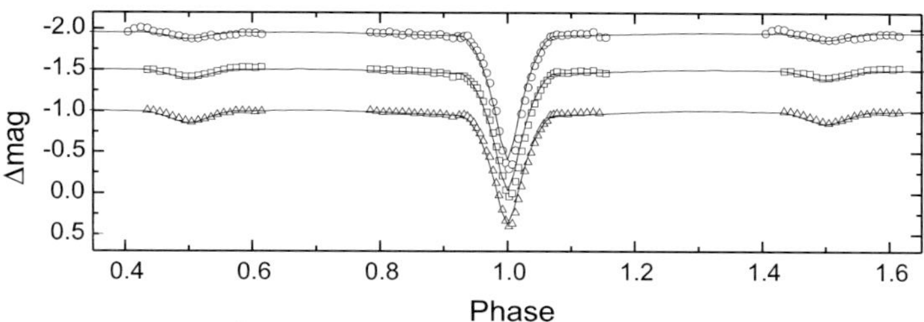

Figure 1. Light curves of KZ Dra. Measurements in V, R_C and I_C passbands are denoted by circles, squares and triangles, respectively. Wilson–Devinney theoretical light curves are plotted as solid lines.

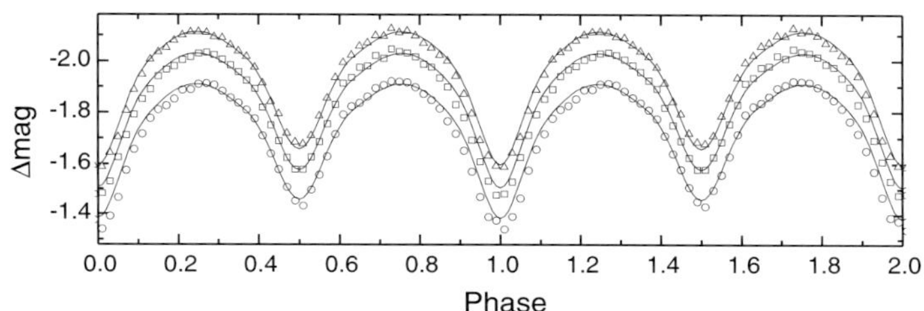

Figure 2. Light curves of LR Cam, meaning of symbols same as in Figure 1.

are some variations in the $O-C$ diagram (observational material is too scarce to discuss the nature of the changes in more thorough detail). As an initial constraint on temperatures we used the 2MASS colour $J - K_S = 0.40$ mag, which corresponds to late G or early K spectral type. LR Cam is also an X-ray source, which suggests coronal activity in the system. The light curves are shown in Figure 2.

3.3. IM Vul

Discovered by Romano (1957), IM Vul ($=$ GSC 1646 1588) remained unstudied until it was included in the programme *Prosper*. As an initial constraint on temperatures we used the 2MASS colour $J - K_S = 0.22$ mag, which corresponds to mid-F spectral type. Since only V and I_C measurements of IM Vul were available, temperatures given in Table I are highly correlated and thus have large uncertainties. Light curves are shown in Figure 3.

4. Conclusions

We determined new ephemerides of the Algol-type star KZ Dra and of the two overcontact binaries LR Cam and IM Vul. We also computed models of these systems using *Nightfall* and the Wilson–Devinney code. Our results provide starting

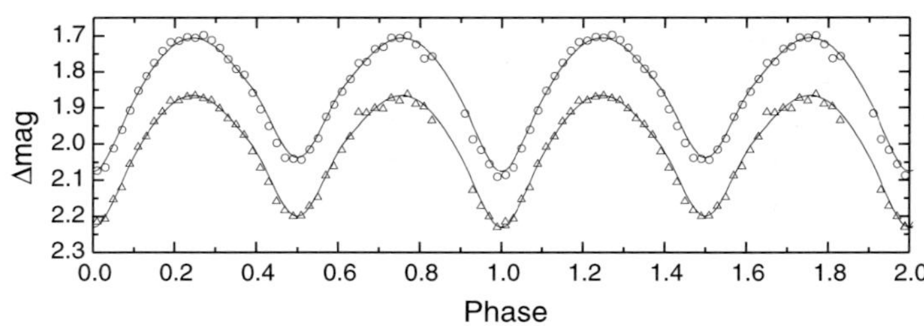

Figure 3. Light curves of IM Vul, meaning of symbols same as in Figure 1.

points for future analysis of these systems. Especially LR Cam deserves further attention, as it is an X-ray source and exhibits variations in the O–C diagram.

Acknowledgements

This investigation was supported by the Grant Agency of the Czech Republic, grant No. 205/04/2063. This research has made use of the SIMBAD database, operated at CDS, Strasbourg, France, and of NASA's Astrophysics Data System Bibliographic Services.

References

Hroch, F., Novák, R., Král, L.: 2001, Available at http://munipack.astronomy.cz.
Pejcha, O., Lehký, M., Sobotka, P., Brát, L., Haltuf, M. and Šmelcer, L.: 2001, *IBVS* **5132**, 1.
Romano, G.: 1957, *Coelum* **25**, N3–N4.
Terrell, D. and Wilson, R.E.: 2004, *Astrophys. Space Sci.* **295**.
Wichmann, R.: 2003, Available at www.lsw.uni-heidelberg.de/˜rwichman/Nightfall.html.
Wilson, R.E.: 1993, in: K.C. Leung and I.S. Nha (eds.), ASP Conference Series, Vol. 38. *New Frontiers in Binary Stars Research*, Astron. Soc. Pac., San Francisco, p. 91.

AN OBSERVATIONAL STUDY OF CONTACT BINARIES WITH LARGE TEMPERATURE DIFFERENCES BETWEEN COMPONENTS

MICHAŁ SIWAK

Astronomical Observatory of the Jagiellonian University, Kraków, Poland;
E-mail: siwak@oa.uj.edu.pl

(accepted April 2004)

Abstract. In this paper we investigate a sample of contact eclipsing binary systems, which exhibit a large temperature difference (at least 1000 K) between the components. Considering the effective temperature of the primary star, the systems were divided into three groups. We applied a Monte Carlo method as a more suitable procedure for the search of the system configurations previously known as contact binaries with a large temperature difference and with a negligible value of the filling factor. Using only data presented in the literature, we found that the geometrical configuration of almost all systems from the second group is near-contact rather than contact.

Keywords: close binaries, V1010 Oph, V747 Cen, DO Cas, BL And

1. Introduction

Objects of special interest are contact binary systems with a large temperature difference (ΔT) between components. How large? Lucy's calculations showed a difference of 880 K in the evolution stage following the loss of contact (Lucy, 1976). According to Flannery (1976), during the semidetached phase of the binary oscillations, the secondary star can be 500 K cooler than the primary, but no more than 1000 K. On the basis of Kähler's results, the value of the order of 1350 K is an upper ΔT limit for stable contact binaries, but only for those with small values of the *fill-out* parameter (Kähler, 2004).

Therefore, on the basis of the above estimations, one can infer that from a theoretical point of view, unlike binary systems with $\Delta T < 1000$ K, predicted by TRO theory (Lucy, 1976; Flannery, 1976; Robertson and Eggleton, 1977) and also really observed, contact binaries with a shared convective envelope and with ΔT more than 1000–1350 K should not exist. However, on the basis of the computations performed by means of light curve synthesis models, we know that these systems, well in physical contact and with a ΔT significantly larger than 1000 K, actually are observed.

In the second section, we present all short period binary systems (also of early type), for which at least one author has obtained a contact configuration with ΔT more than 1000 K between components. In the third section we critically reconsider the light curve solution of four binaries using a Monte Carlo search method (Barone et al., 1988; Zoła et al., 1997) in connection with the Wilson–Devinney model.

Astrophysics and Space Science **296**: 289–292, 2005.
© Springer 2005

2. The Sample

Considering the effective temperature of the primary star, the systems were divided into three groups: the first one includes objects, in which both the primary and the secondary component have radiative envelopes (*SV Cen, V729 Cyg, UW CMa, V606 Cen, TU Mus, V382 Cyg, V1061 Tau,* and *V758 Cen*), the second group includes objects, where the primary component has a radiative envelope, while the secondary has a convective envelope (*V1010 Oph, V747 Cen, DO Cas, BL And,*

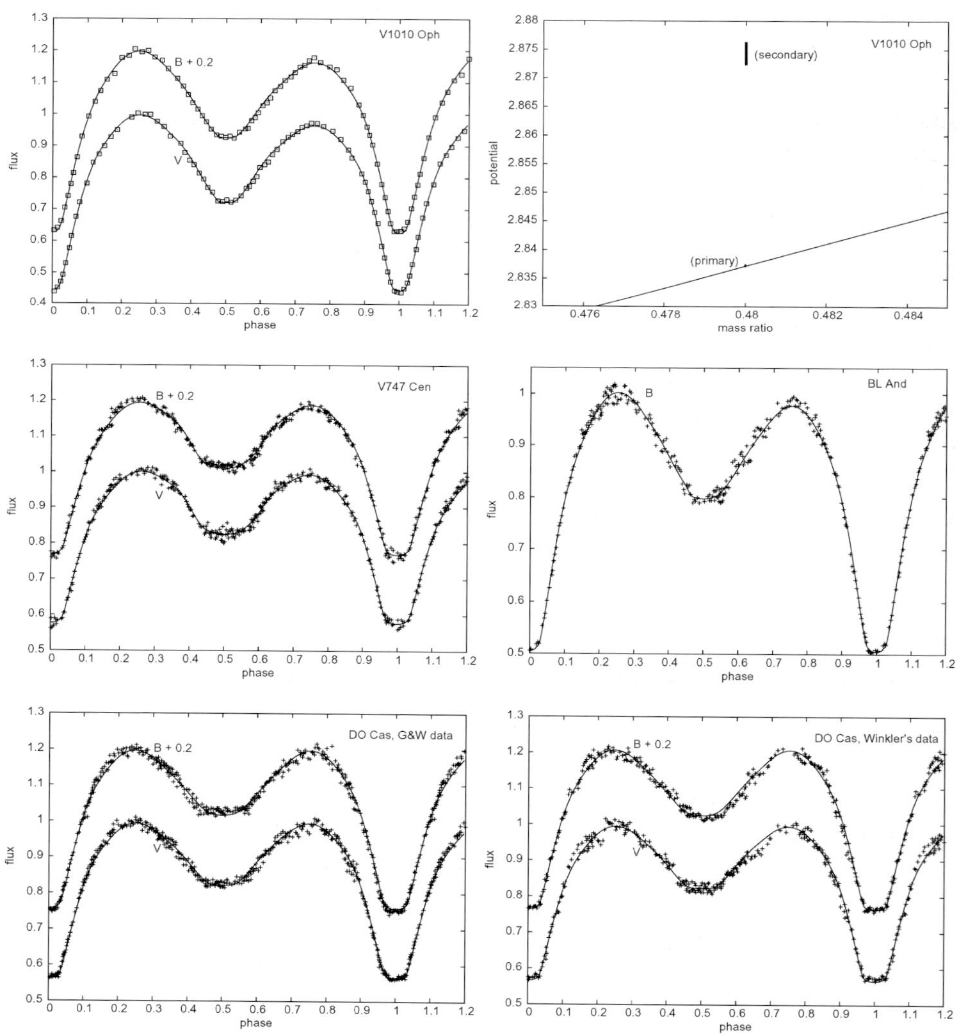

Figure 1. Light curves and the representation of the search array in the q–Ω plane.

and VV Cet). Systems, where both stars have a convective envelope, were put into the third group (*BV Eri, CN And, BX And, CX Vir, FT Lup*, and *FS Lup*).

3. Modeling and Results

It is a widely known fact that for the near-contact or the marginal-contact systems it is very difficult to distinguish between the contact and the near-contact configuration—the solutions with *a priori* assumed either contact or (semi)detached configurations can both give comparable results with practically undistinguishable resulting light curves. Therefore, we applied a Monte Carlo method, which searches for the best configuration around both sides of the line, which is defined as the function of the critical Roche lobe potential in L_1 point versus the mass ratio (see Figure 1). We performed a search for the true configuration for those systems listed above, for which inconsistent results (i.e. contact or near-contact configurations) were presented in the literature: *V1010 Oph, V747 Cen, BL And, DO Cas, CN And*, and *FT Lup*.

For *V1010 Oph* we used *B* and *V* data gathered by Leung (1974). A hot spot on the secondary star was adopted during computations in order to remove an O'Connell effect. A semidetached configuration with the primary filling its Roche lobe was obtained. In this case, only the spectroscopic mass ratio is known (Shaw et al., 1990) and was fixed as constant parameter during computations.

For *V747 Cen* we used Chambliss' data in *B* and *V* filters (Chambliss, 1970). The best fit was obtained with a hot spot on the secondary star. In the solutions of *DO Cas*, two sets of *B* and *V* data (Winkler, 1966; Gleim and Winkler, 1969) were solved simultaneously. For *BL And* we used the data gathered in the B filter by Vetešnik (1967). The best fit was obtained by assuming a hot spot on the secondary

TABLE I

Results derived from the computations

Parameter	V1010 Oph	V747 Cen	DO Cas	BL And
$T_1(K)$	7500*	8200**	8600**	7600**
$T_2(K)$	5090 (25)	4250 (74)	4796 (24)	4190 (97)
$i(°)$	86.49 (62)	86.19 (91)	88.65 (40)	89.15 (72)
$q(M_2/M_1)$	0.48 (3)	0.2983 (50)	0.3198 (16)	0.356 (20)
Ω_1	2.8372	2.4632 (96)	2.5116 (31)	2.5871 (384)
Ω_2	2.8743 (63)	2.4631 (191)	2.5123 (59)	2.5905 (806)
Ω_{inn}	2.8372	2.4625	2.5096	2.5869

Note: Theoretical values of the gravity darkening and the albedo were used: $g_1 = 1$, $A_1 = 1$, $g_2 = 0.32$, $A_2 = 0.5$.

[a]The temperature of the primary was set according to Corcoran et al. (1991).

[b]The temperature of the primary was set according to Harmanec (1988).

component. For these three systems we obtained near-contact configurations, but it is also possible that *BL And* might be a semidetached binary.

Configuration of *CN And* is most probably either detached or semidetached, but our solutions are not finished yet. We found *FT Lup* as a contact system repeatedly, contrary to Lipari and Sisteró (1987), who found it to be semidetached. We used the same data and we performed many types of computations, which always gave a contact configuration (f $= 14\%$) (Table I).

4. Conclusions

We used the WD model and a Monte Carlo algorithm in order to investigate a subgroup of contact binaries, which exhibit a temperature difference of more than 1000 K. Most systems from the second group (defined in Section 2) were found to be near-contact. Unfortunately, because of the lack of data for *VV Cet*, this group has not yet been reanalyzed completely. We are going to do so after the next observational season, based on new observations of this binary star.

Genuine contact systems with a large temperature difference will be subject of our further study.

References

Barone, F., Maceroni, C., Milano, L. and Russo, G.: 1988, *A&A* **197**, 347.
Chambliss, C.R.: 1970, *AJ* **75**, 731.
Corcoran, M.F., Siah, M.J. and Guinan, E.F.: 1991, *AJ* **101**, 1828.
Flannery, B.P.: 1976, *ApJ* **205**, 217.
Gleim, J.K. and Winkler, L.: 1969, *AJ* **74**, 1191.
Harmanec, P.: 1988, *BAC* **39**, 329.
Kähler, H., 2004, *A&A* **414**, 317.
Leung, K.-Ch.: 1974, *AJ* **79**, 852.
Lucy, L.B.: 1976, *ApJ* **205**, 208.
Robertson, J.A. and Eggleton, P.P.: 1977, *MNRAS* **179**, 359.
Shaw, J.S., Guinan, E.F. and Garasi, C.J., 1990, *BAAS* **22**, 1296.
Vetešnik, M.: 1967, *BAC* **18**, 208.
Winkler, L.: 1966, *AJ* **71**, 40.
Zoła, S., Kolonko, M. and Szczech, M.: 1997, *A&A* **324**, 1010.

PHOTOMETRIC SOLUTION OF THE ECLIPSING BINARY
V351 PEGASI

B. ALBAYRAK[1], G. DJURAŠEVIĆ[2], S. SELAM[1], M. YILMAZ[1], S. ERKAPIĆ[2],
O. AKSU[1] and T. TANRIVERDI[1]

[1]*Department of Astronomy and Space Sciences, Ankara University Faculty of Science, TR-06100,
Ankara, Turkey; E-mail: albayrak@astro1.science.ankara.edu.tr*
[2]*Astronomical Observatory, Volgina 7, 11160 Belgrade, Serbia and Montenegro, and Isaac Newton
Institute of Chile, Yugoslav Branch*

(accepted April 2004)

Abstract. *BVR* light curves of the recently discovered eclipsing binary V351 Peg were studied to derive the preliminary physical parameters of the system. The light curves were obtained at the TÜBİTAK*– Turkish National Observatory (TUG) during three nights in August, 2003. The solutions were made using Djurašević's inverse problem method. V351 Peg is a system in an overcontact configuration ($f_{over} \sim 21\%$) with a relatively small temperature difference between the components $\Delta T \approx 20$ K. The results suggest a significant mass and energy transfer from the more massive primary onto the less massive secondary. The hot area on the less massive star, near the neck region, can be considered as a consequence of this mass and energy exchange between the components through the connecting neck of the common envelope.

Keywords: binaries, close-binaries, eclipsing-V351 Peg

1. Introduction

V351 Peg (HIP 115627, BD $+ 14°$ 4990; $8^m.0$) was discovered as a variable star by the HIPPARCOS satellite project (ESA, 1997). The HIPPARCOS photometric observations of the system show a light curve variation with an amplitude of $0^m.30$, and the star was initially classified as an RRc-type pulsating variable with a 0.20 day period. Gomez-Forrellad et al. (1999) then correctly classified the system as a W Ursa Majoris type eclipsing binary. They published a light curve obtained in the *V* band and determined the initial light elements of the system. The light curve shows two almost equally deep minima with an amplitude of $0^m.32$. Rucinski et al. (2001) obtained radial velocity curves of both components and determined a mass ratio of $q = m_c/m_h = 0.36$. The indices (h, c) refer to the hotter (more massive) and cooler (less massive) component, respectively. Rucinski et al. (2001) pointed out that the spectral type of the system is A8 V, and that it belongs to the W sub-class of W UMa systems. More recently, Selam et al. (2003) published a minimum time of V351 Peg. We observed and analyzed the system in order to determine its parameters.

*TÜBİTAK: The Scientific and Technical Research Council of Turkey.

Astrophysics and Space Science **296:** 293–296, 2005.
© Springer 2005

2. Observations and Analysis

New observations of V351 Peg in B, V, and R were obtained at TUG on the nights of August 29, 30, and 31, 2003, by using an SSP-5A photometer attached to a 40 cm Cassegrain telescope. $BD + 15° 4830$ and $BD + 14° 4974$ were chosen as comparison and check stars, respectively. A total of 888 observations were secured during the observing run in each filter. The probable error of a single observation point was estimated to be ±0.012, ±0.010, and ±0.009 in B, V, and R colours, respectively.

To estimate the parameters of V351 Peg, we used the Djurašević (1992b) programme generalized for the case of an overcontact configuration (Djurašević et al., 1998). The programme is based on the Roche model and the principles arising from the paper by Wilson and Devinney (1971). The light curve analysis was performed by applying the inverse-problem method (Djurašević, 1992a) based on the Marquardt (1963) algorithm. The observed and theoretical light curves, and the view of the system at orbital phase 0.25 are shown in Figure 1. The non-linear limb darkening coefficients were taken from Claret's (2000) table.

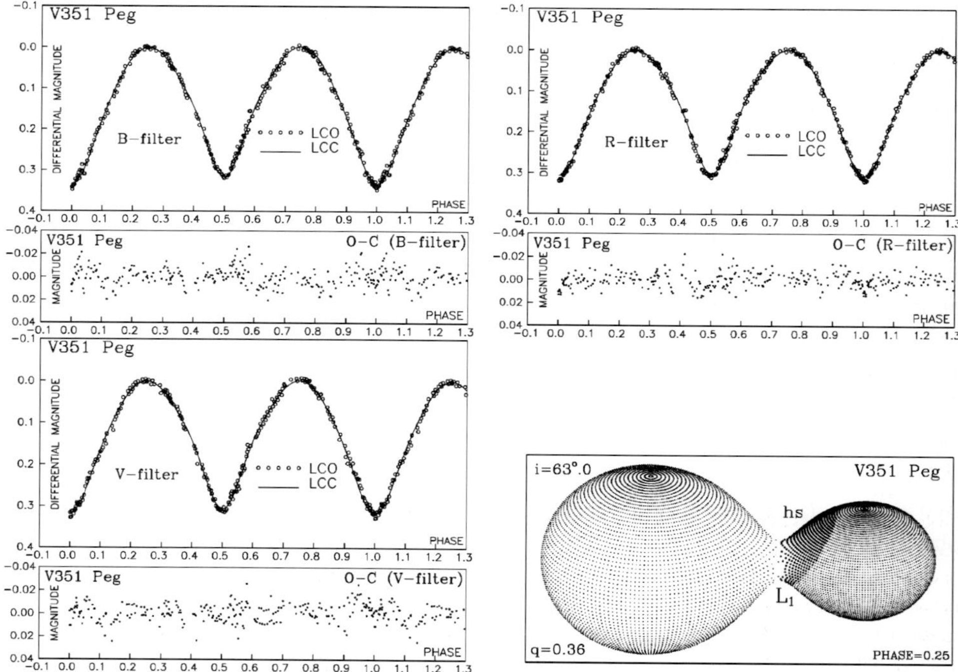

Figure 1. Observed (LCO) and final synthetic (LCC) light curves of V351 Peg with final O-C residuals obtained by analyzing B, V and R observations, and view of the system at orbital phase 0.25, obtained with the parameters of the photometric solution.

3. Conclusion

A summary of our results, given in Table I, proves that a Roche model with a hot active region on the less massive component of V351 Peg can successfully simulate the observed light curves. Synthetic light curves, obtained by solving the inverse problem, fit the observations very well, and we find quite good agreement between the solutions for individual light curves in different BVR filters of the photometric system. All this suggests the consistency of our Roche model with the hot area on the less massive star in simulating the real observations. Without assuming such an active region one can only find fits of much poorer quality.

The results describe the V351 Peg system as an overcontact configuration ($f_{over} \sim 20.6\%$) with a relatively small temperature difference between the components of $\Delta T = T_h - T_c \sim 20$ K. They also suggest a significant mass and energy transfer from the more massive primary onto the less massive secondary. The hot area on the less massive star, near the neck region, is a consequence of this mass and energy exchange.

TABLE I

Results of the simultaneous analysis of the V351 Peg BVR light curves obtained by solving the inverse problem for the Roche model with the hot spot area in the neck region on the less-massive (cooler) component

Parameter	Value	Parameter	Value
$q = m_h/m_c$	0.36 (fixed)	β_h	0.20 ± 0.01
$f_{over}[\%]$	20.64	β_c	0.18 ± 0.01
$i[°]$	63 ± 0.1	$T_h[K]$	7580 (adopted)
$\Omega_{h,c}$	2.5496	$T_c[K]$	7559 ± 15
Ω_{in}	2.5953	$L_h/(L_h + L_c)[B;V;R]$	0.700; 0.702; 0.703
Ω_{out}	2.3738	$A_{hs} = T_{hs}/T_c$	1.066 ± 0.005
$R_h[D = 1]$	0.450	θ_{hs}	41.3 ± 0.9
$R_c[D = 1]$	0.284	λ_{hs}	182.7 ± 0.8
F_h	1.020 ± 0.002	φ_{hs}	23.8 ± 2.2
$f_{h,c}$	1.0	$\Sigma(O - C)^2$	0.0511
$A_{h,c}$	1.0	σ	0.0076

Note: $q = m_c/m_h$: mass ratio of the components, f_{over} [%]: degree of overcontact, i: orbit inclination (in arc degrees), $\Omega_{h,c}$, Ω_{in}, Ω_{out}: dimensionless surface potentials of the components and of the inner and outer contact surfaces respectively, $R_{h,c}$: polar radii of the components in units of the distance between the component centers, F_h: filling factor for the critical Roche lobe of the hotter (more-massive) star, $f_{h,c}$, $A_{h,c}$, $\beta_{h,c}$: non-synchronous rotation coefficients, albedo of the components and gravity-darkening exponents, $T_{h,c}$: temperature of the hotter primary and cooler secondary, $L_h/(L_h + L_c)[B;V;R]$: luminosity of the hotter star (including hot spot on the cooler one), A_{hs}, θ_{hs}, λ_{hs} and φ_{hs}: hot spot temperature coefficient, angular dimension, longitude and latitude (in arc degrees), $\Sigma(O - C)^2$: final sum of squares of residuals between observed (LCO) and synthetic (LCC) light curves, σ: standard deviation of the observations.

                           B. ALBAYRAK ET AL.

Acknowledgements

The authors would like to thank the Turkish National Observatory (TUG) for the observing time. This research was supported by the Turkish Academy of Sciences in the framework of the Young Scientist Award Program (BA/TÜBA-GEBIP/2001-2-2), by the Ministry of Sciences and Technology of Serbia through the project 1191 *Stellar Physics* and by the Research Fund of Ankara University (BAP) under the research project nos: 20040705089 and 20040705090.

References

Claret, A.: 2000, *A&A* **363**, 1081.
Djurašević, G.: 1992a, *Ap&SS* **197**, 17.
Djurašević, G.: 1992b, *Ap&SS* **196**, 241.
Djurašević, G., Zakirov, M., Hojaev, A. and Arzumanyants G.: 1998, *A&AS* **131**, 17.
ESA: 1997, *The Hipparcos and Tycho Catalogues*, ESA SP-1200.
Gomez-Forrellad, J.M., Garcia-Melendo, E. and Guarro-Flo, J.: 1999, *IAU-IBVS* No. 4702.
Marquardt, D.W.: 1963, *J. Soc. Ind. Appl. Math.* **11**, No.2, 431.
Rucinski, S.M., Lu, W., Mochnacki, S.W., Ogloza, W. and Stachowski, G.: 2001, *AJ* **122**, 1974.
Selam, S.O., Albayrak, B., Şenavcı, H.V., et al.: 2003, *IAU-IBVS* No.5471.
Wilson, R.E. and Devinney, E.J.: 1971, *ApJ* **166**, 605.

V2213 Cyg IS A NEW W UMa-TYPE SYSTEM

ELENA PAVLENKO and ANNA LITVINCHOVA

Crimean Astrophysical Observatory, Tavrichesky National University, Crimea, Ukraine;
E-mail: pavlenko@crao.crimea.ua

(accepted April 2004)

Abstract. V2213 Cyg was discovered as a variable star by Pavlenko (1999) in 1998. We present our photometry of V2213 Cyg from 1998–2003 based on CCD observations with the K-380 Cassegrain telescope of CrAO and the 60 cm Zeiss telescope of SAI. Observations have been carried out mostly in R and sometimes in B and V Johnson system. The total amount of data is 2270 points, covering $\sim$50 nights. We classify this binary as a W UMa-type contact system. Using all data we determined the orbital period to be 0.350079 ± 0.000007 day. The mean brightness varies between $R = 14.35$ and 14.05. The mean 1999–2003 orbital light curve has two humps and a primary minimum (I), which is 0.04 mag brighter than the deeper secondary one (II). The mean humps have slightly different height. The difference between two individual maxima varies within 0.1 mag, which may indicate an activity of the components. The highest hump is an asymmetrical one: it has sort of a shoulder at phases 0.75–0.80, before entering the less deep primary minimum (phase 0.0). The system is rather reddened, its colour indices are: $B - V \sim 0.8$ and $V - R \sim 0.7$, and give a spectral class of V2213 Cyg earlier than K.

Keywords: stars, binaries, W UMa-type stars

1. Introduction

V2213 Cyg as a variable star was discovered incidentally by Pavlenko (1999) during observations of the cataclysmic variable V1504 Cyg. The new variable was located within 1 arc min from this star. It showed photometric variations of 0.3 mag with a period of 0.175 d. In a first guess this kind of variability could be attributed to δ Scuti or W UMa-type stars. If the latter case is true the period should be doubled to 0.35 d.

2. Observations

In order to classify this new variable we combined the available data obtained at different telescopes. Most observations of V2213 Cyg have been carried out in the R Johnson system with the Cassegrain K-380 telescope of the Crimean Astrophysical Observatory (CrAO) and the Zeiss-600 telescope of the Crimean Laboratory of the Sternberg Astronomical Institute (SAI). We used a CCD SBIG ST-7 camera. Less data were obtained in V (Johnson). Some brightness estimations in B and V (Johnson) were made in 1998 with the TV-complex instrument of the 0.5 m meniscus Maksutov telescope of the CrAO (Abramenko et al., 1988). Since 1998 to present time we acquired 2270 observations, which span $\sim$ 50 nights. We used the same comparison star for all observations (Pavlenko, 1999). The data reduction was made using the aperture photometry package developed by Goranskij. The accuracy of observations in different nights was 0.02–0.04 mag.

3. Identification

In order to define more exactly the photometrical period we combined all R data and constructed the power spectrum (Figure 1). In 1998–2003 the variable varied between $R = 14.35$ and 14.05 and kept the mean brightness constant within the accuracy of observations, so no correction for a trend had to be used. The most significant peak points to the period 0.175 d. In the case of an SU UMa system the orbital period should be doubled, and the phase diagram should display two maxima and two minima per period. In the case of a δ Scuti-type variable we would not expect a systematic and constant difference between neighbouring maxima and minima. All R and V data folded with the 0.35 d period are shown in Figure 2. Note that only R data covered all phases well enough. They do show the stable difference of the neighboring maxima and minima. The primary minimum at phase 0.0 is $0.^{m}04$ brighter than the secondary one. The mean humps have different shape: the maximum at phase 0.25 is more rounded than the one at phase 0.75 and has sort

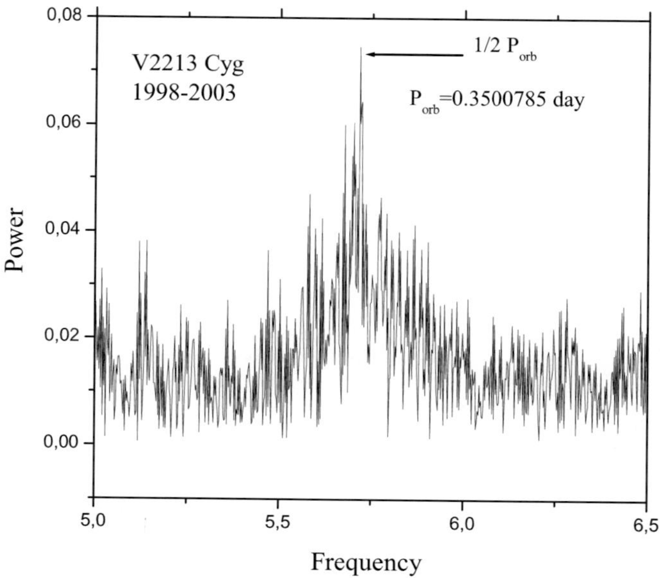

Figure 1. Power spectrum of V2213 Cyg.

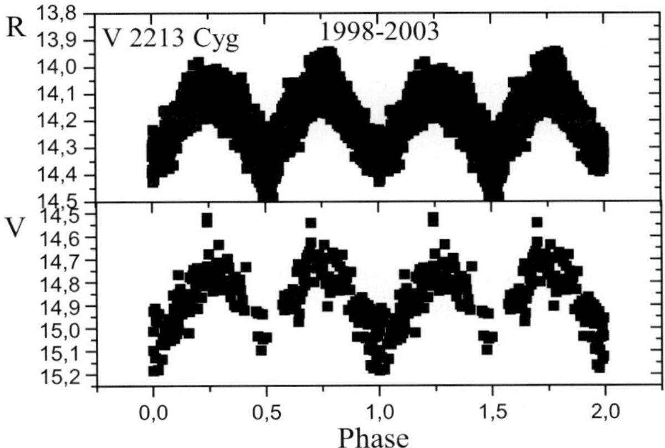

Figure 2. All *R* and *V* data of V2213 Cyg folded with the 0.35 d period.

of a "shoulder" in the phase range 0.75–0.80, before entering the less deep primary minimum at phase 0.0. We derived the following orbital ephemeris for V2213 Cyg:

$$\text{HJD (Min I)} = 2451029.5207 + (0.350079 \pm 0.000007)E, \tag{1}$$

where Min I is the minimum with smaller depth.

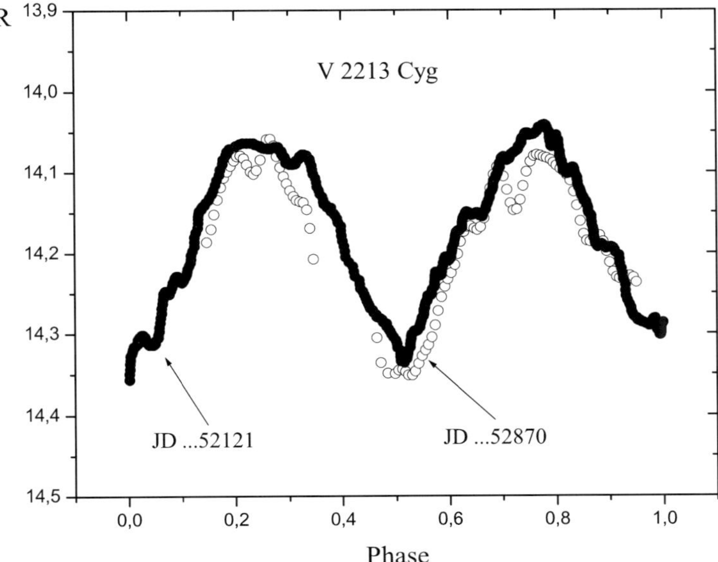

Figure 3. Two selected smoothed light curves of V2213 Cyg.

We determined the mean colour indices of the star as $B - V \sim 0.8$ and $V - R \sim$ 0.7, so the variable is rather reddened. We therefore could define the spectral class of V2213 Cyg as earlier than K.

Taking into account the stable photometric period and amplitudes, the light curve shape and colour indices, we could classify the variable as a new W UMa-type binary. The orbital period and colour indices satisfy the limiting period-colour relation (Rucinski, 1994) for stars of lowered metallicities.

Although the mean light curve is stable from year to year, the individual light curves display a slight difference of their shape within 0.1 mag. In Figure 3 an example of two smoothed curves is shown. The dip at phase 0.72 (open circles) significantly exceeds the errors of observations. Such effect could be caused by a possible activity of the components.

Acknowledgements

This work was partially supported by Ukrainian Fund For Fundamental Researches 02/07/00451.

References

Abramenko, A.N., Prokoféva, V.V. and Bondar, N.I.: 1988, *Izv. CrAO* **78**, 182.
Pavlenko, E.P.: 1999, IBVS No. 4671.
Rucinski, S.M.: 1994, *AJ* **107**, 738.

LIGHT CURVE ANALYSIS OF TWO NEW W UMa STARS IN M15

S.M. SAAD[1,2]

[1]*Astronomicky ustav, Akademie věd Česke republiky, CZ-251 65 Ondřejov, Czech Republic;*
E-mail: somaya@sunstel.asu.cas.cz
[2]*National Research Institute of Astronomy and Geophysics, NRIAG, 11421 Helwan, Cairo, Egypt*

(accepted April 2004)

Abstract. We present the first light curve analysis for two newly discovered W UMa eclipsing binaries in the field of the globular cluster M15. As spectroscopic observations are not available, we used the FOTEL program to obtain the light curve solution of both systems. The fundamental properties of each system are obtained. Several tests concerning values of mass ratio and effective temperatures were carried out. Independent distance moduli for each system are obtained, and the membership to the parent cluster is discussed. It is shown that W2 is a member of M15, whereas W1 is not.

1. Introduction

Based on time series of CCD photometry in both blue and visual filters of M15 from 1997 to 2000, two new variable stars, [W1 ($\alpha_{2000} = 21^h\ 30^m\ 21.06^s$, $\delta_{2000} = +12°9'9''$)] and [W2 ($\alpha_{2000} = 21^h\ 29^m\ 52.25^s$, $\delta_{2000} = +12°6'11''$)] were initially discovered and classified as possible eclipsing systems in M15 by Jeon et al. (2001a). The designations W1 and W2 as assigned in the original paper are used here. In the course of three and nine nights for W1 and W2, respectively, they obtained two compelete light curves for W1 and W2. Periods of $0^d.23306$, $0^d.23576$, mean magnitudes and colour indices of $20^m.246$, $\langle B \rangle - \langle V \rangle = 1.014$ and $19^m.791$, $\langle B \rangle - \langle V \rangle = 0.560$ were reported for W1 and W2, respectively. For more detailes about data reduction and standardization see Jeon et al. (2001a).

The analysis of the light curve is performed using the FOTEL code of Hadrava (1990) with the advantage of simultaneous solution of different bandpasses at individual zero points for each data set. Linear limb darkening coefficients $u_{\lambda,1}$ and $u_{\lambda,2}$ were adopted from Claret (2000), values of (0.88, 0.72) were adopted for W1 and (0.77, 0.63) for W2, respectively for B and V filters. The bolometric albedos ($k_{\lambda,1} = k_{\lambda,2} = 0.5$) and the gravity darkening exponents ($g_1 = g_2 = 0.32$) were fixed through different runs. The light curves were analyzed by simultaneous computations in both filters, and another two solutions were obtained independently for each filter. Since there are no spectroscopic mass ratios of these systems known, several values of mass ratios q between 0.10 and 1.0 with steps 0.05 were allowed to be adjusted. The final adopted solution is where there is a minimum deviation of individual observations from the solution, i.e. at that $(O - C)$, which represents the minimum of the sum of squares. The shape of the primary and secondary

Astrophysics and Space Science **296**: 301–304, 2005.
© Springer 2005

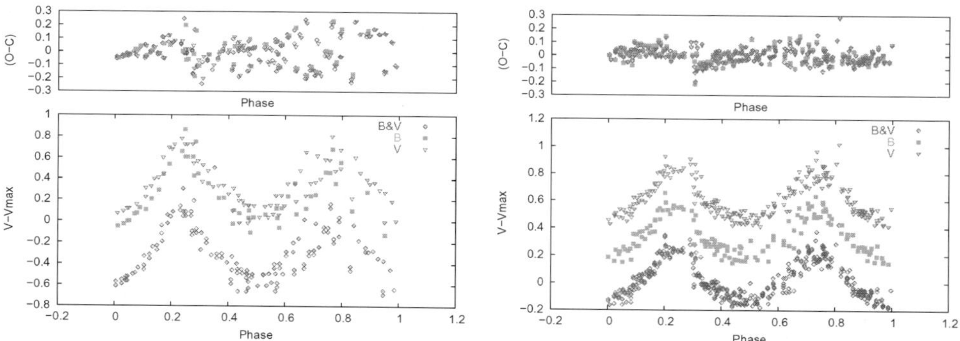

Figure 1. FOTEL results for W1 and W2, *Upper panel*: the residuals of different solutions. *Lower panel*: the difference between individual magnitudes and the zero-level (both folded with the orbital period).

considered is characterized by triaxial ellipsoids with semiaxes $a_{i,j}$, $b_{i,j}$ and $c_{i,j}$, where $i, j = 1, 2$, and the effective radius of each component is taken as the mean value of these three semiaxes $(r_{\mathrm{eff}})_{i,j}$. Masses of the primaries were adopted from Harmanec's (1988) mass vs. T_{eff} relations based on modern binary data. Combining the obtained photometric elements with the fundamental relations of mass, temperature and radius we can roughly estimate the absolute dimensions for each component in both systems.

$$q = (M_2/M_1) \tag{1}$$

$$A/\mathrm{R}_\odot = [74.55(M_1/\mathrm{M}_\odot)(1+q)P^2]^{1/3} \tag{2}$$

$$R_{1,2}/\mathrm{R}_\odot = Ar_{1,2}(\mathrm{side}) \tag{3}$$

$$L_{1,2}/\mathrm{L}_\odot = (R_{1,2}/\mathrm{R}_\odot)^2(T_{1,2}/\mathrm{T}_\odot)^4 \tag{4}$$

$$(M_{\mathrm{bol}})_{1,2} = 42.36 - 10\log T_{1,2} - 5\log(R_{1,2}/\mathrm{R}_\odot) \tag{5}$$

where q, A, $R_{1,2}$, $L_{1,2}$, $M_{1,2}$ and $M_{(\mathrm{bol})_{1,2}}$ represent the mass ratio, projected separation between the two components, absolute radii, absolute luminosities, masses, and bolometric magnitudes of the primary and secondary components.

2. Results and Discussion

2.1. System W1

The light curves in both B and V filters indicate unequal depths and show an asymmetry (i.e., a difference in the levels of maximum light is observed). This effect is known and attributed to the presence of star spots on the surface of one or both components (Bradstreet, 1985; Corcoran et al., 1991). The two components of spectral type K 2 and K 5, with $T_1 = 4700$ K, $T_2 = 4200$ K, $q = 0.50$, and effective radii $r_1 = 0.47$ and $r_2 = 0.34$ are suggested by the solution. The $(O - C)$ residuals from the three solutions show a poor fit, which is a direct reason not to consider the probable effect of the spots in our analysis.

2.2. SYSTEM W2

The shape of the light curve indicates a typical classical EW W UMa system, which we can infer from two nearly equal eclipse depths. The obtained elements suggest two components of spectral type F 8 and G 2 of nearly similar size, with $\langle q \rangle = 0.92$, $T_1 = 6100$ K, $T_2 = 5900$ K, and effective relative radii $r_1 = 0.41$, $r_2 = 0.39$. Figure 1 represents the solutions for W1 (left) and for W2 (right). Table I summarizes the photometric elements of both systems from different solutions, and Table II gives the absolute dimensions of each component in both systems.

Independent distance moduli are obtained by adopting BC corrections of -0.476 and -0.088 (Popper, 1980), respectively, to the obtained M_{bol} of the W1 and W2 primaries. Distances of 6.08 and 11.86 kpc are obtained for W1 and W2, respectively. The best known distances to M15 are 13.55 kpc (Reid, 1997), 11.85 kpc (Harris, 1996), and 12.53 kpc (Jeon et al., 2001b). Considering the criterion for membership

TABLE I

Photometric parameters for W1 and W2 from B, V and $B\&V$ solutions

Parameter	System W1			System W2		
	B	V	$B\&V$	B	V	$B\&V$
i	$65°.5 \pm 1°.72$	$60°.8 \pm 1°.26$	$60°.34 \pm 0°.96$	$79°.2 \pm 2°.86$	$81°.8 \pm 3°.43$	$76°.9 \pm 1°.14$
T_1 (K)	4660 ± 194	4690 ± 130	4700 ± 118	5983 ± 205	6143 ± 173	6174 ± 145
T_2 (K)	4199 ± 152	4185 ± 122	4230 ± 105	5854 ± 345	5883 ± 243	5904 ± 213
q	0.50 ± 0.1	0.50 ± 0.1	0.50 ± 0.1	0.87 ± 0.3	0.91 ± 0.5	0.98 ± 0.15
l_1	0.71	0.71	0.71	0.56	0.56	0.56
l_2	0.27	0.27	0.27	0.43	0.43	0.43
$u_{\lambda,1} = u_{\lambda,2}$	0.88	0.72	0.88 & 0.72	0.77	0.63	0.77 & 0.63
$g_1 = g_2$	0.32	0.32	0.32	0.32	0.32	0.32
$k_{\lambda,1} = k_{\lambda,2}$	0.5	0.5	0.5	0.5	0.5	0.5
r_1(pole)	0.41	0.41	0.41	0.36	0.34	0.35
r_1(side)	0.57	0.57	0.57	0.51	0.48	0.50
r_1(back)	0.44	0.44	0.44	0.38	0.36	0.37
r_2(pole)	0.30	0.30	0.30	0.34	0.36	0.35
r_2(side)	0.43	0.43	0.43	0.48	0.51	0.50
r_2(back)	0.31	0.31	0.31	0.36	0.38	0.37

TABLE II

Absolute physical parameters for W1 and W2

	System	W1	System	W2
Element	Primary	Secondary	Primary	Secondary
M ($M_\odot$)	0.73	0.37	1.19	1.10
R ($R_\odot$)	0.93	0.71	1.08	1.06
T ($T_\odot$)	0.81	0.74	1.05	1.02
L ($L_\odot$)	0.37	0.14	1.44	1.12
M_{bol}	5.85	6.21	4.34	4.53

proposed by Rucinski (2000), and distance moduli of Jeon et al., values of 1.57 mag and 0.12 mag for ($\Delta M_V = M_V^{obs} - M_V^{cal}$) are obtained. Therefore we can confirm the membership of W2 to M15, whereas the distance to W1 suggests to locate it in the foreground field of the cluster.

Acknowledgements

This research has made use of NASA's ADS. The work was supported by grants GAČR 02/0445. My sincere thanks go to Dr. P. Hadrava, who kindly allowed me to use his code "FOTEL", to Dr. J. Kubat for many valuable suggestions, and to the system administrators at Ondřejov Observatory.

References

Bradstreet, D.H.: 1985, *ApJS* **58**, 413.

Claret, A.: 2000, *A&A* **393**, 1081.

Corcoran, M.F., Siah, M.J. and Guinan, E.F.: 1991, *AJ* **101**, 1828.

Hadrava, P.: 1990, *Contrib. Astron. Obs. Skalnate Pleso* **20**, 23.

Harmanec, P.: 1988, *Bull. Astron. Inst. Czech.* **39**, 329.

Harris, W.E.: 1996, *AJ* **112**, 1487.

Jeon, Y.-B., Lee, H., Kim, S.-L. and Lee, M.G.: 2001a, *IBVS*, 5189.

Jeon, Y.-B., Lee, H., Kim, S.-L. and Lee, M.G.: 2001b, *AJ* **121**, 2774.

Popper, D.: 1980, *Ann. Rev. Astron. Astrophys.* **18**, 115.

Reid, I.N.: 1997, *AJ* **114**, 161.

Rucinski, S.M.: 2000, *AJ* **120**, 319.

FIRST GROUND-BASED PHOTOMETRY AND LIGHT CURVE ANALYSIS OF THE RECENTLY DISCOVERED CONTACT BINARY HX UMa

S. SELAM, B. ALBAYRAK, M. YILMAZ, H.V. ŞENAVCI, İ ÖZAVCI
and C. ÇETINTAŞ

Department of Astronomy and Space Sciences, Faculty of Science, Ankara University, TR-06100, Ankara, Turkey; E-mail: selim@astro1.science.ankara.edu.tr

(accepted April 2004)

Abstract. Photoelectric *UBV* light curves of the recently discovered eclipsing binary HX UMa were obtained and studied to determine the preliminary physical parameters of the system for the first time. The observations were taken at the TÜBİTAK[1] – Turkish National Observatory (TUG) on three nights in April 2003. A simultaneous analysis of the light and radial velocity curves yields a typical A-type contact binary with a high degree of overcontact. The influence of the close visual companion to the total light of the system was taken into account during the analysis.

Keywords: binaries, close-binaries, eclipsing-HX UMa

1. Introduction

The variability of HX UMa (HIP 058648, HD 104425, BD + 43° 2177) was discovered during the HIPPARCOS mission (ESA, 1997). The character of the light curve is of typical W UMa-type, with an amplitude of 0.17 mag. A close companion at the distance of 0.63 arcsec was also discovered by HIPPARCOS. The initial light elements were given in the HIPPARCOS catalogue as:

$$\text{Min } I = BJD\ 2448500.3720 + 0^{\text{d}}.379156 \times E \tag{1}$$

The first radial velocity curve measurements of the system were obtained and analyzed by Rucinski et al. (2003). They estimated the spectral type of the system as F8V. Their radial velocity curve solution yields an A-type contact binary with a mass ratio of $q = 0.291 \pm 0.009$. The system has the potential of an excellent combined light and radial velocity curve solution. Thus, we obtained new photoelectric *UBV* light curves of HX UMa and analyzed them simultaneously with the radial velocity curves to obtain a consistent set of parameters of the system.

2. Observations and Analysis

HX UMa was observed in *UBV* pass-bands at TUG on April 6, 27, and 28, 2003, by using an OPTEC SSP-5A photoelectric photometer attached to the 40 cm

---

[1]TÜBİTAK: The Scientific and Technical Research Council of Turkey.

Astrophysics and Space Science **296**: 305–308, 2005.
© Springer 2005

Cassegrain telescope. HD 103954 and HD 104882 were used as comparison and check stars, respectively. A total number of 306 observations were secured in each pass-band. The probable error of a single observation point was estimated to be ± 0.015, ± 0.010, and ± 0.008 in U, B, and V filters, respectively.

Newly determined primary and secondary minima based on our observations were used together with the spectroscopically determined time of primary minimum by Rucinski et al. (2003) to calculate the updated light elements of the system:

$$\text{Min } I = HJD\ 2452757.5357(2) + 0^{d}.3791574(1) \times E \tag{2}$$

The light curves were phased by using this updated light element. The observed levels of both maxima are almost equal, which made the light curve analysis process rather uncomplicated. The observed slight difference in minima led to a very small temperature difference between the components.

To get a preliminary set of parameters for the system and find a good approximation to the third light contribution in each passband, we initially preferred to use the *Nightfall* code by Wichmann (1998), because the third light parameter L_3 is an adjustable quantity in this programme, but not in the WD-code. *Nightfall* is also based on Roche geometry, and all the principles to calculate theoretical light curves of eclipsing binaries was taken from the WD code (Wilson and Devinney, 1971), however did not yield errors of the adjusted parameters. Therefore, after having found a reasonable solution and good approximation to the L_3 parameter with the *Nightfall* programme, we obtained the final results of the analysis by using the WD code. The magnitude difference between HX UMa and its close companion was estimated to be $\Delta m = 3.27 \pm 0.10$ in V-band by Rucinski et al. (2003). We estimated the corresponding third light contribution in light units as $L_3 = 0.047$ in V-band and fixed it during the *Nightfall* analysis. We also fixed the primary

TABLE I

Results derived from the light curve modeling of HX UMa

Parameter	Value	Parameter	Value
f	58.7%	$\Omega_1 = \Omega_2$	2.3395 ± 0.0069
$q = m_2/m_1$	0.291 (fixed)	T_1 (K)	6650 (adopted)
i (degrees)	48.85 ± 0.60	T_2 (K)	6601 ± 47
r_1(pole)	0.481 ± 0.001	$L_1/(L_1 + L_2 + L_3)[U]$	0.731 ± 0.004
r_1(side)	0.525 ± 0.001	$L_3/(L_1 + L_2 + L_3)[U]$	0.013
r_1(back)	0.559 ± 0.001	$L_1/(L_1 + L_2 + L_3)[B]$	0.718 ± 0.003
r_2(pole)	0.284 ± 0.001	$L_3/(L_1 + L_2 + L_3)[B]$	0.032
r_2(side)	0.299 ± 0.001	$L_1/(L_1 + L_2 + L_3)[V]$	0.708 ± 0.003
r_2(back)	0.361 ± 0.002	$L_3/(L_1 + L_2 + L_3)[V]$	0.047

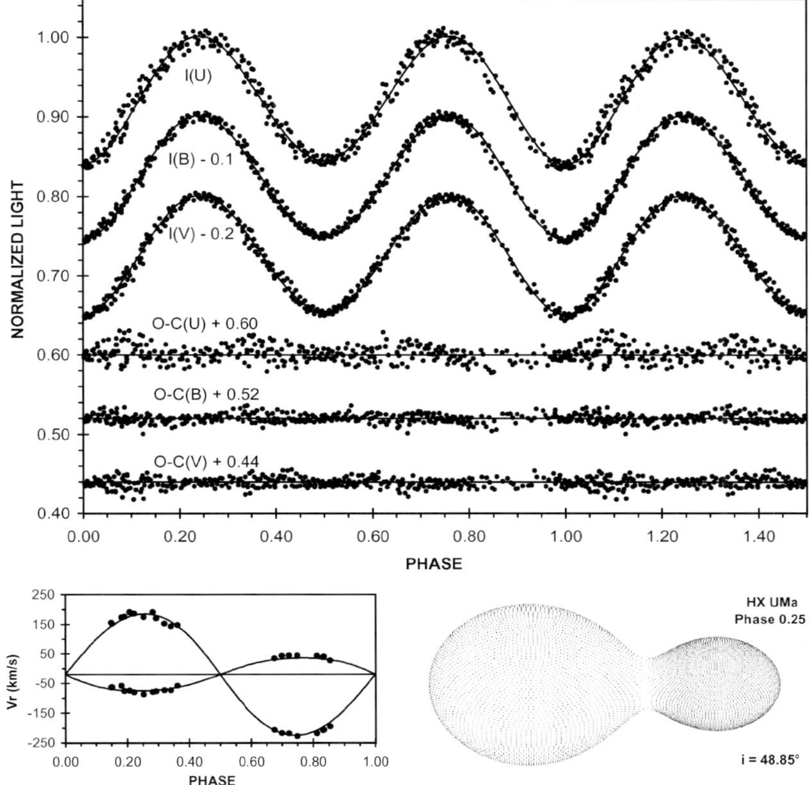

Figure 1. Top panel: Comparison between theoretical and observed light curves of HX UMa. O-C residuals from the theoretical curves in each pass-band are also shown at the bottom of the panel. Bottom left panel: Fits to the radial velocity data. Bottom right panel: Geometrical 3D model of HX UMa as seen at orbital phase 0.25.

component temperature T_1 at 6650 K according to the effective temperature versus spectral type calibration of dwarf stars by Gray and Corbally (1994). The mass ratio $q = 0.291$ was fixed at the spectroscopically determined value by Rucinski et al. (2003). Bolometric albedos and gravity darkening coefficients of the components were fixed at their theoretically adopted values for stars with convective envelopes. Limb darkening coefficients were taken from the work by Van Hamme (1993). The adjustable parameters were the orbital inclination i, the mean temperature of the secondary star T_2, the potentials of the components Ω_1 and Ω_2, the monochromatic luminosity of the primary star L_1 (Planck approximation is used to compute the luminosities), and the L_3 parameter for U and B passbands. The converged solution reveals that the third light contribution from the close companion to the total light of the system is almost negligible. After the Nightfall solution the last version of the Wilson-Devinney Code WD-2003 (Wilson and Devinney, 1971) was used to perform the final analysis. We transferred the resulting parameters from the *Nightfall* analysis to the WD code as input parameters and made several runs to get the final solution by fixing the L_3 parameters to the values obtained during the Nightfall analysis. The fixed and adjustable parameters were the same. After several runs we obtained the best fitting model in mode 3 of the WD code. The parameters of the best fitting model are given in Table I. A comparison between theoretical and observed light curves in *UBV* passbands, fits to the radial velocity data and a 3D geometrical model of the system are shown in Figure 1.

3. Conclusions

Photoelectric UBV light curves of the eclipsing binary HX UMa were obtained and studied to determine the preliminary physical parameters of the system for the first time. The simultaneous analysis of the light curves together with the radial velocity curves showed that the temperature difference between the components is only about $\Delta T = T_h - T_c = 50$ K. Thus, the system appears to be a typical A-subtype W UMa system with a high-degree of over-contact configuration ($f = 58.7\%$). The slightly hotter and more massive component is eclipsed at primary minimum. The third light contribution from the close companion to the total light of the system was found to be almost negligible.

Acknowledgements

The authors would like to thank the Turkish National Observatory (TUG) for the observing time. This research was supported by the Turkish Academy of Sciences in the framework of the Young Scientist Award Program (BA/TÜBA-GEBIP/2001-2-2) and by the Research Fund of Ankara University (BAP) under the research project no: 20040705090.

References

ESA: 1997, *The Hipparcos and Tycho Catalogues*, ESA SP-1200.
Gray, R.O. and Corbally, C.J.: 1994, *AJ* **107**, 742.
Rucinski, S.M., Capobianco, C.C. and Lu, W.: 2003, *AJ* **125**, 3258.
Van Hamme, W.: 1993, *AJ* **106**, 2096.
Wichmann, R.: 1998, *Nightfall User Manual* (http://www.lsw.uni-heidelberg.de/users/rwichman/Nightfall.html).
Wilson, R.E. and Devinney, E.J.: 1971, *ApJ* **166**, 605.

$UBV(RI)_C$ PHOTOMETRY OF THE CONTACT W UMa BINARY BD + 14°5016

A. KARSKA[1], M. MIKOLAJEWSKI[1], G. MACIEJEWSKI[1], C. GALAN[1]
and P. LIGEZA[2]

[1]*Uniwersytet Mikolaja Kopernika, Centrum Astronomii, ul. Gagarina 11, 87-100 Torun, Poland;*
E-mail: mamiko@astri.uni.torun.pl
[2]*Astronomical Observatory of the Adam Mickiewicz University, ul. Sloneczna 36, 60-286*
Poznan, Poland

(accepted April 2004)

Abstract. New $UBV(RI)_C$ observations and radial velocities of the W UMa-type system BD + 14°5016 were collected. Changes of the properties of a hot spot, located on the surface of the primary component, have been noticed.

Keywords: binaries, eclipsing – stars, individual, BD + 14°5016

1. Introduction

BD + 14°5016 (GSC 01720-00658, SAO 108714) was discovered as a variable by Maciejewski et al. (2002) during the semi-automatic variability search (SAVS) sky survey at Torun observatory (Niedzielski et al., 2003). These CCD observations (collected in *B* and *V* bands) indicated a variability of W UMa type and served to determine a preliminary ephemeris of the system. The characteristic shift in phase of secondary minimum and the noticeable O'Connell effect (different brightness maxima heights) suggested the existence of a spot on the surface of BD + 14°5016.

The Wilson–Devinney (WD) light and radial velocity curves analysis code was applied to the photometric and very few spectroscopic data (Maciejewski et al., 2003). The solution showed that BD + 14°5016 is an A-type W UMa system, which is in contact with a filling factor of about 54%. Assumption of a hot spot on the surface of the more massive component, near the neck connecting both stars, is best suited to explain the asymmetry of the light curve of the system.

2. New Observations

Several new spectra have been collected at David Dunlop Observatory (University of Toronto, Canada) using the 1.9 m telescope and the Cassegrain spectrograph with dispersion of 10.8 Å mm^{-1}. They allowed to compute a more accurate mass ratio of the system: $q = 0.23 \pm 0.02$ (Maciejewski and Ligeza, 2004).

New multicolor $UBV(RI)_C$ photoelectric observations of BD + 14°5016 were collected during three nights on September 18, 19, and October 17, 2003, using the

Astrophysics and Space Science **296**: 309–310, 2005.
© Springer 2005

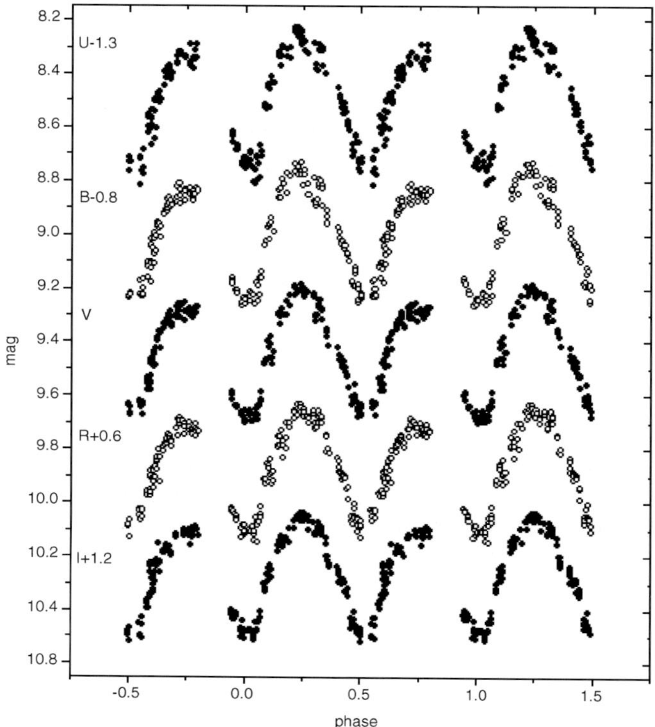

Figure 1. The light curves in $UBV(RI)_C$ filters.

0.6 m reflector of Torun observatory. The telescope was equipped with a single-channel photometer and cooled C31034 photomultiplier. The comparison star BD + 15°4849 ($V = 9^m45$) and the check star BD + 14°5011 ($V = 10^m13$) were chosen. Adopting the $UBV(RI)_C$ magnitudes for the comparison star from Ducati et al. (2001), the obtained light curves are presented in Figure 1. All curves were phased using the ephemeris from Maciejewski et al. (2002).

The observations show that the hot spot on the surface of the more massive component has not changed its location during the year. However, its size and/or temperature factor increased significantly.

References

Ducati, J.R., Bevilacqua, C.M., Rembold, S.B. and Ribeiro, D.: 2001, *ApJ* **558**, 309.
Maciejewski, G., Karska, A. and Niedzielski, A.: 2002, *IBVS* **5343**.
Maciejewski, G., Ligeza, P. and Karska, A.: 2003, *IBVS* **5400**.
Maciejewski, G. and Ligeza, P.: 2004, *IBVS* **5504**.
Niedzielski, A., Maciejewski, G. and Czart, K.: 2003, *Acta Astron.* **53**, 281.

UV LEO: THE BINARY WITH THE TWO SUNS

G. DJURAŠEVIĆ[1], P. ROVITHIS[2], H. ROVITHIS-LIVANIOU[3]
and E. FRAGOULOPOULOU[3]

[1]*Astronomical Observatory, Volgina 7, 11160 Belgrade, Serbia and Montenegro;*
E-mail: gdjurasevic@aob.bg.ac.yu
[2]*Astronomical Institute, National Observatory of Athens, Athens, Greece*
[3]*Department of Astrophysics, Astronomy and Mechanics, Faculty of Physics, University of Athens,*
Panepistimiopolis, Athens, Greece

(accepted April 2004)

Abstract. *BV* light curves of the eclipsing binary UV Leo obtained at the Kryonerion Astronomical Station of the National Observatory of Athens, Greece, are analyzed. The analysis is based on a Roche configuration with two spots on the secondary surface. The elements of the two components of the system are calculated and the spot characteristics are given.

Keywords: binary stars, photometry, individuals, UV Leo

1. Introduction

UV Leo (BD+15° 2230) was discovered as a variable star by Hoffmeister (1934). Its period was estimated as 0.30 d by Jensch (1935); the star was finally confirmed to be a binary by Schneller (1936). According to Broglia (1961), UV Leo showed some kind of solar-like eruptions. On the other hand, spectral analysis of the system carried out by Popper (1965, 1993) did not reveal any emission in its rotationally broadened spectral features, similar to those found in chromospherically active stars. Special attention has been paid to the nature of primary eclipse (occultation or transit?) as well as to the relative luminosity of both components (e.g., Perek, 1952; Wellmann, 1954; McCluskey, 1966; Cester et al., 1978; Popper, 1993; Frederic and Etzel, 1996).

We observed UV Leo and analyzed its light curves since it adds to our knowledge about stars of the lower part of the main sequence. This very interesting system UV Leo has recently attracted the attention of other investigators too (e.g., Mikuz et al., 2002; Zwitter et al., 2003).

2. Light Curves and Analysis Procedure

Photoelectric observations of UV Leo were carried out with the 48-in. Cassegrain reflector at the Kryonerion Astronomical Station of the Athens National Observatory, Greece, mainly during 1997. The individual observations together

Astrophysics and Space Science **296**: 311–314, 2005.
© Springer 2005

with other observational details and with a study of the orbital period behavior of UV Leo will be given elsewhere (Rovithis-Livaniou and Rovithis, 2004). Here, we limit ourselves to the solution of the obtained light curves. The analysis was carried out using the Djurašević (1992) code. The primary star's temperature was taken

TABLE I

Results of the analysis of the UV Leo light curve in V passband

Quantity	V-filter	Quantity	V-filter
n	153		
$\Sigma(O-C)^2$	0.0386		
σ	0.0159		
$q = m_c/m_h$	0.917		
$f_{h,c}$	1.0	$a_1^{h,c}$	$+0.5035, +0.5092$
$\beta_{h,c}$	0.08	$a_2^{h,c}$	$+0.1274, +0.0955$
$A_{h,c}$	0.5	$a_3^{h,c}$	$+0.4433, +0.4917$
T_h	5916	$a_4^{h,c}$	$-0.2827, -0.3024$
$A_{S1} = T_{S1}/T_c$	0.83 ± 0.01	$\Omega_{h,c}$	$4.261, 4.089$
θ_{S1}	40.1 ± 0.9	$R_{h,c}[D=1]$	$0.296, 0.297$
λ_{S1}	275.5 ± 2.9	$L_h/(L_h + L_c)$	0.495
φ_{S1}	57.5 ± 1.9	$\mathcal{M}_h[M_\odot]$	1.21 ± 0.1
$A_{S2} = T_{S2}/T_c$	0.82 ± 0.01	$\mathcal{M}_c[M_\odot]$	1.11 ± 0.1
θ_{S2}	30.5 ± 0.8	$\mathcal{R}_h[R_\odot]$	1.21 ± 0.06
λ_{S2}	81.5 ± 3.2	$\mathcal{R}_c[R_\odot]$	1.22 ± 0.06
φ_{S2}	-6.3 ± 8.2	$\log g_h$	4.36 ± 0.06
T_c	5880 ± 18	$\log g_c$	4.31 ± 0.06
F_h	0.814 ± 0.009	M_{bol}^h	4.28 ± 0.11
F_c	0.851 ± 0.007	M_{bol}^c	4.28 ± 0.12
$i\,[°]$	83.1 ± 0.2	$a_{orb}[R_\odot]$	3.96 ± 0.09

Notes: n: number of observations; $\Sigma(O-C)^2$: final sum of squares of residuals between observed (LCO) and synthetic (LCC) light curves; σ: standard deviation of the observations; $q = m_c/m_h$: mass ratio of the components; $f_{h,c}, \beta_{h,c}, A_{h,c}$: nonsynchronous rotation coefficients, gravity-darkening exponents and albedo of the components; $T_{h,c}$: temperature of the hotter primary and cooler secondary; $A_{S1,2}, \theta_{S1,2}, \lambda_{S1,2}$ and $\varphi_{S1,2}$: spots' temperature coefficient, angular dimension, longitude and latitude (in arc degrees); $F_{h,c}$: filling factors for the critical Roche lobe of the hotter (more-massive) and cooler (less-massive) star; $i = [°]$: orbit inclination (in arc degrees); $a_1^{h,c}, a_2^{h,c}, a_3^{h,c}, a_4^{h,c}$: nonlinear ($V$-filter) limb-darkening coefficients of the components (Claret's formula); $\Omega_{h,c}$: dimensionless surface potentials of the primary and secondary; $R_{h,c}$: polar radii of the components in units of the distance between the component centres; $L_h/(L_h + L_c)$: (V-filter) luminosity of the hotter star (including spot on the cooler one); $\mathcal{M}_{h,c}[M_\odot], \mathcal{R}_{h,c}[R_\odot]$: stellar masses and mean radii of stars in solar units; $\log g_{h,c}$: logarithm (base 10) of the mean surface acceleration (effective gravity) for system stars; $M_{bol}^{h,c}$: absolute bolometric magnitudes of UV Leo components; and $a_{orb}[R_\odot]$: orbital semi-major axis in units of solar radius.

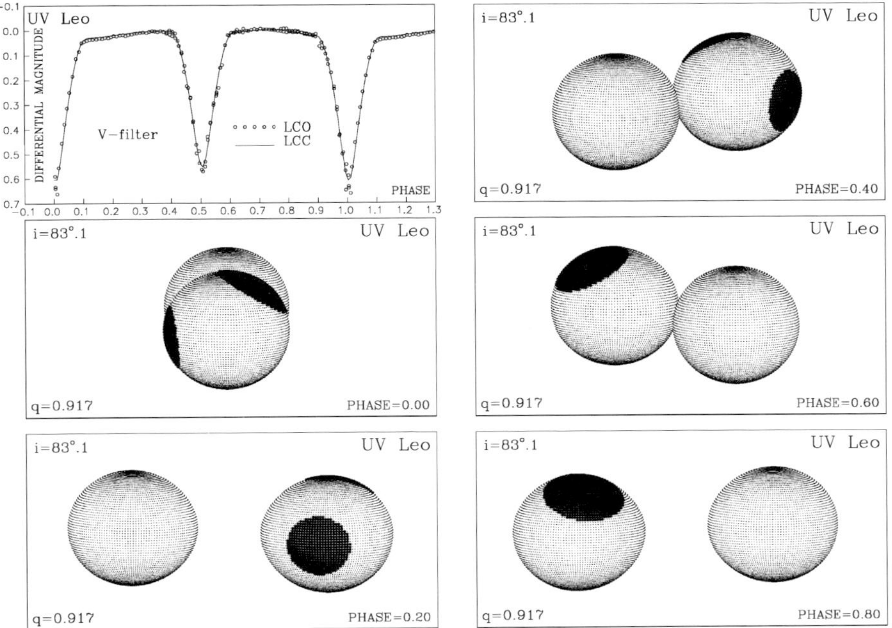

Figure 1. Observed (LCO) and final synthetic (LCC) light curve of UV Leo and view of its Roche configuration at different orbital phases.

as $T_h = 5916$ K, according to Popper (1980), and the spectroscopically estimated mass ratio $q = m_c/m_h = 0.917$ (Zwitter et al., 2003) was adopted. For the gravity darkening coefficients and bolometric albedos of the two components the values for convective envelopes were adopted, in accordance with the spectral types.

The obtained results are given in Table I and shown in Figure 1, where the points stand for our individual observations and the continuous line denotes the theoretical light curve. In the same figure the Roche model for the two components of UV Leo with the spotted areas is also presented, as viewed at different phases.

3. Conclusions and Discussion

The partial eclipses of UV Leo and the fact that the two components are very similar in size and spectral type introduce a difficulty not only with respect to the nature of primary eclipse, but also concerning the question which of the two components is the hotter or more luminous one. Regarding relative luminosities, Popper (1965) found from early photographic spectra that the secondary is slightly less luminous than the primary. Moreover, using high resolution CCD spectra and applying a cross-correlation function, Popper (1993) concluded that the hotter primary component is the more luminous one. On the other hand, Frederic and

Etzel (1996) analyzed the light curves of UV Leo using spotted and unspotted models. They deduced from the unspotted solution that the cooler secondary should be the slightly more luminous component. From the spotted solution, however, they found the two components to be equally bright in the V, R, and I passbands, and the primary to be brighter in the U and B bands.

In the present analysis, Popper's findings were adopted (Popper, 1993), assuming the primary component to be the hotter and more luminous of the two. Since UV Leo is a very active binary, the model with two dark spots on the secondary surface was used in our light curve analysis. The proposed solution is based on the latest estimates of the mass ratio and the primary temperature.

Acknowledgements

This work was partly supported by Athens University (grant 70/4/3305).

References

Broglia, P.: 1961, *Mem. Soc. Astron. Ital.* **32**, 43.
Cester, B., Fedel, B., Giuricin, G., Mardirossian, F. and Mezzetti, M.: 1978, *A&AS* **32**, 351.
Djurašević, G.: 1992, *Ap&SS* **197**, 17.
Frederik, M.C.G. and Etzel, P.B.: 1996, *AJ* **111**, 2081.
Jensch, A.: 1935, *Astron. Nachr.* **257**, 139.
Hoffmeister, C.: 1934, *Astron. Nachr.* **253**, 195.
McCluskey, G.E., Jr.: 1966, *AJ* **71**, 536.
Mikuz, H., Dintinjana, B., Prša, A., Munari, U. and Zwitter, T.: 2002, *IBVS* **5338**, 1.
Perek, L.: 1952, *Contr. Astron. Inst. Brno* **10**, 1.
Popper, D.M.: 1965, *ApJ* **141**, 126.
Popper, D.M.: 1980, *Ann. Rev. Astron. Astrophys.* **18**, 115.
Popper, D.M.: 1993, *ApJ* **404**, L67.
Rovithis-Livaniou, H. and Rovithis, P.: 2005 (in preparation).
Schneller, H.: 1936, *Astron. Nachr.* **260**, 404.
Wellmann, P.: 1954, *Zeitschr. f. Ap.* **34**, 99.
Zwitter, T., Munari, U., Marrese, P.M., Prša, A., Milone, E.F., Boschi, F., Tomov, T. and Siviero, A.: 2003, *A&A* **404**, 333.

INFLUENCE OF INTERSTELLAR AND ATMOSPHERIC EXTINCTION ON LIGHT CURVES OF ECLIPSING BINARIES

A. PRŠA and T. ZWITTER

Department of Physics, University of Ljubljana, Jadranska, Ljubljana, Slovenia;
E-mail: andrej.prsa@fmf.uni-lj.si and tomaz.zwitter@fmf.uni-lj.si

(accepted April 2004)

Abstract. Interstellar and atmospheric extinctions redden the observational photometric data and they should be handled rigorously. This paper simulates the effect of reddening for the modest case of two main sequence $T_1 = 6500$ K and $T_2 = 5500$ K components of a detached eclipsing binary system. It is shown that simply subtracting a constant from measured magnitudes (the approach often used in the field of eclipsing binaries) to account for reddening should be avoided. Simplified treatment of the reddening introduces systematics that reaches $\sim$0.01 mag for the simulated case, but can be as high as $\sim$0.2 mag for, e.g., B8 V-K4 III systems. With rigorous treatment, it is possible to *uniquely* determine the colour excess value $E(B - V)$ from multi-colour photometric light curves of eclipsing binaries.

Keywords: binaries: eclipsing, ISM: dust, extinction, atmosphere: scattering, stars: fundamental parameters, methods: numerical

1. Introduction

Although interstellar extinction has been discussed in many papers and quantitatively determined by dedicated missions (IUE, 2MASS, and others), there is a lack of proper handling in the field of eclipsing binaries. The usually adopted approach is to calculate the amount of reddening from the observed object's coordinates and its inferred distance and to subtract it uniformly, regardless of phase, from photometric observations. This paper shows why this approach may be inadequate, especially for objects where interstellar extinction and the colour difference between both components are significant. Atmospheric extinction is a better-posed problem: similarly as interstellar extinction depends on $E(B - V)$, atmospheric extinction depends on air-mass, which is a measurable quantity, whereas $E(B - V)$ has to be estimated.

2. Simulation

To estimate the effect of reddening on eclipsing binaries, we built a synthetic binary star model, consisting of two main sequence G9 V-F5 V stars with $T_1 = 5500$ K, $R_1 = 0.861$ R$_\odot$ and $T_2 = 6500$ K, $R_2 = 1.356$ R$_\odot$ and 1 day orbital period. The simulation logic is as follows: for the given phase, we calculate the effective spectrum

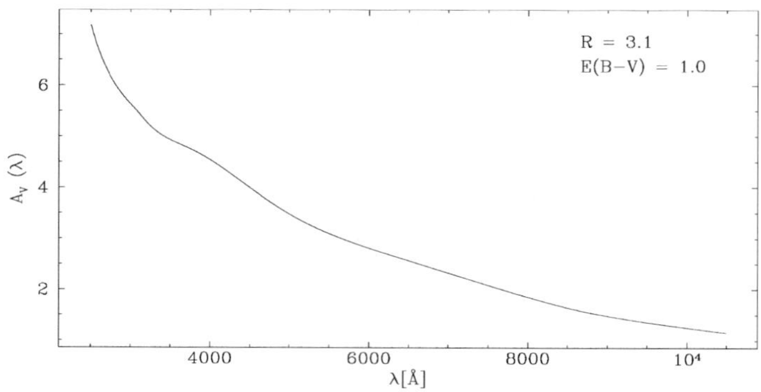

Figure 1. The reddening law adopted from Cardelli et al. (1989).

of the binary by convolving Doppler-shifted individual spectra of the visible surfaces of both components. To this intrinsic spectrum we rigorously apply interstellar and atmospheric extinctions (both as functions of wavelength). We then convolve this reddened spectrum with the overall response function (composed of the filter transmittivity and detector response functions) and integrate over the bandpass wavelength range to obtain the flux. In contrast, we use the same intrinsic spectrum without rigorously applying the reddening. To simulate the subtraction[1] of a reddening *constant* from photometric observations, we simply divide the intrinsic spectrum by the flux that corresponds to this constant. Finally, we calculate the flux in the same manner as before and compare it to the flux obtained by applying rigorous reddening.

For building synthetic light curves we use PHOEBE[2] (Prša and Zwitter, 2005; in preparation). Each light curve consists of 300 points uniformly distributed over the whole orbital phase range. To be able to evaluate the impact of reddening on photometric light curves exclusively, all second-order effects (limb darkening, gravity brightening, reflection effect) have been turned off.

We take Kurucz's synthetic spectra ($R = 20,000$) from precalculated tables by Munari et al. (2004; in preparation). The used $(UBV)_J(RI)_C$ response data (filter × detector) are taken from ADPS (Moro and Munari, 2000), where we apply a cubic spline fit to obtain the overall response function.

For interstellar extinction, we use the Cardelli et al. (1989) empirical formula (Figure 1), where $\mathcal{R}_V = 3.1$ was assumed throughout this study. Schlegel et al. (1998) interstellar dust catalogue was used to obtain the maximum colour excess $E(B - V)$ values for different lines of sight.

---

[1] To emphasize that this constant is *subtracted* from photometric observations (given in magnitudes), we shall use the term *constant subtraction* throughout this study, although we are *dividing* when operating with fluxes.

[2] PHOEBE stands for Physics Of Eclipsing Binaries and it is based on the Wilson-Devinney code. See http://www.fiz.uni-lj.si/phoebe for details.

For atmospheric extinction we use the equation triplet for Rayleigh-ozone-aerosol extinction sources given by Forbes et al. (1996) and summarized by Pakštienė and Solheim (2003). The observatory altitude $h = 0\,\mathrm{km}$ and the zenith air-mass are assumed throughout the study.

To rigorously deredden the observations for the given $\mathcal{R}_V$ and $E(B - V)$, it is necessary to determine the reddening for each wavelength of the spectrum. Correcting differentially and integrating over the filter bandpass then yields the dereddened flux of the given filter. However, without spectral observations, it is difficult to calculate properly the flux correction. Since Cardelli et al. (1989) formula depends on the wavelength, the usually adopted approach found in literature is to use the effective wavelength λ_{eff} of the filter transmittivity curve to calculate the reddening correction. We demonstrate the implications in the following section.

3. Implications

3.1. Interstellar Extinction

By comparing the rigorously calculated fluxes against intrinsic fluxes with a simple constant subtracted, we come to the following conclusions. (1) Taking the effective wavelength of the filter bandpass should be avoided. Since the flux is the integral over the filter bandpass, λ_{eff} has a *conceptually* different meaning. Furthermore, λ_{eff} of the given filter depends heavily on the effective temperature of the observed object and on the colour excess $E(B - V)$ (Figure 2). To determine the subtraction constant, one has to make sure that *the integral* (rather than any particular wavelength) of the both curves is the same. Figure 3 shows the discrepancy between the properly calculated light curve and the one obtained by subtracting a λ_{eff}-calculated constant. Table I summarizes the differences between the proper treatment and other approaches. (2) Even if the subtraction constant is properly calculated, the light curves still exhibit measurable differences in both minima (Figures 3 and 4). This is due to the effective temperature change of the binary system during eclipses. For

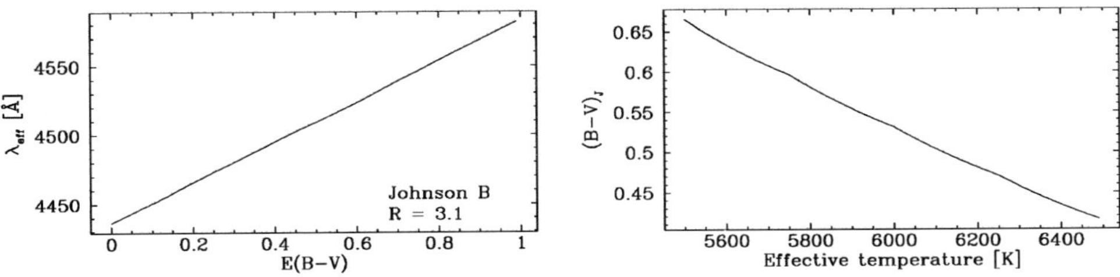

Figure 2. Left: The change of the effective wavelength of the Johnson *B* filter due to reddening of the simulated G9 V-F5 V binary. Right: The $(B - V)_J$ colour index on the G9 V-F5 V (5500–6500 K) temperature interval, calculated by integrating the spectrum over both filter bandpasses.

TABLE I

The summary of different approaches to calculate the wavelength to
be used for the dereddening constant

Approach:	λ_{eff}	Δm_B	$\varepsilon_{\Delta m_B}$
Rigorously calculated value:	4470.2Å	4.03	0.00
Filter transmittivity:	4410.8Å	4.09	0.06
Filter transmittivity + reddening law:	4452.1Å	4.05	0.02
Intrinsic spectrum:	4436.3Å	4.06	0.03
Effective (reddened) spectrum:	4583.6Å	3.90	−0.13

λ_{eff} is determined by requiring that the area under the spectrum on
both sides is equal. Δm_B is the value of extinction in B filter and
$\varepsilon_{\Delta m_B}$ is the deviation from the rigorously calculated value. All values
are calculated for $E(B - V) = 1$ at quarter phase. Note that Δm_B is
smaller than $\mathcal{R}_V + E(B - V) = 4.1$, since our simulated binary is
cooler than 10,000 K.

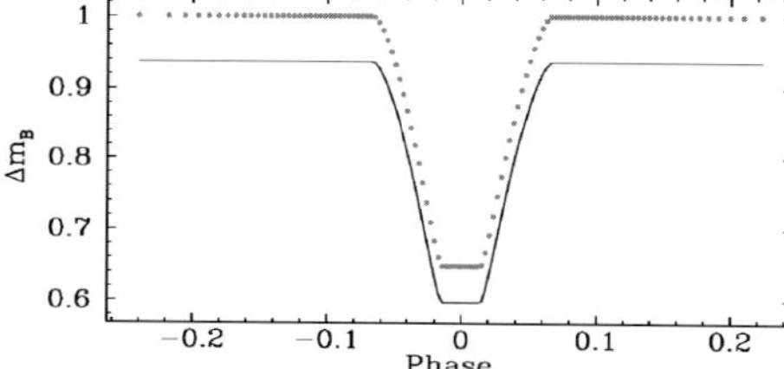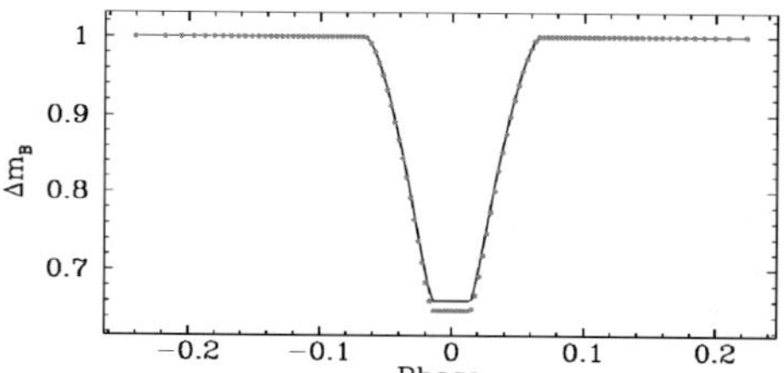

Figure 3. Left: The discrepancy due to simplified constant subtraction approach (solid line) compared
to the rigorously applied reddening (points) for a G9 V-F5 V binary. The subtracted constant was
obtained from the effective wavelength ($\lambda_{\text{eff}} = 4410.8$Å) of the Johnson B transmittivity curve.
Right: Overplotted light curves with the subtraction constant calculated so that the magnitudes in
quarter phase are aligned. There is still a *measurable* difference in eclipse depth of both light curves.
$E(B - V) = 1$ is assumed.

the analysed case, the difference in B magnitude is ∼0.01 mag, which is generally
observable. (3) If light curves in three or more photometric filters are available, it
is possible to *uniquely* determine the colour excess value $E(B - V)$ by comparing
different colour indices in-and-out of eclipse. The reddening may thus be properly
introduced to the fitting scheme of the eclipsing binary analysis program. This was
done in PHOEBE.

3.2. ATMOSPHERIC EXTINCTION

Atmospheric extinction is comprised of three different sources: the Rayleigh scattering, the aerosol scattering and ozone absorption (Forbes et al., 1996). It depends on
the wavelength of the observed light and on the air-mass of observations, which is of

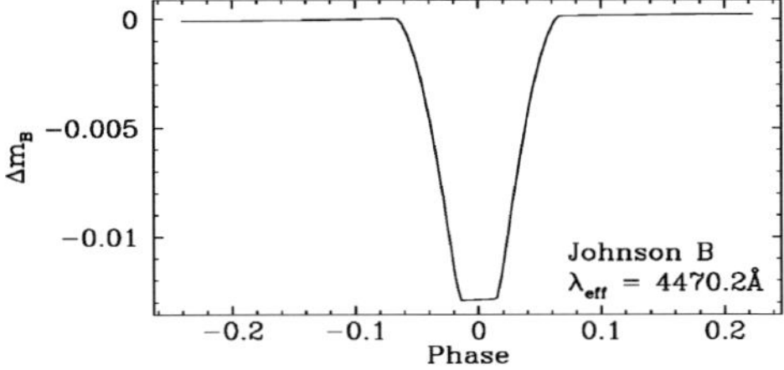
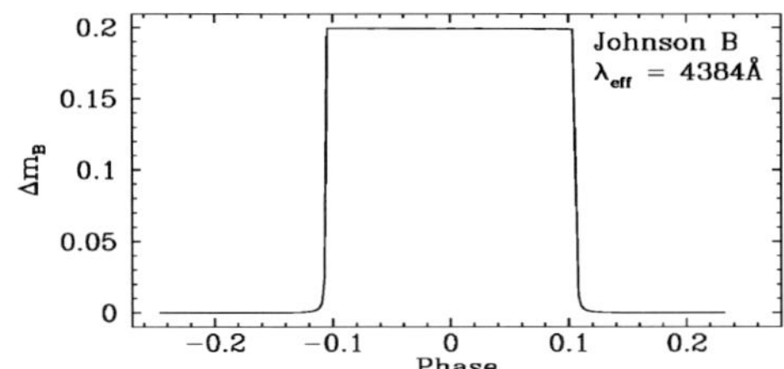

Figure 4. The difference between the rigorously calculated reddening and a constant subtraction in Johnson *B* filter for the G9 V-F5 V binary (left) and a B8 V-K4 III binary (right) during primary minimum. $E(B - V) = 1$ is assumed.

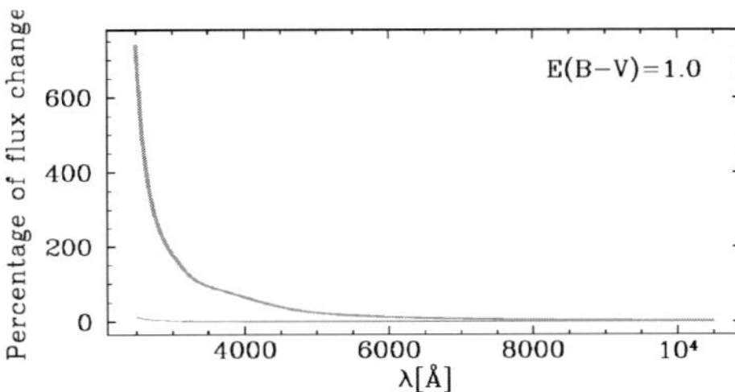
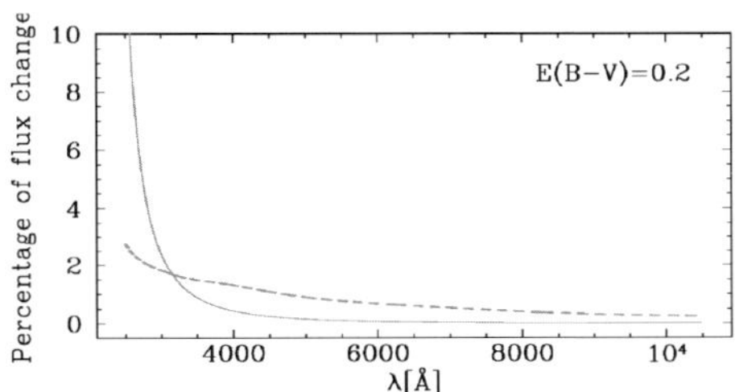

Figure 5. The percentage of the flux change imposed on the intrinsic spectrum by reddening (thick line) and atmospheric extinction (thin line). If the colour excess is large, the reddening completely dominates the spectrum (left), but if the colour excess is moderate, the blue part of the spectrum is dominated by the Rayleigh scattering (right).

course the same for both binary components; we are thus left only with wavelength dependence at some inferred air-mass. To assess the impact on the photometric data, we compare the flux change imposed on the intrinsic spectrum by reddening and by atmospheric extinction (Figure 5). We conclude that for weakly and moderately reddened eclipsing binaries the atmospheric extinction dominates the blue parts of the spectrum due to Rayleigh scattering, but for larger colour excesses $(E(B - V) \gtrsim 0.5)$ the reddening is dominant throughout the spectrum.

4. Discussion

By not properly taking reddening into account, we introduce systematics of ~ 0.01 mag into the solution for our simulated binary. One may ask: is this difference worth bothering with? To demonstrate that it is, let us simulate a bit more exotic eclipsing binary with blue B8 main-sequence ($T_1 = 12{,}000$ K, $R_1 = 3.221 \, R_\odot$) and red K4 type III giant ($T_2 = 4000$ K, $R_2 = 30.0 \, R_\odot$) components with orbital period of 15 days. Figure 4 (right) shows that the discrepancy between the rigorously calculated light curve and the constant-subtracted one is as large as ~ 0.2 mag, which means $\sim 10\%$ error in the determined distance to the observed object. On the other hand, the reddening effect may be reduced by using narrower filter-sets, e.g. Strömgren *ubvy* set. Furthermore, typical observed colour excesses rarely exceed few tenths of the magnitude, which additionally diminishes this effect. However, future space scanning missions such as Gaia (Perryman et al., 2001) will acquire measurements of eclipsing binaries without spatial bias. Thus interstellar extinction should be carefully and rigorously introduced into the reduction pipeline for eclipsing binaries.

References

Moro, D. and Munari, U.: 2000, *A&AS* **147**, 361.
Cardelli, J.A., Clayton, G.C. and Mathis, J.S.: 1998, *ApJ* **345**, 245.
Schlegel, D.J., Finkbeiner, D.P. and Davis, M.: 1998, *ApJ* **500**, 525.
Forbes, M.C., Dodd, R.J. and Sullivan, D.J.: 1996, *Balt Astron* **5**, 281.
Pakštienė, E. and Solheim, J.E.: 2003, *Balt Astron* **12**, 221.
Perryman, M.A.C., de Boer, K.S., Gilmore, G., Hòg, E., Lattanzi, M.G., Lindgren, L., Luri, X., Mignard, F., Pace, O. and de Zeeuw, P.T.: 2001, *A&A* **369**, 339.

RESULTS OF LIGHT CURVE ANALYSIS OF THE MASSIVE BINARY CYG X-1/V1357 CYG IN PRIMARY MINIMUM

IRINA VOLOSHINA

Sternberg Astronomical Institute, Universitetskij prospect, 13, Moscow, Russia;
E-mail: vib@sai.msu.ru

(accepted April 2004)

Abstract. The results of new *UBV* photometric observations of V1357 Cyg in primary minimum are presented. Observations were carried out from 1996 up to now with 60 cm telescope in Crimea with the goal to study additional radiation that was detected in the mean light curve of this system near orbital phase 0.0. The properties of this additional radiation are also considered.

Keywords: Cyg X-1/V1357 Cyg, massive binary systems, UBV photometry

The massive close X-ray binary system Cyg X-1/V1357 Cyg was discovered in 1971. It is attracting the attention of observers for a quarter of century, and probably remains the best black-hole candidate to date. This system contains an optical star – a blue O9.7 Iab supergiant – and a compact object. The basic optical variability of this binary system is connected with the ellipsoidal effect of the optical star. The orbital period of the system was derived by different authors and is equal to $5^d.6$. Despite a number of different periods reported for this system in different ranges, the orbital period is still the only one that was confirmed in all bands – optical, infrared, X-ray and also in the radio range. The values of the masses of the two components have been refined a number of times, most recently by Herrero et al. (1995), giving 17.8 $M_\odot$ for the supergiant companion star and 10.1 $M_\odot$ for the black hole.

In 1985 Lyuty reported detection of some additional radiation in the mean light curve of Cyg X-1, based on numerous optical observations during 12 years and published data. This additional radiation appears as a narrow peak (with a width of $\Delta\varphi \approx 0.03$) near orbital phase $\varphi = 0.0$. But the effect was found from the analysis of the mean light curve from all data (1–2 measurements during a night) published until then and needed an independent confirmation. Subsequent photometric observations (Voloshina and Lyuty, 1995) from 1988 to 1993 permitted to confirm the existence of additional radiation near the primary minimum in the light curve of this system. The analysis shows that the effect itself is variable. Observations made in eight nights revealed the effect in four minima. To study this effect in detail and to understand its nature it was necessary to collect a more extended observational database.

The new observations of Cyg X-1 at primary minimum were carried out with a *UBV* photometer attached to the 60 cm telescope of the Sternberg Astronomical

Astrophysics and Space Science **296**: 321–324, 2005.
© Springer 2005

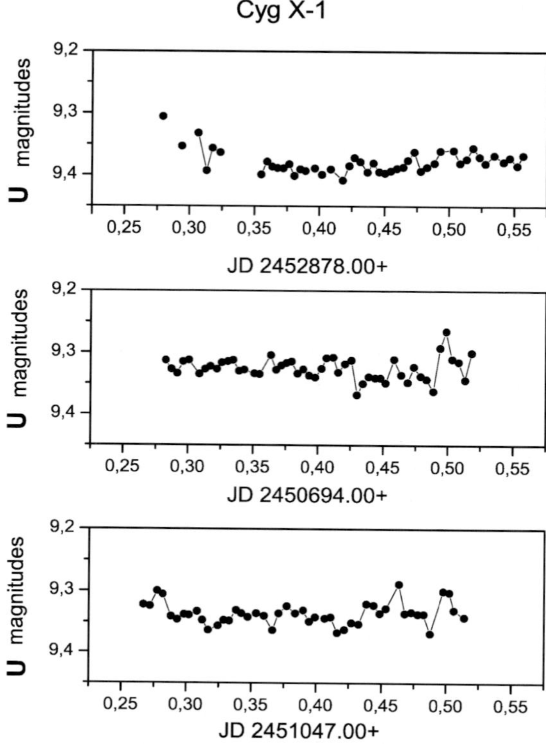

Figure 1. Daily light curves of V1357 Cyg/Cyg X-1.

Institute in Crimea. All observations were made with reference to the standard star BD+35°3816 with magnitudes $V = 9^m\!.976$, $B - V = +0^m\!.590$, $U - B = +0^m\!.064$. The usual method of differential observations was used. The individual measurements are accurate to $0^m\!.005$ in V and B bands and $0^m\!.008$ in U band. From 1996 on up to now 23 new observational sets from different nights with durations between 2.5 and 6 h were obtained. As an example, some of them are shown in Figure 1 (U band). The journal of all our observations is given in Table I. The orbital phases were calculated with elements of Brocksopp et al. (1998):

$$\text{Min}\,I = \text{JD}\ 2441163.529 + 5^d\!.599829\,E.$$

As one can see in Figure 1, most observations of Cyg X-1 show rapid fluctuations significantly exceeding the errors of measurements, but sometimes there are no fluctuations, e.g. during the nights of JD 2450246 and 2452700. The maximum amplitude of the rapid light fluctuations was observed on JD 2450694, up to 10% in several minutes. The existence of such rapid fluctuations with an amplitude of several percent significantly complicates the study of small-amplitude effects, such

TABLE I

Journal of observations

JD	Date	Orbital phases	N
2450246	11.06.1996	0.990–0.010	19
2450252	17.06.1996	0.050–0.080	18
2450274	09.07.1996	0.975–0.010	33
2450302	06.08.1996	0.970–0.010	30
2450319	23.08.1996	0.015–0.040	27
2450627	27.06.1997	0.020–0.040	25
2450655	25.07.1997	0.010–0.030	27
2450694	02.09.1997	0.970–0.015	50
2450700	08.09.1997	0.040–0.080	38
2450728	06.10.1997	0.050–0.070	17
2450739	17.10.1997	0.988–0.021	45
2451019	24.07.1998	0.015–0.025	21
2451047	21.08.1998	0.010–0.050	46
2451398	07.08.1999	0.690–0.740	14
2451401	10.08.1998	0.220–0.270	11
2451406	15.08.1998	0.120–0.160	5
2451769	12.08.2000	0.940–0.990	40
2451772	15.08.2000	0.480–0.530	15
2451775	18.08.2000	0.010–0.040	28
2451831	13.10.2000	0.000–0.040	43
2452150	28.08.2001	0.997–0.984	11
2452867	15.08.2003	0.990–0.020	45
2452878	26.08.2003	0.969–0.019	53

as the additional radiation at the time of the superior conjunction of the X-ray source. Figure 2 displays the results of our observations obtained during the recent years in the U band (different symbols indicate different nights).

Only 15 observational sets of Cyg X-1 near primary minimum were used for our analysis. Observations of JD 2451722, specially obtained at the secondary minimum (orbital phase range 0.48–0.53) were also used for comparison.

Our analysis shows that though additional radiation was not observed in some individual light curves of Cyg X-1, it appears as excess on the mean light curve determined by all our observations since 1971 (see Figure 2) near orbital phase $\sim$0.01 (superior conjunction of the relativistic component, when the X-ray source is behind the optical star).

This radiation possesses the following properties: the width of the additional radiation beam is very small (less than 15°), its center lies in the phase region of

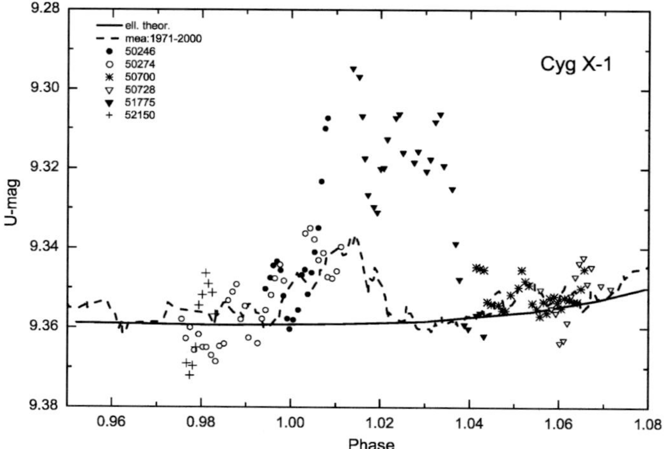

Figure 2. Light curve of Cyg X-1 near primary minimum. The solid line indicates the ellipsoidal curve. The dashed line shows the mean light curve for all our observations since 1971.

$\varphi = 0.01$–0.02, the amplitude is about 1% of total luminosity in the V band and 2% in the U band, and its temperature exceeds the temperature of the optical companion in this system.

We can also state the following:

- The additional radiation was not observed in each orbital cycle.
- Additional radiation is not observed during secondary minimum.
- Apparently the peak of the additional radiation has a rather complicated structure.

Acknowledgements

I am grateful to N. Metlova and V. Sementsov for assistance with observations and data processing, and V. Lyuty for valuable discussions. This study was supported by the grant of Leading Scientific Schools of Russia (project 388.2003.2) and Russian Fund of Basic Research (project 02-02-16462).

References

Brocksopp, C., Tarasov, A.E., Lyuty, V.M. and Roche, P.: 1999, *A&A* **343**, 861.
Herrero A., Kudritzki, R.P., Gabler, R. et al.: 1995, *A&A* **297**, 556.
Lyuty, V.M.: 1985, *Astron. Zh.* **62**, 731.
Voloshina, I. and Lyuty, V.M.: 1995, *Astron. Nachr.* **316**, 85.

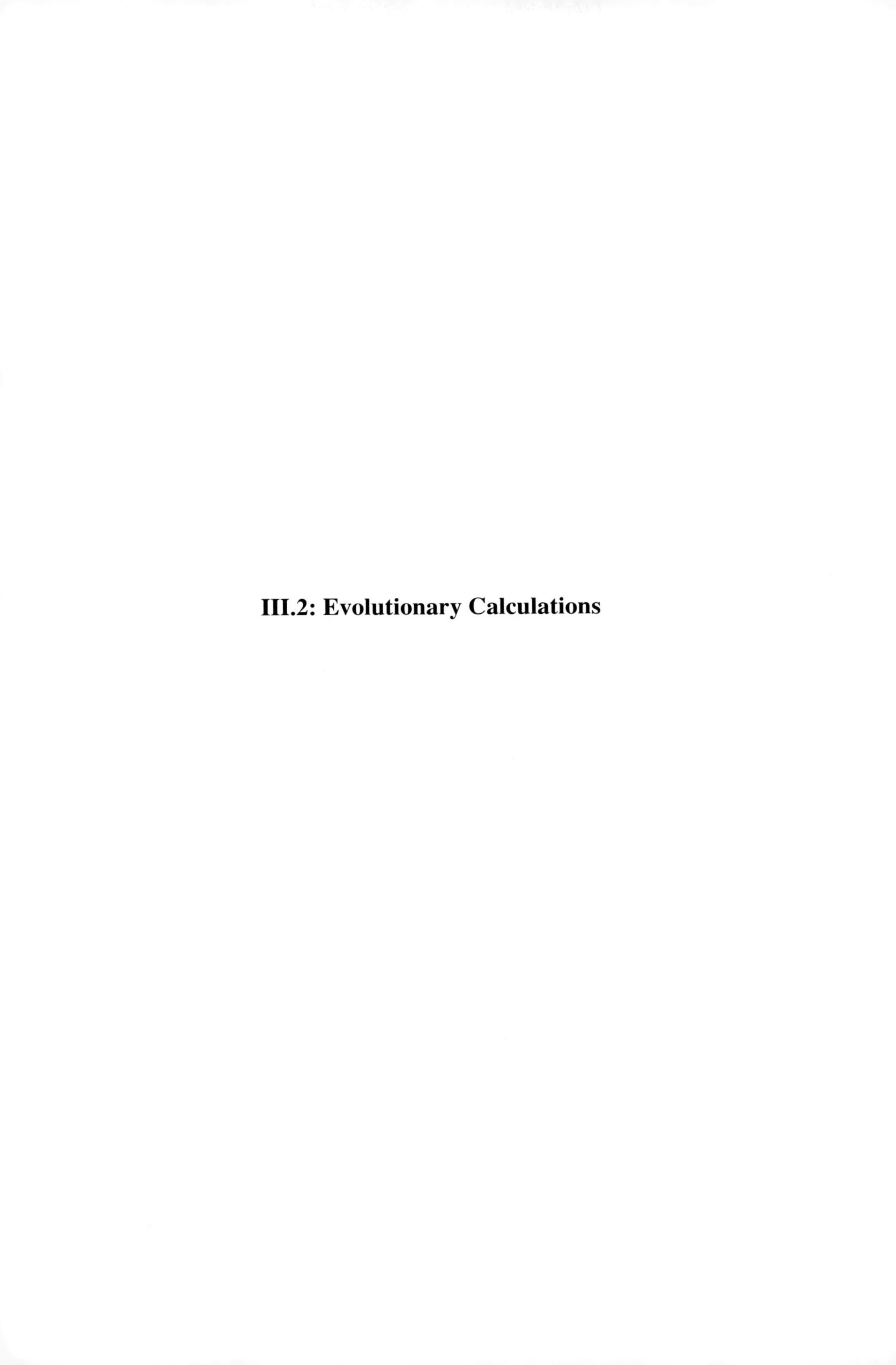

III.2: Evolutionary Calculations

EVOLUTIONARY PROCESSES IN LOW-MASS BINARY SYSTEMS

PETER P. EGGLETON[1] and LUDMILA KISSELEVA-EGGLETON[2]

[1]*Lawrence Livermore National Laboratory, East Avenue, Livermore, CA, USA;*
E-mail: ppe@igpp.ucllnl.org
[2]*Department of Mathematics, University of the Pacific, Pacific Avenue, Stockton, CA, USA*

(accepted April 2004)

Abstract. We consider the evolution of certain low-mass binaries, incorporating models of (a) internal evolution, (b) tidal friction, (c) dynamo activity driven by an elementary α, Ω dynamo, (d) stellar wind driven by the activity, and (e) magnetic braking as a consequence of wind and poloidal dynamo-generated magnetic field. In some circumstances the stellar wind is found to remove mass on a nuclear timescale, as is necessary to explain some observed systems.

We can hope that various uncertainties in the model may be clarified by a careful comparison of the models with such observed quantities as rotation periods. These are modified by processes (a), (b) and (e). Assuming that stellar evolution is slow, rotation rate should in some circumstances represent a balance between magnetic braking trying to slow the star down and tidal friction trying to spin it up. Preliminary attempts are promising, but indicate that some fine tuning is necessary.

When there is a third body present, in an orbit which is inclined but not necessarily of short period, the eccentricity of a close binary can be strongly modified by 'Kozai cycles'. We show that this may complicate attempts to account for spin rates of stars in close binaries.

Keywords: double stars, dynamo activity, winds, magnetic braking, tidal friction

1. Introduction

Many low-mass binaries, by which we mean binaries where the two masses are $\lesssim 4M_\odot$ each, show evidence that they have lost either angular momentum, or mass, or both. This is particularly true if one or both components are of spectral type G/K/M, whether giant, subgiant or dwarf. It seems likely that dynamo activity, somewhat as seen in the Sun but considerably enhanced, is the explanation. The enhancement is probably due to the fact that stars in fairly close binaries are forced to spin faster than they would if they were single.

2. Discussion

Table I gives some data on a selection of systems. The first two are semidetached binaries (Algols). It has long been realized (Refsdal et al., 1974) that AS Eri has remarkably little angular momentum. Any semidetached binary must evolve through a stage where the two masses are equal; if AS Eri has conserved its mass and angular momentum since then, its orbital period at that time would have been ~ 0.1 days,

Astrophysics and Space Science **296**: 327–336, 2005.
© Springer 2005

TABLE I

Name	Spectra	P^a	e	M_1	M_2	R_1	R_2	$P_{\rm rot}$	$P_{\rm PS}^b$
AS Eri	K0 + A3	2.66		0.2	1.9	2.2	1.8		
R CMa	G8IV + F1	1.14		0.17	1.07	1.15	1.5		
Z Her	K0IV + F4	3.99		1.31	1.6	2.7:	1.9:		
RZ Eri	K2III + F5m	39.3	.35	1.62	1.68	7.0	2.8	31.4	23
RS CVn	K0IV + F4IV-V	4.80		1.44	1.41	4.0	2.0		
λ And	G8III-IV + ?	20.5	.04	0.0006^c				54	20
AE Lyn	G5IV + F8IV-V	11.1	.13	0.97^c	0.96^c			10.2	10.2
V1379 Aql	SDB + K0III-IV	20.7	.09	0.30	2.27	0.05	9.0	24	19.6
o Leo	F9IIIm + A7m	14.5		2.12	1.85	5.7	2.6		
HR2030	K0IIb + B8IV	66.5	.02	4.0	4.0	41	5.9		
TZ For	G8III + F7III	75.7		2.05	1.95	8.3	4.0		
α Aur	G8III + G0III	104		2.61	2.49	11.4	8.8		
SU Cyg	F2-G0I-II + (B + A)	549	.34	5.9^c	5.5^c				
BY Dra	K6Ve + M0Ve	5.98	.30	0.59:	0.52:			3.83	3.86
BY Dra	(K + M) + M5V	17″		1.1:					
41 Dra	F7V + F7V	1247	.9754	1.39	1.30	1.93	1.57	13:	4.2
40 Dra	F7V + F7V	10.5	.374	1.32	1.20	1.63	1.32	6:	5.5
41/40Dra	(F + F) + (F + F)	19″		2.69	2.52				
AB Dor	K0-2e + ?	10:yr	.5:	.76:	.09:			.51	
AB Dor	(K + ?) + M4Ve	9.3″							

[a]Period in days (or years where indicated); or visual separation in seconds of arc.
[b]Pseudosynchronous rotation period, subject to tidal friction alone.
[c]Mass function, or if two values $M \sin^3 i$.
References: AS Eri – Popper, 1980; R CMa – Sarma et al., 1996; Z Her – Popper, 1988a; RZ Eri – Popper, 1988b; RS CVn – Popper, 1988a; λ And – Boyd et al., 1983; AE Lyn – Fekel et al., 1998; V1379 Aql – Jeffery and Simon, 1997; o Leo – Griffin, 2002; HR 2030 – Griffin and Griffin, 2000; TZ For – Andersen et al., 1991; α Aur – Barlow et al., 1993; SU Cyg – Evans and Bolton, 1990; BY Dra – Boden and Lane, 2001, Zuckerman et al., 1997; 41/40 Dra – Tokovinin et al., 2003; AB Dor – Guirado et al., 1997; Lim, 1993.

far too short to have contained two stars of $\sim 1 M_\odot$ each. R CMa is similar, but with the extra problem that it must surely have lost mass as well. A star (of roughly solar composition) has to be $\gtrsim 0.9 M_\odot$ in order to evolve at all, and therefore the initial mass ratio would have had to be $\gtrsim 3$. Such an extreme mass ratio would almost certainly have led to a very rapid burst of mass transfer at the onset of Roche-lobe overflow (RLOF), and a more likely outcome would have been a brief common-envelope phase followed by a merger into a single star. Although we cannot assert that AS Eri *must* have lost mass, it seems probable to us that both binaries have lost both angular momentum and mass in the course of their evolution. Both kinds of loss must have occurred on a nuclear timescale; probably at least 50% of the initial angular momentum was lost, and 40% of the initial total mass, at least in the case of R CMa.

The second group of stars in Table I are members of the RS CVn class, in which at least one component is seen to be unusually active. It can be seen in Z Her that the more evolved star is 20% less massive than its companion, and yet is by no means filling its Roche lobe. Here is further evidence that stellar activity, enhanced by the relatively rapid rotation of the giant as it is forced to corotate with the binary, can cause mass loss on a *nuclear* timescale. The same phenomenon is evident to a lesser extent in RZ Eri. In this latter system star 1 is observed to be rotating in 31.4 days, which is significantly slower than the pseudosynchronous period (23 days) expected on the basis of tidal friction alone (Hut, 1981). RS CVn itself does not show a mass anomaly, except that the substantially different radii of the two components suggest a mass difference of $\sim 12\%$ rather than the observed 2%, and so may imply that star 1 has lost $\sim 10\%$ of its initial mass. λ And has a giant rotating substantially less rapidly than its pseudosynchronous rate, but on the other hand AE Lyn appears to be rather exactly pseudosynchronized.

One of us (PPE) has suggested a very crude model for dynamo activity, with concomitant mass loss and angular momentum loss (Eggleton, 2001). In this model it is conjectured that activity in a star is a function solely of its mass M, its radius R, its luminosity L, and its rotational period $P_{\rm rot}$. For given values of these four quantities the model gives a value of B and dM/dt (the former being the mean poloidal magnetic field), and hence also of the Alfvén radius $R_{\rm A}$. Certain constants were calibrated to ensure that the Sun would have roughly its observed rotation rate, magnetic field, mass-loss rate and Alfvén radius at age 4.6 Gyr, starting from a rotation period of 3 days. Further observational results used in the calibration were the analysis by Stępień (1995) of rapidly rotating solar-type stars, and by Brandenburg et al. (1998) of solar-type cycles in cool giants and dwarfs.

In order to relate the rate of spin of the components of a binary to the orbital rotation it is necessary to include a model for tidal friction. We used the 'equilibrium-tide' model of Hut (1981), as generalized by Eggleton et al. (1998) to include spins which are oblique to the orbit. A stellar evolution code (Eggleton, 1972; Pols et al., 1995) was modified so as to solve the interior structure of both stars *simultaneously*, and in addition extra boundary conditions were added and solved simultaneously with the structure to include both the dynamo model and the tidal friction model. The latter is restricted to spins parallel to the orbit, however. The components were assumed to be in uniform rotation.

In addition to a code which follows internal evolution in detail, we find it useful to have a second code, which follows the orbital dynamics in detail, but uses only a very simple approximation to internal evolution. This allows us to follow, for example, (a) the process by which spin and orbit become parallelized, and (b) the influence of a third body in a possibly non-parallel orbit.

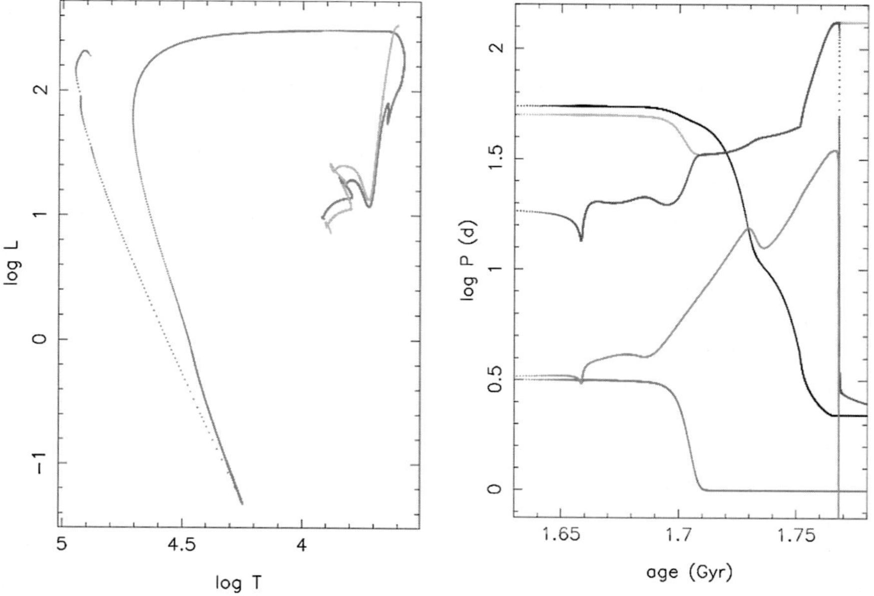

Figure 1. Possible evolution of RZ Eri, taking account of enhanced wind, magnetic braking, and tidal friction in both components. The system was started with parameters $(1.75 + 1.68M_\odot; e = 0.5; P = 49$ days); both stars were started with an initial rotation period of 2 days. The present epoch is at $\sim$1.70 Gyr. (a) The theoretical HRD. Star 1 became a white dwarf, and then flared rapidly in luminosity (and broke down) because of partial accretion of wind from star 2 as star 2 climbed the First Giant Branch. (b) The evolution of mass (solar units), eccentricity, orbital period and rotational period of star 1 (log days) and primary radius (log solar radii), during the time interval when they were varying most rapidly. RLOF began at 1.75 Gyr.

Figure 1 is our first attempt to model the evolution of RZ Eri. We started with an arbitrary value of 0.5 for the eccentricity, and by trial and error found initial masses and an initial orbital period that led to roughly the observed values of masses, radii and orbital period. The age implied by this procedure was 1.70 Gyr. Only the rotation period was unsatisfactory at this epoch: our model reached the pseudosynchronous rate at $\sim$1.69 Gyr. This suggests that either tidal friction was too strong (in the model), or magnetic braking was too weak. Stronger magnetic braking could keep star 1 in slower rotation, maintaining a pseudosynchronism at a slower rate than tidal friction alone.

It is difficult to find a good fit to a binary, in which non-conservative processes have been significant. In conservative evolution we are helped by the fact that the total mass and the orbital angular momentum are known from observation, and only the initial mass ratio has to be guessed. Good fits can be found to semidetached systems (Nelson and Eggleton, 2001), provided they are of sufficiently early spectral type for dynamo activity to have been unlikely. Not only do we have to guess more initial quantities, including spin periods and eccentricity as well, but we also have considerable uncertainty in the coefficients governing the tidal friction and dynamo models. We are only beginning what we hope will be a systematic attempt to model the many systems in which these processes play a part.

It is already clear that this may be difficult, and the difficulties may point to the rather disappointing, though hardly surprising, possibility that there is something inherently chaotic in dynamo activity, even when averaged over a long stretch of time. For example, our model can actually deal quite well with Z Her, giving more-or-less the correct mass anomaly in star 1 (but our model also gives significant mass loss in star 2, so that *both* initial masses must have been larger than present masses). However, we then find it difficult to model the much smaller (or rather non-existent) mass anomaly in the somewhat similar system RS CVn.

V1379 Aql is an interesting RS CVn-like system, which is presumably a *post-Algol*. What is worrying here is that the orbit is quite significantly eccentric: $e = 0.09 \pm 0.01$. It is hard to see how such an eccentricity can either have persisted through the RLOF, or have been generated subsequently. We believe that most probably it has been generated subsequently, through the agency of a third body that has not yet been detected. We discuss third bodies shortly, but for the moment note that if third bodies are capable of affecting orbits such as those in Table I, then the business of modeling these systems becomes substantially more uncertain.

The four systems o Leo, HR 2030, TZ For, and α Aur all have (nearly) circular orbits, which should set some lower limit to tidal friction. The first two systems are unusual in that star 1 appears to be in the rapid evolutionary stage of crossing the Hertzsprung gap for the first time, and not yet at helium ignition. In the last two systems it is likely that at least one component was previously larger, at helium ignition, and that most of the circularisation of the orbit took place then. However, this means that it will be harder than we would like to pin down the timescale of tidal friction, because the maximum radius at helium ignition depends importantly on two other uncertain parameters: the convective mixing-length ratio and the degree of convective overshooting.

These uncertainties also enter into the possibility of using the remarkable Cepheid binary SU Cyg (Table I) to give an upper limit to tidal friction. Here, substantial eccentricity persists, despite the fact that the Cepheid must have approached quite close to its Roche-lobe radius at helium ignition. We would like to be able to blame this on the fact that the companion is itself a close pair, so that the system is in fact triple. But although a distant third body, the Cepheid, can have a marked influence on the close orbit (see below), it does not appear that the converse can happen.

The low-mass short-period system BY Dra (Table I) seems like a good laboratory for testing the combination of tidal friction and magnetic braking. This system has often been described as a very young system, possibly with one or both components still larger than, and contracting to, the ZAMS. However, this is now untenable, since Zuckerman et al. (1997) determined that the binary has a CPM companion

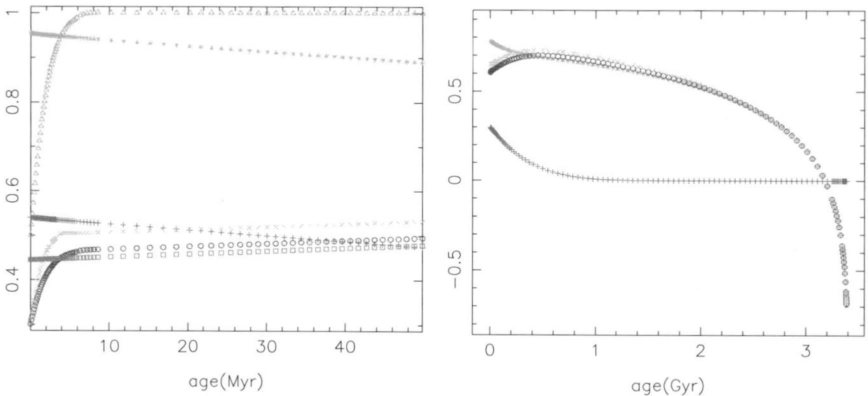

Figure 2. Orbital and spin evolution of a model for BY Dra. (a) The short term, starting 200 Myr ago and ending 150 Myr ago. The initial period and eccentricity were chosen to give the present period and eccentricity after 200 Myr. The six curves, starting from the top at $\sim$10 Myr, are (1) cosine of inclination of star 1's spin to orbit, (2) log orbital period (days), (3) eccentricity, (4) log $P_{\rm rot}$ (days), for star 2, (5) the same for star 1, and (6) the pseudosynchronous period (log, days) for either star, subject to tidal friction alone. (b) The long term, starting from present conditions. The four curves, starting from the top at age zero, are (1) orbital period (log days), (2) and (3) log $P_{\rm rot}$ (days) of star 2 and star 1, and (4) eccentricity.

which is a normal M5 V star. Such a star should be at least 200 Myr old. The system does not eclipse, but an inclination ($151.8 \pm 3.5°$) has been determined by VLTI (Boden and Lane, 2001), leading to the masses quoted. However the uncertainty in $\sin^3 i$ is still quite substantial, and as the quoted masses appear rather low for the spectral types (Popper, 1980) we experimented with masses that were larger but still within the error bars. Figure 2 shows the evolution of some parameters in our models.

We assume initial parameters ($0.73 + 0.64 M_\odot$, $P = 9$ days, $e = 0.54$). After $\sim$200 Myr (of which only the first 50 Myr is illustrated in Figure 2a) it reaches its present parameters ($P = 6$ days, $e = 0.3$). We choose this age because of the CPM companion, as above. Because we cannot be clear how the binary formed in the first place, we cannot be sure that evolution 'started' with synchronous, parallel rotation in both components, and so we assume arbitrarily that the initial spin periods were both 2 days, and that the two axes were at 60° to the orbital axis, and 120° to each other in the same plane. In Figure 2a, we see that pseudosynchronisation and parallelisation take $\sim$6 Myr. Our model gives the *spin* period of star 1 as 4 days, not very different from the observed value of 3.83 days. In Figure 2b we follow the evolution much further, and find that the orbit shrinks to RLOF at $\sim$3.4 Gyr, by which time the masses have dropped 15 and 10%, respectively. They might form a contact binary, but also might merge quickly in a hydrodynamic burst of RLOF to form a single rapidly rotating star, as we conjecture for AB Dor below.

Unfortunately, any conclusions that we may draw can, in principle, be entirely vitiated by the presence of the third body, even though its period may be $\sim$1000 year. This can happen if the outer orbit is inclined at more than $39°$ $(\sin^{-1} \sqrt{2/5})$ – Kozai (1962); Mazeh and Shaham (1979); Kiseleva et al. (1998). We illustrate this with the remarkable quadruple system ADS 11061 (Tokovinin et al., 2003). This consists of four rather similar late F dwarfs, marking out the turnoff region of a cluster of $\sim$2.5 Gyr. One orbit is long and thin ($P = 1274$ days, $e = 0.9754$), the other smaller and rounder ($P = 10.5$ days, $e = 0.374$); but both orbits have much the same angular momentum, rather surprisingly.

The outer orbit is not known, though it can reasonably be estimated to be $\sim$10^4 year. Its inclination is almost certainly different from those of the two sub-binaries, which differ from each other. Assuming a range of initial parameters for all three orbits we get a wide variety of possible scenarios. In some, neither orbit is much affected by Kozai cycles, but in others one or both orbits are affected, perhaps seriously. Figure 3 is a possible model of the long, thin orbit. Starting with parameters as listed in the caption, the eccentricity cycles powerfully (Figure 3a) with e fluctuating between $\sim$0 and $\sim$0.98, and the inclination between $70°$ and $86°$ (Figure 3b). But in the close periastron passages at the peaks of eccentricity, tidal friction after $\sim$2 Gyr reduces the range considerably (but not its *upper* limit); and by 2.7 Gyr, the eccentricity though still large starts to diminish rapidly. By 3 Gyr the orbit is much smaller and only moderately eccentric, as is observed in the *other* sub-binary. The inclination (Figure 3b) cycles intimately with the eccentricity, and the period (Figure 3c) drops, though

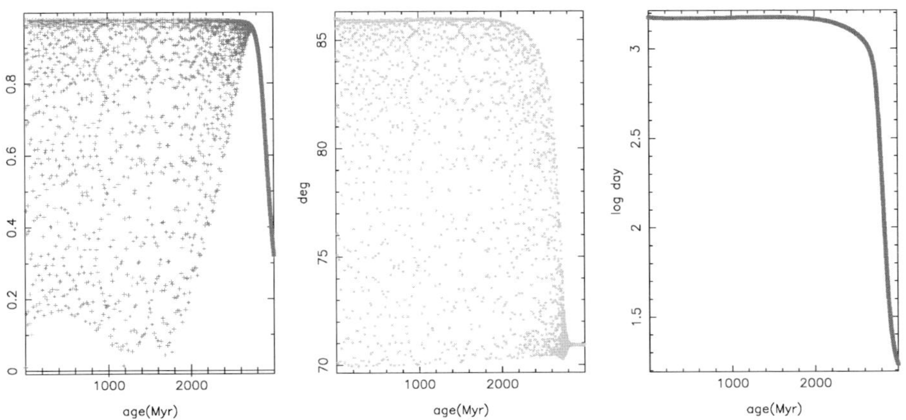

Figure 3. Possible orbital evolution for 41 Dra, the more eccentric sub-binary of ADS 11061. Initial conditions: $P_1 = 1500$ days, $e_1 = 0.01$ (inner orbit); $P = 15,000$ years, $e = 0.73$ (outer orbit); mutual inclination $86°$. Left panel: eccentricity. Centre panel: inclination of inner to outer orbit. Right panel: log period (days). Individual Kozai cycles are about 20 Myr long, and are severely undersampled by the plotting processs.

to 20 days rather than 10 days as seen in the other system. Either or both sub-binaries may have suffered, or be suffering, such evolution, for all one can tell at present.

We included the triple companion of BY Dra in the calculation of Figure 2a, and it made little difference. However, it may be possible to start with a very different inner orbit, and still end with much the same orbit as observed; this is what Figure 3a illustrates, for 41/40 Dra. Thus it is not clear that BY Dra can really be used to calibrate the model of magnetic braking that we propose.

AB Dor (Table I) is an unusually rapidly rotating K dwarf (0.514 d), which like BY Dra shows rotational modulation, erratic variability and emission lines. It does have two companions, as listed, but neither appears to be near enough to have significant influence. The nearer component is not seen, but is inferred from an astrometric orbit. The further (M4 Ve) component is itself active, and a rapid rotator. We might suppose that the rapid rotation of AB Dor is due to youth; it is believed to be associated with the Pleiades and other young clusters in the Local Association, with inferred ages of 50–80 Myr. And yet by no means all young K dwarfs are rotating as rapidly. More typically, even very young but effectively single K dwarfs have rotation periods of 2–5 days. A possible expla-nation is that the K dwarf was formed fairly recently (within 1–5 Myr) by the merger of two M dwarfs; much as we expect for BY Dra in the future (Figure 2b). But we imagine that the 'initial' period in proto-AB Dor was probably $\lesssim 1$ days, in order to allow the merger to take place much more rapidly. It could easily be that Kozai cycling induced in the proto-AB-Dor binary by the companion in the $\sim$10 year orbit first reduced the inner orbit from several days to something of the order of one day, and then magnetic braking reduced it further until the recent merger. Note that the magnitude of the eccentricity fluctuation induced by the third body does not depend *at all* on the mass of the third body. This mass only dictates the cycle time of the fluctuation. For an outer orbit of 10 yr, an inner orbit of 10 days, and a mass ratio of 8, the cycle time would be $\sim$30,000 year, so that hundreds or thousands of cycles could have taken place before our assumed circularisation at short period, followed by our assumed merger driven by magnetic braking.

There is probably, and justifiably, a considerable degree of scepticism in the as-trophysical community regarding evolutionary scenarios that require several stars to interact significantly in a fairly short interval of time, e.g. tens of Myrs. Never-theless, we derive comfort from the fact that one improbable dynamical encounter, involving three stars, was actually observed to take place in about 1996. T Tau (Ghez et al., 1991; Loinard et al., 2003), the prototypical pre-main-sequence star, is in a multiple, formerly quadruple system, though it was not known to be other than single 30 years ago. This appears recently to have broken up. There is a cool infrared companion (T Tau S), probably dominated by an accretion disc rather than a star, 0.7″ to the south of the main G5 V:e component (T Tau N). T Tau S is at least two components, one of which (T Tau Sb) appears to have been in an eccentric 'visual' (actually, radio VLBI) orbit of period $\sim$20 year around the other, but to have been ejected in $\sim$1996 into a hyperbolic orbit with velocity $\sim$20 km s^{-1} towards the east. This suggests that T Tau Sa is itself a closer binary, with an estimated period $\sim$2 year, and that an interaction at the periastron of the eccentric outer orbit between Sb and the binary Sa led to an ejection, as is commonly seen in N-body gravitational simulations of non-hierarchical systems (Anosova, 1986). It is likely that the entire solar neighbourhood has been populated by stars or systems ejected in a somewhat similar manner from star-forming regions.

3. Conclusions

Binaries containing low-mass, cool stars are subject to several non-conservative evolutionary processes, involving dynamo activity, mass loss, magnetic braking, and tidal friction. These can influence masses and orbits on a nuclear timescale, even though the stars may be dwarfs or subgiants. It is not clear that these processes are completely deterministic, but it seems reasonable to attempt to model them in a deterministic way, by assuming that they depend only on such bulk parameters as the masses, luminosities, radii and rotational periods of the components. A considerable number of observed systems can potentially constrain such a deterministic model. The same processes may be important in several other classes of binaries, notably contact binaries, cataclysmic binaries and low-mass X-ray binaries, and we would hope that consistency would be achieved among all these disparate systems.

However, care must be taken to either exclude triple and higher multiple systems, or else to include their potentially important orbital-dynamical interactions. An inconspicuous third body, in a surprisingly wide orbit (perhaps up to $\sim$10^4 years) can have an important influence on shorter-period sub-orbits with periods of days to years.

At the Lawrence Livermore National Laboratory we are developing a 3-D stellar structure code, which should in the medium term be able to model not only hydrodynamic, but also magnetohydrodynamic processes by resolving stellar interiors with $\gtrsim$10^8 cells, and in the longer term we hope $\sim$10^{11} cells. Such a code will be necessary for investigating the complex interaction of turbulent convection with rotation and magnetic fields.

Acknowledgements

We would like to thank Dr. A.A. Tokovinin for helpful discussions. This work was performed under the auspices of the U.S. Department of Energy, National Nuclear Security Administration by the University of California, Lawrence Livermore National Laboratory under contract No. W-7405-Eng-48.

References

Andersen, J., Clausen, J.V., Nordström, B., Tomkin, J. and Mayor, M.: 1991, *A&A* **246**, 99.

Anosova, J.P.: 1986, *Ap&SS* **124**, 217.

Barlow, D.J., Fekel, F.C. and Scarfe, C.D.: 1993, *PASP* **105**, 476.

Boden, A.F. and Lane, B.F. 2001, *ApJ* **547**, 1071.

Boyd, R.W., Eaton, J.A., Hall, D.S., Henry, G.W., Genet, R.M., Lovell, L.P., Hopkins, J.L., Sabia, J.D., Krisciunas, K. and Chambliss, C.R.: 1983, *Ap&SS* **90**, 197.

Brandenburg, A., Saar, S.H. and Turpin, C.R.: 1998, *ApJ* **498**, L51.

Eggleton, P.P.: 1972, *MNRAS* **156**, 361.

Eggleton, P.P.: 2001, in: P. Podsiadlowski, S. Rappaport, A.R. King, F. D'Antona and L. Burderi (eds.), Evolution of Binary and Multiple Star Systems, ASP Conference 229, p. 157.

Eggleton, P.P., Kiseleva, L.G. and Hut, P.: 1998, *ApJ* **499**, 853.

Evans, N.R. and Bolton, C.T.: 1990, *ApJ* **356**, 630.

Fekel, F.C., Eitter, J.J., de Medeiros, J.-M. and Kirkpatrick, J.D.: 1998, *AJ* **115**, 1153.

Ghez, A.M., Neugebauer, G., Gorham, P.W., Haniff, C.A., Kulkarni, S.R. and Matthews, K.: 1991, *AJ* **102**, 2066.

Griffin, R.E.M.: 2002, *AJ* **123**, 1001.

Griffin, R.E.M. and Griffin, R.F. 2000, *MNRAS* **319**, 1094.

Guirado, J.C., Reynolds, J.E., Lestrade, J.-F., Preston, R.A., Jauncey, D.L., Jones, D.L., Tzioumis, A.K., Ferris, R.H., King, E.A., Lovell, J.E.J., McCulloch, P.M., Johnston, K.J., Kingham, K.A., Martin, J.O., White, G.L., Jones, P.A., Arenou, F., Froeschle, M., Kovalevsky, J., Martin, C., Lindegren, L. and Söderhjelm, S.: 1997, *ApJ* **490**, 835.

Hut, P.: 1981, *A&A* **99**, 126.

Jeffery, C.S. and Simon, T.: 1997, *MNRAS* **286**, 487.

Kozai, Y.: 1962, *AJ* **67**, 591.

Kiseleva, L.G., Eggleton, P.P. and Mikkola, S.: 1998, *MNRAS* **300**, 292.

Lim, J.: 1993, *ApJ* **405**, L33.

Loinard, L., Rodríguez, L.F. and Rodríguez, M.I.: 2003, *ApJ* **587**, L47.

Mazeh, T. and Shaham, J.: 1979, *A&A* **77**, 145.

Nelson, C.A. and Eggleton, P.P.: 2001, *ApJ* **552**, 664.

Pols, O.R., Tout, C.A., Eggleton, P.P. and Han, Z.: 1995, *MNRAS* **274**, 964.

Popper, D.M.: 1980, *ARA&A* **18**, 115.

Popper, D.M.: 1988a, *AJ* **95**, 1242.

Popper, D.M.: 1988b, *AJ* **96**, 1040.

Refsdal, S., Roth, M.L. and Weigert, A.: 1974, *A&A* **36**, 113.

Sarma, M.B.K., Vivekananda Rao, P. and Abhyankar, K.D.: 1996, *ApJ* **458**, 371.

Stępień, K.: 1995, *MNRAS* **274**, 1019.

Tokovinin, A.A., Balega, Y.Y., Pluzhnik, E.A., Shatsky, N.I., Gorynya, N.A. and Weigelt, G.: 2003, *A&A* **409**, 245.

Zuckerman, B., Webb, R.A., Becklin, E.E., McLean, I.S. and Malkan, M.A.: 1997, *AJ* **114**, 805.

THE EVOLUTION OF MASSIVE CLOSE BINARIES

C. DE LOORE and W. VAN RENSBERGEN

Astrophysical Institute, Vrije Universiteit Brussel, Pleinlaan 2, B-1050 Brussel;
E-mail: cdeloore@nets.ruca.ua.ac.be

(accepted April 2004)

Abstract. We present results of evolutionary computations for massive close binaries with the Brussels simultaneous evolution code for conservative and non-conservative Roche lobe overflow (RLOF). We discuss mass transfer in massive close binaries during phases of RLOF, common envelope, spiral-in and merging. We examine the effects of stellar wind during successive stellar evolution phases and the final fate of primaries. We show how our library can be used to explain well-known binaries such as the WR + OB system V444 Cyg, HMXBs Vela X-1 and Wray 977, LMXBs like Her X-1, and binary pulsars. More details on the evolution of massive close binaries can be found in "The Brightest Binaries" (Vanbeveren et al., 1998).

Keywords: massive stars, binaries, Roche lobe overflow (RLOF), X-ray binaries

1. Massive Stars: Definition

A massive star performs all nuclear burning phases non-degenerately until the formation of a Fe–Ni core. It is different for single stars and components of binaries. Certain evolutionary phases show properties corresponding to observed characteristics of stellar subclasses, such as the hottest O and B stars of luminosity classes V–II, the hottest OBA stars, LBVs, Yellow Supergiants, Red Supergiants, Hypergiants, and WR stars. Figure 1 shows the galactic HRD of O3-M5 stars, members of a cluster or association (Humphreys and McElroy, 1984) and evolutionary tracks for 15, 20 and 40 $M_\odot$.

2. Stellar Wind Phases

2.1. OB STARS AND THE LBV PHASE

LBVs are unstable and very luminous OB supergiants. LBVs with $M_{\text{init}} \geq 40M_\odot$ have stellar wind mass loss rates of 10^{-7}–$10^{-4} M_\odot$/year) and lose mass by eruptions at rates of 10^{-3}–$10^{-2} M_\odot$/year.

2.2. THE RSG-PHASE

Reid et al. (1990) published stellar wind (SW) mass loss rates for luminous Red Supergiants (RSGs) in the LMC, using a formalism proposed by

Astrophysics and Space Science **296**: 337–352, 2005.
© Springer 2005

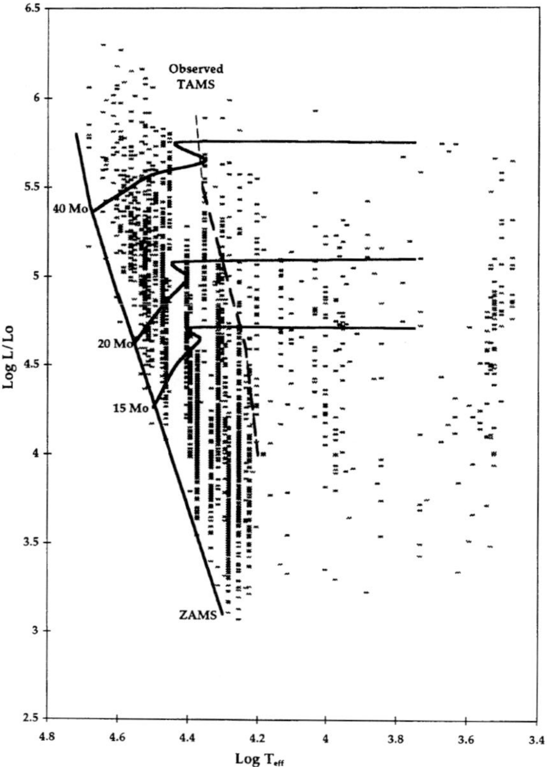

Figure 1. HRD of stars with spectral type O3-M5 together with some evolutionary tracks labeled with initial mass.

Jura (1987):

$$\log(-\dot{M}) = 0.8 \log L - C \quad \text{with} \quad C \in [8.7\text{--}9] \tag{1}$$

2.3. THE WR-PHASE

Typical mass loss rates for WR stars: $-5.1 < \log(-\dot{M}) < -4.1$.
   A possible $\dot{M} - L$ relation for galactic WR stars: $\log(-\dot{M}) = \log L - 10$.

3. Massive Star Evolution: Stellar Wind and Rotation

Evolutionary computations show that stellar wind mass loss during core H-burning is unimportant for stars of $M_{\text{initial}} \leq 40 M_{\odot}$. Stars with $M_{\text{initial}} > 40 M_{\odot}$ lose most of their H-rich layers by OB + LBV-type stellar winds. For close binaries with a period P so large that RLOF does not occur before the start of the LBV phase

($P \geq 10$ days) mass transfer may be prevented. Stellar winds change the binary period drastically. Meynet and Maeder (2000) stressed the importance of rotation for stellar evolution and the composition of the outer layers. Faster rotation means larger convective cores: overshooting of convective cores mimics the effect of rotation. During RLOF the mass-losing component loses the larger part of its H-rich layers. The remnant corresponds more or less to the He core left after core H-burning.

4. WR Stars

Emission lines of He and N (WN stars), He and C (WC stars) or He and O (WO stars) dominate the spectrum of WR stars. The majority are hot H-deficient core He-burning stars. WN stars show CNO equilibrium abundances. In WC and WO types we see the products of the 3α process. A few WR stars show WC and WN characteristics; together: they are WNC stars. A further subdivision ranges from WN2 to WN9, WC4 to WC9 (Smith et al., 1996). The larger part of the late type WN stars show H in their spectra; most of the early types are H-deficient. WNE stars are WN stars without H. WNL stars are WN stars with H. WC and WO stars do not have H in their atmospheres. Figure 2 shows the overall position of galactic massive stars in the HRD.

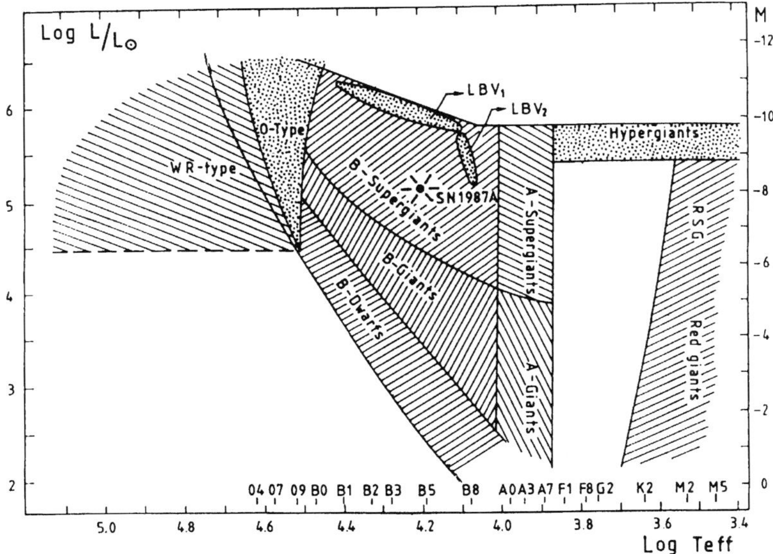

Figure 2. Overall HRD of the massive stars in the Galaxy.

                    C. DE LOORE AND W. VAN RENSBERGEN

5. Massive Close Binaries

Binaries are characterized by the masses M_1 and M_2 of their components, the orbital period P, the eccentricity e, and the mass ratio $q = M_2/M_1$. The semi-major axis A is then determined from Kepler's law: $P^2/A^3 = $ constant. The orbital angular momentum J is given by:

$$J = \frac{M_1 \times M_2}{M_1 + M_2} \times \frac{2\pi A^2 \sqrt{1 - e^2}}{P} \tag{2}$$

A binary is called massive, if at least one of its components will collapse into a neutron star (NS) or a black hole (BH).

6. The Roche Model

The evolution of a component of a binary differs from that of a single star of same mass and chemical composition by the existence of a point between the two components, where the effective gravity vanishes and consequently matter can flow freely from one component to the other.

6.1. ROCHE EQUIPOTENTIALS

Computations show that early during core hydrogen burning (CHB) massive stars are centrally condensed. The gravitational field of a binary can hence be approximated by that of two point sources. In a Cartesian coordinate system: origin at the center of mass, z-axis along the spin axis, x-axis through the stellar centers, the equipotential surfaces are given by:

$$\Phi = -\frac{G \times M_1}{\sqrt{[x - \mu a]^2 + y^2 + z^2}} - \frac{G \times M_2}{\sqrt{[x - (1 - \mu)a]^2 + y^2 + z^2}}$$
$$-\frac{1}{2} \times \Omega_B^2 \times [x^2 + y^2] = \text{const} \tag{3}$$

with

$$\mu = \frac{M_2}{M_1 + M_2} \quad \text{and} \quad \Omega_B = \frac{2\pi}{P} \tag{4}$$

Figure 3 depicts the intersection of the equipotential surfaces with the orbital plane ($z = 0$). The function Φ has five saddle points (Lagrangian points L_i), where centrifugal and gravitational forces modified by the Coriolis acceleration are in balance. L_2 and L_3 are potential minima on the x-axis. The most important

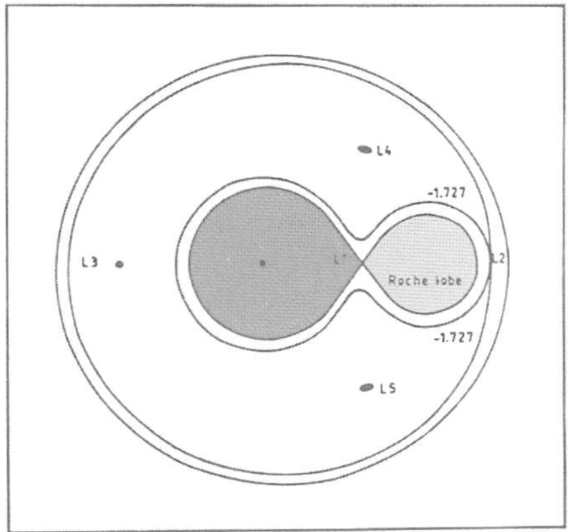

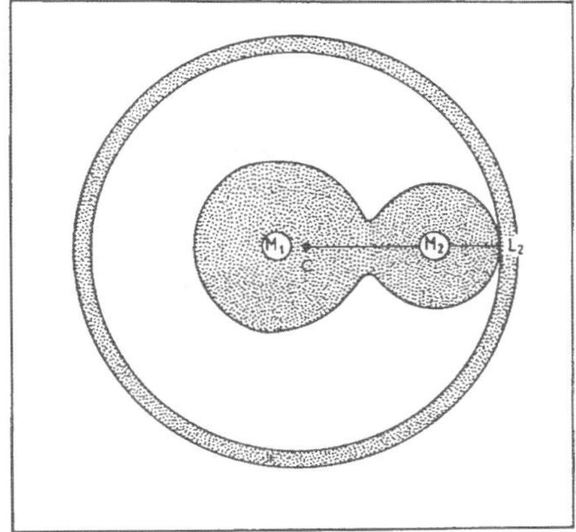

Figure 3. Left: Intersection of equipotential surfaces with the orbital plane. Right: The center of gravity C and the distance $CL_2 \approx 1.25A$. The Keplerian ring that carries away mass and angular momentum from the system is located at $\approx 2.25A$.

saddle point is L_1, the point of gravitational balance on the x-axis between the two components. The equipotential surface across L_1 is the critical or Roche surface (Roche lobe). The Roche radius R_c is the radius of a sphere with the same volume as the Roche lobe. With $q = M_2/M_1$, Eggleton (1983) presented a formula giving the Roche radius of star "1" as a function of the mass ratio:

$$\frac{R_c}{A} = \frac{0.49}{0.6 + q^{2/3}\ln(1 + q^{-1/3})} \tag{5}$$

R_c of star "2" is found using $1/q$ instead of q in Eq. (5).

6.2. THE ROCHE LOBE OVERFLOW PROCESS

In the case that rotational motion of the envelope of the primary and orbital motion are synchronous, we have hydrostatic equilibrium in the corotating frame, and the pressure equation is then:

$$\nabla P = -\rho \nabla \Phi \tag{6}$$

Surfaces of equal pressure and density are equipotential surfaces. The shape of a binary component is given by an equipotential. During the evolution the expanding outer layers may reach L_1, where $\nabla \Phi = 0$. Hydrostatic equilibrium is no longer valid. For an isothermal gas near L_1, the equation of motion is:

$$\left(v^2 - c_s^2\right)\frac{dv}{dx} = \nabla \Phi \tag{7}$$

v is the velocity of the gas, c_s the local isothermal sound speed. In order to retain a monotonic increasing velocity field near and across L_1, ($\nabla\Phi = 0$), $v \approx c_s$. Matter funnels near L_1 and flows hypersonically towards the companion. The process of filling (and overflowing) the Roche lobe is called RLOF. The component losing matter is called the loser or donor, the other is the gainer or accretor.

6.3. MASS LOSS RATE OF THE MASS LOSER DURING RLOF

Scheme for the computation of the mass loss rate is as follows.

(a) Calculate the internal structure of the two stars using the set of equations for stellar structure for single stars with spherical symmetry.
(b) Compare the radius of the primary R_1 to the Roche radius R_c.
(c) If $R_1 < R_c$: accept model. Calculate new model for next time step.
(d) If $R_1 > R_c$: determine the mass loss rate by equalling the stellar radius and the Roche radius. Combination of this rate and the time step allows the determination of the masses of loser and gainer.

This procedure works well in situations where the response of the star to mass loss implies a radius decrease. This is the case for a radiative envelope, not for deep convective envelopes. When the star suffers mass loss (time scale < thermal time scale), its adiabatic response is a radius increase, leading to violent mass loss on a dynamical time scale of hours. The secondary will be engulfed by the primary. This phase is known as the common envelope phase.

6.4. DIFFERENT TYPES OF UNEVOLVED MASSIVE CLOSE BINARIES

Massive stars are characterized by various expansion phases: CHB, H-shell burning and He-shell burning. Primaries of close binaries fill their Roche lobe during one of these phases. Kippenhahn and Weigert (1967) and Lauterborn (1969) divided the class of unevolved massive close binaries (MCBs) into:

Case A: the primary fills the Roche lobe during CHB;
Case B: the primary fills the Roche lobe during H-shell burning;
Case C: the primary fills the Roche lobe during He-shell burning.

Most MCBs with P smaller than 2–4 days: case A; P between 4 and ≈ 1000 days: case B; P between ≈ 1000 and 2000 days: case C.

The observed period distribution $\Pi(P)$ of binaries before RLOF follows the relation of Popova et al. (1982): $\Pi(P) \sim (1/P)$.

6.5. MASS TRANSFER PROCESS DURING RLOF. DIRECT HIT OR THE FORMATION OF A KEPLERIAN DISK

Matter leaving the primary via L_1 can hit the secondary directly or it can first form a disk around it. Lubow and Shu (1975) studied the latter case. Given M_1, M_2 and P, computation of particle trajectories reveals that a disk is formed if $R_2 < R_{\text{disk}}$, R_{disk} is given by:

$$\ln \frac{R_{\text{disk}}}{A} = 0.367 \ln q - 2.93 \tag{8}$$

Before RLOF both components evolve like single stars. If gas streams towards a star, a shock front forms near the surface (Ulrich and Burger, 1976). The gravitational energy released by this inflowing matter is transformed into thermal energy and is radiated away as X-rays and UV radiation. If all the gravitational energy is transformed into radiation, the radiated luminosity (L_{acc}) is given by: $L_{\text{acc}} = G \times M \times \dot{M}/R$. This exceeds the critical Eddington luminosity only when $\dot{M} > 10^{-2} M_\odot$/year, much larger than the mass transfer rates during RLOF.

6.6. THE STANDARD ACCRETION MODEL

For MCBs we may assume that the envelope of the gainer is in radiative equilibrium and has a positive entropy gradient. In the "standard accretion model" one assumes the outer layers to be in radiative equilibrium. This is the case when the entropy of the impinging matter is larger than the entropy of the envelope of the mass gainer.

$$\dot{M} \geq 3.2 \times 10^{-8} \times \frac{RL}{M} \tag{9}$$

where $\dot{M}$ is in $M_\odot$/year and R, L, M are in solar units.

A mass gainer of $15 M_\odot$ at the end of CHB becomes overluminous, when the accretion rate is larger than $10^{-3} M_\odot$/year. Due to significant overluminosity, the outer stellar layers expand, and the mass gainer may as well fill its critical equipotential: the binary is then a contact binary.

6.7. THE ACCRETION-INDUCED MIXING MODEL

Mass accretion implies accretion of angular momentum; consequently a mass gainer spins up. Rotation induces mixing, i.e. radiative equilibrium is destroyed. Furthermore, if the entropy of the accreted matter is significantly lower than the entropy of the envelope of the gainer, convection will smear out the entropy profile. This convection zone will develop inwards as long as its specific entropy exceeds that of the matter, which is accreted on top of it. The overluminosity will be smaller than in the standard accretion model. The limiting situation is a mixing of the whole star.

This is "accretion-induced mixing". We studied the evolution of mass gainers by adopting both models. Both can be used to examine agreement with observations.

7. Conservative and Non-Conservative Mass Transfer

In case of conservative RLOF, all mass leaving the primary is accreted by the secondary. The rotational angular momentum of the binary components is always much smaller than the orbital angular momentum. Conservation of mass implies conservation of total orbital angular momentum, and this means in the circularized case:

$$\frac{M_1^2 M_2^2}{M_1 + M_2} \times A = \text{const} \tag{10}$$

or using Kepler's law

$$\frac{P}{P^0} = \left(\frac{M_1^0 M_2^0}{M_1 M_2} \right)^3 \tag{11}$$

The superscript "o" in Eq. (11) denotes values at the start of RLOF. For non-conservative RLOF, we define a parameter β, the fraction of the lost mass accreted by the gainer. If $\dot{M}_1$ (respectively $\dot{M}_2$) denotes the mass loss (respectively mass gain) rate of the mass loser (respectively mass gainer):

$$\dot{M}_2 = -\beta \dot{M}_1; \quad 0 \le \beta \le 1.$$

In most MCBs, SW mass loss is not large enough to allow the primary to avoid RLOF. The primary loses matter by RLOF at very high rates ($\approx 10^{-3} M_\odot$/year). The secondary expands and both stars may enter into contact. Matter can leave a case A/case B_r MCB in an efficient way at rates similar to the mass loss rate of the primary by RLOF. Mass escapes through L_2 located at $\approx 1.25 \times A$ from the center of gravity C of the system shown in Figure 3 (right). This matter stabilizes outside L_2 into a Keplerian ring located at $\approx 2.25 \times A$ (Soberman et al., 1997). This ring carries a lot of mass and angular momentum, resulting in a very important decrease of the orbital period.

8. Evolution of a Massive Primary Before and During RLOF

8.1. PRIMARIES WITH $M_{\text{initial}} \le 40\text{–}50 M_\odot$: CASES A AND B_r

Conservative evolution of a case A $15 M_\odot$ primary (galactic composition) with $q = 0.9$ and $P = 2$ days is shown in Figure 4 as dashed track. Case B_r means:

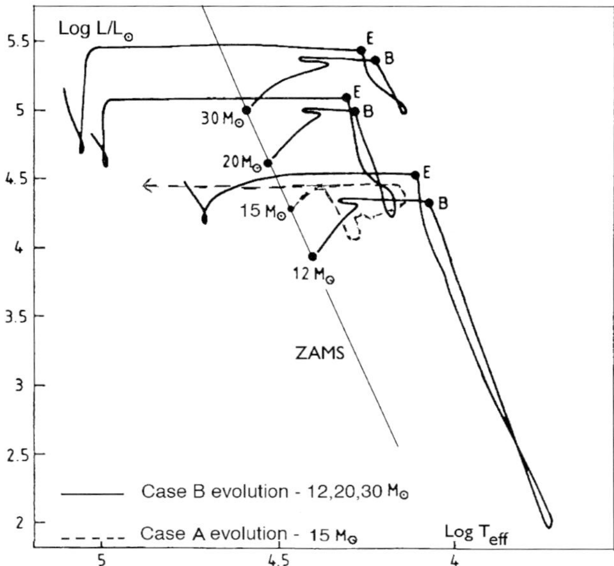

Figure 4. In full lines: tracks of massive primaries in a case B_r binary assuming conservative RLOF. The points B and E denote the beginning and the end of RLOF, respectively. The dashed line shows the conservative evolution of a 15 $M_\odot$ primary with initial $q = 0.9$ and initial $P = 2$ days.

RLOF starts when the primary is a H-shell burning star with a radiative envelope. Figure 4 shows typical tracks of massive primaries in a case B_r binary assuming conservative RLOF.

The duration of RLOF and the behaviour of the primary during RLOF are determined by the expansion of the H-shell. During the first $\approx 10\%$ of this phase, the star evolves on the Kelvin-Helmholtz timescale and loses mass at very high rates, approximately equal to the right hand side of Eq. (9).

The maximum values are a few times $10^{-3} M_\odot$/year: the star is in the rapid phase of RLOF. The entropy term in the equation of energy causes a rapid decrease of the stellar luminosity. When the central temperature and density are sufficiently high to start core He burning (CHeB), the expansion rate of a H-shell burning star decreases and the mass loss rate diminishes: the star is in the slow phase of RLOF. During this phase the stellar luminosity increases again; the star regains thermal equilibrium. This is attained when most of its H-rich layers have been removed. Evolutionary computations reveal that a massive H-shell burning star stops its expansion when He is burning in the core and when the atmospheric H-abundance $X_{atm} \approx$ 0.2–0.3. The star regains its thermal equilibrium and starts contracting: RLOF is finished.

Figure 5 shows the overall evolutionary behaviour of primaries of case B_r close binaries (small convective core overshooting) with galactic and SMC initial

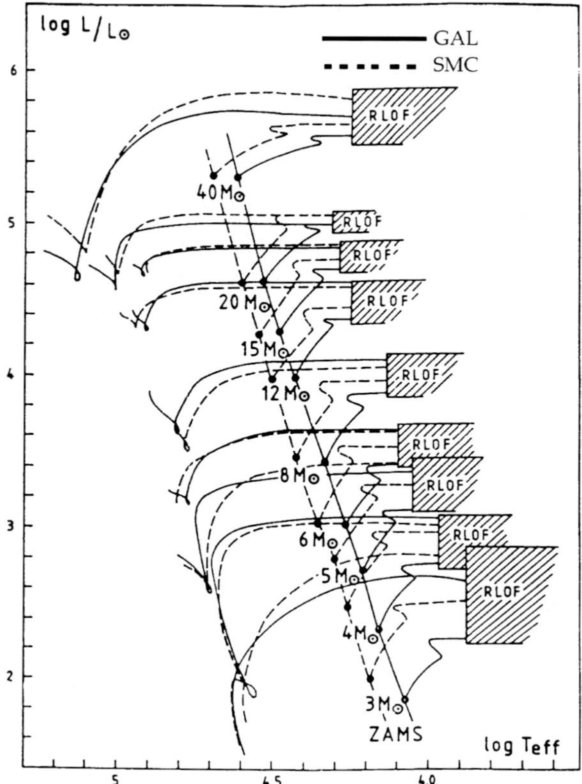

Figure 5. Evolution of primaries of case B_r close binaries with galactic and SMC initial chemical composition.

chemical composition (LMC being intermediate). The evolution of a primary after RLOF does not depend drastically on the details during RLOF; the RLOF is covered by a black box. Relations between pre-M_b and post-M_a RLOF mass were derived: for the Galaxy: $M_a = 0.093 M_b^{1.44}$; for the LMC: $M_a = 0.085 M_b^{1.52}$; for the SMC: $M_a = 0.048 M_b^{1.7}$.

8.2. PRIMARIES WITH $M_{\text{initial}} \leq 40\text{--}50 M_\odot$: CASES B_c AND C

The larger the radius of the H-shell burning star the smaller the thermal time scale, or, the larger the initial period of the binary, the more violent the rapid RLOF phase and the larger the mass loss rates. This becomes critical when the period is sufficiently large to start RLOF when the primary has a convective envelope: case B_c. Mass loss by RLOF will go very fast, on a dynamical time scale (hours), and a common envelope phase sets in (see Section 6.3).

9. The Final Fate of Primaries of MCBs

Primaries with $M_{init} \leq 40 M_\odot$ develop a CO core mass $\leq 3.8 M_\odot$. This accords to a Fe–Ni core mass of $\approx 1.8 M_\odot$ (Woosley, 1986). Primaries with $M_{init} > 40 M_\odot$ have massive CO cores. Since the corresponding mass of the Fe core is sufficiently large they are potential candidates for BH formation. In all He stars with mass $> 2.2 M_\odot$, the core becomes larger than the Chandrasekhar mass; it will undergo a SN leaving a compact star. A 2.2 $M_\odot$ CHeB star is the remnant of a $10 M_\odot$ primary, after RLOF in a case B or late case A binary. The $10 M_\odot$ mass limit for NSs and the $40 M_\odot$ limit for BHs are perhaps smaller for large convective core overshooting during CHB. Since the effect of rotation is similar to that of large convective overshooting, the lower mass limits also apply to rapid rotators.

10. Evolutionary Computations of MCBs and Observations

Both components first evolve as independent single stars. The most massive star expands faster than its companion, it first reaches its critical radius and RLOF starts. When most of the H-rich envelope is transferred an overall contraction phase sets in: RLOF stops. The system consists now of a H-deficient CHeB star and an OB-type star that has accreted part or all the matter lost by the loser. Paczynski (1967) suggested first that this phase corresponds to the observed WR + OB binaries. Shortly after core-He exhaustion, the He star explodes. If the SN does not disrupt the binary, the post-SN system resembles an OB + compact star binary. Van den Heuvel and Heise (1972) suggested that this phase could represent the observed massive X-ray binaries. The OB-type gainer expands, its radius also may reach a critical value, and a second RLOF starts. However, due to the extreme mass ratio of the binary, the low-mass compact star will be dragged into the envelope of the OB star and spirals inwards. A very narrow system will be formed. At the end of the evolution of the He star, a second SN occurs. In most cases, the system will be disrupted, and two single runaway pulsars are left. In the (exceptional) case that the binary remains bound a binary pulsar like, e.g., PSR 1855–09 is formed. Our library of evolutionary sequences explains further in this section some well-known binaries such as the WR + OB system V444 Cyg, the massive X-ray binaries Vela X-1 and Wray 977, and low-mass X-ray binaries like Her X-1.

10.1. V444 Cyg

The observed mass of the WNE component of $\approx 9 M_\odot$ requires a progenitor of $\approx 30 M_\odot$. The age of the binary is thus ≈ 7 million years. Its OB companion is an O6 star. Since the age of a normal O6 star is ≈ 1–2 million years, the OB star must have been rejuvenated: mass transfer must have occurred. The observed mass of the O6 component is $\approx 25 - 26 M_\odot$. However, a normal O6 star has a mass of $\approx 37 M_\odot$, if it is of class V, and $\approx 48 M_\odot$, if it is of class III. Either the O-type component is not an O6-type star but rather O7–O8, or the O6 component is undermassive compared to its spectral type, in other words, overluminous compared to its mass. Agreement with observations can only be obtained by assuming that the system evolved non-conservatively, with $\beta \approx 0.5$. In a next phase the WR star explodes leaving a NS. If the binary is disrupted, the single OB-type runaway becomes a RSG. SW removes the H-rich layers, and the star will become a WR star of 10–$14 M_\odot$. In the non-disrupting case, the binary evolves through a spiral-in phase. Since the present period of the WR binary is only 4.2 days, the NS will spiral-in completely leading to a Thorne-Zytkov object (TZO). Further SW mass loss during the RSG phase of this TZO removes the H-rich layers: a WR + Thorne-Zytkov binary (WRTZ) is formed.

10.2. Vela X-1

The atmosphere shows an overabundance of He, $\epsilon = 0.28$ ($\epsilon_\odot \approx 0.1$). From orbit and X-ray eclipse Rappaport and Joss (1983, 1984) obtain $R \in [28\text{–}35]R_\odot$, $M \in [21.5\text{–}26.5]M_\odot$. $\log L/L_\odot \in [5.5\text{–}5.7]$; distance $d \in [1.8\text{–}2]$ kpc. Using the annual proper motions and the radial velocity, Van Rensbergen et al. (1996) investigated the origin of Vela X-1 and concluded that only Vela OB1 is hit by the past path, and that Vela X-1 left the association some 2 ± 1 million years ago. The distance of Vela OB1 is ≈ 1900 pc (Blaha and Humphreys, 1989), a value close to that of Vela X-1. The progenitor of Vela X-1 was a member of Vela OB1. Figure 6 shows tracks of mass gainers. At the left standard accretion was adopted. After accretion, the evolution is continued up to the beginning of H-shell burning. Models have a final mass $\approx$ the observed mass of the optical star in Vela X-1. The middle of Figure 6 shows tracks calculated with accretion-induced mixing. Numbers along the tracks refer to masses. The tracks are compared to the observed HRD position of the optical component. The figure at the right shows tracks of a MCB explaining the observed properties of the HMXB Vela X-1. The HRD position of the optical component is compared to the members of the Vela OB1 association. Only accretion-induced mixing models agree with the observations, as well for the position in the HRD as for the ϵ-values. Further evolution: spiral-in of the compact component into the B supergiant. A TZO will be formed. When SW removes the H-rich layers, the star might be seen as a WRTZ.

10.3. Wray 977

This X-ray source is a pulsar; a NS. The optical companion is a hypergiant of $[38\text{–}48]M_\odot$ (Sato et al., 1986; Kaper et al., 1995).

If we treat the system in the non-conservative way, $\beta = 0.5$, we find that the initial mass of the primary $\in [39\text{–}46]M_\odot$. For conservative RLOF, these initial

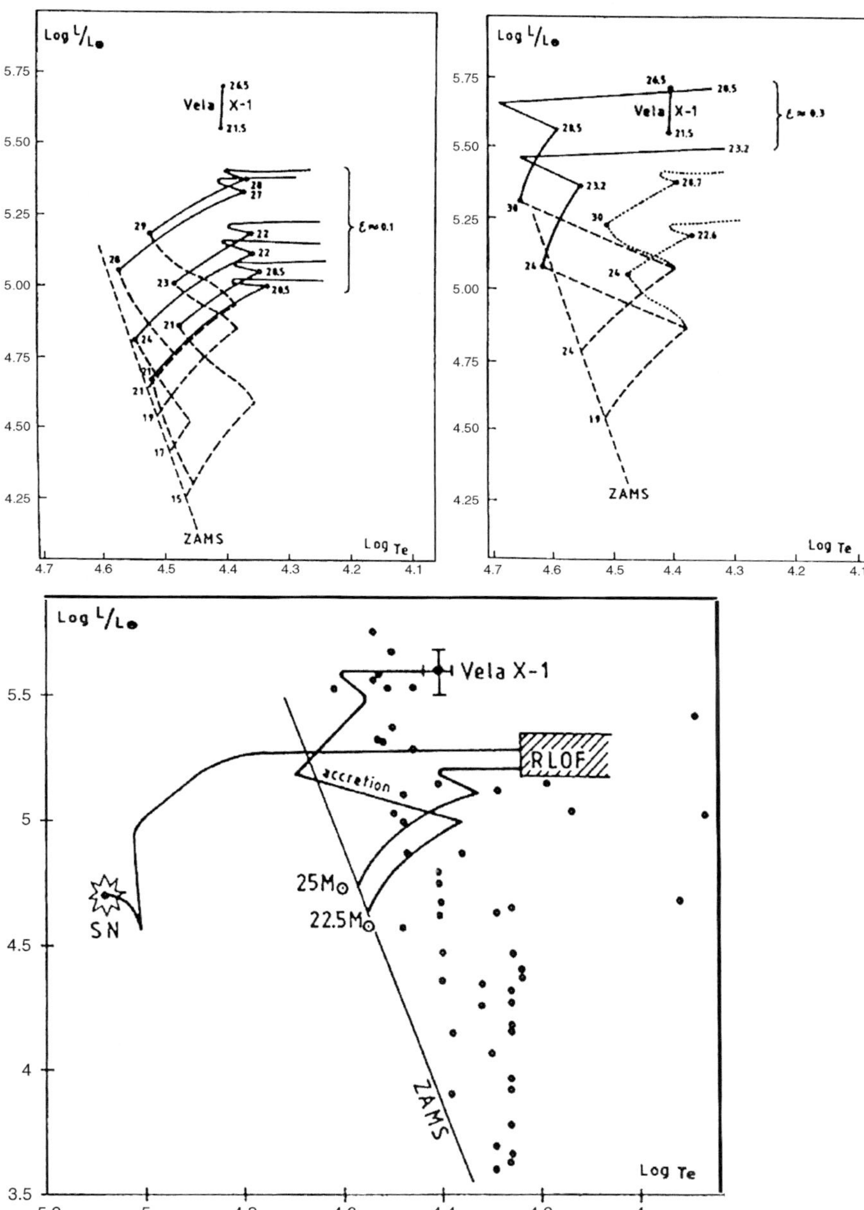

Figure 6. Starting from upper left and clockwise: evolutionary tracks of mass gainers with standard accretion. Tracks calculated with accretion-induced mixing compared to the observed HRD position of the optical component of Vela X-1. Tracks of a MCB explaining the observed properties of Vela X-1. The HRD position of the optical component of Vela X-1 is compared to that of the members of the cluster Vela OB1.

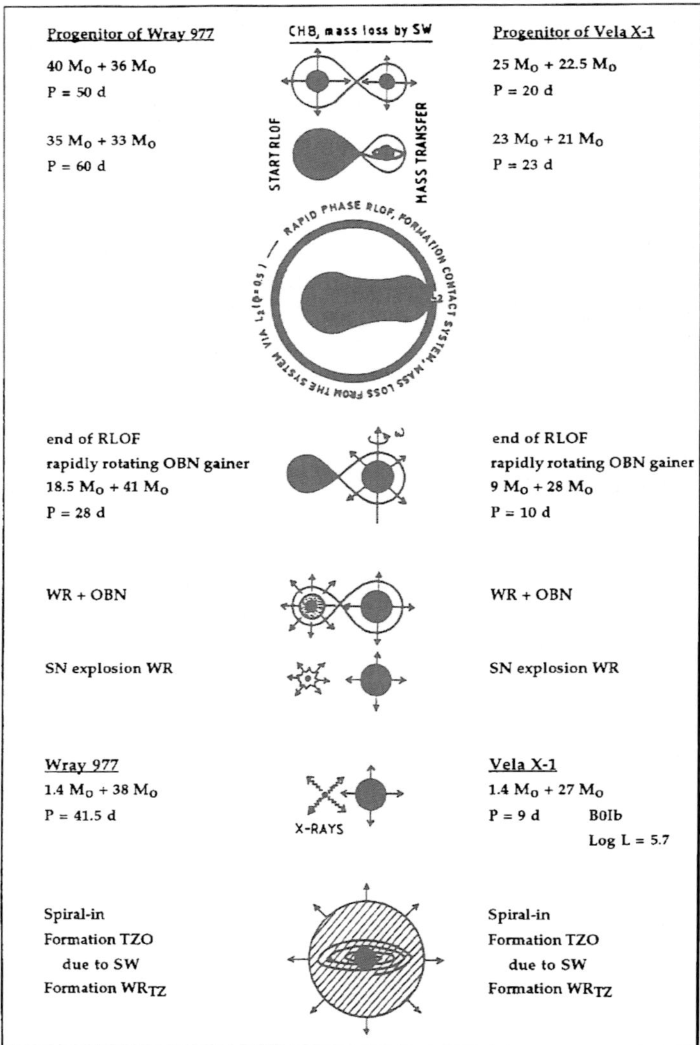

Figure 7. The overall evolutionary scenario of the HMXBs Wray 977 and Vela X-1.

masses reduce to $\in [33\text{–}40]M_\odot$. Hence a primary in an interacting binary with initial mass up to $\approx 30M_\odot$ (perhaps $\approx 40M_\odot$) ends its life as a NS after a SN explosion. Further evolution of Wray 977 is compared to that of Vela X-1 in Figure 7.

10.4. HER X1

This X-ray pulsar consists of a NS of $\approx 1.4 \pm 0.3 M_\odot$ and an A-type companion of $\approx 2 M_\odot$ (Van Kerkwijk et al., 1995). Her X-1 sits out of the galactic plane. The age of a $2M_\odot$ star is 800 million years, so starting from within the galactic disc, its z-velocity component at the explosion must have been larger than 120 km s^{-1} to reach its present position. A binary model explaining the observed properties of the LMXB Her X-1 was suggested by Sutantyo (1975) (see also Verbunt et al., 1990). As initial system Sutantyo proposed a massive star of $\approx 15 M_\odot$ and a companion of $2M_\odot$ with a period $P \geq 1$ year. The system evolves via a common envelope phase into a system of a $\approx 4 M_\odot$ CHeB star and a $2M_\odot$ A-type star with a period ~ 1 day. The CHeB star explodes and forms a neutron star of $1.4 M_\odot$, in an eccentric orbit around the A-star. As a consequence of the explosion the system gets a space velocity of 100 km s^{-1}. The A-type companion evolves, fills its Roche lobe and mass transfer starts. The inflowing matter produces the X-rays.

11. Summary: The Massive Close Binary Evolution Model

The synthesis of our library of evolutionary computations for massive close binaries prior to the final collapse of the primary comparing computations and observations is depicted in Figure 8. A flow chart with the various evolutionary possibilities for MCBs after the final collapse of the primary is shown in Vanbeveren et al. (1998).

If $M_1 \geq 40\text{-}50\ M_\odot$		If $M_1 < 40\text{-}50\ M_\odot$	
If $q \leq 0.2$		If $q \leq 0.2$	
Case A spiral-in mergers possibly populating the BHG	Larger period binaries LBV-scenario WR+OB dwarf or low mass (solar type) companion (P≥10 days)	Case A, early case B$_r$ spiral-in mergers possibly populating the BHG	Larger period binaries spiral-in and/or RSG stellar wind sdO/WD/WR like star/WR+low mass (solar type) companion depending on importance of RSG stellar wind mass loss, either small period or period ≈ year
If $q > 0.2$		If $q > 0.2$	
Case A (quasi)-conservative RLOF WR + OB	Larger period binaries LBV-scenario WR + OB (P≥10 days)	Case A, case B$_r$ (quasi)-conservative-RLOF WR + OB (1 day ≤ P ≤ 120 days)	Larger period binaries common-envelope and/or RSG stellar wind sdO/WD/WR like star/WR+OB depending on importance of RSG stellar wind mass loss, either a period of 1-100 days or period of the order of a year

Figure 8. Brussels MCB model, prior to the final collapse of the primary.

References

Blaha, C. and Humphreys, R.: 1989, *AJ* **98**, 1598–1608.

Eggleton, P.: 1983, *AJ* **268**, 368–369.

Humphreys, R. and McElroy, D.: 1984, *AJ* **284**, 565–577.

Joss, P. and Rappaport, S.: 1984, *ARA&A* **22**, 537–592.

Jura, M.: 1987, *AJ* **313**, 743–749.

Kaper, L., Lamers, H., Ruymaekers, E., Van den Heuvel, E. and Zuiderwijk, E.: 1995, *A&A* **300**, 446–452.

Kippenhahn, R. and Weigert, A.: 1967, *Zeitschr. F. Astrophys* **65**, 251.

Lauterborn, D.: 1969, in: M. Hack (ed.), Mass Loss From Stars, Astrophysics and Space Science Library, p. 269.

Lubow, S. and Shu, F.: 1975, *AJ* **198**, 383–405.

Meynet, G. and Maeder, A.: 2000, *A&A* **361**, 101–120.

Paczynski, B.: 1967, *Acta Astron* **17**, 355–380.

Popova, E., Tutukov, A. and Yungelson, L.: 1982, *Ap&SS* **88**, 55–80.

Rappaport, S. and Joss, P.: 1983, in: W. Lewin and E. Van den Heuvel (eds.), Accretion Driven Stellar X-Ray Sources, Astrophysics and Space Science Library.

Reid, N., Tinney, C. and Mould, J.: 1990, *AJ* **348**, 98–119.

Sato, N., Nagase, F., Kawai, N., et al.: 1986, *AJ* **304**, 241–248.

Smith, L., Shara, M. and Moffat, A.: 1996, *MNRAS* **281**, 163–191.

Soberman, G., Phinney, E. and Van den Heuvel, E.: 1997, *A&A* **327**, 620–635.

Sutantyo, W.: 1975, *A&A* **41**, 47–52.

Ulrich, R. and Burger, H.: 1976, *AJ* **206**, 509–514.

Van den Heuvel, E. and Heise, J.: 1972, *Nat Phys Sci* **239**, 67.

Van Kerkwijk, A., Van Paradijs, J. and Zuiderwijk, E.: 1995, *A&A* **303**, 497–501.

Van Rensbergen, W., Vanbeveren, D. and De Loore, C.: 1996, *A&A* **305**, 825–834.

Vanbeveren, D., Van Rensbergen, W. and De Loore, C.: 1998, The Brightest Binaries, Astrophysics and Space Science Library, p. 232.

Verbunt, F.: 1990, in: W. Kundt (ed.), Neutron Stars and Their Birth Events, Astrophysics and Space Science Library, p. 179.

Woosley, S.: 1986, in: B. Hauck (ed.), Nucleosynthesis and Chemical Evolution, 16th Saas-Fee Course, Geneva Observatory, p. 1.

BINARY EVOLUTION COMPARED TO OBSERVED ALGOLS

C. DE LOORE and W. VAN RENSBERGEN

Astrophysical Institute, Vrije Universiteit Brussel, Pleinlaan 2, B-1050 Brussel;
E-mails: cdeloore@nets.ruca.ua.ac.be, wvanrens@vub.ac.be

(accepted April 2004)

Abstract. We present a grid of evolutionary tracks of 240 binaries with a B-type primary at birth and initial mass ratios: 0.4, 0.6 and 0.9. Conservative calculations are done for cases A and A/B RLOF. For pure RLOF B, conservative and liberal evolutionary sequences have both been computed. In order to compare statistically our computations with the observed distributions of orbital periods and mass ratios of Algols, we enlightened the Algol appearance in every evolutionary sequence. Conservative RLOF reproduces the observed distribution of orbital periods well, but it underestimates the observed mass ratios in the range $q \in [0.4\text{--}1]$.

Keywords: conservative RLOF, observed Algols

1. Introduction

The Brussels simultaneous evolution code is described in detail in Vanbeveren et al. (1998). We used this to compute the evolution of about 240 close binaries with a B-type primary at birth and small initial orbital periods so that case A evolution will occur. The criterion of Peters (2001) allows to determine for each of the evolutionary sequences the beginning and ending of various Algol stages. This criterion requires that in a semi-detached system the less massive star fills its Roche lobe. The most massive star does not fill its Roche lobe and is still on the main sequence. The less massive star is the cooler, less luminous and larger component.

2. The Grid

We calculated some 240 evolution sequences for binaries with a B-type primary at birth ($M_0^1 \in [2.5\text{--}16.7]\ M_\odot$) and with an initial orbital period so that RLOF A will occur. Three representative values of the mass ratio at birth have been considered: $q^0 = M_0^2/M_0^1 = 0.4$, 0.6 and 0.9, respectively. With increasing initial orbital periods the cases start from contact systems at ZAMS until a detached system showing RLOF A occurs. A fraction of the RLOF duration is occupied with Algol appearance. For larger periods the RLOF A is followed by RLOF B. Above a certain limiting period only RLOF B can cause the Algol appearance of the binary.

Astrophysics and Space Science **296**: 353–356, 2005.
© Springer 2005

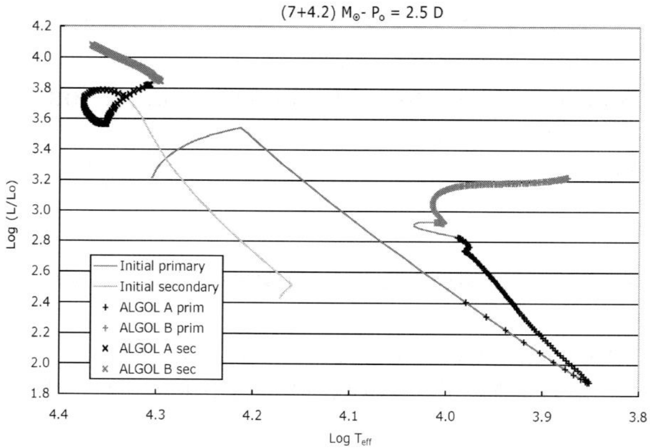

Figure 1. The evolution of a $(7 + 4.2)$ $M_\odot$ system with an initial period of 2.5 days. Subsequent Algol phases are enlightened.

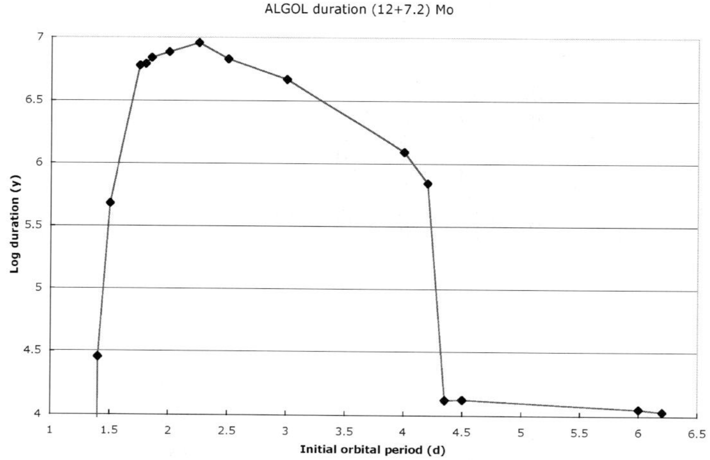

Figure 2. Algol duration for a $(12 + 7.2)$ $M_\odot$ binary. RLOF A leads to Algol appearance for $P >$ 1.4 days. Algol duration increases with increasing initial period. From $P \approx 2.1$ days on, Algol A is succeeded by an Algol B phase. Above $P \approx 4.3$ days, the Algol B duration lasts only a little over ten thousand years.

3. HRD Tracks and Algol Duration

As an example we show the evolutionary tracks of a system of $(7 + 4.2)$ $M_\odot$ with an initial period of 2.5 days in Figure 1. The Algol appearance is enlightened in this figure. Generally all systems become semi-detached at a certain moment and will fulfill the requirements of "Algolism" during a certain time. As an example we show the Algol duration for a $(12 + 7.2)$ $M_\odot$ system in Figure 2. In general the case A Algol durations increase with increasing periods, reach a maximum, and then decrease. The shortest initial orbital periods herein ingnite two subsequent phases of Algol A, whereas A/B sequences show up with increasing periods. The pure Algol B cases occur starting from still larger initial periods, but have not been considered here, since this stage of Algol appearance was already described by Van Rensbergen (2003).

4. Conservative Simulation and Conclusions

From our grid of conservative calculations we simulated the distribution of orbital periods and mass ratios of Algols using a Monte Carlo algorithm. We started with

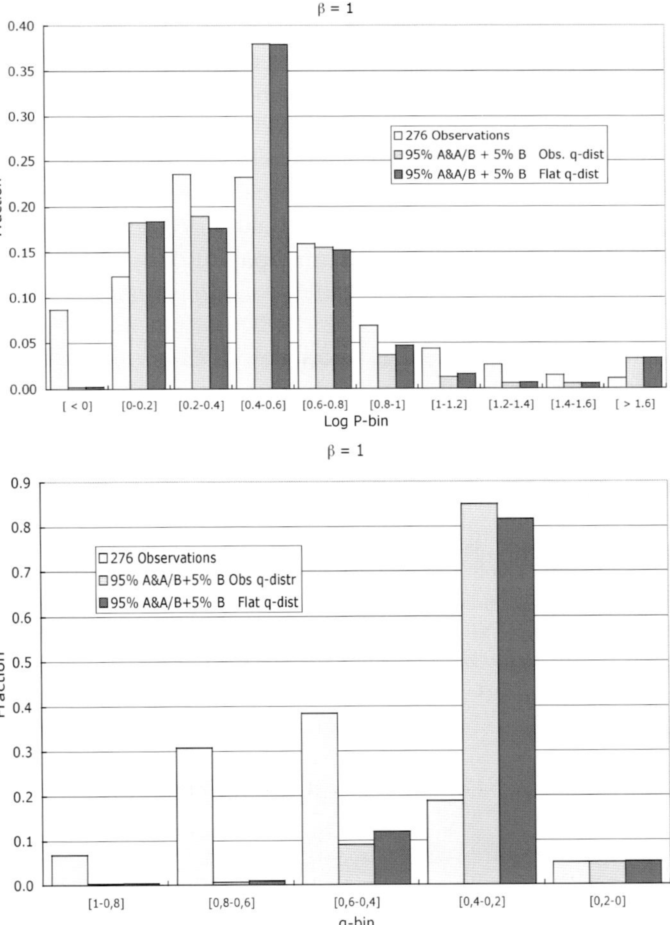

Figure 3. Comparison of the distribution of periods (top) and mass ratios (bottom) of 276 observed Algols with conservative RLOF calculations. Whereas the P-distribution fits well, the q range [0.4–1] is underpopulated by any conservative scenario.

an IMF for the primary given by Salpeter (1955) ($\zeta(M) \sim M^{-2.35}$), an initial period distribution from Popova et al. (1982) (($\Pi(P) \sim 1/P$), and an initial mass ratio distribution from Van Rensbergen (2001) $\Psi(q) \sim (1+q)^{-\alpha}$ with $\alpha = 3.37$ for early B primaries and $\alpha = 1.47$ for late B primaries). Next, we compared the simulation to the observations (Budding, 1984). The fraction of Algols, which are formed through RLOF B, can be estimated to be 10 % at most. RLOF B occurs indeed in a large range of initial orbital periods, but the duration is much shorter compared to RLOF A. The top of Figure 3 shows that the observed orbital periods are well represented by the conservative model with, for instance, an admixture of 5% pure RLOF B cases. The bottom of the figure shows that the observed mass ratios are not well represented by the model. Whereas RLOF B reproduces most Algols in the q-bin [0–0.2] (which is not observed), RLOF A still overpopulates the q-bin [0.2–0.4]. The use of another initial mass ratio distribution ($\Psi \sim 1$) is also shown in Figure 3: the q-range [0.4–1] remains to be underpopulated by the simulation. We will investigate the effect of liberal mass transfer (as well in Case A as in Case B RLOF) hoping to solve this discrepancy. This possibility might very well be at work, especially in cases with a low q-value at birth, where the less massive and small gainer might very well be incapable to capture all the matter that is sent to it by the massive and large donor. The consequences of this liberal mass exchange will be considered in a forthcoming paper.

References

Budding, E.: 1984, *CDS Bull.* **27**.

Peters, G.: 2001, *Ap&SS* **264**, 79–94.

Popova, E., Tutukov, A. and Yungelson, L.: 1982, *Ap&SS* **88**, 55.

Salpeter, E.: 1955, *ApJ* **121**, 161.

Vanbeveren, D., Van Rensbergen, W. and De Loore, C.: 1998, *The Brightest Binaries*, Astrophysics and Space Science Library, vol. 232.

Van Rensbergen, W.: 2001, *The Influence of Binaries on Stellar Population Studies*, Astrophysics and Space Science Library, vol. 264, pp. 21–35.

Van Rensbergen, W.: 2003, *Stellar Astrophysics "A tribute to Helmut Abt"*, Astrophysics and Space Science Library, vol. 298, pp. 117–125.

III.3: Exploring Interaction Effects – Modern Tools

CONTACT BINARIES: A STUDY OF THE PROXIMITY EFFECTS AND GRAVITY DARKENING BASED ON KOPAL'S FOURIER METHOD

P.G. NIARCHOS

Department of Astrophysics, Astronomy and Mechanics, Faculty of Physics, University of Athens, GR 157 84 Zographou, Athens, Greece; E-mail: pniarcho@cc.uoa.gr

(accepted April 2004)

Abstract. A new method for the determination of the proximity effects and gravity darkening exponents in contact binaries of W UMa type is presented. The method is based on Kopal's method of Fourier analysis of the light changes of eclipsing variables in the Frequency Domain. The method was applied to 36 W UMa systems for which geometric and photometric elements have been derived by the most powerful techniques. The derived values are very close to those predicted by the existing theory of radiative transfer or convective equilibrium.

Keywords: binary stars, W UMa-type stars, proximity effects, gravity darkening

1. Introduction

The analysis of eclipsing binary light curves involves a large number of parameters and previous knowledge of some of them is required. To avoid multiple solutions, as many parameters as possible are kept fixed either at the observed values (e.g. mass ratio) or at theoretical ones (limb darkening coefficients, gravity darkening exponents and bolometric albedos). The light curve generation models are able to directly adjust such parameters, but this may introduce correlations among the adjusted parameters and the convergence to physically accepted solutions is difficult or impossible. Experience has shown that the fit of geometrical or radiative parameters to actual observations by means of synthetic light curve modeling is quite difficult for second order effects like those of limb and gravity darkening. The common practice in solving the light curves is to adopt these parameters from realistic theoretical computations (Van Hamme, 1993; Díaz-Cordovés et al., 1995; Claret et al., 1995 and references therein; Vaz and Nordlund, 1985; Nordlund and Vaz, 1990) instead of trying to derive them from observations. Despite these difficulties several attempts have been made to compute the second order parameters from observations. For the gravity darkening exponents the currently adopted values are still the old ones based on the work of von Zeipel (1924) and Lucy (1967).

Astrophysics and Space Science **296**: 359–370, 2005.
© Springer 2005

2. Evaluation of the Proximity Effects

2.1. BASIC EQUATIONS

The basic equation in Kopal's method of Fourier analysis of light curves of eclipsing variables is

$$A_{2m} = \int_0^{\pi/2} [\ell(\pi/2) - \ell(\psi)] \, d(\sin^{2m} \psi), \quad m = 1, 2, 3, \ldots, \tag{1}$$

where $\ell(\pi/2)$ is the light intensity at quadrature and $\ell(\psi)$ is the the light at any phase angle ψ. The term A_{2m} on the left-hand side of Eq. (1) denotes the *empirical moments* which can be evaluated from the observations and is expressed as

$$A_{2m} = A_{2m}^{\text{ecl}} + A_{2m}^{\text{prox}} + P_{2m}, \tag{2}$$

where the first term A_{2m}^{ecl} on the right-hand side stands for the theoretical expression of the *eclipse moments* in terms of the elements of the system. It assumes different expressions for different types of the eclipse. All these expressions have been given by Kopal (cf. Kopal, 1975, 1977, 1979, 1986) and can be regarded as known.

The second term A_{2m}^{prox} in the same equation represents the *proximity effects* outside eclipses. Their evaluation can be obtained by a suitable modulation of the part of the light curve, which is unaffected by eclipses. This term can be expressed as

$$A_{2m}^{\text{prox}} = -m \sum_{j=1}^{n} B\left[m, \frac{j}{2} + 1\right] C_j, \quad m = 1, 2, 3, \ldots, \quad j = 1, 2, 3, \ldots, n, \tag{3}$$

where C_j are empirical coefficients, which can be expressed in terms of the elements and other physical parameters of the system. The C_j's can be evaluated from the observations through the formula

$$C_j = \int_{-\alpha}^{\alpha} [\ell(\pi/2) - \ell(\psi)] F_j^{(\alpha,n)}(x) \, dx, \tag{4}$$

where $\cos \psi \equiv x$, $\cos \psi_1 \equiv \alpha$, and ψ_1 the angle of the first contact, and $F_j^{\alpha,n}(x)$ are appropriate Legendre polynomials.

The method for the evaluation of the proximity effects has been fully described by Kopal (1976), and a generalization of this method has been done by Niarchos (1981). A new method for the evaluation of the proximity effects is given in the present work.

The last term P_{2m} expresses the photometric perturbations arising from the distortion of both components. The explicit forms of the P_{2m} terms for any type of eclipses have been given by Kopal (1986).

The light changes of a close eclipsing system can be expressed (Kopal, 1976, 1977, 1986) by a series of the form

$$L(\psi) = \frac{A_0}{2} + \sum_{j=1}^{n} A_j \cos(j\psi), \tag{5}$$

where $L(\psi)$ is the observed light of the system at any phase angle ψ; $A_0, A_1, \ldots, A_n$ are properly determined coefficients, and n denotes the order of the approximation considered.

Moreover the variation of light between minima, where no eclipses occur and the light changes are due only to proximity effects, can be approximated (Kopal, 1986) by a series of the form

$$L_{\text{prox}}(\psi) = \sum_{j=0}^{n} C_j \cos^j \psi, \tag{6}$$

where C_j's constitute the so called *coefficients of the proximity effects* and n denotes the order of approximation ($n = 4$ for first order).

The coefficient C_2, which is the dominant one among the C_j's, can be considered (Kopal, 1979) as a function of several parameters associated with the particular system, namely

$$C_2 = f(q, i, r_1, r_2, T_1, T_2, x_1, x_2, \beta_1, \beta_2, \lambda), \tag{7}$$

where $q = m_2/m_1$ is the mass-ratio, where the subscripts 1 and 2 refer to the eclipsed star and the eclipsing one at primary minimum, respectively; i is the orbital inclination; r_1, r_2 are the fractional radii of the two components; T_1, T_2 are the effective temperatures; x_1, x_2 the linear coefficients of limb darkening; β_1, β_2 denote the exponents of gravity darkening, and λ the respective wavelength of observations.

All the quantities (except the two exponents β_1, β_2) appearing on both sides of Eq. (7) can be determined from photometric and spectroscopic observations by using proper methods of analysis. Thus, the only unknown quantities in Eq. (7) are the two exponents β_1, β_2. Particularly for contact systems of W UMa type, with common convective envelopes, the two gravity exponents can be considered to be equal, so that in Eq. (7) there is only one unknown $\beta = \beta_1 = \beta_2$, for which it can be solved.

2.2. THE LIGHT CURVE AND ITS DERIVATIVES

Equation (1) constitutes the expression for the whole light of the system at any phase angle ψ. A least squares method was used to calculate the coefficients A_j

in Eq. (1) by using the observational points (light intensities ℓ_i and phase angles $\psi_i, i = 1 \ldots N1$, where N1 is the number of observations). The number n of terms used in Eq. (5) is defined for each case by the requirement that the coefficient A_j should not be smaller than the respective uncertainty. For all the cases we studied (except for some special cases) 6–7 terms were enough for the previous criterium to be fulfilled. For reasons of smoothing the data and saving computational time normal points were used.

It is well known that the light curve of a W UMa-type system is produced by pure *eclipse effects* and the so-called *proximity effects*. A necessary step in our method is to separate the eclipse from the proximity effects, e.g. to find the phase angle, where the eclipses begin (or end). We propose a new method for the determination of the angle of the first contact by simply studying the light curve, an approximation of which is given by Eq. (5). Assuming that the coefficients A_j are known, the derivatives of any order of Eq. (5) can be evaluated. In Figure 1 the graphic representation of the light curves and of the first and second derivatives, derived from Eq. (5), are shown for the test systems AB And and V535 Ara, respectively. In both cases six terms in Eq. (5) were used.

In the phase interval $0 - \pi/2$ and at the point, where the eclipse ends, we expect an abrupt increase of the second derivative and an abrupt diminution of the first derivative of the light curve. Similar conclusions can be drawn for the rest phase intervals: $\pi/2 - \pi$, $\pi - 3\pi/2$ and $\pi/2 - 2\pi$. These sudden changes in the values of the derivatives can be used to locate the points, where the eclipse begins or ends, and, consequently, to define the phase interval of the light curve, which is free from eclipse effects. This method has the advantage that it is based on the observations alone, and no previous knowledge of the system parameters is required.

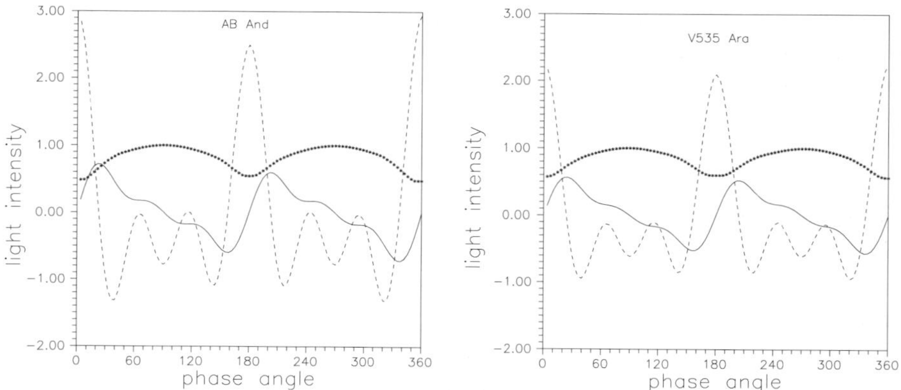

Figure 1. Graphic representation of the light curve and of its first and second derivative of AB And (left) and V535 Ara (right). The asterisks are theoretical V observations; the solid line represents the 1st derivative and the dashed line the 2nd derivative, derived from Eq. (5).

2.3. EVALUATION OF THE COEFFICIENTS
OF THE PROXIMITY EFFECTS

According to the previous analysis, the part of the light curve, which is unaffected by eclipses, can be specified from a study of the second derivative of the light curve. When the eclipse ends, the second derivative shows a sudden increase, while previously was being reduced. Therefore, the end points of the phase interval, where no eclipses occur, can be defined by the points, where the third derivative becomes zero. By using this interval we obtained the best results, e.g. we succeeded to calculate gravity darkening exponents almost identical to those used to produce theoretical light curves of the test systems AB And and V535 Ara (see Section 4).

An analytic expression for the approximation of the proximity effects curve should rely on phase intervals, which are free from eclipse effects. Such an expression is given by Eq. (6), which is a periodic function with period 2π. The coefficients C_j, $j = 0, 1, 2, 3, 4$ can be evaluated by using the observations of the part of the light curve, which is not affected by eclipses. The coefficient C_0 can be found directly from Eqs. (5) and (6) and it is equal to $C_0 = L(\pi/2)$, while the coefficients C_j, $j = 1, 2, 3, 4$ follow from the relation

$$\sum_{i=1}^{N} (L_{\mathrm{prox}}(\psi_i) - l_i)^2 = \min, \tag{8}$$

where l_i, ψ_i, are the N observational points (light intensity, phase angle) lying inside the phase interval, where no eclipses occur.

In order for the proximity effects curve to be reliable in phase regions, where eclipses occur, we must exploit most of the information included in the regions, where no eclipse effects are present. Assuming that the functions $L(\psi)$ and $L_{\mathrm{prox}}(\psi)$, defined by Eqs. (5) and (6), are known from observations, we can secure the identity of the two functions in the phase interval outside eclipses by minimizing, in addition to the quantity defined by Eq. (8), the quantity

$$\sum_{i=1}^{N} (L'_{\mathrm{prox}}(\psi_i) - L'(\psi_i))^2 = \min \tag{9}$$

defined in the phase interval outside eclipses. The $'$'s stand for the respective derivatives of 1st order. A simultaneous solution of Eqs. (8) and (9) yield the coefficients C_j's. Similar systems of equations can be derived by taking derivatives of higher order. But, as it is shown below, derivatives of first order lead to a satisfactory solution of the problem. The proximity effects curves together with the theoretical light curves for the two test systems are shown in Figure 2.

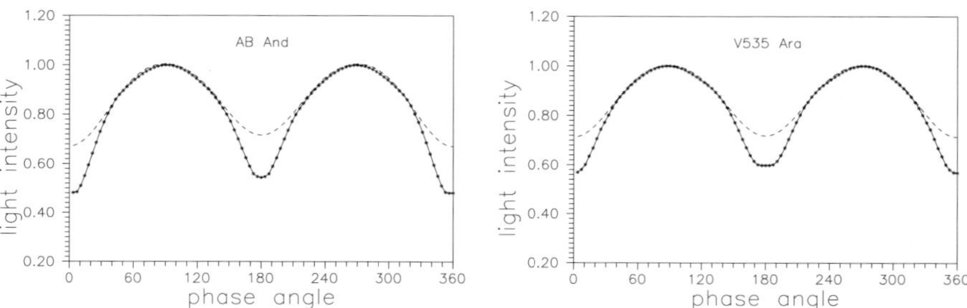

Figure 2. Theoretical V light curve (solid line) and proximity effects curve (dashed line) of AB And (left) and V535 Ara (right).

3. The Gravity Darkening Effect

3.1. THEORETICAL AND OBSERVATIONAL APPROACHES

One of the most important observable phenomena, which enables us to probe the stellar structure, is the gravity darkening exhibited by the distorted stars in close binary systems. The general law of gravity darkening is written as

$$H \propto g^{\beta},\tag{10}$$

where H is the bolometric surface brightness at any arbitrary point on the stellar surface; g is the local gravity and β the exponent of gravity darkening.

Von Zeipel (1924) proved that the emergent flux of total (bolometric) radiation at any point over the surface of a rotationally or tidally distorted star in radiative equilibrium varies proportionally to the local gravity ($\beta = 1$ in Eq. (10)). The validity of Zeipel's law is limited to the case, in which the energy transfer in sub-surface layers is purely radiative. The gravity darkening law with $\beta = 1$ was used after the work of von Zeipel in all light curve solutions for the next 45 years and some-times led to results, which were difficult to interprete.

The application of von Zeipel's law is of doubtful validity for W UMa stars, whose subphotospheric layers are mostly convective. This problem was first re-alized by Lucy (1967), who predicted a β value of about 0.32, based on fitting together convective envelopes having different surface gravities at a depth, where the temperature gradient becomes adiabatic.

Recently, Alencar and Vaz (1997) studied the gravity darkening exponent β in non-illuminated convective grey and non-grey atmospheres for $3700 < T_{\text{eff}} < 7000\,K$. Their results confirmed the value $\beta = 0.32$ only for $T_{\text{eff}} \approx 6500\,K$.

In a recent paper Claret (1998) developed a method to compute the gravity darkening exponent using interior models. His method embraces convective and radiative envelopes. For the first time, the gravity darkening exponents are presented as functions of mass and degree of evolution. According to Claret (1998) the old values of β(0.32 and 1.0) for convective and radiative envelopes, are thus superseded by the new results, and a smooth transition is achieved between both energy transport mechanisms.

Several observational approaches for the gravity darkening determination have been made so far by a quantitative analysis of the ellipticity effect in the light curves of W UMa systems. The observed values of the gravity darkening were found to be larger than the theoretical values at the respective wavelength of observation.

A different approach based on light curve synthesis has also been made by Rafert and Twigg (1980). From an analysis of a dozen contact binaries, the individual determinations for cooler, convective stars showed a scatter within a range of about $0.16 < \beta < 0.48$, on average, confirming the prediction of Lucy.

As Rucinski (1989) pointed out, despite the efforts that have been made so far, our knowledge about gravity darkening is still incomplete. Therefore, we need new determinations of gravity darkening exponents, especially in the case of convective stars, both theoretically and observationally. As a result of Rucinski's (1989) appeal, a theoretical determination of gravity darkening was made by Claret (1998), while a new method for the empirical determination of gravity darkening in contact binary systems was proposed by Pantazis and Niarchos (1998) and Niarchos (2000). A brief description of the method is given in the following sections.

3.2. THE GRAVITY DARKENING EFFECT AND ITS WAVELENGTH DEPENDENCE

It has already been stated that the gravity darkening exponent β in Eq. (10) refers to the bolometric light of the star. Since photometric observations are made at a specific wavelength λ, it is necessary to relate the exponent β with the gravity darkening effect determined from observations at any wavelength λ. The bolometric exponent β is found to be connected with the coefficient τ of gravity darkening (see Pantazis and Niarchos, 1998; Niarchos, 2000) through the relation

$$\beta = 4\lambda k T_0 \tau \left(1 - e^{\frac{-hc}{\lambda k T_0}}\right)/hc \tag{11}$$

The exponent β follows from this equation, provided that the coefficient τ is derived from observations. It should be pointed out that the coefficient τ depends on the wavelength of radiation λ, while the exponent β is independent of λ and refers to the bolometric radiation.

3.3. EVALUATION OF THE GRAVITY DARKENING EXPONENT

The coefficient C_2 in Eq. (7) is given (Kopal, 1975) by

$$C_2 = C_2^* + \gamma_2, \tag{12}$$

where

$$C_2^* = \left(\frac{3}{2} \sin^2 i \, L_1 q (15 + x_1)(1 + \tau)/(15 - 5x_1) \right) r_1^3$$
$$+ \left(\frac{3}{2} \sin^2 i \, L_1 q \, 45(1 - x_1)(1 + \tau/3)/(24 - 8x_1) \right) r_1^5$$
$$+ \left(\frac{3}{2} \sin^2 i \, L_2 (15 + x_1)(1 + \tau)/(q(15 - 5x_2)) \right) r_2^3$$
$$+ \left(\frac{3}{2} \sin^2 i \, L_2 \, 45(1 - x_1)(1 + \tau/3)/(q(24 - 8x_2)) \right) r_2^5, \tag{13}$$

where $L_{1,2}$ are the fractional luminosities of the two components in the respective wavelength of observations; $r_{1,2}$ the fractional radii of the two stars in terms of their separation taken as unit of length; q is the mass-ratio; i is the inclination of the orbit with respect to the normal of the line of observation; $x_{1,2}$ are the coefficients of limb darkening, and

$$\gamma_2 = L_2 \gamma_{2,1} + L_1 \gamma_{2,2}, \tag{14}$$

where

$$\gamma_{2,k} = \left(\frac{r_k^2}{3\pi} + \frac{3r_k^3}{8} \right) f_k \sin^2 i, \quad k = 1, 2 \tag{15}$$

and

$$f_k = (T_{3-k}/T_k)^4 (L_k/L_{3-k})(r_{3-k}/r_k)^2. \tag{16}$$

In the above Eq. (13) all the quantities occurring on the right-hand side, except for the coefficient τ, can be found from observations or (like $x_{1,2}$) from theory. On the other hand, C_2^*, on the left-hand side, can be evaluated from Eq. (12), since C_2 and γ_2 can also be determined from observations through the Eqs. (5)–(6) and (14)–(16), respectively. Thus, Eq. (13) can be solved for the only unknown τ, and then from Eq. (11) we finally get the value of the gravity darkening exponent β.

4. Application of the Method

4.1. APPLICATION TO THEORETICAL (TEST) LIGHT CURVES

Our new method for the determination of the gravity darkening exponent can be fully automatized following the steps described in the previous sections. The method was

TABLE I

Results for the test systems AB And and V535 Ara

Coefficient	AB And	V535 Ara
A_0	0.81375 ± 0.00500	0.83822 ± 0.00500
A_1	-0.02431 ± 0.00005	-0.00269 ± 0.00005
A_2	-0.22405 ± 0.00005	-0.19586 ± 0.00005
A_3	-0.01208 ± 0.00005	-0.00455 ± 0.00005
A_4	-0.06070 ± 0.00005	-0.04854 ± 0.00005
A_5	-0.00382 ± 0.00005	-0.00122 ± 0.00005
A_6	-0.02371 ± 0.00005	-0.01685 ± 0.00005
C_2	-0.19810 ± 0.00005	-0.17751 ± 0.00005
τ	0.39 ± 0.02	0.86 ± 0.08
β	0.33 ± 0.01	1.08 ± 0.06
β_{th}	0.32^*	1.00^*

*Theoretical value.

first applied to synthetic light curves of the W UMa systems AB And and V535 Ara. These light curves were produced by the *Binary Maker 2.0* programme (Bradstreet, 1993), where the gravity darkening exponents were assigned theoretical values (0.32 and 1.00, respectively), and input parameters were taken from the *Pictorial Atlas of Binary Stars* prepared by Terrell et al. (1992). The first system is a typical example of a system with convective envelopes (cool system), while the second one is a system with radiative envelopes (hot system). The results of our method are given in Table I.

Figure 2 shows the normal points and the corresponding theoretical light curves, obtained by our method (Eqs. (5) and (6)), of the systems AB And and V535 Ara. It is obvious that the fitting is extremely good, while the evaluated gravity darkening exponents are almost identical, within the limits of their errors, to the theoretical values 0.32 and 1.00.

4.2. APPLICATION TO W UMa SYSTEMS

Having shown that our new method works very well, we applied it to a large enough sample of W UMa systems carefully selected from the literature to have very accurately determined parameters. The majority of the systems were chosen to possess common convective envelopes, since, according to Rucinski (1989), the problem of gravity darkening is still open, mainly for convective stars. A dividing line is drawn at the temperature 7200 K, and all systems with $T < 7200\,K$ are considered to have convective envelopes, while those with $T > 7200\,K$ are supposed to possess radiative envelopes. The results are given in Table II together with the values of some basic parameters of the systems. The necessary input parameters

P.G. NIARCHOS

TABLE II

Results for W UMa systems

n	System	r_1	r_2	$q\left(\frac{m_2}{m_1}\right)$	i (deg.)	T_1 (K)	T_2 (K)	β
1	AB And	0.320	0.451	2.03	86.80	5821	5450	0.29 ± 0.01
2	OO Aql	0.412	0.380	0.84	90.00	5700	5635	0.18 ± 0.04
3	VW Boo	0.470	0.322	0.43	75.64	5700	5190	0.27 ± 0.01
4	TY Boo	0.309	0.457	2.14	77.50	5864	5469	0.38 ± 0.02
5	TU Boo	0.471	0.342	0.50	87.50	5800	5787	0.32 ± 0.02
6	V523 Cas	0.332	0.436	1.75	83.70	4407	4200	0.53 ± 0.02
7	CW Cas	0.326	0.438	1.84	73.40	5440	5090	0.32 ± 0.01
8	AD Cnc	0.341	0.428	1.60	64.90	5164	4825	0.69 ± 0.01
9	CC Com	0.326	0.447	1.93	90.00	4300	4140	0.26 ± 0.01
10	FS Cra	0.360	0.411	1.32	86.50	4700	4567	0.39 ± 0.05
11	V677 Cen	0.580	0.243	0.15	89.50	5700	5700	0.33 ± 0.03
12	V700 Cyg	0.335	0.452	1.88	81.60	6890	6334	0.67 ± 0.02
13	BV Dra	0.318	0.472	2.43	76.28	6345	6245	0.40 ± 0.01
14	BW Dra	0.281	0.499	3.57	74.40	6164	5980	0.33 ± 0.01
15	YY Eri	0.303	0.470	2.49	82.50	5600	5362	0.30 ± 0.01
16	AK Her	0.517	0.276	0.26	80.47	6400	6033	0.33 ± 0.01
17	DF Hya	0.306	0.463	2.36	84.30	6000	5851	0.22 ± 0.01
18	ST Ind	0.438	0.350	0.60	71.30	6430	6414	0.47 ± 0.02
19	XY Leo	0.317	0.444	2.00	65.80	4850	4575	0.53 ± 0.02
20	RT LMi	0.306	0.481	2.59	83.38	6000	5800	0.29 ± 0.02
21	V502 Oph	0.299	0.477	2.65	71.67	6200	5954	0.49 ± 0.04
22	V508 Oph	0.441	0.324	0.53	86.13	6000	5830	0.12 ± 0.01
23	V566 Oph	0.534	0.271	0.34	79.80	6700	6618	0.35 ± 0.02
24	BX Peg	0.300	0.479	2.72	87.50	5528	5300	0.25 ± 0.03
25	AE Phe	0.305	0.480	2.22	86.00	6100	5809	0.12 ± 0.01
26	FG Sct	0.359	0.403	1.27	89.90	4800	4662	0.37 ± 0.01
27	V781 Tau	0.310	0.476	2.47	65.36	5950	5861	0.50 ± 0.03
28	W UMa	0.480	0.325	0.49	86.00	6500	6400	0.16 ± 0.01
29	BI Vul	0.363	0.431	1.44	78.80	4600	4598	0.19 ± 0.01
30	V535 Ara	0.432	0.287	0.36	82.14	8750	8572	1.04 ± 0.08
31	XY Boo	0.558	0.253	0.18	68.38	7200	7100	0.81 ± 0.08
32	RR Cen	0.555	0.247	0.18	78.70	7250	7188	0.87 ± 0.01
33	XZ Leo	0.412	0.353	0.73	72.03	7850	7044	1.27 ± 0.04
34	MW Pav	0.561	0.256	0.18	85.07	7620	7570	0.97 ± 0.01
35	TY Pup	0.559	0.257	0.18	81.80	7800	7567	1.36 ± 0.02
36	RZ Tau	0.502	0.320	0.37	82.88	7200	7146	0.86 ± 0.04

and observations were taken from the literature (see Niarchos and Pantazis, 1998). The given values of β are mean values obtained from the analysis of light curves in different colours.

5. Conclusions

The crucial point in our method is to define the phase interval of the light curve, which is free from eclipse effects. After several numerical experiments we found that the interval giving the best results is that defined by the points, where the third derivative of the equation of the light curve becomes zero. A successful test of the method using theoretical light curves was made, before we analyzed the light curves of W UMa systems.

As far as the limitations of our method are concerned, it should be noted: (i) the effectiveness of the method depends strongly on the quality of the observations, (ii) the method in its present form can be used only for W UMa systems, where we assume $\beta_1 = \beta_2$, as a result of the common envelope of the system, and (iii) for systems with unequal maxima in their light curves (O'Connell effect), the part of the light curve around the higher maximum (unspotted part) is used for the determination of the gravity darkening exponent.

The present results show that for radiative (hot) systems the computed values of β are close to the theoretical value $\beta = 1$, as proposed by von Zeipel's theory, while for systems with convective envelopes (cool systems) the computed values of β cluster around the theoretical value $\beta = 0.32$ proposed by Lucy (1967). It should be noted here that the jump of β from both radiative to convective atmospheres was awkward from the observational and theoretical point of view: the two processes of energy transport can exist simultaneously in a particular stellar envelope. One of the advantages of Claret's method is that a smooth transition is achieved between both energy transport mechanisms. This result is also confirmed by our method, where the computed values of β for the sample of W UMa systems used range from 0.12 to 1.36.

In conclusion, the importance of the results of the present work is two-fold: (a) they confirm to a great degree the validity of the existing theories of von Zeipel and Lucy and are consistent with the model calculations of Claret (1998), and (b) the computed values of the gravity darkening exponents can be used as input parameters in light curve synthesis programs for a more precise solution of the observed light curves.

References

Alencar, S.H.P. and Vaz, L.P.R.: 1997, *Astron. Astrophys.* **326**, 257.
Bradstreet, D.H.: 1993, Binary Maker 2.0 User Manual.
Claret, A.: 1995, *Astron. Astrophys. Suppl.* **109**, 441.

Claret, A.: 1998, *Astron. Astrophys. Suppl.* **131**, 395.

Claret, A., Díaz-Cordovés, J. and Gimenez, A.: 1995, *Astron. Astrophys. Suppl.* **114**, 247.

Díaz-Cordovés, J., Claret, A., and Gimenez, A.: 1995, *Astron. Astrophys. Suppl.* **110**, 329.

Kopal, Z.: 1975, *Astrophys. Space Sci.* **38**, 191.

Kopal, Z.: 1976, *Astrophys. Space Sci.* **45**, 269.

Kopal, Z.: 1977, *Astrophys. Space Sci.* **51**, 439.

Kopal, Z.: 1979, Language of the Stars, D. Reidel, Dordrecht and Boston.

Kopal, Z.: 1986, *Vistas Astron.* **29**, 295.

Lucy, L.B.: 1967, *Z. Astrophys.* **65**, 89.

Niarchos, P.G.: 1981, *Astrophys. Space Sci.* **76**, 503.

Niarchos, P.G.: 2000, in: C. Ibanoglu (ed.), Variable Stars as Essential Astrophysics Tools, NATO science series, Vol. 544, p.631.

Nordlund, A. and Vaz, L.P.R.: 1990, *Astron. Astrophys.* **228**, 231.

Rafert, J.B. and Twigg, L.W.: 1980, *MNRAS* **193**, 79.

Pantazis, G. and Niarchos, P.G.: 1998, *Astron. Astrophys.* **335**, 199.

Rucinski, S.M.: 1989, *Comm. Astrophys.* **14**, 79.

Terrell, D., Mukherjee, J. and Wilson, R.E.: 1992, *BINARY STARS, A Pictorial Atlas*, Krieger, Florida.

Van Hamme, W.: 1993, *Astron. J.* **106**, 2096.

Vaz, L.P.R. and Nordlund, A.: 1985, *Astron. Astrophys.* **147**, 281.

Zeipel, H.V.: 1924, *MNRAS* **84**, 702.

MULTI-FACILITY STUDY OF THE ALGOL-TYPE BINARY δ LIBRAE

E. BUDDING[1,2], V. BAKIS[1], A. ERDEM[1], O. DEMIRCAN[1], L. ILIEV[3],
I. ILIEV[3] and O.B. SLEE[4]

[1]*Physics Department, Canakkale Onsekiz Mart University (COMU), TR 17020, Turkey;
E-mail: ebudding@comu.edu.tr*
[2]*Carter National Observatory, P.O. Box 2909, Wellington, New Zealand*
[3]*Institute of Astronomy, BAS, 72 Tsarigradsko Shosee Blvd., BG-1784, Sofia, Bulgaria*
[4]*ATNF, CSIRO Radiophysics, P.O. Box 76, Epping, NSW 2121, Australia*

(accepted April 2004)

Abstract. Various branches of observational information, and related physical problems concerning the bright, relatively close, Algol binary δ Lib are discussed. Times of minimum light confirm the classical Algol status, which, combined with the optical and IR photometry of Lazaro et al. (2002), provide the basis for a good understanding of the system's basic parameters. New spectroscopy from the Coudé spectrograph of the Rozhen Observatory (Bulgaria) allows further information on the model and the mass transfer process. Very high resolution radio observations would also have much to reveal about astrophysics of the semi-detached configuration, not only directly, by informing about the outermost envelopes of the components, but indirectly through high accuracy positional information. We show how this may be related to careful period and astrometric studies. δ Lib thus provides a rich source of information on the astrophysics of the Algol configuration.

1. Introduction

δ Librae is one of the nearest ($\sim$90 pc) Algol systems: a close, interacting binary made up of A0V + K0IV stars with $V \sim 4.9$ mag and an orbital period close to 2.33 days. The classical Algol configuration, of which δ Lib may be typical, consists of an early-type, near MS primary in a semi-detached pairing with a cool subgiant secondary, i.e. the secondary star fills its Roche surface of limiting dynamic stability (c.f. Plavec, 1973). Matter is expected to overflow from this Roche lobe and flows towards the primary. In fact, the angular momentum in this stream would carry at least some proportion of its matter past the region of closest approach, particularly in wide systems, where an accretion structure may form around the primary. Otherwise, there should be a region of enhanced temperatures around that part of the primary closest to the infalling stream (hot spot). Spectroscopic studies of such effects are well-known (c.f. e.g. Batten, 1989).

δ Lib may be compared with the Algol prototype β Per, which has a similar overall mass, little longer period and is about one third as distant; or V505 Sgr, which has smaller mass and period, but is at a comparable distance. These Algol binaries appear to be in gravitationally bound systems with three, or more stars.

Astrophysics and Space Science **296**: 371–389, 2005.
© Springer 2005

Worek (2001), for example, reported the likely presence of a third star in the δ Lib system, with a three-body configuration quite similar to β Per.

Algol secondaries are cool and relatively rapidly rotating stars with convective atmospheres, so they may well also have regions of local magnetic flux concentrations in their surface layers, producing phenomenological 'starspots' that could be very much larger than those of the Sun. Algols are known sources of microwave radioemission (Slee et al., 1987) at levels of, typically, a few mJy for the nearer examples, where signal-to-noise ratios permit useful measurements. Stellar radio emission is usually associated with high brightness temperatures, often related to regions of magnetic field enhancements, electrons accelerated by flare-type mechanisms and gyrosynchrotron emission. In this respect, the cool secondaries of Algol systems resemble the active cool stars known as RS CVns (c.f. Strassmeier et al., 1993). Active-star radio emission (including that from Algols) has been found to correlate with rotation rate (Slee and Stewart, 1989).

Feasible models of the gyrosynchrotron emission (c.f. Slee et al., 2004) then lead to moderate coronal magnetic fields ($\sim$10–100 G) that have regions whose size is on the scale of tens of percent of the stellar diameter and very high (brightness) temperatures $\geq 10^9$ K. The radio emission constrains magnetic field values, and so its study helps build up a picture of the field distribution, and perhaps its origin. A recent and exciting development in the study of Algols by radioastronomy is the possibility of source resolution using VLBA (c.f. Mutel et al., 1998). More direct information on coronal magnetic fields may be possible from such work. It would be interesting to check, for instance, if the double-lobed structure with opposite senses of polarization, seen in Algol itself, is repeated in a system with very similar components, i.e. δ Lib.

The existence of the mass-transferring stream (a high-speed conductive plasma with a particular, known geometry) suggests different radio emission models than those of normal RS CVn stars (which do not have such streams). Algol streams can be modelled in some detail from analysis of high-dispersion spectroscopy (Richards et al., 1996), but so far no such model development has been proved to have a clear relevance for radioemission. Gunn et al. (1999) speculated about an inter-component source for V505 Sgr, however. The possibly similar (but relatively much more energized) streams of polars show a phase dependence, pointing to a geometry-related emission scenario. Further detailed observations may help clarify the relevance of the orbital geometry (c.f. also Kalomeni et al., 2003).

2. Period Analysis

The observed times of minimum light for Algol systems, compared with the predicted ones (O–C), sometimes provide clear evidence supporting mass transfer. The atlas of O–Cs against primary minimum epoch (E) of Kreiner et al. (2000) shows many Algols with an apex-down parabola, indicating a steady separation

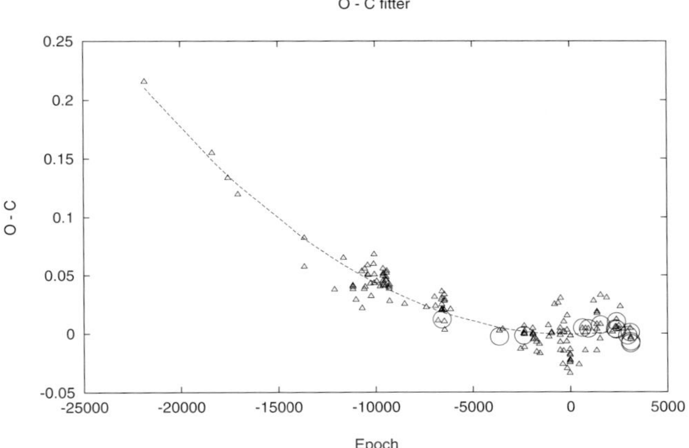

Figure 1. Observed and predicted times of light minima for δ Lib from Kreiner et al. (2000). A parabola has been fitted to the O–C data.

of components, in agreement with the mechanics of mass transfer from the less massive secondary to the primary. The situation for δ Lib is shown in Figure 1, where the predominating parabolic variation is clearly visible. Kreiner et al.'s data were here fitted with the form:

$$\text{O–C} = aE^2 + bE + c, \tag{1}$$

where $a = 4.44 \times 10^{-10}$, $b = 3.01 \times 10^{-8}$ and $c = 0.00075$. The rate of period lengthening can be set against a corresponding rate of mass transfer from the loser, of mass x, expressed as a fraction of the total mass. We then write

$$\frac{\Delta P}{P} = 3g(x)\frac{\Delta x}{x}, \tag{2}$$

where, for example, with conservation of orbital angular momentum, the function $g(x) = (2x - 1)/(1 - x)$. In this way, 'conservative' mass transfer for δ Lib, (using Tomkin's (1978) spectroscopic mass ratio) works out at about $1.1 \times 10^{-7} M_\odot$ per year, which is typical for recession stages of the Case B model of Algol evolution involving mass-losers with original mass a few times that of the Sun (c.f. e.g. Özdemir et al., 2000).

The residuals from the fit of Figure 1 can be examined for further dynamical effects. In view of the considerations about additional components in the system apart from the eclipsing binary (Worek, 1998; Lazaro et al., 2002), we have fitted a light time effect associated with a third body in an eccentric orbit (c.f. e.g. Irwin, 1963) to the O–C residuals after removing the main parabolic term. Results are

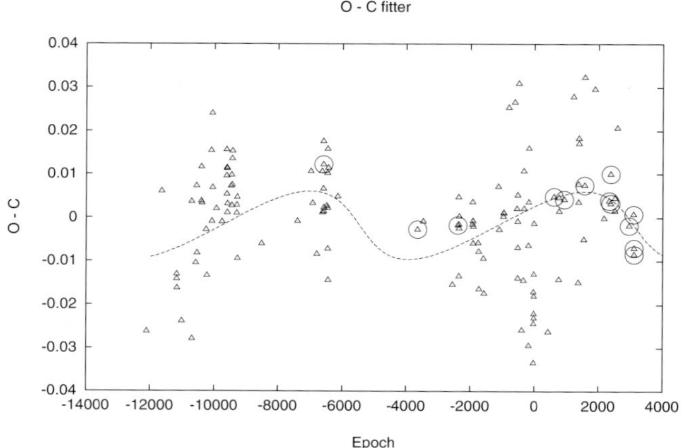

Figure 2. Fit of residuals from Figure 1 to an eccentric orbit. Photoelectric times of minima are shown as full circles.

shown in Figure 2, where extra weight has been given to the photoelectrically observed minima.

Although the inclusion of this orbital effect has improved the total goodness of fit parameter χ^2, it is only a small improvement over a straight line fit and sensitive to the weights given to the relatively few photoelectric times of minima. Thus, even giving these 10 times the weight of visual observations, the net value of χ^2 decreases by only about 10% with the orbit inclusion, and, whilst of interest, cannot be considered decisive. The question of other components will be reconsidered in a later section. The main upshot from this part of the times of minima analysis, however, concerns obtaining more photoelectric times of minima in the future, to check on the tentative 13-min amplitude oscillation. The parameters of the fit shown in Figure 2 are as follows: $a \sin i = 1.52$ a.u., $P = 23.49$ y, $e = 0.49$, $\varpi = 317°$.

3. Light Curve Analysis

The variability of δ Lib has been known for a long time, with a photoelectric light curve produced as long ago as 1920 (Stebbins, 1934). The near-UBV light curves presented by Koch (1962) (Figure 3) are often cited for their definitive quality. Spectroscopy has tended to assign an A0V classification to the primary (Roman, 1956; Sahade and Hernandez, 1963; Tomkin, 1978), although Koch's photometry suggested a late B spectral type. Tomkin (1978) identified secondary features in the spectrum, whose type he inferred to be early G. Tomkin thence measured radial velocities of both stars, so that a spectroscopic mass ratio of 0.345 ± 0.012 could be assigned. He gave masses as 4.9 and 1.7 $M_\odot$ with errors of about 5%. This points to an earlier type than A0, if the primary is close to the Main Sequence.

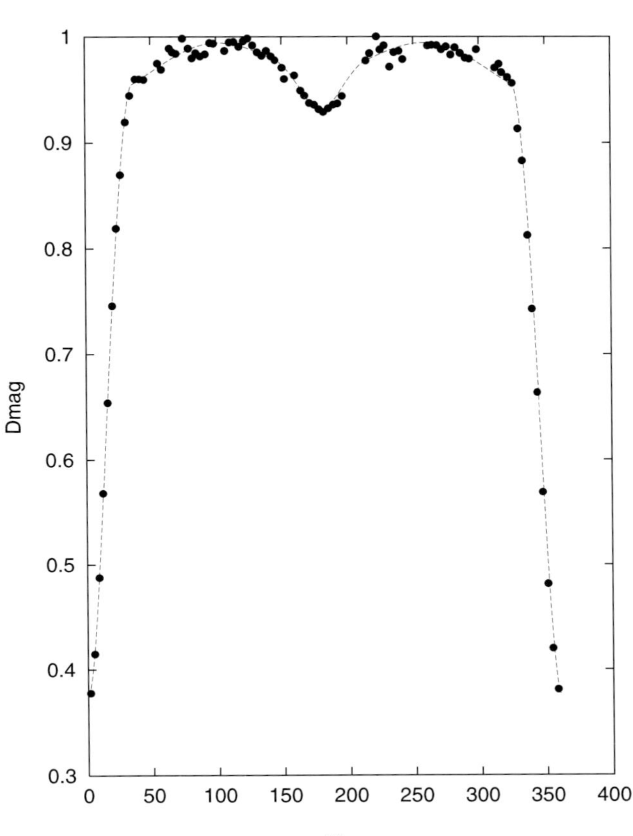

Figure 3. ILOT fitting to Koch's (1962) *B* light curve of δ Lib.

A recent detailed study of the light curves of δ Lib was carried out by Lazaro et al. (2002), who included their own new light curves in the J, H and K bands. These good quality data should permit well-defined photometric parameters for both stars, given that the secondary minimum becomes much deeper in the infrared. Combining this with available spectroscopy allows absolute parameters to be checked. The data have been here independently analysed using a recent version of the ILOT light-curve fitter program (Budding et al., 2004), and values generally similar to those of Lazaro et al. were found. They are listed in Table I. (using standard symbols), and a corresponding curve-fit shown in Figure 3. Methods of finding parameters like those of Table I were discussed by Budding et al. (2004), where the importance of knowing accuracy levels was noted. Parameter accuracy, in the case of Table I, is approximately reflected in the number of retained digits, although, in general, such presentations tend to be over-optimistic Figure 4.

TABLE I

Light curve fittings for δ Lib

Parameter	U	B	V	J	H	K	Error
L_1	0.865	0.877	0.803	0.651	0.585	0.541	0.01
L_2	0.047	0.057	0.094	0.269	0.344	0.368	0.01
L_3	0.088	0.066	0.103	0.080	0.071	0.091	0.01
r_1	0.279	0.276	0.275	0.276	0.279	0.277	0.007
r_2	0.317	0.313	0.309	0.313	0.313	0.315	0.007
i	80.4	80.0	81.0	80.0	80.3	80.5	1.0
$\Delta\phi_0$	0.1	0.1	0.1	0.3	1.9	0.7	0.2
q	0.35	0.35	0.35	0.35	0.35	0.35	0.05
τ_1	1.01	0.86	0.72	0.43	0.38	0.36	–
τ_2	1.87	1.58	1.26	0.62	0.52	0.44	–
E_1	2.73	1.36	0.64	012	0.09	0.07	–
E_2	0.50	0.50	0.71	1.09	0.75	0.80	0.1
χ^2/ν	1.05	0.88	0.94	1.20	1.07	1.13	–
Δl	0.008	0.006	0.006	0.006	0.008	0.005	–

TABLE II

Magnitudes of stars in δ Lib

	U	B	V	J	H	K	Error
m_1	4.99	5.08	5.19	5.20	5.20	5.26	0.02
m_2	8.15	8.04	7.52	6.17	5.78	5.68	0.1
m_3	7.47	7.88	7.42	7.48	7.49	7.20	0.1

Koch gives the V magnitude of δ Lib at quadrature as 4.92, which corrects to 4.95, to give the distortion-free sum of contributions of both stars. Adopting the fractional luminosities of Table I, the primary's V magnitude then becomes 5.19. With the Hipparcos distance of 93 pc, we find the absolute magnitude as $M_V = 0.35$. This appears significantly brighter than typical Main Sequence stars of around A0 spectral type. That the primary should not be so far from the standard star Vega in temperature can be confirmed by calculating its magnitudes in the various bands, using the quadrature measures provided by Koch (1962) and Lazaro et al. (2002). We give the results in Table II, together with those of the secondary and the putative third star. The slightly bluer colour of the primary relative to Vega across the whole wavelength range, using the approximate proportionality to reciprocal temperature difference for slight changes of colour (Budding, 1993), indicates a temperature about 5% higher than Vega's 9550 K (Castelli and Kurucz, 1994), i.e. 10,000 K. We had adopted the value 9800 K, close to the preferred one of Lazaro et al.

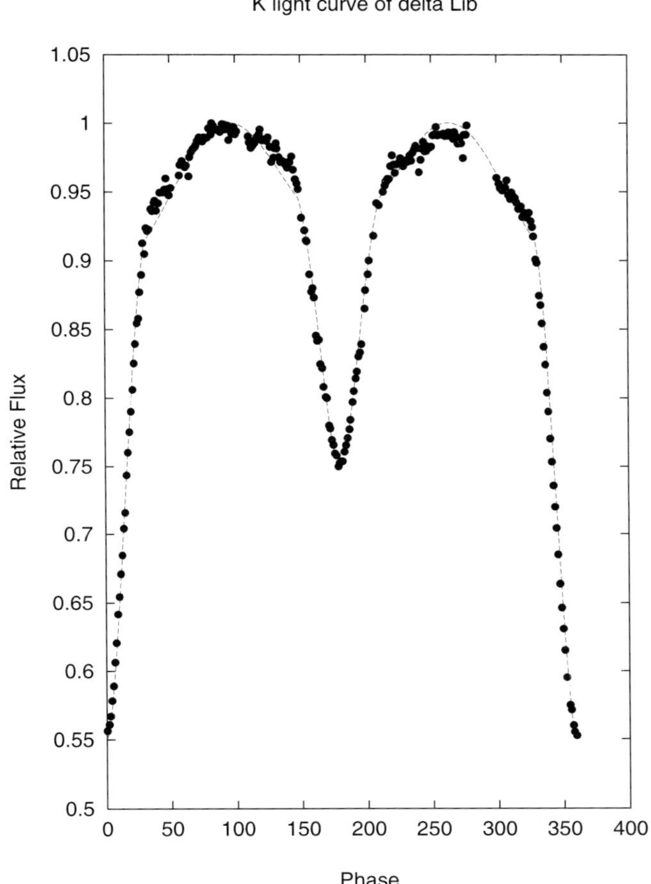

Figure 4. ILOT fitting to Lazaro et al.'s (2002) K light curve of δ Lib.

(9760 K), in preparing the auxiliary parameters (e.g. gravity-brightening and re-radiation coefficients) of Table I, but these are not changed significantly by a shift to 10,000 K.

Kepler's law, combined with the eclipse solution for the radii, allows us to write

$$3 \log R_1 - \log M_1 = 1.062, \tag{3}$$

but if we use, for example, the empirical data of Popper (1980) on typical masses and radii for this region of the Main Sequence and then eliminate the variables, we will find a very high mass for the corresponding spectral type. This was noted already by Lazaro et al. (2002), who used evolutionary models in the $T{:}g$ plane to find that the primary mass cannot be far from 3 $M_\odot$ and adopted 2.85 $M_\odot$ as

TABLE III

Adopted parameters for δ Lib

Parameter	Value
Period	2.32735297 d
Epoch (HJD)	2422852.3598
A	11.84 $R_\odot$
$R_{1,2}$	3.28; 3.72 $R_\odot$
i	80.3 deg
$M_{1,2}$	2.9; 1.1 $M_\odot$
MM_V	-0.09
V	4.92
$B - V$	-0.01
$V - J$	0.23
$T_{1,2}$	10000; 5200 K
Distance	100 pc
v_{rot} (meas., calc.)	69; 71 km s^{-1}

providing the best match (since R, and therefore g, can be derived from (3) for a given M). This produces a radius 3.25 $R_\odot$, which is significantly larger than the typical MS values of Popper (1980). Lazaro et al. (2002) pointed out that the light curves could be consistent with Tomkin's higher masses, but then much larger luminosities would be implied, and the distance to the star becomes much too large. It should also be noted that the rotation velocity, if synchronous, as is usual for such close binaries, supports the 3.25 $R_\odot$ radius. A 2% increase in temperature, to give better agreement with the colours, taken together with the mass–luminosity relation coupled to (3), would entail increases of 3% in mass and 1% in radius over the values adopted by Lazaro et al. These are shown in the absolute parameters of Table III.

Lazaro et al. (2002) remarked that their solutions were still not entirely self-consistent across the full range of wavelengths studied. They gave various reasons for this, including the role of extra absorption from circumstellar matter or the presence of a third body, although their fitting procedure did not set this as a free parameter, as with Table I. Our solutions, given the goodness of fit, low deviation of observations and relatively high determinacy of the photometry, give persuasive evidence about a third star. The magnitudes given in Table II point to the third star being a late A or early F-type object, with a feasible (MS) mass of about 1.4 $M_\odot$.

Lazaro et al.'s discussion shows the desirability of further spectroscopy, both to check on the previous analysis of Tomkin (1978), and also to provide evidence for their remarks on unresolved photometric issues. Extra stars in a system can produce complications for a Hipparcos distance, so various issues are interrelated for this star, and it is worthwhile to recheck the astrometry. In their published solutions, both Hipparcos and Tycho surveys indicate that δ Lib is a reliable astrometric standard, with well-defined positional parameters (a 'reference' star), suggesting the parallax, at least, should not be greatly at variance with the published value. The evidence provided by the O–C curve (see above) should also be kept in mind, however, since this can provide an independent check on a third component.

4. Spectroscopic Data

Figure 5 shows a high-dispersion normalized profile for the solar Hα line, taken from the BASS solar reference spectrum (BASS, 2001), which was analysed using an ILOT procedure involving a Voigt function (c.f. Olah et al., 1998). The curve-fitting procedure for such high-dispersion spectrograms is relevant for those of stars obtained with the Coudé spectrograph on the 2m telescope of the Rozhen Observatory in southern Bulgaria, from where a series of 4 Å/mm spectrograms of the Hα region for δ Lib has been recently gathered.

The Voigt function convolves an inverse-square damping effect and Gaussian contributions to the spectral line. The formula can be expressed thus:

$$V(\Delta\lambda, \gamma, \delta) = \int_0^\infty \frac{\delta/4\pi^2}{(\Delta\lambda' - \Delta\lambda)^2 + (\delta/4\pi)^2} \frac{1}{\pi^{1/2}\gamma} \exp\{-(\delta\lambda'/\gamma)^2\} \, d\Delta\lambda', \quad (4)$$

where the function V expresses the decrement at separation $\Delta\lambda$ from the line centre, with δ the damping half-width, γ the Gaussian width, and the convolution is performed with the integration variable $\Delta\lambda'$.

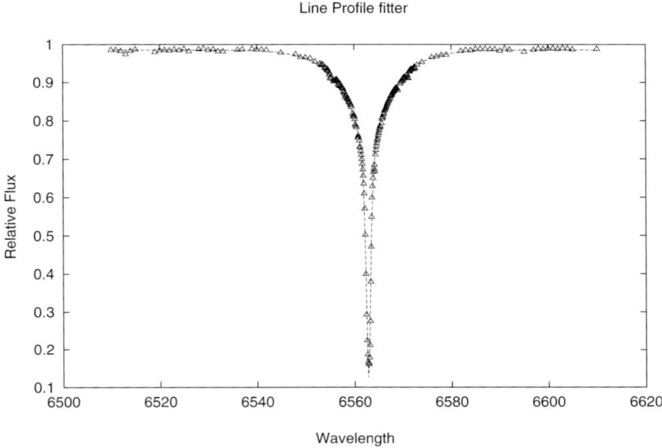

Figure 5. Opacity-scaled Voigt model profile fit to the solar Hα line.

The Voigt function is dominated by the gaussian component at small $\Delta\lambda$, and by damping due to radiation or collisional broadening at large $\Delta\lambda$. The lower frequency Balmer lines also show strong wings, related to inter-particle interactions statistically distorting atomic energy levels. One needs to know the atmospheric pressure, together with various other physical details to interpret these effects properly.

The absorption lines from real atmospheres are not simply Voigt profiles according to the foregoing formula (4), however: there is also the relative opacity (β) effect, that brings out the wings relative to the line centre. This becomes particularly significant for those lines that involve large numbers of atoms in the line-of-sight, i.e. the hydrogen lines. More detailed considerations (e.g. Mihalas, 1978) imply that a model for $\beta(\Delta\lambda)$ is required to rescale the Voigt form in an appropriate way for the Balmer lines. From the point of view of finding a convergent fitting procedure, the observed profile itself may not be so suitable, due to the low parameter discrimination of the wings. One may better approach the opacity model by examining the logarithmic form $\eta(\Delta\lambda)$ of the observed profile $u(\Delta\lambda)$ given by

$$\eta(\Delta\lambda) = -\ln\{1 - u(\Delta\lambda)\}/\beta_0. \tag{5}$$

Here, we expect the opacity variation $\beta(\Delta\lambda)$ to appear as a linear factor in $\eta(\Delta\lambda)$, and this appears able to be approximated by a relatively simple form $\beta = \beta(\Delta\lambda, w, a)$, w and a being simple width and depth terms, respectively. In Figure 5 we show the form of $\eta(\Delta\lambda)$, which has been scaled by taking $\beta_0 = 4.3$, so that when the relative line depth approaches 0.01, the ordinate approaches unity. The line was taken to effectively merge with the continuum at relative flux 0.99 and higher points truncated.

Examination of standard reference spectra of the Sun (BASS, 2001) and also Vega (Pickles, 1998) indicated this logarithmic form of the $H\alpha$ line, to have, in general, such a structure as that shown in Figure 5, although there are somewhat different shapes for the different observed sources, probably reflecting different observing and resolution conditions. On this basis, an empirical fitting function for the observed $H\alpha$ line can be given as

$$f(\Delta\lambda) = 1 - \exp(-\beta(\Delta\lambda, a, w)).(1 + bV(\Delta\lambda, \gamma, \delta)), \tag{6}$$

where β is of the form $\beta = a\{1 + (\Delta\lambda/w)^2\}^{\frac{1}{2}}$. The presence of the radical recalls classical expressions for the emergent flux. In practice, the width terms w and δ are essentially similar.

A feature of interest on some of the δ Lib high dispersion spectra is the spur on the wing of the main (primary) absorption formation (Figure 6). This does not, by itself, prove emission is present, because, with suitable combinations of primary and secondary parameters, at the orbital elongations (which is when the spur is

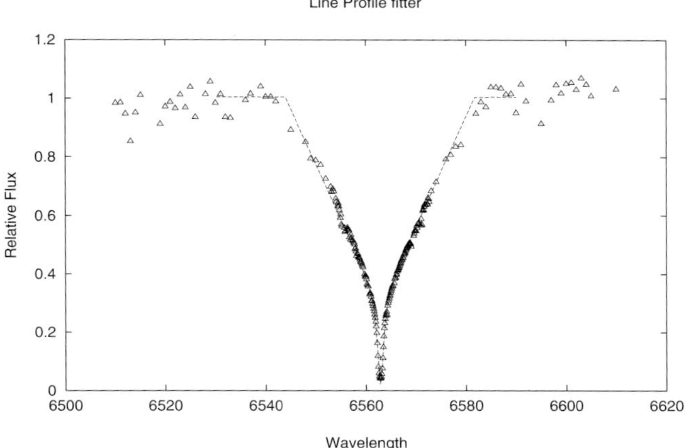

Figure 6. Logarithmic form of the solar H_α line scaled to unity at depth 0.01 from the continuum.

seen), an 'up-down' effect can be created by adding two normal absorption lines, corresponding to primary and secondary, but suitably displaced due to their relative radial velocities. In fact, further study of the δ Lib profiles and those of Vega and the Sun (to compare with primary and secondary contributions – c.f. Lazaro et al., 2002) suggested an additional narrow absorption source, close to the wavelength of the primary.

In Figure 6 we show a three absorption source fit to an observed $H\alpha$ line of δ Lib. It is tempting to associate the third source with the third star identified in Section 3. We should stress that this work is still in progress, but it appears that the fitting is not well-defined as an optimization problem, i.e. several combinations of parameters, or possible alternative forms of the empirical fitting function, are capable of matching the data. The absence of damping, though with a sizeable Gaussian component in the central core, however, is unexpected for an A-type dwarf, so we may also think about circumbinary material.

To probe further into this, we may look at the difference between model and the observed line in Figure 7. The absorption core appears here superposed on a narrow emission feature. This feature could perhaps be explained by an accretion structure around the primary, together with a surplus of emission from near the L_1 point. More detailed analysis will be necessary to establish the relevance of this picture.

5. δ Lib as a Radio Source

Most of our knowledge of the statistical properties of the cm-band emission from δ Lib was acquired during a large-scale survey of active stars from 1981 to 1988 (Slee

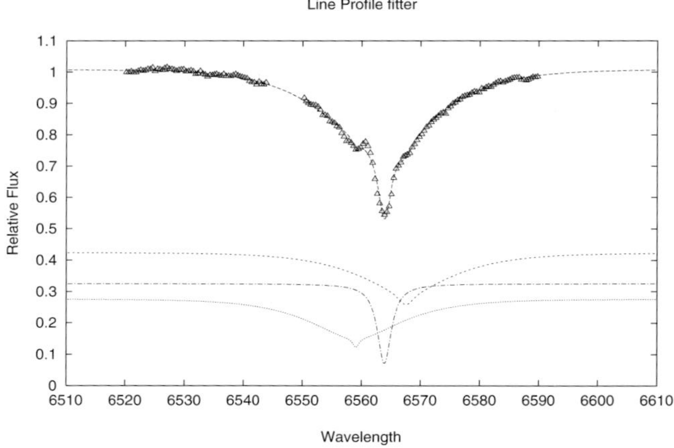

Figure 7. Three absorption source model fit to the $H\alpha$ line of the δ Lib system.

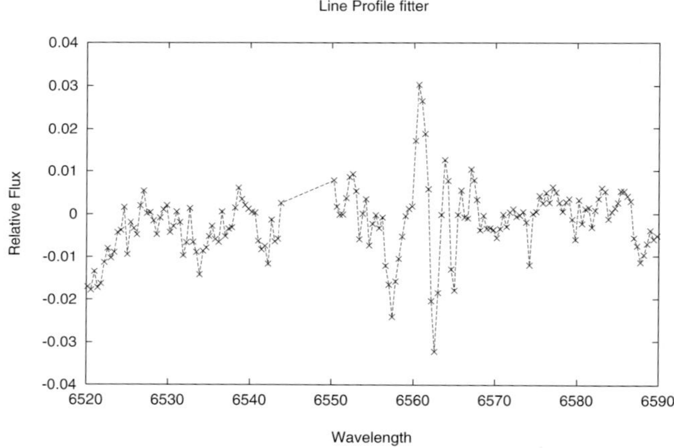

Figure 8. Difference between the observed and absorption model of the $H\alpha$ line for δ Lib.

et al., 1987). The 8.4 GHz measurements of δ Lib discussed here were acquired with the dual-beam, low-noise system on the Parkes 64-m reflector, measuring flux densities on 26 days in the interval September 1985 to July 1988. The FWHM beamwidth was 2.7 arcmin, and the 3σ detection level was 6.9 mJy (Figure 8).

Table IV shows the distribution of 26 daily measurements of flux density. The measurements in the lowest flux bin are $<2\sigma$, and are not significant detections. Using the Hipparcos distance of 93.3 pc, the highest emitted 8.4 GHz power was 7.3×10^{17} erg s^{-1} Hz^{-1}, which is very similar to the maximum power measured from the well-known RS CVn star HR 1099. The median value of flux density (using all 26 measurements) was 7.6 mJy, and the detection rate was 0.73 (from the sources in the last four bins of Table IV). It is clear from Table IV, however, that

TABLE IV

Distribution of 8.4 GHz flux

Flux range	No. of days	Median
0.0–4.9	7	2.3
5.0–9.9	8	2.3
10.0–19.9	5	3.0
20.0–39.9	4	4.2
40.0–79.9	2	2.8

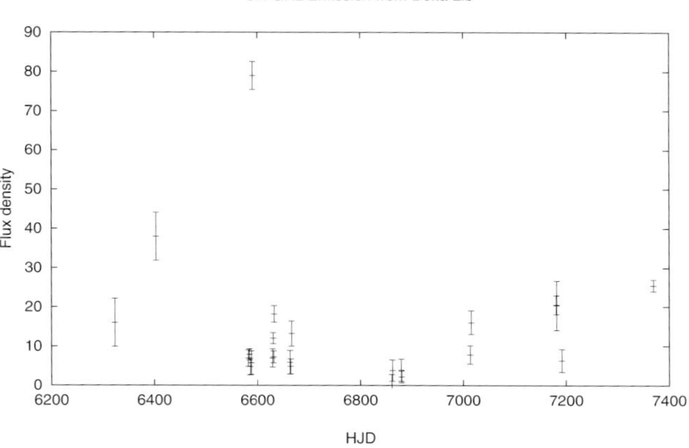

Figure 9. Averaged daily 8.4 GHz flux densities for 26 days during the interval from September 1985 to July 1988. Further details are given in the text.

weak emission is much the more prevalent, falling below 10 mJy for 58% of the observing days.

Table IV gives no information on the temporal variation in radio activity, but this is partially clarified in Figure 9, which traces flux density variations over the interval September 1985 to July 1988. Each plotted point represents the averaged daily flux density. On those days when the emission level was low, or below the detection threshold, the integration was confined to just 12 min, but on the two successive days with strongest emission we made nine and six 12-min observations, respectively, each spread over $\sim$4 h. On the second of these days, the flux density increased steadily from 40 to 84 mJy over the 4-h interval of observation. Circularly polarized emission of 19% in the right-handed sense was present on the first of these days, but was below the detection threshold of $\sim$5% on the second day.

The variation of detection rate with orbital phase is a quantity of interest regarding possible eclipsing of active regions in the corona of the K0 IV star, or the possible presence of a radio source in the inter-component plasma stream. Our data on δ Lib are insufficient to construct a meaningful plot of detection rate against phase, and even gathering together all the 8.4 GHz detections for the 8 EA2 systems in the Parkes sample (Figure 10), produces a distribution that is not significantly different from a uniform one. When, however, one constructs a phase distribution from all our 8.4 GHz detections of Algols and RS CVn's (Figure 11), the distribution is found to be non-uniform at the 99% confidence level. Of particular interest in this plot is the significantly increased detection rate in the phase range 0.2–0.4, and, to a lesser extent, in the range 0.8–1.0. It seems more than pure chance that

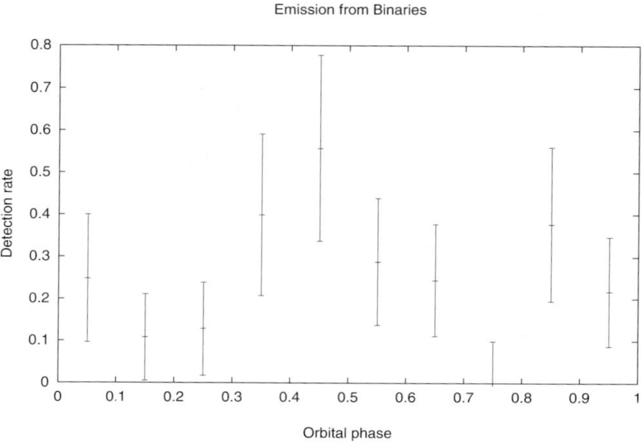

Figure 10. Dependence of detection rate on orbital phase using data from the Parkes Active Star Survey (Slee et al., 1987): EAS Algols only.

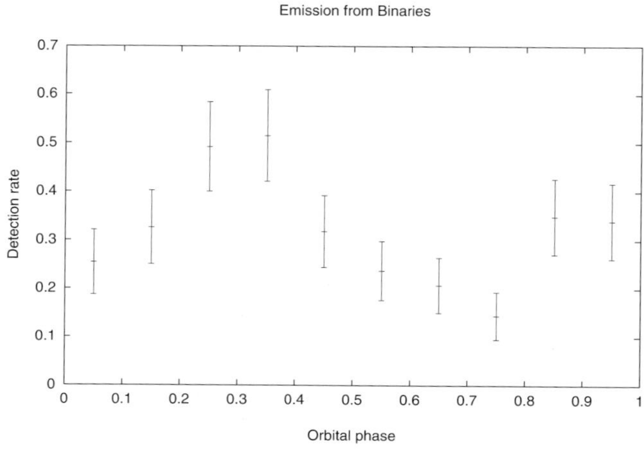

Figure 11. Dependence of detection rate on orbital phase using data from the Parkes Active Star Survey (Slee et al., 1987): combined data from 31 EAD, EAS and RS CVn binaries.

the distributions for the 8 EA2 Algols and (all Algols + RS CVns) are similar, despite the latter containing 5 times the number of detections as the former (see also Budding, Slee and Jones, 1988).

Slee et al. (2003) have presented a simple coronal 'slab' model, employing the microwave emission formulae of Dulk (1985), to interpret emission from cool star coronae. These models require fairly dense, relatively large structures, containing magnetic fields of moderate field strength (typically $\sim$100 Gauss) and high local kinetic temperatures. Budding et al. (2002) reported a tendency to a phase delay in cross-correlations of higher and lower frequency microwave emission from such sources, such that lower frequency emission features tend to arrive later than corresponding higher frequency ones. The interpretation put on this was in terms of a coronal propagation tending (usually) to progress outward from lower more dense layers.

The simple slab model does not correspond very obviously with high-resolution radio observations of Algols. Mutel et al. (1998) used the VLBA's good imaging capability on β Per to show the prolonged presence of a double-lobed structure having an E–W spacing of 1.6 mas. The lobes were strongly circularly polarized, but in opposite senses. Although the authors were unable to register the VLBI and optical images to better than $\pm$1.6 mas, they assumed that the K2IV secondary lay at the centroid of the lobes. With the help of other, shorter baseline, measurements that apparently showed no evidence of radio eclipses, Mutel et al. produced a model, in which emitting lobes are placed in the polar regions of an assumed dipolar field. The opposite senses of polarization in the two lobes were interpreted as due to different field components in opposite directions along the line-of-sight.

Note also that microwave emission from polars like V834 Cen indicate a phase-dependent emission (Chanmugam and Dulk, 1982; Wright et al., 1988, Ferrario et al., 1989). This suggests a role for the mass transfer process in generating the radio emission. This might be related to an interaction scenario, involving the magnetospheres of both components (Uchida and Sakurai, 1983). Observational exposition of such issues is technically difficult, but it is a good challenge for current observational technology.

Modeling improvements could be applied to a new investigation of δ Lib. The magnetic field and electron density model could include both stellar envelopes and take into account coronal distortions associated with the Roche lobe filling of the cool component. Consistency of the assumed magnetic field and density models with X-ray spectra (Singh et al., 1995) should be checked. A three-dimensional inhomogeneous magnetic field and electron density treatment could permit the study of possible emission associated with the Roche lobe overflow process. While Mutel et al. (1998) solved the radiative transfer equation separately for each polarization component (X and O-modes), a more comprehensive discussion could adopt more general radiative transfer equations including mode-coupling effects (Melrose, 1980).

6. Astrometric Checks

Mention has already been made of the third body postulated by Worek (2001). This was to account for irregularities found in the primary's radial velocity curve. Worek's model has the relatively short period for the third orbit as 2.762 y, which can, in principle, be checked from Hipparcos observations that covered a $\sim$4 y time interval. There is possible ambiguity arising from radial velocity data, since what is called the 'longitude of periastron' (ϖ) is measured from the plane of the sky, and not the line of nodes in the equator, as in the standard 3-dimensional specification for double stars. If we use the local plane of the sky as the reference, the angle Ω becomes referred to the direction of increasing right ascension. There are then ranges of possible values of Ω and inclination i of the (assumed) plane of the wide orbit to the plane of the sky, that go with a given value of ϖ. An example of a plot (with $\Omega = \pi/2$, and the wide and close orbits coplanar, i.e. $i = 80.5°$) corresponding to Worek's parameters is shown in Figure 12.

Apart from the very poor definition of any possible orbit from such trials, there are other reasons why we demur from Worek's solution. Firstly, the Hipparcos/Tycho programme investigators failed to find any indication of duplicity, although it was known that a third body was suspected and checks were made (ESA, 1997). Secondly, there are a number of ambiguities in Worek's (2001) paper, including, for example, the selected frequency in the periodogram analysis (his Figure 3) appears as only one of several possibles shown (and perhaps others not shown), which, in any case are at very low S/N resolution. Again, the 8 points used (his Figure 4) to define his third body orbit radial velocities (from which group some other measures were discarded), really define only the maximum region of a cyclical curve. The corresponding minimum is almost undefined.

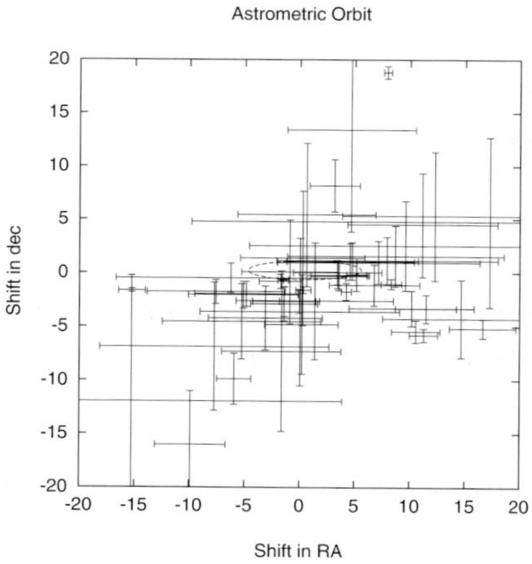

Figure 12. Plot of Hipparcos angular measurement components in RA and dec (in mas) on the assumption of displacements due to Worek's (2001) 2.8 year period third body.

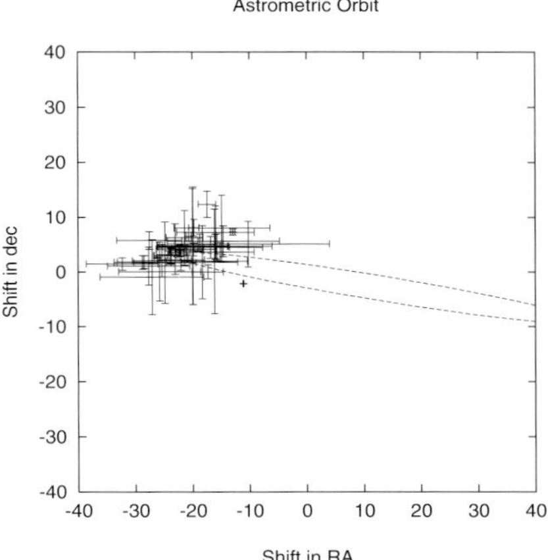

Figure 13. Plot of Hipparcos data trial fit to an orbit of δ Lib about a barycentre with a third body, corresponding to the period analysis of Section 2 (see text).

Our photometric solution for the third body indicates a significantly more massive and brighter star than Worek's. If we accept Worek's γ-velocity as of the right order, there would have to be an order of magnitude increase in the period of the third body orbit to increase the mass function to an appropriate value. Such an increased period, while having the possibility to reconcile the mass with the radial velocity displacements, could also allow that Hipparcos would fail to detect significant changes of position, particularly if the third star was close to its apse at the time of observation. The longer period is also supported by the O–C analysis given in Section 2.

In Figure 13 we show the result of a trial optimization run in which the period, projected semi-major axis, eccentricity and longitude values were taken from the preceding tentative O–C fitting, and inclination and nodal angle values were optimized, also allowing for a mean shift in the centroid of the orbit from the centroidal position given for δ Lib in the Hipparcos catalogue. This approach is essentially similar to that of Ribas et al. (2001) for R CMa. The solution found that corresponds with Figure 13 has $i = 86.5$ and $\Omega = 279.2°$. Shifts of 19.0 and 1.6 mas in right ascension and declination, respectively, are also required for this fit. The s.d. error of an individual residual is 1.8 mas in this fitting, which is close to typical Hipparcos solutions for fixed stars. Hence, this tentative orbit is as consistent as the no-motion solution.

The third body mass that would go with the light–time effect mass function, however, would be only 0.48 $M_\odot$ with such a high inclination, and this is inconsistent with the photometric third body of Table II. There are several areas of uncertainty about this: firstly, the O–C analysis for the third body orbit is very provisional. A longer period and greater amplitude could be found after more reliable photoelectric photometry will become available in the future. Secondly, the spectroscopic information appears very ambiguous: our $H\alpha$ profiles are not convincing about a third star and have other complications, while the difficulties with Worek's solution were already discussed. Thirdly, the fact that both orbital and non-orbital fittings can fit Hipparcos measurements to comparable accuracy (provided we allow for small, but significant, shifts in the reference positions) indicates inherent non-uniqueness in the four-parameter fittings to such data. We found one such orbit after some initial trials, but others could be possible.

A possible check on the tentative orbit may come from proper motion comparisons. The Hipparcos proper motions of $\mu_\alpha = -66.9 \pm 0.9$ mas/y and $\mu_\delta = -3.4 \pm 0.8$ mas/y may be checked against historic values. Some 43 measurements in RA and 38 in dec, spanning an interval of a little over a century, gave rise to average proper motions of $\mu_\alpha = -61.2 \pm 5.9$ mas/y and $\mu_\delta = -5.4 \pm 1.5$ mas/y. However, it is difficult to bring this to bear strongly on the tentative astrometric orbit, given the combined errors of the measures. Even if there is no significant disagreement between these values (within their error limits), it does not necessarily rule out a third body orbit, since the Hipparcos observations could be near a stationary point in RA of the wide orbit (as with R CMa), while the motion in declination is small. The real difference between the orbital motion and the mean proper motion is likely to be only a few mas per year at most, while the Hipparcos and Tycho mean proper motions in declination (for a similar time interval) already differ by 9 mas/year, which is greater than the quoted errors of either measure. The rather more negative Tycho value of $\mu_\delta = -12.3 \pm 3.9$ mas/y could be in keeping with the geometry of Figure 13, however. In summary, there are various issues about the astrometry that must be fully checked before a tentative orbit, such as that presented, could become definite. Confirmation for some of these issues may have to wait until the precision astrometry of the GAIA era becomes available.

7. Conclusions

The bright classical Algol δ Lib invites attention using the power of combined astronomical techniques. There is also a long historical background of observation for this bright star. The Algol configuration is a relatively simple example of binary physical interaction, for which detailed checks of theory of structure and evolution should be possible.

Microwave radio observations will undoubtedly help probe astrophysics of the semi-detached configuration, not only directly by informing about the outermost envelopes of the components, but indirectly through its high accuracy positional information. In turn, this can relate back to studies of the period of the binary and careful astrometry.

Algol systems like δ Lib thus contain interesting, still unsolved problems of physical parameter evaluation, stream and accretion structure modeling, radio emission mechanisms and plasma physics, that can support much further observational attention.

References

Batten, A.H.: 1989, *Algols; Proc. IAU Coll. 107, Space Sci. Rev.* **50**.

Budding, E.: 1993, Introduction to Astronomical Photometry, Cambridge University Press, Cambridge.

Budding, E.: 2003, *Proc. Summer School Astron. Math.*, Kültür University, Yayinlari **35**, Istanbul., 21.

Budding. E., Slee, O.B. and Jones, K.: 1998, *Publ. Astron. Soc. Aust.* **15**, 183.

Budding, E., Lim, J., Slee, O.B. and White, S.M.: 2002, *New Astron.* **7**, 35.

Castelli, F. and Kurucz, R.L.: 1994 *A&A* **281**, 817.

Chanmugam, D. and Dulk, G.A.: 1982, *ApJ*, **255**, L107.

Demircan, O.: 2000, *NATO ASI Ser. C*, Dordrecht, Kluwer, 615.

Drake, S.A., Simon, T. and Linsky, J.L.: 1989, *ApJ Suppl.* **71**, 905.

Ferrario, L., Wickramasinghe, D.T. and Tuohy, I.R.: 1989, *ApJ* **341**, 327.

Gunn, A.G., Brady, P.A., Migenes, V., Spencer, R.E. and Doyle, J.G.: 1999, *MNRAS* **304**, 611.

Kalomeni, B., Pekünlü, E.R. and Yakut, K.: 2003, *Publ. COMU Ap. Research Center* **3**, 257.

Koch, R.H.: 1962, *AJ* **67**, 130.

Kreiner, J.M., Kim, C-H. and Nha, I.-S.: *An Atlas of O–C Diagrams of Eclipsing Binary Systems*, Wydawnictwo Naukowe Akademi Pedagogicnnej, Krakow.

Lazaro, C., Arevalo, M.J. and Claret, A.: 2002, *MNRAS*, **334**, 542.

Melrose, D.B.: 1980, *Plasma Astrophysics* Vol. I, Gordon and Breach, New York, p.195.

Mihalas, D.: 1978, *Stellar Atmospheres*, W.H. Freeman & Co., San Fransisco.

Mutel, R.L. Molnar, L.A., Waltman, E.B. and Ghigo, F.D.: 1998, *ApJ* **507**, 371.

Oláh, K., Marik, D., Houdebine, E.R., Dempsey, R.C. and Budding, E.: 1998, *A&A* **330**, 559.

Özdemir, S. et al.: 2001, *Publ. Astron. Soc. Aust.*, **18**, 151.

Pickles, A.J.: 1998, *PASP*, **110**, 863.

Plavec, M.: 1973, *Proc. IAU Symp. 51*, O. Struve and A.H. Batten (eds.), Dordrecht, p. 216.

Radhakrishnan, K.R., Sarma, M.B.K. and Abhyankar, K.D.: 1984, *Ap%SS* **99**, 229.

Ribas, I., Arenou, F. and Guinan, E.F.: 2002, *AJ* **123**, 2033.

Richards, M.T., Jones, R.D. and Swain, M.A.: 1996, *ApJ* **459**, 249.

Robinson, P.A. and Melrose, D.B.: 1985, *Aust. J. Phys.* **37**, 675.

Roman, N.G.: 1956, *ApJ* **123**, 246.

Sahade, J. and Hernandez, C.A.: 1963, *ApJ* **137**, 845.

Singh, K. P., Drake, S. A. and White, N.E.: 1995, *ApJ* **445**, 840.

Slee, O.B. et al.: 1987, *MNRAS* **229**, 659.

Slee, O.B. et al.: 1987b, *Publ. Astron. Soc. Aust.* **7**, 558.

Slee, O.B. et al.: 1988, *ApJ* **L27**, 247S.

Slee, O.B. and Stewart, R.T.: 1989, *MNRAS* **236**, 129.

Slee, O.B., Budding, E., Carter, B.D., Mengel, M.W., Waite, I. and Donati, J.-F.: 2004, *Publ. Astron. Soc. Aust.* **21**, 72.

Strassmeier, K., Hall, D.S., Fekel, F.C. and Scheck, M.: 1993, *A&AS* **100**, 173.

Tomkin, J.: 1978, *AJ* **221**, 608.

Uchida, Y. and Sakurai, T.: 1983, Activity in Red Dwarf Stars, *Proc. IAU Colloq. 71*, in: P.B. Byrne and M. Rodono (eds.), p. 629.

Umana, G., Leto, P., Trigillio, C., Hjellming, R.M. and Catalano, S.: 1999, *A&A* **342**, 709.

Worek, T.F.: 2001, *PASP* **113**, 964.

Wright, A. E., Cropper, M., Stewart, R.T., Nelson, G.J. and Slee, O.B.: 1988, *MNRAS* **231**, 319.

NUMERICAL MODELING OF MASS TRANSFER IN CLOSE BINARIES

D.V. BISIKALO

Institute of Astronomy, Russian Academy of Sciences, Moscow, Russia;
E-mail: bisikalo@inasan.rssi.ru

(accepted April 2004)

Abstract. The review of gasdynamic models used for the description of the mass exchange in close binaries is presented. Main features of the flow structure are summarized. Special attention is paid to physics of accretion discs in binary systems. It is shown that in self-consistent considerations of gas dynamics of mass transfer in close binaries the interaction of the stream from L_1 with the forming accretion disc is shock-free, and, hence, a "hot spot" does not form at the outer edge of the disc. To explain the presence of the observed zones of high luminosity in semidetached binaries a self-consistent "hot line" model is proposed. According to this model the excess energy is released in a shock wave formed due to interaction of the circumdisc halo and the gas stream flowing out of the donor-star through the vicinity of the inner Lagrangian point.

Keywords: binary stars, accretion disk, gasdynamic numerical modeling

1. Introduction

The main objective of this paper is the study of gas dynamics of mass transfer in close binary systems (CBS), being at the stage when a system becomes a semidetached one according to the Kopal (1959) classification. The study of the flow structure is of great importance, and the results can be used both for consideration of the evolutionary status of binary stars and for the interpretation of observational data. Summarizing the main arguments in favour of the actuality of the study of gas dynamics of mass transfer in CBS, one can distinguish the three most important reasons.

- The evolution of a CBS at the stage of mass transfer is determined by several main processes, namely: mass loss by the donor-star, gaining of matter by the accretor, formation of a circumbinary envelope as well as mass and angular momentum loss by the system. As a rule, the study of the evolution of binary systems involves the time-averaged characteristics of mass transfer and does not consider the details of flow patterns in the system. At the same time, the study of the flow structure allows one to determine the mass transfer parameters at a specific stage of star evolution and hence to define more exactly the averaged parameters of mass transfer.
- A number of observations prove the complex flow structure in CBSs. Starting from the study of Struve (1941), who first conceived the idea of a gas stream appearing between components in β Lyrae to explain the peculiar behaviour of

Astrophysics and Space Science **296:** 391–401, 2005.
© Springer 2005

the spectrum at eclipse, the effects of mass transfer resulting in formation of gas flows, streams, discs, circumbinary envelopes and other structures were observed in a number of CBSs (see, e.g., Struve, 1941; Batten, 1973; Sahade and Wood, 1978; Goncharskii et al., 1985; Shore et al., 1994; Warner, 1995).

The observations reflect the current state of a binary system, and for the interpretation of them one should consider the gas dynamics of flow patterns. The analysis of the model considering the features of the system caused by mass transfer between the components and comparison of the result with observational data allows one to define the physical processes taking place in the binary system more exactly.

– The importance of study of binary systems becomes even more obvious when one considers the energy spectrum of the available observations. Even the simplest analysis involving the estimation of the total energy of a single star proves that some additional energy sources should be taken into account in order to explain the observations in short-wavelength bands. Since a single star cannot produce the observed amounts of energy, the researchers suggest that the majority of the observed far ultra-violet, X-ray, and γ-ray phenomena are not related to single stars, but are caused by the processes in binaries (mass transfer, accretion processes, accretion discs, etc.). Therefore, the study of mass transfer in binaries is of great importance for the understanding of the nature of many (or the majority) of non-stationary stars and peculiar objects.

The importance of the study of mass transfer in CBSs has been noted in pioneer works by Struve (1941), Kuiper (1941), and Crawford (1955). The gas dynamics of mass transfer through the inner Lagrangian point L_1 has been investigated by many authors. Prendergast (1960) and Gorbatsky (1964, 1977) first attempted to consider the gas flow structure in binaries. Lubow and Shu (1975) successfully analysed the gas flow in the vicinity of L_1 and estimated its basic characteristics using the perturbation method.

To describe mass transfer in a system, one should consider not only parameters of the stream, but the further motion of matter from L_1 to the accretor as well. It is the gas motion that forms a general flow pattern in the system that is observed and, hence, great attention was focused on this problem. Warner and Peters (1972), Lubow and Shu (1975), and Flannery (1975) first considered the fate of particles that left L_1 and moved in the gravitational field of a binary system. The results obtained in these studies as well as in subsequent ones were widely accepted. However, the application of the ballistic approach to concrete binaries shows that the results failed to agree with observational data. These disagreements are explained by the simplifications of the ballistic approach, since this approach ignored gasdynamic effects caused by the circumbinary envelope. The presence of a circumbinary envelope in a number of close binaries was revealed from observations and was thoroughly studied by many authors. To consider the impact of the forming circumbinary envelope on gas flows and, hence, to describe gas flows correctly, we should solve a complete system of gasdynamic equations. This system can be solved only within the framework of rather complex mathematical models.

The application of numerical methods to studies of mass transfer in binaries was limited for a long time by insufficient computer power, and hence two-dimensional models were widely used for the analysis of gas flows. Despite the simplifications, the two-dimensional approach allowed one to consider some details of the flow patterns and obtain some interesting results (see, e.g., Sawada et al., 1986, 1987; Taam et al., 1991; Blondin et al., 1995; Murray, 1996). In the last few years gas dynamics of mass transfer was numerically studied in the framework of more realistic three-dimensional models (see, e.g., Nagasawa et al., 1991; Hirose et al., 1991; Molteni et al., 1991; Lanzafame et al., 1992, 1994; Belvedere et al., 1993; Armitage and Livio, 1996). In these studies, in particular the formation of accretion discs in semidetached binaries (see, e.g., Nagasawa et al., 1991; Sawada and Matsuda, 1992; Yukawa et al., 1997) and the stream–disc interaction (see, e.g., Hirose et al., 1991; Armitage and Livio, 1996) were considered. In this paper we summarize the results of 3D numerical simulations of mass transfer in semidetached binaries that were mainly obtained by Bisikalo et al. (1997, 1998, 2000, 2003), and Boyarchuk et al. (2002).

2. The Model Description

Let us consider a semidetached binary system with masses of accretor and mass-donating star M_1 and M_2, respectively, separation A, and angular velocity of orbital rotation Ω. Gaseous flows in the system can be described by Euler equations with addition of the ideal gas equation of state. The shape of the secondary is defined by Roche lobe geometry, so there are only few dimensionless parameters determining the flow structure: mass ratio $q = M_2/M_1$, Lubow–Shu parameter $\epsilon = c(L_1)/A\Omega$ (Lubow and Shu, 1975) where $c(L_1)$ is the sound speed in L_1, relative size of accretor R_1/A, and adiabatic index γ. We also should expand this list by adding the value of viscosity (expressed as dimensionless parameter α introduced by Shakura, 1972; Shakura and Sunyaev, 1973), since in approximate solutions some numerical viscosity is present. The model implies that the magnetic field does not influence the gas motion in the system.

The analysis shows a sort of stability of the obtained solutions against perturbations of the input parameters, if their values lie in intervals pertinent to cataclysmic variables (CVs) and low mass X-ray binaries (LMXBs). The calculations prove the qualitative similarity in the structure of flow in the studied LMXB and CV systems (see, e.g., Boyarchuk et al., 2002). Taking into account the qualitative similarity of the results, below we will present the general properties of the flow structure.

To solve the system of gasdynamic equations we use the Roe–Osher TVD scheme of a high approximation order (Roe, 1986; Chakravarthy and Osher, 1985) with Einfeldt modification (Einfeldt, 1988). This numerical method allows one to study flows with a significant density contrast, and, hence, to investigate the morphology of gaseous flows in binaries and consider the impact of the forming circumbinary envelope on the flow patterns.

The modelling was carried out in a non-inertial reference frame rotating with the binary, in Cartesian coordinates in a rectangular three-dimensional grid. Since the problem is symmetrical about the equatorial plane, only half of the space occupied by the disc was modelled. To increase the accuracy of the solution the grid was made denser in the zone of interaction between the stream and disc, making it possible to resolve well the formed shock wave. The grid was also denser towards the equatorial plane, so that the vertical structure was resolved. The numerical simulations of mass transfer in semidetached binaries have been conducted on large time intervals that allowed us to consider the main features of flow structure in the steady-state regime.

Other details of the numerical model can be found in Boyarchuk et al. (2002) and Bisikalo et al. (2003).

3. Mass Transfer in a System with Hot Accretion Disc

These solutions were obtained for temperatures of the outer parts of the accretion disc of 200,000–500,000 K.

The morphology of gaseous flows in the considered binary system can be evaluated from Figure 1. In this figure the distribution of density over the equatorial plane and velocity vectors are presented. We also show a gasdynamic trajectory of a particle moving from L_1 to the accretor (a white line with circles) and a gasdynamic trajectory passing through the shock wave along the stream edge (a black line with squares).

The sketch of main peculiarities of the morphology of gaseous flows in semidetached binaries for the "hot" case is given in Figure 2. This scheme is based on the results of 3D gasdynamic simulations. In Figure 2 the fragment of the mass-losing star that fills its Roche lobe, the location of the inner Lagrangian point L_1, the stream of matter from L_1 as well as the location of the accretor are shown. The dashed line marks the Roche lobe.

The morphology of gaseous flows in semidetached binaries is governed by the stream of matter from L_1, a quasi-elliptical accretion disc, a circumdisc halo, and a circumbinary envelope. This classification of the main constituents is based on their physical properties: (i) if the motion of a gas particle is not determined by the gravitational field of the accretor then this particle belongs to the circumbinary envelope filling the space between the components of the binary; (ii) if a gas particle revolves around the accretor and after that mixes with the matter of the stream then it does not belong to the accretion disc, but forms the circumdisc halo (zone "C" in Figure 2); (iii) the accretion disc is formed by the matter of the stream, which

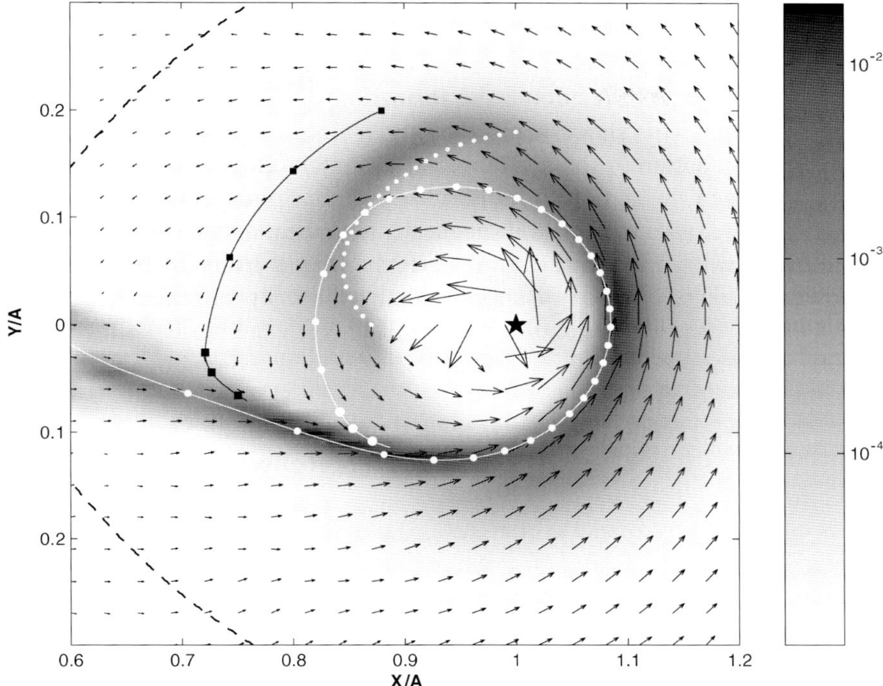

Figure 1. Distribution of density over the equatorial plane. The X and Y coordinates are expressed in terms of the separation A. Arrows are the velocity vectors. The black asterisk is the accretor. The dashed line bounds the Roche lobe. The dotted line is the tidally induced spiral shock. The gasdynamic trajectory of a particle moving from L_1 to the accretor is shown by a white line with circles. Another gasdynamic trajectory is shown by a black line with squares. Symbols of larger size correspond to the passing of the trajectory through the shock wave located at the edge of the stream.

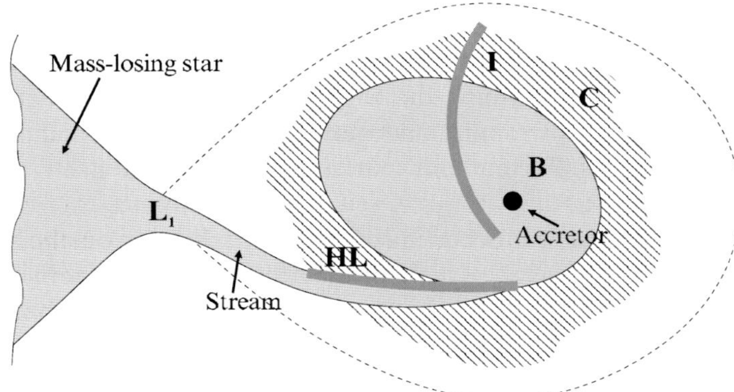

Figure 2. Sketch of main peculiarities of the morphology of gaseous flows in semidetached binaries for the case of high gas temperature.

is gravitationally captured by the accretor and hereinafter does not interact with the stream, but moves to the accretor losing the angular momentum (zone "B" in Figure 2). The interaction of matter of the circumdisc halo and circumbinary envelope with the stream results in the formation of the shock located along the edge of the stream. This shock is referred to as "hot line" and is marked by "HL" in Figure 2. Tidal action of the mass-losing star results in the formation of a spiral shock marked by "I" in Figure 2. Our 3D gasdynamic simulations for the "hot" case have shown only a one-armed spiral shock, while in the place where the second arm should be the flow structure is determined by the stream from L_1. It also should be stressed that in this case the spiral shock penetrates deeply into the inner part of the disc.

4. Morphology of the Interaction Between the Stream and the Cool Accretion Disc

Analysis conducted by Bisikalo et al. (2003) has shown that for realistic values of parameters ($\dot{M} \simeq 10^{-12} \div 10^{-7} M_\odot$/year and $\alpha \simeq 10^{-1} \div 10^{-2}$) the gas temperature in the outer parts of the disc is between $\sim 10^4$ and $\sim 10^6$ K. This implies that cool accretion discs can form in some close binaries.

Let us consider the morphology of gaseous flows in a system with a cool accretion disc. In the model used the temperature decreases to 13,600 K over the entire computation domain due to the radiative cooling. The basic problem here is whether the interaction between the stream and the disc remains shockless, as was shown for relatively hot discs. Figure 3 depicts the density distribution and velocity vectors in the equatorial plane of the system (the XY plane). Two panels of Figure 4 show the density distribution and velocity vectors (left panel) and visualization of the velocity field (right panel) in the zone of stream–disc interaction (see dashed rectangular in Figure 3). Figure 4 shows that in the "cool" case the interaction between the circumdisc halo and the stream displays all features typical of an oblique collision of two streams. We can clearly see two shock waves and a tangential discontinuity between them. The gases forming the halo and stream pass through the shocks corresponding to their flows, mix, and move along the tangential discontinuity between the two shocks. Further, this material forms the disc itself, the halo, and the envelope.

Let us consider the changes occurring during the transition from the hot accretion disc to the cool one. The sketch of main peculiarities of the morphology of gaseous flows in semidetached binaries for the "cool" case, when non-adiabatic processes of radiative heating and cooling result in dropping of the gas temperature, is given in Figure 5. Our 3D gasdynamic simulations presented in Bisikalo et al. (2003) have shown that for the "cool" case, when the radiative cooling decreases the gas temperature to $\sim 10^4$ K, the solution has the same qualitative features as in the "hot" case, namely: the interaction between the stream and disc is shockless, the region

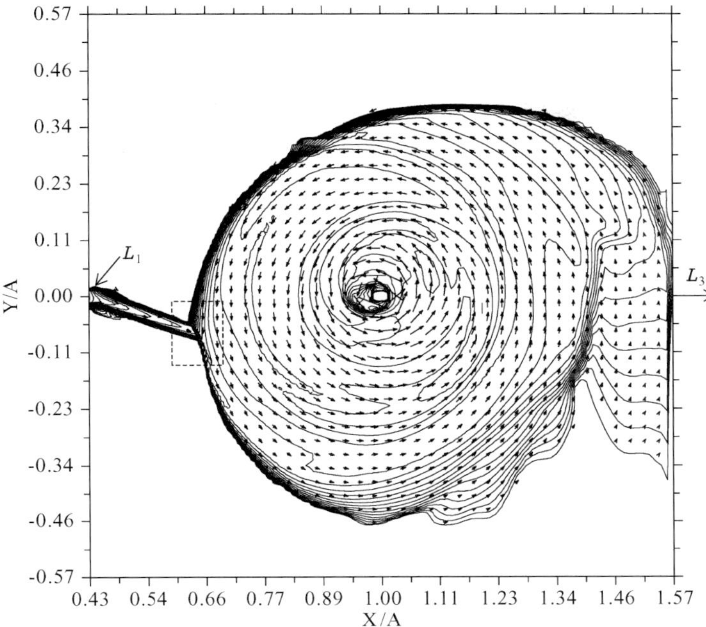

Figure 3. Contours of constant density and velocity vectors in the equatorial plane *XY* of the system. The dashed rectangle indicates the zone of interaction between the stream and disc, shown in Figures 4a and 4b. The point L_1 and the direction towards L_3 are marked.

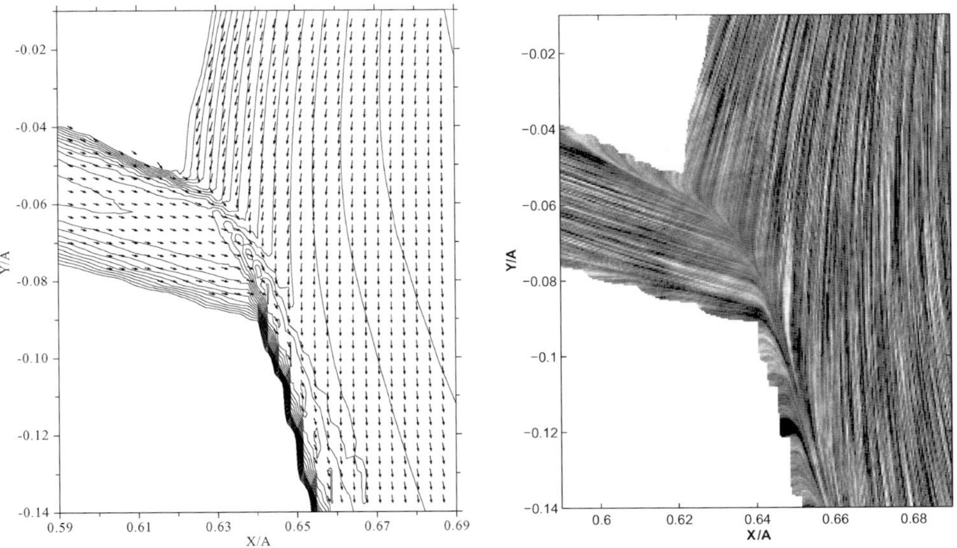

Figure 4. Contours of constant density and velocity vectors (left panel) and visualization of the velocity field (right panel) in the zone of interaction between the stream and disc (the dashed rectangle in Figure 3).

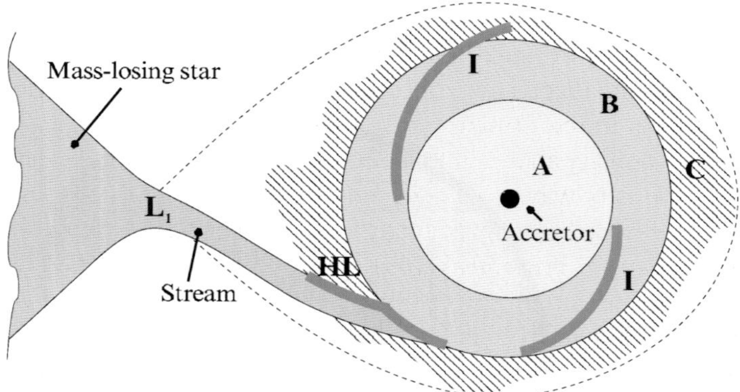

Figure 5. Sketch of main peculiarities of the morphology of gaseous flows in semidetached binaries for the case of low gas temperature.

of enhanced energy release is formed due to the interaction between the circumdisc halo and the stream and is located beyond the disc. The resulting shock – the "hot line" ("HL" in Figure 5) is fairly extended, that is particularly important for explaining the observations. However, unlike the solution with a high temperature in the outer regions of the disc, in the cool case, the shape of the zone of shock interaction between the stream and halo is more complex than a simple line. This is due to the sharp increase of the halo density as the disc is approached. Those parts of the halo that are far from the disc have low density, and the shock, due to their interaction with the stream, is situated along the edge of the stream. As the halo density increases, the shock bends, and eventually stretches along the edge of the disc. In the "cool" case the accretion disc (zones "A" and "B" in Figure 5) is significantly denser than the matter of the stream, the disc is thinner and has not quasi-elliptical, but circular form. The size of the circumdisc halo is smaller as well. The second arm of the tidal spiral shock is formed, and both arms do not reach the accretor, but are located in the outer part of the disc.

Taking into account that the stream influences the dense inner part of the disc weakly as well as that all the shocks ("hot line" and two arms of tidal wave) are located in the outer part of the disc, we can introduce a new element of the flow structure for the "cool" case: the inner region of the accretion disc (zone "A" in Figure 5), where the influence of gasdynamic perturbations mentioned above is negligible.

Formation of a gasdynamically non-perturbed region in the inner part of the disc allows to consider the latter as a slightly elliptical disc with typical size of ~ 0.2–$0.3\,A$ embedded in the gravitational field of binary. It is known (see, e.g., Warner, 1995), that the influence of the companion star results in precession of orbits of particles rotating around the binary's component. The precession is retrograde and its period increases when approaching the accretor. We have shown (Bisikalo et al., 2004a) that the retrograde precession with a specified law of precession rate results in the formation of the density spiral wave of a new, "precessional" type in the inner part of the disc. This wave is formed by apoastrons of flowlines, and its appearance leads to the growth of the radial component of matter flux due to increasing of both density and radial velocity. Increasing of the radial flux of matter after passing the density wave results in growth of the accretion rate and formation of a compact zone of energy release on the accretor surface. This zone can be seen as a periodic increase of brightness in light curves of semidetached binaries. The features of the "precessional" spiral wave allow us to propose a new mechanism explaining superoutbursts in binaries of SU UMa type (Bisikalo et al., 2004b). This mechanism explains both the energy release during the outburst and all its observational manifestations. The distinctive characteristics of a superoutburst in an SU UMa-type star is the appearance of the superhump in the light curve. The proposed model reproduces well the formation of the superhump as well as its observational features, such as the period that is 3–7% longer than the orbital one and the detectability of superhumps regardless of the binary inclination.

5. Conclusions

The morphology of gaseous flows in semidetached binaries was investigated. It was shown that in steady-state 3D solutions the rarefied gas of the circumbinary envelope influences the flow patterns significantly. The gas of the circumbinary envelope interacts with the stream of matter and deflects it. This leads, in particular, to the shock-free (tangential) interaction between the stream and the outer edge of the accretion disc, and, consequently, to the absence of a "hot spot." At the same time the interaction of the gas of the circumdisc halo with the stream results in the formation of an extended shock wave located along the stream edge ("hot line").

Currently, one of the best sources of information about the flow structure in binary systems is the analysis of light curves of CVs. These curves are characterized by the presence of well-known humps in their quiescent state. The positions of the observed humps and changes of their amplitudes allow us to describe the flow structure in CVs. We can get even more information from an analysis of Z Cha, OY Car, V2051 Oph, HT Cas, and IP Peg. For these five CVs in their quiescent state one can see not only an orbital hump in the light curves, but also a so-called "double eclipse" (i.e., the eclipse of the central white dwarf and the hot region at the outer edge of the accretion disc by the red dwarf). Over many years the presence of a hump in a light curve was considered as proof of the existence of a "hot spot" on the accretion disc. However, as new information appeared, it became clear that the standard model precludes the explanation of a number of effects.

In spite of the fact that in the "hot line" model the region of shock energy release is located outside the accretion disc, there are good grounds to believe that this region can be considered in observations as an equivalent of a "hot spot" in the disc. In order to make certain that this assumption is valid and, hence, the "hot line" gasdynamic model is adequate, we synthesize the light curves and compare them with observational data. The proposed "hot line" model was confronted with observations and confirmed by virtue of comparison of synthetic and observational light curves for CVs (Bisikalo et al., 1998b; Khruzina et al., 2001, 2003a,b) and by analysis of synthetic Doppler tomograms (Kuznetsov et al., 2001).

As follows from the present study, the application of numerical approaches to investigations of gas dynamics of mass transfer in close binaries has allowed us to obtain a body of new interesting results that, to some extent, change the standard point of view on flow patterns in semidetached binary systems. We hope that the results presented above can be of use in astrophysical practice and utilized as a basis for subsequent more sophisticated numerical models. Summarizing, we may conclude that numerical simulation of mass transfer in close binary systems is a rapidly progressing branch of astrophysics, and the use of these results makes it possible to make some progress both in the interpretation of observed data and the analysis of the evolutionary status of binary systems.

Acknowledgement

The work was partially supported by Russian Foundation for Basic Research (projects NN 02-02-16088, 03-02-16622), by Science Schools Support Program (project N 162.2003.2), by Federal Programme "Astronomy," by Presidium RAS Programs "Mathematical modeling and intellectual systems," "Nonstationary phenomena in astronomy," and by INTAS (grant N 00-491).

References

Armitage, P.J. and Livio, M.: 1996, *AJ* **470**, 1024.

Batten, A.: 1973, Binary and Multiple Systems of Stars, Pergamon Press, Oxford.

Belvedere, G., Lanzafame, G. and Molteni, D.: 1993, *A&A* **280**, 525.

Bisikalo, D.V., Boyarchuk, A.A., Kuznetsov, O.A. and Chechetkin, V.M.: 1997, *Astron Rep* **41**, 786.

Bisikalo, D.V., Boyarchuk, A.A., Chechetkin, V.M., Kuznetsov, O.A. and Molteni, D.: 1998a, *MNRAS* **300**, 39.

Bisikalo, D.V., Boyarchuk, A.A., Kuznetsov, O.A., Khruzina, T.S. and Cherepashchuk, A.M.: 1998b, *Astron Rep* **42**, 33.

Bisikalo, V., Harmanec, P., Boyarchuk, A.A., Kuznetsov, O.A. and Hadrava, P.: 2000, *A&A* **353**, 1009.

Bisikalo, D.V., Boyarchuk, A.A., Kaigorodov, P.V. and Kuznetsov, O.A.: 2003, *Astron Rep* **47**, 809.

Bisikalo, D.V., Boyarchuk, A.A., Kaigorodov, P.V., Kuznetsov, O.A. and Matsuda, T.: 2004a, *Astron Rep* **48** [astro-ph 0403053].

Bisikalo, D.V., Boyarchuk, A.A., Kaigorodov, P.V., Kuznetsov, O.A. and Matsuda, T.: 2004b, *Astron Rep* **48** [astro-ph 0403057].

Blondin, J.M., Richards, M.T. and Malinowski, M.L.: 1995, *AJ* **445**, 939.

Boyarchuk, A.A., Bisikalo, D.V., Kuznetsov, O.A. and Chechetkin, V.M.: 2002, Mass Transfer in Close Binary Stars, Taylor and Francis, London.

Chakravarthy, S.R. and Osher, S.: 1985, *AIAA Pap.*, N 85-0363.

Crawford, J.A.: 1955, *AJ* **121**, 71.

Einfeldt, B.: 1988, *SIAM J Numer Anal* **25**, 294.

Flannery, B.: 1975, *MNRAS* **170**, 325.

Goncharskii, A.M., Cherepashchuk, A.M. and Yagola, A.G.: 1985, Ill-Posed Problems of Astrophysics, Nauka Academic Press, Moscow (in Russian).

Gorbatskii, V.G.: 1964, *Astron Zh* **41**, 849; *Soviet Astron* **8**, 680.

Gorbatskii, V.G.: 1977, Space Gasdynamics, Nauka Academic Press, Moscow (in Russian).

Hirose, M., Osaki, Y. and Mineshige, S.: 1991, *PASJ* **43**, 809.

Khruzina, T.S., Cherepashchuk, A.M., Bisikalo, D.V., Boyarchuk, A.A. and Kuznetsov, O.A.: 2001, *Astron Rep* **45**, 538.

Khruzina, T.S., Cherepashchuk, A.M., Bisikalo, D.V., Boyarchuk, A.A. and Kuznetsov, O.A.: 2003a, *Astron Rep* **47**, 164.

Khruzina, T.S., Cherepashchuk, A.M., Bisikalo, D.V., Boyarchuk, A.A. and Kuznetsov, O.A.: 2003b, *Astron Rep* **47**, 768.

Kopal, Z.: 1959, Close Binary Systems, Chapman & Hall, London.

Kuiper, G.P.: 1941, *A J* **93**, 133.

Kuznetsov, O.A., Bisikalo, D.V., Boyarchuk, A.A., Khruzina, T.S. and Cherepashchuk, A.M.: 2001, *Astron Rep* **45**, 872.

Lanzafame, G., Belvedere, G. and Molteni, D.: 1992, *MNRAS* **258**, 152.

Lanzafame, G., Belvedere, G. and Molteni, D.: 1994, *MNRAS* **267**, 312.

Lubow, S.H. and Shu, F.H.: 1975, *A J* **198**, 383.

Molteni, D., Belvedere, G. and Lanzafame, G.: 1991, *MNRAS* **249**, 748.

Molteni, D., Bisikalo, D.V., Kuznetsov, O.A. and Boyarchuk, A.A.: 2001, *MNRAS* **327**, 1103.

Murray, J.R.: 1996, *MNRAS* **279**, 402.

Nagasawa, M., Matsuda, T. and Kuwahara, K.: 1991, *Numer Astrophys Jpn* **2**, 27.

Prendergast, K.H.: 1960, *A J* **132**, 162.

Roe, P.L.: 1986, *Ann Rev Fluid Mech* **18**, 337.

Sahade, J. and Wood, F.B.: 1978, Interacting Binary Stars, Pergamon Press, New York.

Sawada, K. and Matsuda, T.: 1992, *MNRAS* **255**, 17P.

Sawada, K., Matsuda, T. and Hachisu, I.: 1986, *MNRAS* **219**, 75.

Sawada, K., Matsuda, T., Inoue, M. and Hachisu, I.: 1987, *MNRAS* **224**, 307.

Shakura, N.I.: 1972, *Astron Zh* **49**, 921; *Soviet Astron* **16**, 756.

Shakura, N.I. and Sunyaev, R.A.: 1973, *A&A* **24**, 337.

Shore, S., Livio, M. and van den Heuvel, E.P.J.: 1994, Interacting Binaries, Kluwer Academic Publishers, Dordrecht.

Struve, O.: 1941, *AJ* **93**, 104.

Taam, R.E., Fu, A. and Fryxell, B.A.: 1991, *A J* **371**, 696.

Warner, B.: 1995, Cataclysmic Variable Stars, Cambridge University Press, Cambridge.

Warner, B. and Peters, W.L.: 1972, *MNRAS* **160**, 15.

Yukawa, H., Boffin, H.M.J. and Matsuda, T.: 1997, *MNRAS* **292**, 321.

DOPPLER TOMOGRAPHY

T.R. MARSH

Department of Physics, University of Warwick, Coventry CV4 7AL, U.K.;
E-mail: t.r.marsh@warwick.ac.uk

(accepted April 2004)

Abstract. I review the method of Doppler tomography which is widely used to understand the complex emission line profile variations displayed by accreting binary stars. Doppler tomography uses the information encoded in line profiles as a function of orbital phase to calculate the strength of emission as a function of velocity, using a process closely related to medical X-ray tomographic imaging. I review applications which have revealed spiral structures in accretion discs, the accretion flows within magnetically-dominated systems and irradiation induced emission in X-ray binary stars. I also cover some of the more recent extensions to the method which variously allow for Roche geometry, modulation of the fluxes and motion at angles to the orbital plane.

Keywords: accretion

1. Introduction

The modeling of the light curves and spectra of binary stars has reached a high level of refinement. If one can understand the physics well enough, it is possible to produce a parameterised model, and then given data, compute the region of parameter space that it supports. While degeneracies are always a problem, the process is nevertheless familiar and well-understood. However, we cannot always apply this process with confidence. The structures of accreting binary stars are many and various, and very often we have little idea of how their brightness varies with position. Consider an accretion disc. We can perhaps assume that it is flat, just as we can assume a star to be spherical or tidally distorted (although a flat disc is not nearly as secure an approximation). The difficulty comes with the surface brightness. Stars have limb and gravity darkening and irradiation, which although not perfectly understood, are pillars of certainty compared to the surface brightness distribution over accretion discs. While parameterisation is still possible (e.g. a power law in radius), it can be misleading when there is no clear reason to suppose any particular pattern *a priori*. A power law in radius for instance can only fit axi-symmetric distributions. While one can add more complexity, it becomes hard to know what to add or when there is enough flexibility. This is especially the case in accreting systems where the usual χ^2 goodness-of-fit is compromised by the erratic variability called flickering – even a correct model may not fit perfectly.

An alternative approach, pioneered for cataclysmic variable stars by Keith Horne (Horne, 1985), is to create a model of almost complete flexibility. For instance, one

Astrophysics and Space Science **296**: 403–415, 2005.
© Springer 2005

can divide up a disc into thousands of small elements and seek to determine the surface brightness of all the elements. Given a fine enough grid, such a model can fit arbitrary brightness distributions, but it will also be highly degenerate. To cope with this, Horne (1985) introduced the crucial ingredient of a regularising function, picking the image of maximum entropy consistent with the data.

Doppler tomography grew from Horne's work, but applies to spectra rather than light-curves. The essential basis of the method is that information about the distribution of line emission is encoded in the emission lines of a system through Doppler shifting. The paper presenting the method (Marsh and Horne, 1988) has been cited by over 200 other papers, and Doppler tomography is now commonly applied to the interpretation of the spectra of accreting binary stars. Doppler tomography has been applied to all types of cataclysmic variables and low mass X-ray binaries and Algols. It has been reviewed by Marsh (2001) and more recently in a series of short papers by Morales-Rueda (2004), Richards (2004), Schwope et al. (2004), Steeghs (2004), and Vrtílek et al. (2004). In this review I will cover the method briefly, but focus more upon the main results of its application.

2. Fundamental Principles

Consider the schematic plot of Figure 1. This shows lines of equal radial velocity over a disc in a close binary, as seen from the perspective of an observer located to the right of the picture. Two orbital phases are shown to make the point that the pattern of equal radial velocity lines is fixed in the observer's frame rather than that of the binary. From the perspective of the rotating frame of the binary, the dipole-field-like pattern rotates with the observer. If Doppler shifting is the dominant broadening mechanism, all emission within a given region of more-or-less equal radial velocity will end up at one particular part of the line profile (see for instance Figure 1 of Marsh (1986) for how this forms the well-known double peaks from accretion discs). Hence spectra tell us the integrated flux over particular regions of

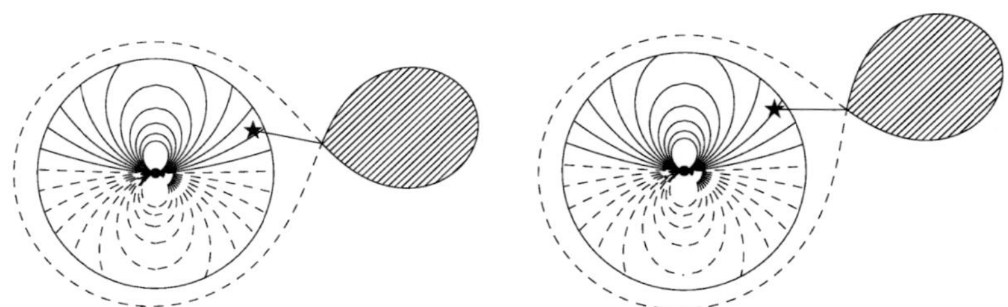

Figure 1. Schematic plots of a cataclysmic variable star, with the observer located off to the right of the images. Two orbital phases are shown, 0.03 (left) and 0.06. The filled circle is a white dwarf and the circle surrounding it is an accretion disc. Lines of equal line-of-sight speed are plotted. The solid lines are red-shifted, while the dashed lines are blue-shifted. The lines are stepped by $100\,\mathrm{km\,s^{-1}}$.

the disc. These regions continuously change orientation, and this can be used to obtain an image.

2.1. VELOCITY SPACE

Viewed in the spatial coordinates of Figure 1, the process of inverting data to obtain an image can seem complex. Things are a great deal simpler if viewed in terms of velocity coordinates. That is, instead of considering intensity as a function of (x, y, z) coordinates, we use (V_x, V_y, V_z) coordinates, where these are components of velocity as measured in an inertial frame that coincides with the rotating frame at orbital phase zero (the latter is usually defined by superior conjunction of the white dwarf when it is furthest from us). The coordinate axes are standardly defined by the direction from the white dwarf to the mass donor (x), the direction of motion of the mass donor (y) and the direction along the orbital axis (z), such that a right-handed set of axes is created. Figure 2 shows the equivalent of Figure 1, but now in velocity coordinates. Each structure plotted in Figure 1 has a well-defined velocity and can therefore be plotted in Figure 2. Take the mass donor for instance. We assume that it is co-rotating with the binary, and therefore is effectively in solid-body rotation with the binary, $\mathbf{v} = \omega \wedge \mathbf{r}$. This transform preserves the shape of the mass donor. Since it moves in the positive y-direction by definition, the mass donor ends up on the positive y axis. The important point about Figure 2 is that the dipole pattern of lines of equal radial velocity becomes a series of straight lines. Moreover, these lines can be plotted over all components, not just the disc. To do the same in Figure 1 would have required multiple sets of lines for the different components. I plot the stream twice in Figure 2: the lower curve, leading from the inner Langrangian point and ending with a star symbol to indicate the impact point with the disc, shows the velocity of the stream directly; the upper curve, also ending in a star, shows the velocity of

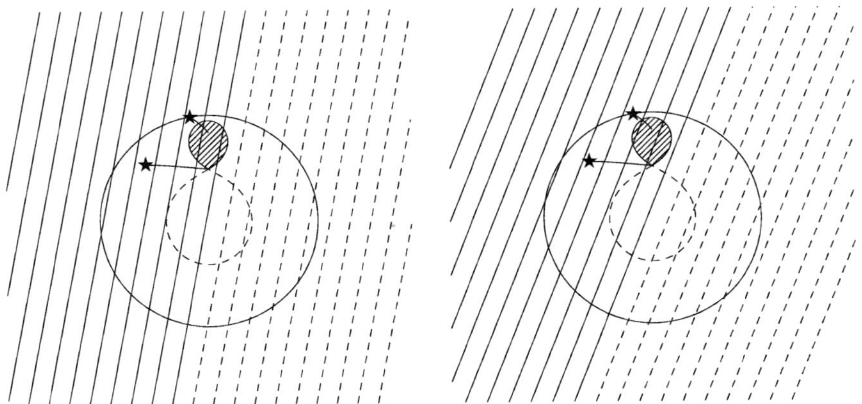

Figure 2. The equivalent of Figure 1 in terms of velocity coordinates.

the disc (assuming it has a Keplerian velocity field) along the path of the stream (somewhat unrealistically, since we don't expect the disc to extend as far as the inner Lagrangian point). There appears to be ambiguity here, however the ambiguity is felt most keenly if one attempts to reconstruct into *position* space, because spectra are measured directly in terms of velocity. Spots appear as sharp peaks which vary sinusoidally in velocity. Such a component can be assigned a unique position in velocity space, but when it comes to its position *we cannot say whether it might be directly from the stream or the disc along the stream without additional information.* This is a fundamental restriction which has lead me always to reconstruct in velocity space, despite starting from position space reconstructions in my thesis.

Provided that Doppler broadening dominates, all emission sources lying between two of the straight lines in Figure 2 will contribute to the same part of the emission line. Effectively the velocity space image is collapsed or 'projected' along the direction defined by these lines, which continuously alters as the binary rotates. A series of emission lines is thus equivalent to a series of projections at different angles of the same image. This is analogous to medical X-ray imaging in which we have a series of projections, measured by integrated optical depth to X-rays, of someone's head. This was a problem solved in the 1970s and goes back to work by Radon (1917). It goes under the name of 'Computerized Axial Tomography' (CAT scanning), or sometimes 'Computed Tomography' for short, and hence 'Doppler Tomography'.

2.2. COMPUTING DOPPLER TOMOGRAMS

There are two main methods used for implementing Doppler tomography. These have been described in some detail elsewhere (Marsh and Horne, 1988; Marsh, 2001), and so I will discuss them only briefly here. The main one presented by Marsh and Horne (1988) is based upon the same maximum entropy regularisation used by Horne (1985) in his 'eclipse mapping'. One divides velocity space into many elements and seeks the image of maximum entropy for a given goodness-of-fit measured with χ^2. This has the advantage of producing a model fit which can be compared directly with the data.

There is also a linear method which directly inverts the integrals defining the emission line formation by projection (the Radon transform). The usual method employed is that of 'filtered back-projection'. This is fast to compute, although speed is usually only an issue if hundreds of maps are being computed. Filtered back-projection is a two-step process in which first each spectrum is filtered. In Fourier space, the filter is proportional to $|k|$, where k is the wave number, and hence this filter enhances high frequencies. The second step is to compute

$$I(V_x, V_y) = \int_0^{0.5} f(\gamma - V_x \cos 2\pi\phi + V_y \sin 2\pi\phi, \phi)\, d\phi, \tag{1}$$

where ϕ is the orbital phase, γ is the systemic velocity and $f(V, \phi)$ is the filtered profile as a function of velocity and orbital phase. This operation is known as 'back-projection,' because it can be viewed as smearing each profile along a direction defined by phase, which is almost the reverse of projection; see Marsh (2001) for details. One of the most useful aspects of the linear method is the intuition one gains from it in trying to understand artifacts. For example, a cosmic ray, if not removed, will lead to a streak across the image at an angle dependent upon the orbital phase of the spectrum affected. Small numbers of spectra also characteristically produce such structures. This is an obvious consequence of back-projection.

2.3. PRINCIPLES OF STANDARD DOPPLER TOMOGRAPHY

Doppler tomography allows for arbitrary brightness distributions, but not arbitrary data. A stationary component displaced from the systemic velocity γ for instance is impossible under the standard assumptions of Doppler tomography. These assumptions are as follows:

1. The visibility of all elements remains constant.
2. The flux of each element is constant.
3. Motion is parallel to the orbital plane.
4. All velocity vectors rotate with the binary.
5. The intrinsic width (e.g. thermal) of the profile is negligible.

It is item 4 which makes a stationary component displaced from γ an impossibility within standard Doppler tomography, for if such a component has velocity $\gamma + V$ at phase ϕ, it should have velocity $\gamma - V$ at phase $\phi + 0.5$.

It is *very common* for one or more of these assumptions to be wrong. A simple variation of emission line flux is one indication of possible problems. Item 1 is perhaps most commonly a problem. I will discuss later an attempt to deal with it. The result of such discrepancies are artifacts in the image. In the interpretation of Doppler tomograms one must always consider this possibility.

3. Applications of Doppler Tomography

In this section, I will review some of the applications of Doppler tomography, concentrating upon those which would have been difficult or perhaps impossible without it. Before looking at these maps some of which are complex, I show in Figure 3, a fairly simple example to illustrate some of the features shown by Doppler maps. The accretion disc produces a broad annulus. The *inner* edge of this annulus corresponds to the *outer* edge of the disc. Normally one does not expect emission from within the annulus except from the component stars. In the particular case shown there is emission from the (slowly rotating) white dwarf, which is not normally seen. As mentioned earlier, the mass donor is shown with its shape

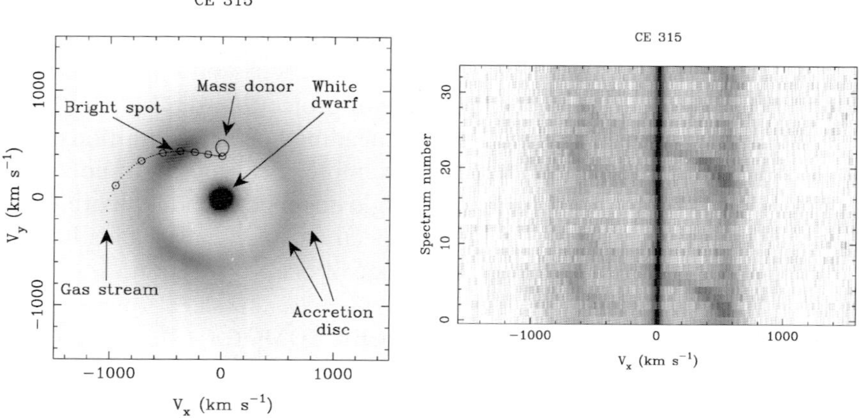

Figure 3. The left panel shows a Doppler map of the star CE 315 which shows many of the structures seen in Doppler images (labelled). The right panel shows the data (phase-folded). This system has an extreme mass ratio and thus the white dwarf is almost at the centre of mass at (0, 0). Both data and map are courtesy of Danny Steeghs.

preserved; we will soon see an example where emission is seen at its location. The gas stream however is not in solid-body rotation, and so is distorted compared to the usual shape in position coordinates. It is important to note that, whether or not distortion occurs, it is possible to make quantitative predictions for the position of any given component. The convention used in this figure is to mark radii along the gas stream by points every 0.01 (dots) and 0.1 (circles) of the inner Lagrangian distance, R_{L1}. In this instance one can see that the gas stream hits the disc at $0.7 R_{L1}$, a measure of the radius of the disc.

3.1. RESULTS

I now go through a few key results from the application of Doppler tomography. I focus upon cataclysmic variables; the reviews mentioned in the introduction cover other types of systems.

3.1.1. *Spiral Structure*

The discovery of spiral structure in the dwarf nova IP Peg during outburst by Steeghs et al. (1997) is a key discovery only made possible by Doppler tomography. As shown in the left-hand panel of Figure 4, the profile changes in this system are complex and hard to interpret. This is caused by significant asymmetry in the brightness pattern, as can be seen in the centre panel, which shows the Doppler map computed from the data. Such structures were predicted in the 1980s to come from tidally driven shocks by Sawada et al. (1986) and Spruit (1987), and so this was the first interpretation of them (Steeghs et al., 1997). More recently, doubt has

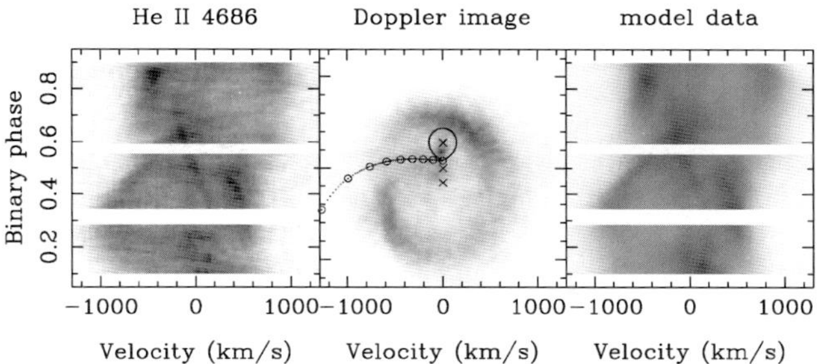

Figure 4. The figure shows trailed spectra (left), with time running upwards, of the dwarf nova IP Peg during outburst. The other panels show the Doppler image (middle) and the fit from the Doppler image (right). Figure taken from Harlaftis et al. (1999).

been cast upon the shock model (Ogilvie, 2002; Smak, 2001) with a combination of variable disc thickness and irradiation being the proposed alternative. As yet there is no clear solution to this problem.

3.1.2. *Emission from the Mass Donor in WZ Sge*

WZ Sge is a key object amongst cataclysmic variable stars. First of all, it is probably the closest of all CVs at only 43 pc (Thorstensen, 2003). It is also of interest for its very long inter-outburst period, which, until the most recent in 2001, was about 33 years. In July 2001, WZ Sge went into outburst only 23 years after its previous high state. Despite its closeness, WZ Sge remains a hard system to pin down, because its short orbital period (81.6 min) implies a very low mass and faint donor star, probably a brown dwarf. As a result, until 2001, the donor had never been detected. This changed during the outburst of 2001, when the donor showed line emission (Steeghs et al., 2001; see Figure 1) presumably as the result of irradiation during the outburst (Figure 5). The distance of the mass donor from the origin is a measure of its radial velocity semi-amplitude, which can be used to place a lower limit on the mass of the white dwarf. This turned out to be much larger than expected.

3.1.3. *Accretion Flows in Polars*

The accreting white dwarfs in some cataclysmic variable stars have magnetic fields strong enough to completely disrupt the accretion disc (surface field strength $\gtrsim 10$ MG). These systems are known as polars. Although in these systems significant motion out of the orbital plane is expected, violating assumption 3 of Section 2.3, Doppler tomography has been applied to them with considerable success. Doppler images of polars have shown emission associated with the gas stream apparently before it is entrained into the magnetic field (the 'ballistic stream'), and also emission from the gas as it flows onto the magnetic poles Schwope et al. (2004).

These features were nicely brought out in the work of Heerlein et al. (1999) (Figure 6), who modelled the accretion in the polar HU Aqr. The problem of

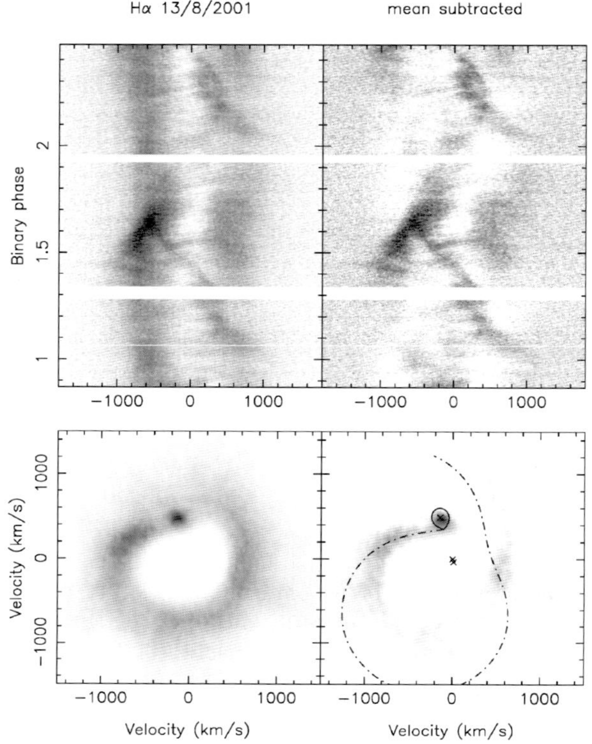

Figure 5. The data (top) and Doppler image (bottom) of WZ Sge taken during its July 2001 outburst. The right-hand panels show the same after subtraction of the symmetric part of the image and its equivalent from the data. Figure taken from Steeghs et al. (2001).

out-of-plane motion remains however. An attempt to account for this is described by Schwope et al. (2004).

3.1.4. *Bowen Fluorescence in Low-Mass X-ray Binary Stars*

The basic parameters of low-mass X-ray binaries (LMXBs) are notoriously difficult to pin down. The mass donor in particular is completely outshone by the accretion disc. It turns out, however, that the mass donors are sometimes visible in the light of the Bowen fluorescence lines near 4640 Å. This lead to the first detection of the mass donor in the famous system Sco X-1 (Steeghs and Casares, 2002). It is arguable whether Doppler tomography was necessary in this case, but it helped with the analysis, and can be invaluable in the case of low signal-to-noise ratios.

Doppler tomography has lead to similar discoveries in cataclysmic variables, and even signs of disc shadowing (Haralaftis et al., 1999; Morales-Rueda et al., 2000), which is seen in the tendency for irradiated emission to concentrate towards the poles of the mass donors. It has also been of great use for study of the gas stream/disc

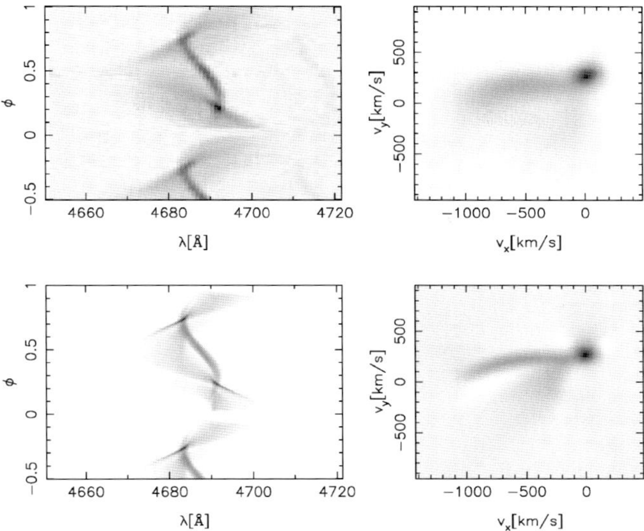

Figure 6. The spectra (top left) and Doppler image (top right) of the polar HU Aqr are plotted together with the equivalent pictures derived from a simple model of ballistic flow followed by magnetic entrainment (bottom panels). Figure taken from Heerlein et al. (1999).

impact in cataclysmic variable stars and black-hole binary stars (Marsh et al., 1990, 1994). Unfortunately there is not space to cover all of these areas, instead I turn to some of the ways in which Doppler tomography is being developed.

4. Extensions to Doppler Tomography

Since the presentation of Doppler tomography by Marsh and Horne (1988), several variations of the method have been presented. These can be split into two classes: those which add more physics, and those which allow for more flexibility. Bobinger et al. (1999) presented a combination of Doppler tomography and eclipse mapping, making use of a Keplerian velocity field to transform between space and position. Schwope et al. (2004) have started to work on simultaneous eclipse and Doppler mapping in polars, where they assume a curtain-like geometry for the magnetically-dominated flow. Both these fall into the first category of adding more physics. It remains to be seen how successful these extensions will be, but they can be criticised for adding additional assumptions, which are hard to be sure of. For instance, it is certainly *not* the case that the velocity field of all components is Keplerian, and the assumption that it is could lead to problems. There is however one case where the assumptions are fairly robust, which is the case of 'Roche tomography' (Rutten and Dhillon, 1994; Watson and Dhillon, 2001), in which the emission or absorption from the mass donor is mapped. The extra physics here is the shape and size of the Roche lobe, which are usually reasonably well constrained. The chief

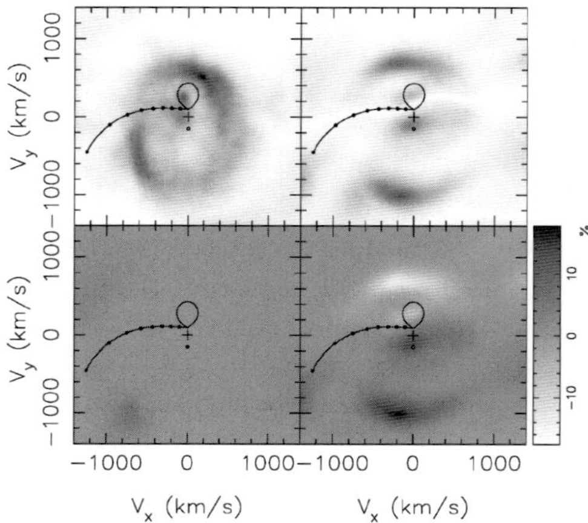

Figure 7. Doppler maps of IP Peg allowing for sinusoidal modulation (Steeghs, 2003). The top-left panel shows the average map (cf. Figure 4), while running clockwise from this are the amplitude of the modulation and the sine and cosine components.

difficulty in this case is obtaining data uncontaminated by emission from accretion flows.

The second approach of more flexibility has been pioneered by Billington (1995), who modelled spectra and eclipses with a fairly general velocity–position relation, and most recently by Steeghs (2003). The latter work allows for sinusoidal variation of the flux from each velocity with orbital phase, thus relaxing the first assumption listed in Section 2.3. This is a very generally applicable method, which Steeghs demonstrates leads to improved fits to data.

Interpretation of the results is tricky, but it does highlight regions of the images, which are likely to be affected by shadowing leading to modulation. For instance, Figure 7, which shows the modulation maps for IP Peg in outburst when spiral structure is present. There appears to be strong modulation close to the spiral arms, indicative of vertical structure causing shadowing.

4.1. ALLOWING FOR OUT-OF-PLANE MOTION

As I have mentioned before, standard Doppler tomography does not allow for motion out of the orbital plane. Can it be modified to cope with this? The short answer is no, because it is possible to explain some line profiles as coming from a disc or equally well from a particular distribution of out-of-plane motion. This problem for example sometimes leads to an ambiguity between emission from a jet or a disc as an explanation for double-peaked emission. However, there are cases where one can deduce that there *must* be out-of-plane motion. I discussed earlier the case of a profile constantly offset from the systemic velocity as something that cannot be fitted using Doppler tomography, but it can easily be explained from

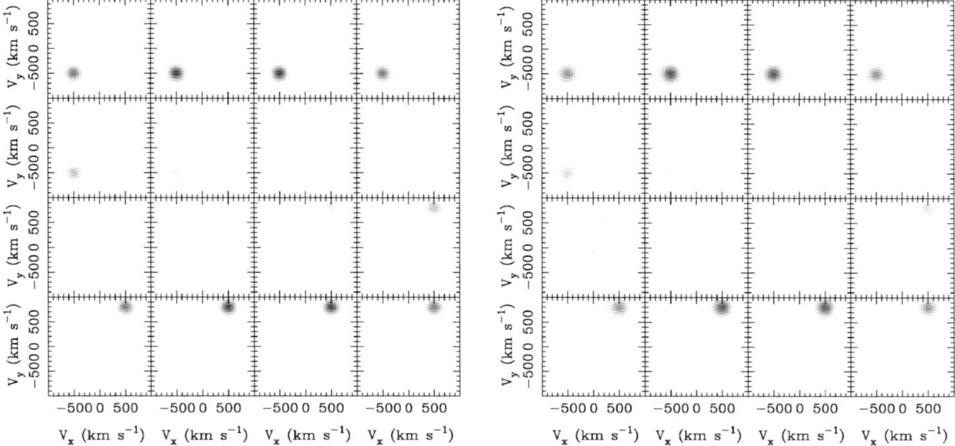

Figure 8. A model and reconstruction allowing for out-of-plane motion. The panels show slices of constant out-of-plane speed, V_z. The model consists of two spots located at $(-500, -500, -300)$ and $(+800, +500, +300)$ in units of km s^{-1}. The slices shown are stepped by 50 km s^{-1} in the V_z component and range from -375 (top-left) to $+375$ km s^{-1} (bottom-right).

out-of-plane motion. This suggests that there is some information on out-of-plane motion.

To test this I have carried out a reconstruction of data computed from a model, which includes out-of-plane motion, as shown in Figure 8. The reconstruction does appear to recover the out-of-plane motion. However, the fake data here had high signal-to-noise, and it is far from clear whether the reconstruction will survive lower signal-to-noise and especially a more realistic distribution of flux. Development of this method may however be useful as an indication of when out-of-plane motion is significant. Ultimately a more prescriptive technique such as that outlined by Schwope et al. (2004) might be preferable, if uncertainties of geometry can be controlled.

5. Observational Requirements for Doppler Tomography

Any review of Doppler tomography would be incomplete without some discussion of the data requirements for Doppler tomography. The key point to appreciate is that to create a map of a given resolution (e.g. 30 km s^{-1}) places requirements upon both spectral *and* time resolution. The time resolution must be such that features do not change their line-of-sight speed by more than the desired resolution during the exposure. For a feature of speed K from the centre of mass, the exposure time Δt must satisfy

$$\Delta t \lesssim \frac{P}{2\pi} \frac{\Delta V}{K}, \tag{2}$$

where ΔV is the velocity resolution desired. If we wish to cut ΔV by a factor 2, then we must cut Δt by the same factor, which would cut the number of photons per pixel by a factor of 4. It is thus very easy to become readout noise limited, even on a large telescope, and also to suffer from slow detector readout speed. Figure 3 is a good example of this: the data were taken with the echelle spectrograph UVES and the 8m VLT, and yet the resolution of the map is limited by the exposure time, which causes significant azimuthal smearing. Even at $V \approx 17.5$, which is bright by some standards, the target, CE 315, was too faint for the VLT and UVES given its orbital period of 65 mins and the high spectral resolution.

6. Conclusion

I have reviewed the method of Doppler tomography, which is now widely used in the interpretation of the spectra of binary stars. Doppler tomography has lead to the discovery of asymmetric structures in accretion discs and revealed details of the gas flow in a variety of types of systems, including magnetically-dominated binaries. There remain areas of both observation and analysis which can be improved. Data collection can be improved both in terms of resolution and time coverage. Analysis is often rather qualitative, and the issue of when a feature is real has rarely been tackled. With a wide range of applicability and much to improve, Doppler tomography will continue to be an essential analysis tool for binary stars for the foreseeable future.

Acknowledgements

I thank Keith Horne and Danny Steeghs for many conversations over the years and Danny for supplying the data for Figure 3. I thank PPARC for the support of a Senior Research Fellowship.

References

Billington, I.: 1995, *Images of Accretion Discs in Cataclysmic Variable Stars.*
Bobinger, A., Barwig, H., Fiedler, H., Mantel, K.H., Simic, D. and Wolf, S.: 1999, *A&A* **348**, 145–153.
Harlaftis, E.T., Steeghs, D., Horne, K., Martín, E. and Magazzú, A.: 1999, *MNRAS* **306**, 348–352.
Heerlein, C., Horne, K. and Schwope, A.D.: 1999, *MNRAS* **304**, 145–154.
Horne, K.: 1985, *MNRAS* **213**, 129–141.
Horne, K. and Marsh, T.R.: 1986, *MNRAS* **218**, 761–773.
Marsh, T.R.: 2001, Doppler Tomography. *LNP Vol. 573: Astrotomography, Indirect Imaging Methods in Observational Astronomy*, pp. 1–26.
Marsh, T.R. and Horne, K.: 1988, *MNRAS* **235**, 269–286.
Marsh, T.R., Horne, K., Schlegel, E.M., Honeycutt, R.K. and Kaitchuck, R.H.: 1990, *ApJ* **364**, 637–646.
Marsh, T.R., Robinson, E.L. and Wood, J.H.: 1994, *MNRAS* **266**, 137+.

Morales-Rueda, L.: 2004, *Astronomische Nachrichten* **325**, 193–196.

Morales-Rueda, L., Marsh, T.R. and Billington, I.: 2000, *MNRAS* **313**, 454–460.

Ogilvie, G.I.: 2002, *MNRAS* **330**, 937–949.

Potter, S.B., Romero-Colmenero, E., Watson, C.A., Buckley, D.A.H. and Phillips, A.: 2004, *MNRAS* **348**, 316–324.

Radon, J.: 1917, *Ber. Verh. Sächs. Akad. Wiss. Leipzig Math. Phys. K1* **69**, 262–277.

Richards, M.T.: 2004, *Astronomische Nachrichten* **325**, 229–232.

Rutten, R.G.M. and Dhillon, V.S.: 1994, *A&A*, **288**, 773–781.

Sawada, K., Matsuda, T. and Hachisu, I.: 1986, *MNRAS* **219**, 75–88.

Schwope, A.D., Catalán, M.S., Beuermann, K., Metzner, A., Smith, R.C. and Steeghs, D.: 2000, *MNRAS* **313**, 533–546.

Schwope, A.D., Mantel, K.H. and Horne, K.: 1997, *A&A* **319**, 894–908.

Schwope, A.D., Staude, A., Vogel, J. and Schwarz, R.: 2004, *Astronomische Nachrichten* **325**, 197–200.

Smak, J.I.: 2001, *Acta Astronomica* **51**, 279–293.

Spruit, H.C.: 1987, *A&A* **184**, 173–184.

Steeghs, D.: 2003, *MNRAS* **344**, 448–454.

Steeghs, D.: 2004, *Astronomische Nachrichten* **325**, 185–188.

Steeghs, D. and Casares, J.: 2002, *ApJ* **568**, 273–278.

Steeghs, D., Harlaftis, E.T. and Horne, K.: 1997, *MNRAS* **290**, L28–L32.

Steeghs, D., Marsh, T., Knigge, C., Maxted, P.F.L., Kuulkers, E. and Skidmore, W.: 2001, *ApJL* **562**, L145–L148.

Thorstensen, J.R.: 2003, *AJ* **126**, 3017–3029.

Vrtílek, S.D., Quaintrell, H., Boroson, B. and Shields, M.: 2004, *Astronomische Nachrichten* **325**, 209–212.

Watson, C.A. and Dhillon, V.S.: 2001, *MNRAS* **326**, 67–77.

INDIRECT IMAGING OF THE ACCRETION DISK RIM IN W CRUCIS

K. PAVLOVSKI[1], G. BURKI[2] and P. MIMICA[3]

[1]*Department of Physics, University of Zagreb, Zagreb, Croatia;*
E-mail: pavlovski@phy.hr
[2]*Observatoire de Genève, Sauverny, Switzerland*
[3]*Max-Planck Institut für Astrophysik, Garching, Germany*

(accepted April 2004)

Abstract. The eclipsing binary W Crucis belongs to the rare group of strongly interacting binaries, which are believed to be shortly after the first and rapid mass transfer between components, prior to the Algol phase. New 7-colour photometric measurements in the Geneva system are presented for this long period binary. Several consecutive cycles were covered, which revealed rather complex light curves with pronounced bumps and asymmetries in the eclipse shoulders, as well in out-of-eclipse brightness. We modelled light curves with a variant of Rutten's 3D eclipse-mapping method, which makes it possible to indirectly image the accretion disk rim. A patchy structure has emerged, which explains the rather erratic and complex light curve and its cycle-to-cycle variations.

Keywords: close binary stars, accretion disk, W Cru

1. Introduction

W Cru (HD 105998) is a member of the rather small group of strongly interacting binaries of the *W Serpentis* class (Plavec, 1980), among which β Lyrae is also a prominent member. It is believed that these binaries are in or shortly after the first and rapid mass transfer process, in which mass reversal between components happens. Many signatures of that almost cataclysmic event are present, but the most important one is the presence of the highly-ionized species in the UV spectra, which can only be explained by an accretion process (cf. Plavec, 1980, and references therein).

W Cru is exceptional among 'serpentids' for its almost 200 days period. Such long period is a great observational challenge, in particular it is difficult to cover the complete light curve. As was shown by Zoła (1996), the light curves could be explained only if an accretion disk obscures the more massive, accreting star. Therefore, it is not visible in the spectrum (Woolf, 1962) and, as a consequence, the deeper minimum is the one caused by the eclipse of the mass losing star by the disk.

2. Observations

Between December 1984 and July 1989 this binary star has been observed with the photometer P7 (Burnet and Rufener, 1979) attached to the 0.7 m Swiss

Astrophysics and Space Science **296**: 417–420, 2005.
© Springer 2005

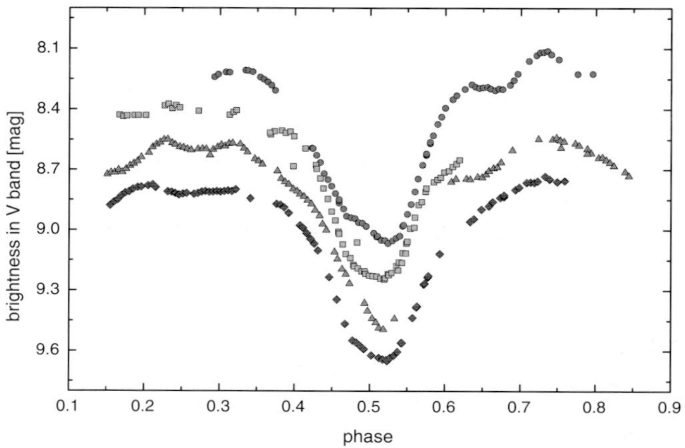

Figure 1. Photometric observations of W Cru in the V passband of Geneva system for four consecutive cycles around the deeper minimum. Peculiarities and complex cycle-to-cycle variations are evident.

Telescope at the ESO La Silla Observatory, Chile, by Dr. Zdeněk Kviz. In total, 388 sets of photometric measurements in the Geneva 7-colour photometric system have been secured. Some initial results from the first season of the photometric follow-up have been reported by Kviz and Rufener (1988). Almost four consecutive cycles of this long-period binary have been covered, and fine details and changes in the light curves are revealed for the first time (cf. portion around the primary minimum as shown in Figure 1). Observations are deposited in the Geneva photometric database and will be published by us in the subsequent paper.

3. Light Curve Modeling and Results

An important progress in the understanding of W Cru has been made by the light curve modeling performed by Zoła (1996) and Daems (1998), respectively. In these two studies a binary model including an accretion disk, which surrounds the more massive component, has been used in the calculations. The synthetic light curves match globally well the shape of the observed light curves, but cannot account for the anomalies which are present. As was already noted by previous observers, and this is also evident by the observations presented in this paper (Figure 1), the light curves are rather complex with unequal maxima, asymmetric eclipse branches and out-of-eclipse variations changing from cycle to cycle. Therefore, the modeling of the light curve(s) has to be done preferably by the eclipse mapping technique (Horne, 1985).

Recently, Mimica and Pavlovski (2003) have applied this technique to reconstruct the surface brightness distribution in V367 Cyg, another *serpentid* binary system. A variant of Rutten's (1998) method of 3D eclipse mapping was applied. The idea of Rutten's approach is to fit the whole light curve instead of only the eclipsed part. In the present work, we have used the same code, but the disk model was replaced by an α–disk model, which is physically more realistic than the previously used torus-like disk model. The optimization routine is based on the genetic algorithm. We found it very efficient, but also very CPU time consuming.

The characteristics of the components and the accretion disk, as derived by our calculations, are in close agreement with previous results (Zoła, 1996; Daems, 1998). The mass ratio we derived, $q = 0.19$, together with the mass function $f(m) = 5.83$ $M_\odot$ (Woolf, 1962), yields the masses $M_1 = 8.2$ $M_\odot$ and $M_2 = 1.6$ $M_\odot$. However, the more massive component is completely hidden by the accretion disk. The disk thickness constrains its radius, $R_1 < 17$ $R_\odot$, and this supports the hypothesis that the more massive component is a mid-B MS star. The disk is not extended to the Roche lobe of the mass-accreting component, but, due to a very large separation between the components, it is quite large ($R_d = 124$ $R_\odot$). After this outline of the binary model, a detailed analysis was performed, i.e. the optimal brightness distribution of the accretion disk elements was calculated. Since the inclination is very high ($i = 88°$), only the accretion disk rim is visible (Figure 2). A rather patchy disk rim emerged from our optimized fitting of the observed light curve in the V passband (Figure 3).

In conclusion, we hope that further work, which is in progress, will shade light on mass transfer processes, which are (still) occurring in this system, as is evident from the cycle-to-cycle variability in its light curve(s).

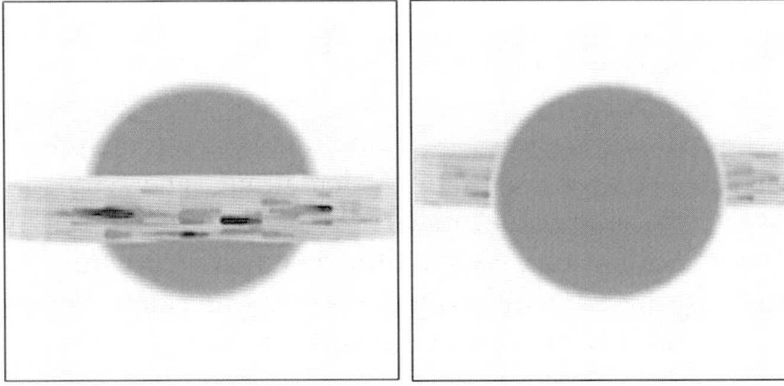

Figure 2. Reconstructed disk rim image of W Cru in the phases of photometric minima; snapshot at left is for the deeper minimum.

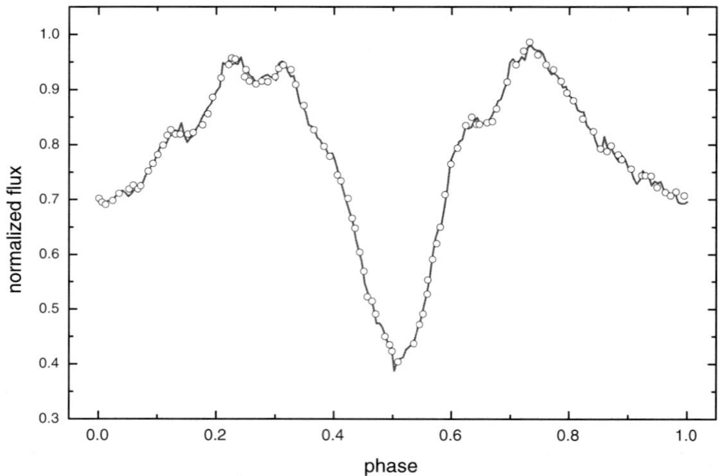

Figure 3. Synthetic light curve (solid line) of W Cru calculated for the optimized disk image compared to the observed light curve (circles) in the V passband of Geneva system.

References

Burnet, M. and Rufener, M.: 1979, *A&A* **74**, 54.

Daems, C.: 1998, PhD Thesis, Katholijke Universiteit Leuven (unpublished).

Horne, K.: 1985, *MNRAS* **213**, 129.

Kviz, Z. and Rufener, F.: 1988, *Inf. Bull. Var. Stars No. 3158*.

Mimica, P. and Pavlovski, K.: 2003, *ASP Conf. Ser.* **288**, 120.

Plavec, M.J.: 1980, *IAU Symp.* **88**, 251.

Rutten, R.G.M.: 1998, *A&AS* **127**, 581.

Woolf, N.J.: 1962, *MNRAS* **123**, 399.

Zoła, S.: 1996, *A&A* **308**, 785.

PRESENT UNDERSTANDING OF THE LIGHT CURVES
OF SYMBIOTIC BINARIES

AUGUSTIN SKOPAL

Astronomical Institute, Slovak Academy of Sciences, 059 60 Tatranská Lomnica, Slovakia;
E-mail: astrskop@ta3.sk

(accepted April 2004)

Abstract. We explain a complex behaviour of the light curves (LC) of symbiotic binaries on the basis of their spectral energy distribution (SED) in the optical domain. During quiescent phases we observe a wave-like variation of the optical light as a function of orbital phase. This is produced by the orbitally-related variation of the emission measure (EM) of the symbiotic nebula. During active phases this type of variability disappears, the continuum level increases by 2–3 magnitudes in U and, in the case of a highly inclined orbit, narrow minima – eclipses – in the LC can occur. These phenomena are connected with the creation of a cool pseudo-photosphere around the hot star, whose contribution is significant in the optical, while the nebula practically disappears.

Keywords: stars, binaries, symbiotic

1. Introduction

Symbiotic stars are long-period interacting binary systems consisting of a cool giant and a hot compact star embedded in a gaseous nebula. Many systems show stages of activity, during which the star's brightness increases by $\approx$2–3 mag.

During so called *quiescent phases* the hot star radiation ionizes a fraction of the neutral giant's wind giving rise to the nebular radiation. As a result the spectrum consists of basically three components of radiation – two of stellar origin (from the hot and the cool star) and a nebular one. In systems with a strong nebula and a late red giant the nebula dominates the UBV passbands (e.g. V1329 Cyg). On the other hand, in systems with earlier type giants (e.g. AG Dra) the nebula becomes more evident only in the U band.

During *active phases* the hot component expands in radius and becomes significantly cooler. As a result the ionized region is suppressed and the source of optical light is restricted mostly to the hot star pseudo-photosphere. The wave-like variation disappears and, in the case of a high orbital inclination, deep narrow minima in the LC, caused by the eclipse of the hot star by its giant companion, are observed.

The aim of this paper is to explain basic characteristics of the LCs of symbiotic binaries on the basis of their composite SED.

Astrophysics and Space Science **296**: 421–425, 2005.
© Springer 2005

2. Light Curves During Quiescent Phases

According to a simple photoionization model for symbiotic binaries (Seaquist et al., 1984) we reconstructed the SEDs for selected systems during quiescent phases to outline the nature of the orbitally-related changes in their LCs. Here we demonstrate this case for the example of V1329 Cyg. Figure 1 shows that the nebular continuum emission (dashed line) is subject to variation with orbital phase of the binary. A maximum/minimum of the nebular emission at $\varphi = 0.57/0.95$ corresponds to a maximum/minimum in the UBV LCs. Skopal (2001) showed that the observed

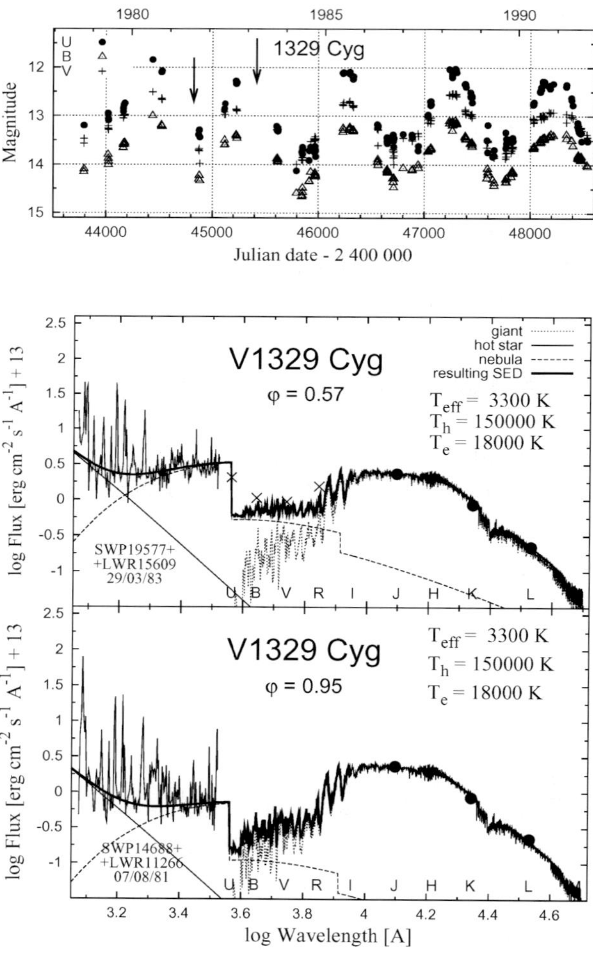

Figure 1. Example of the wave-like orbitally-related variation in the LC of V1329 Cyg during quiescent phase (top). The nature of this type of variability can be understood within an ionization model and its corresponding SED (bottom).

nebular emission is adequately explained by the ionization model and that the variation in EM is fully responsible for the shape of the LCs. Based on these revealings it is possible to reconstruct LCs by converting the observed EM to the scale of magnitudes. The observed nebular flux is connected to EM as

$$F_\lambda^{\rm obs} = \left(\frac{\varepsilon_\lambda}{4\pi d^2} \right) {\rm EM},$$

where d is the distance to the object and ε_λ is the volume emission coefficient. This relation allows us to express the magnitude

$$m_\lambda = -2.5\log({\rm EM}) + {\rm C}_\lambda, \tag{1}$$

where

$$C_\lambda = q_\lambda - 2.5\log\left(\frac{\varepsilon_\lambda}{4\pi d^2} \right),$$

and the constant q_λ defines magnitude zero. Figure 2 shows the comparison of such reconstructed LC in the B band with that directly obtained by standard photometric measurements for the prototypical symbiotic star Z And.

3. Light Curves During Active Phases

Figure 3 shows the LC of CI Cyg covering its major outburst from 1975 and the SED corresponding to a high level of activity. It is an eclipsing system. During the

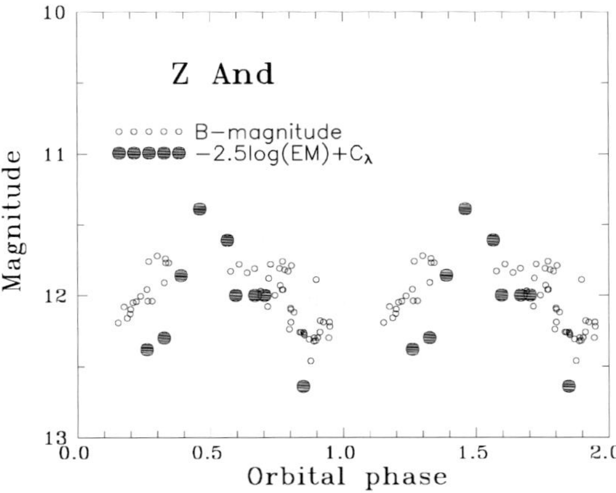

Figure 2. The LCs of Z And in B obtained by standard photometry and from variation in the observed EM (figure taken from Skopal, 2001).

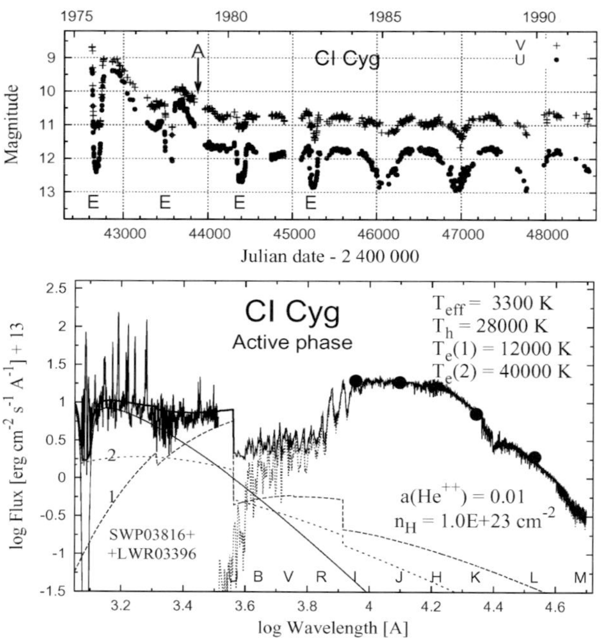

Figure 3. Top: The LC of CI Cyg covering its active phase. Bottom: The SED from the epoch marked by **A** shows a significant contribution of the stellar component of radiation in the optical. As a result we observe narrow minima – eclipses (**E**) – in the active LC at the spectroscopic conjunction of the giant. During the following quiescence profiles of minima are broad. Meaning of the lines as in Figure 1.

active phase (until about 1984) the minima – eclipses of the hot object by the giant – are deep and narrow. From about 1985 the LC displays the periodic wave-like variation, which is typical for quiescent phases of symbiotic stars (Section 2 above). Modeling the SED in activity demonstrates the nature of such behaviour. The stellar component of radiation from the hot object, which is subject to eclipse, has a significant contribution in the optical (Figure 3), which forms a narrow minimum at the spectroscopic conjunction of the giant. The nebular component of radiation is located far from the eclipsed object and its amount does not rival that from the hot object, at least in U. After a few orbital cycles, the cool pseudo-photosphere around the hot star dilutes, and its stellar radiation becomes significantly hotter ($\sim 10^5$ K) with a negligible contribution in the optical – typical properties of quiescent phases (cf. Figure 1, Section 2).

Acknowledgements

This work was supported by Science and Technology Assistance Agency, contract no. APVT-20-014402 and by the VEGA grant no. 2/4014/4.

References

Seaquist, E.R., Taylor, A.R. and Button, S.: 1984, *ApJ* **286**, 202.
Skopal, A.: 2001, *A&A* **366**, 157.

ON THE ASYNCHRONOUS ROTATION OF ACCRETORS IN INTERACTING BINARIES

AUGUSTIN SKOPAL,[1] RICHARD KOMŽÍK[1] and MÁRIA CSATÁRYOVÁ[2]

[1]*Astronomical Institute, Slovak Academy of Sciences, Tatranská Lomnica, Slovakia;*
E-mail: astrskop@ta3.sk
[2]*Faculty of Arts and Natural Sciences, University of Prešov, Prešov, Slovakia*

(accepted April 2004)

Abstract. We show that the gaining component in interacting binaries can rotate faster than its orbital revolution as a consequence of the accretion process. We derive an approximative analytical formula for the Roche lobe radius of asynchronously rotating accretors. We present the case of the semi-detached interacting binary TX UMa, for which we measured directly asynchronous rotation of its accretor. We suggest a method to detect indirectly a fast spinning of accretors in symbiotic binaries based on the the X-ray luminosity of the boundary layer. We demonstrate this possibility for the case of EG And.

Keywords: stars, binaries, asynchronous rotation

1. Introduction

At a certain stage of binary evolution the process of mass transfer from the donor star onto the gaining star (the accretor) can lead to a high activity of the binary including effects of accretion, outbursts, mass outflows, etc. Accretion of part of the matter lost by the donor star can happen by means of Roche lobe overflow or stellar wind. The former mainly concerns short-period binaries (e.g. cataclysmic variables) and the latter takes place in long-period systems (e.g. symbiotic stars). Generally, effects of interaction in a binary system depend on the accretion rate. Part of the energy of the accreted material is used for spinning up the accreting star at its final stage of accretion. As a result, the accretor will rotate faster than the orbital revolution. This situation leads to a shrinking of its critical equipotential surface, which can affect the efficiency of the accretion process.

In this paper, we introduce a possibility of spinning up the accreting star, derive a formula for the approximate Roche lobe radius for asynchronous rotation, and as an example, demonstrate the case of the interacting binaries TX UMa and EG And.

2. Limiting Case for Spinning Up the Accretor

Here we derive the angular velocity of the accretor, $\Omega_*(t)$, as a consequence of mass accretion at the rate $dM/dt = \dot{M}$ for the extreme case, in which the total

Astrophysics and Space Science **296**: 427–430, 2005.
© Springer 2005

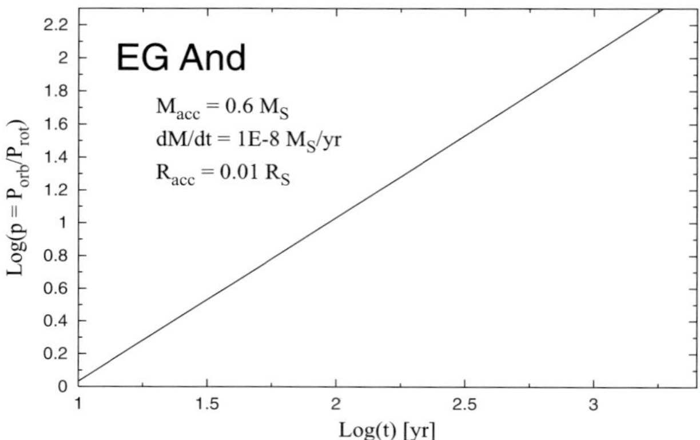

Figure 1. Spinning up of the accretor with time in the symbiotic binary EG And.

kinetic energy of the accreted matter converts into spinning up the accreting star. In addition, we assume that the changes in radius and the star's momentum of inertia due to the mass infall can be neglected. This strong simplification allows us to express the angular velocity of the accretor in the form

$$\Omega_*(t) = \frac{v_K}{R_*}\left[1 - \exp\left(-\frac{5\dot{M}}{2M_*}t\right)\right] + \Omega_0 \exp\left(-\frac{5\dot{M}}{2M_*}t\right),\tag{1}$$

where M_* and R_* are parameters of the accretor and v_K is the Keplerian velocity. Figure 1 shows the case of such spinning up due to a direct-impact accretion for parameters of the symbiotic star EG And for the initial angular velocity $\Omega_0 = 0$ (i.e. the synchronous case).

3. The Roche Lobe Radius for Asynchronos Rotation

In this section, we derive a suitable analytical expression of the Roche lobe radius, R_L, for the case of an asynchronously rotating star in a binary system. For this purpose, we used a generalized binary potential as derived by Kruszewski (1963). For a circular orbit, we found that

$$\frac{R_L}{A} = 0.361 - 0.0904\ln(p) + [0.0710 - 0.0183\ln(p)]\ln(q),\tag{2}$$

which is valid for the mass ratio $0.05 < q < 1$ and the ratio of the orbital to the rotational period $1 < p < 20$. This relation reproduces the numerical values to better than 6%. Example is given in Figure 2.

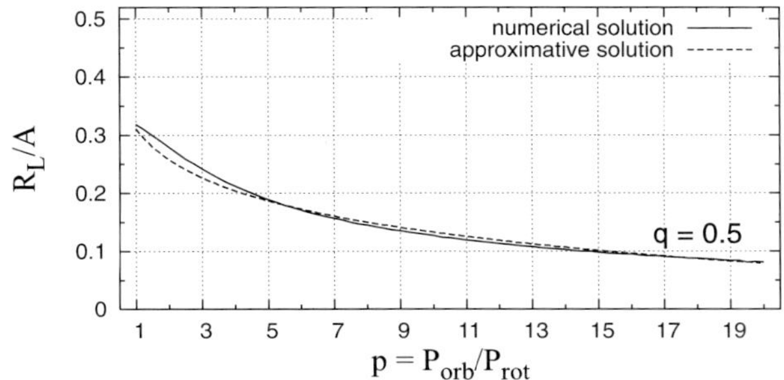

Figure 2. Comparison of the numerically calculated Roche lobe radius with R_L approximated by relation (2); A is the separation of the binary components.

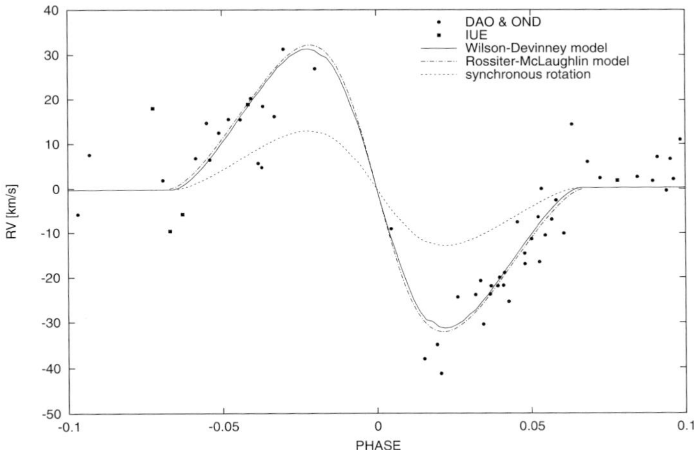

Figure 3. Rotational effect during the TX UMa primary eclipse (Komžík, 1998).

4. TX UMa – A Direct Evidence for Asynchronous Rotation

TX UMa is an Algol-type binary with an orbital period of 3.06 days consisting of a G0 III–IV lobe filling star and a gaining primary of type B8 V (e.g. Albright and Richards, 1993). Based on radial velocities of absorption hydrogen lines, we revealed the effect caused by rotation of the eclipsed component (Figure 3). This result suggests an asynchronous rotation of the primary component in TX UMa with factor $p = 2.39 \pm 0.004$.

5. A Fast Spinning Accretor in EG And?

EG And is a quiet symbiotic binary. Its orbital period is 482 days, and it consists of a red M3 III giant and a white dwarf accreting from the giant's wind with a

                    A. SKOPAL ET AL.

rate of $\dot{M} \sim 10^{-8}\,M_\odot\,\mathrm{yr}^{-1}$ (Kolb et al., 2003). The hot component luminosity is about $60\,(d/590\,\mathrm{pc})^2\,L_\odot$. In this case, there are no direct signatures of a faster than synchronous rotation of the accretor. Therefore, we suggest an indirect method to detect a (possible) fast spinning of the accretor in this long-period system. The approach is based on the luminosity of the boundary layer, L_{BL}, which depends on the angular velocity of the accretor (Popham and Narayan, 1995). For the luminosity of the accretion disk, L_{AD}, we can write the ratio

$$\frac{L_{\mathrm{BL}}}{L_{\mathrm{AD}}} = (1 - u)^2, \tag{3}$$

where $u = \Omega_*/\Omega_{\mathrm{K}}(R_*)$ and Ω_{K} is the Keplerian velocity. For a synchronously rotating star ($\Omega_* = 0$), the ratio $L_{\mathrm{BL}}/L_{\mathrm{AD}} = 1$, but $L_{\mathrm{BL}} < L_{\mathrm{AD}}$ indicates a faster rotation of the accretor ($\Omega_* > 0$). The luminosity of the EG And BL can be estimated from its X-ray luminosity,

$$L_{\mathrm{X}} = 4\pi d^2 \times 1.5\,10^{-10}\eta^{-1}\,\mathrm{erg\,s}^{-1}, \tag{4}$$

derived from a strong He II 1640 Å emission line (we used the HST spectrum Z27E0307N) as proposed by Patterson and Raymond (1985). Assuming that $L_{\mathrm{X}} = L_{\mathrm{BL}}$ and for $L_{\mathrm{AD}} = 60\,L_\odot$ we get

$$\frac{L_{\mathrm{BL}}}{L_{\mathrm{AD}}} = 0.03\eta^{-1} = 0.3{-}0.15 \tag{5}$$

for $\eta = 0.1{-}0.2$ (Vrtilek et al., 1994). With the aid of relation (3), this result implies a very fast rotation of the accreting star in EG And at $\Omega_*(\mathrm{EG\ And}) = 0.3 - 0.5\,\Omega_{\mathrm{K}}(R_*)$.

Acknowledgements

This work was supported by Science and Technology Assistance Agency, contract APVT-20-014402 and by the VEGA grant 2/4014/4.

References

Albright, G.E. and Richards, M.T.: 1993, *ApJ* **414**, 830.
Kruszewski, A.: 1963, *Acta Astron.* **13**, 106.
Kolb, K.J., Miller, J.K. and Sion, E.M.: 2003, AAS Meeting, 202, No. 3908.
Komžík, R.: 1998, Ph.D. Thesis, Tatranská Lomnica (in Slovak).
Patterson, J. and Raymond, J.C.: 1985, *ApJ* **292**, 550.
Popham, R. and Narayan, R.: 1995, *ApJ* **442**, 337.
Vrtilek, S.D., Silber, A., Raymond, J.C. and Patterson, J.: 1994, *ApJ* **425**, 787.

PHOTOMETRIC EVOLUTION OF THE ORBITAL LIGHT CURVES OF THE SLOW NOVA V723 CAS

SERGEI YU. SHUGAROV[1], VITALIJ P. GORANSKIJ[1], NATALY A. KATYSHEVA[1],
ANATOLIJ V. KUSAKIN[1], NATALY V. METLOVA[1], IGOR M. VOLKOV[1],
DRAHOMÍR CHOCHOL[2], THEODOR PRIBULLA[2], EUGENIA A. KARITSKAJA[3],
ALON RETTER[4], OHAD SHEMMER[5] and YIFTAH LIPKIN[5]

[1]*Sternberg Astronomical Institute, Universitetskij Prosp. 13, Moscow, Russia;*
E-mail: sg@sai.msu.ru
[2]*Astronomical Institute, Slovak Academy of Sciences, Tatranská Lomnica, Slovakia*
[3]*Institute for Astronomy, Russian Academy of Sciences, Piatnitskaja str. 48, Moscow, Russia*
[4]*University of Sydney, School of Physics, New South Wales, Australia*
[5]*School of Physics and Astronomy, Wise Observatory, Tel Aviv University, Tel Aviv, Israel*

(accepted April 2004)

Abstract. UBVRI photoelectric and CCD photometry of the slow nova V723 Cas obtained in the years 1995–2003 is presented. The evolution of light curves in 1-year intervals, folded with the orbital period 0.69326 days, shows an increase of the amplitude of the wave-like variations from 0.07 to 1.3 mag during the years 1997–2003. The fact that the shape and amplitude of the orbital light curves does not depend on wavelength is most probably related to the geometry of eclipses combined with the distribution of circumstellar matter in the system.

Keywords: stars, binaries, novae, photometry

1. Introduction

Novae are semi-detached binaries, in which the red dwarf transfers matter to the white dwarf component. The nova outburst is a thermonuclear event on the surface of the white dwarf. As a result, the nova brightens and ejects the envelope.

V723 Cas (Nova Cas, 1995) was discovered by M. Yamamoto on August 1995 (Hirosawa, 1995). It reached the maximum brightness at $V = 7.1$ mag, $R = 6.5$ mag on December 17, 1995. Chochol and Pribulla (1997) classified it as a slow nova with $T_{3,V} = 173^d$, $T_{3,B} = 189^d$. In 1996–1998, the decline was interrupted by small flares on 180-day scale (Chochol and Pribulla, 1998). The orbital period of 0.69325 days was derived by Chochol et al. (2000) by the analysis of photometric observations.

2. Our Photometry and Results

We present our extensive UBVRI photoelectric and CCD monitoring of V723 Cas carried out at the Sternberg Astronomical Institute, Astronomical Institute of the

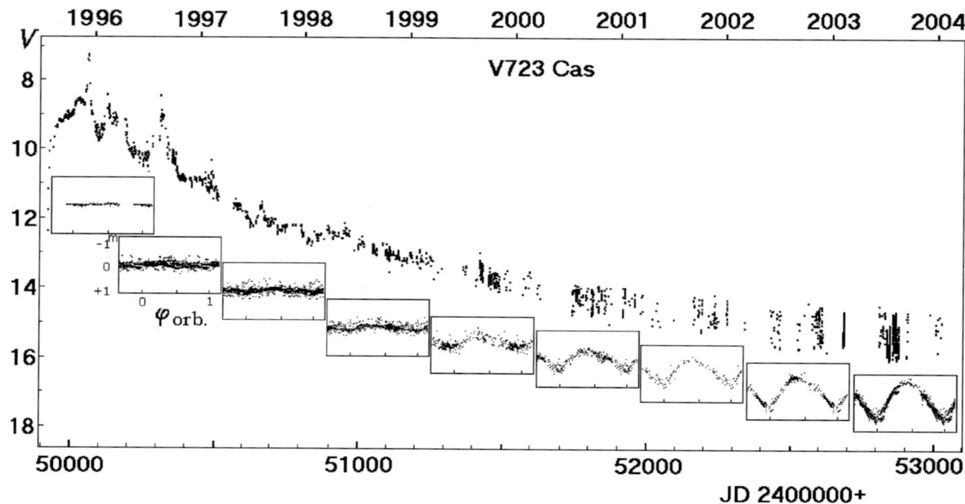

Figure 1. Long-term and orbital light curves of V723 Cas in *V* band.

Slovak Academy of Sciences and Tel Aviv University in the period 1995–2003. The long-term *V* light curve (LC) of V723 Cas is presented in Figure 1. Inserted windows display the evolution of the orbital LCs constructed from non-flare observations in 1-year intervals using the ephemeris:

$$JD_{min.hel.} = 2450421.48 + 0^d.69326 \times E.$$

Regular orbital variability with an amplitude 0.07 mag started to become evident in 1997. The saw-tooth shaped LC was formed in September 1998. The maximum of the orbital LC occurred at orbital phases 0.3–0.4. A dip-like feature resembling a secondary minimum was sometimes detected near phase 0.6, suggesting a high inclination angle of the system. The quasi-periodic oscillation, found by Goranskij et al. (2000), which caused the distortions of the orbital LCs, disappeared in 2000. New observations demonstrated the stable shape of the orbital LC and a gradual increase of its amplitude up to 1.3 mag reached in 2003. At early stages, the luminosity of the envelope exceeded the luminosities of both components and prevented the detection of brightness variations caused by their orbital motion. At later stages, the dissipation of the ejected envelope was responsible for the increase of the amplitude of orbital wave-like variations.

Figure 2 displays the evolution of the multicolor (IRVBU) orbital LCs constructed from non-flare observations in 1-year intervals. They exhibit a slightly asymmetric shape. The fact that the shape and amplitude of the orbital LCs do not depend on wavelength is particularly striking. Such a behaviour cannot be explained by a simple irradiation effect. It is most probably related to the geometry

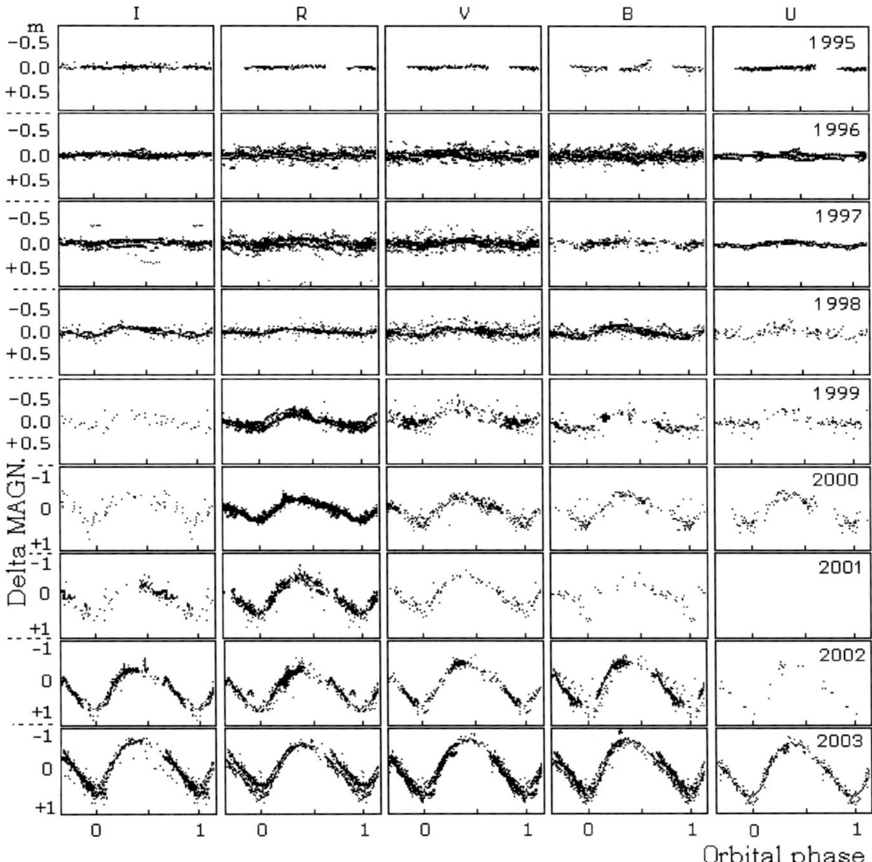

Figure 2. IRVBU orbital light curves of V723 Cas in 1-year intervals.

of eclipses combined with the distribution of circumstellar matter in the system, consisting of a mass transfer stream, a quasi-elliptic accretion disk and a part of the stream moving around the system, as was found by three-dimensional numerical simulations of gaseous flows in semi-detached binaries by Bisikalo et al. (1998).

Acknowledgements

This work was supported by Science and Technology Assistance Agency under the contract no. APVT-20-014402, the VEGA grant no. 2/4014/4, the grants 02-02-16235, 02-02-17524 of Russian Foundation of Basic Researches and Russian President Grant NS-388.2003.3.

References

Bisikalo, D.V., Boyarchuk, A.A., Chechetkin, V.M., Kuznetsov, O.A. and Molteni, D.: 1998, *MNRAS* **300**, 39.

Chochol, D. and Pribulla, T.: 1997, *CoSka* **27**, 53.

Chochol, D. and Pribulla, T.: 1998, *CoSka* **28**, 121.

Chochol, D., Pribulla, T., Shemmer, O., Retter, A., Shugarov, S.Y., Goranskij, V.P. and Katysheva, N.A.: 2000, IAU Circ. No. 7351.

Goranskij, V.P., Shugarov, S.Y., Katysheva, N.A., Shemmer, O., Retter, A., Chochol, D. and Pribulla, T.: 2000, IBVS No. 4852.

Hirosawa, K.: 1995, IAU Cir. No. 6213.

OPTICAL AND X-RAY OBSERVATIONS OF THE SYMBIOTIC SYSTEM AG DRACONIS DURING QUIESCENCE AND OUTBURST*

ROBERTO F. VIOTTI[1], ROSARIO GONZÁLEZ-RIESTRA[2], TAKASHI IIJIMA[3],
STEFANO BERNABEI[4], RICCARDO CLAUDI[5], JOCHEN GREINER[6], MICHAEL
FRIEDJUNG[7], VITO FRANCESCO POLCARO[1] and CORINNE ROSSI[8]

[1]*Istituto di Astrofisica Spaziale e Fisica Cosmica, CNR, Tor Vergata, Roma, Italy;*
E-mail: uvspace@rm.iasf.cnr.it
[2]*XMM Science Operations Centre, ESA, VILSPA, Madrid, Spain*
[3]*INAF–Osservatorio Astronomico di Padova, Sezione di Asiago, Asiago, Italy*
[4]*INAF–Osservatorio Astronomico di Bologna, Loiano Observing Station, Italy*
[5]*INAF–Osservatorio Astronomico di Padova, Vicolo Osservatorio 5, Padova, Italy*
[6]*MPI f. extraterrestrische Physik, Garching, Germany*
[7]*Institut d'Astrophysique, CNRS, 98 bis Bd. Arago, Paris, France*
[8]*Dip. di Fisica, Università La Sapienza, Pz.le Aldo Moro 3, 00185 Roma, Italy*

(accepted April 2004)

Abstract. We review the main characteristics of the symbiotic system AG Draconis, with special emphasis on its optical and X-ray variations. We also discuss the X-ray to visual energy distribution during quiescence and outburst and describe our spectroscopic and X-ray observations during the 2003 outburst.

Keywords: symbiotic binaries, AG Dra, optical spectroscopy, X-rays, outburst

1. Introduction

AG Dra is a spectroscopic binary system with a K2 III primary and a WD whose surface is steadily heated to ~ 1–2×10^5 K by thermonuclear burning on its surface of H-rich matter accreted from the K-star wind (Greiner et al., 1997). The WD emission extends to the soft X-ray range, which was first detected by the Einstein satellite in April 1980 (Anderson et al., 1981). Because of its high X-ray luminosity and very soft spectrum, AG Dra shares the typical properties of the supersoft X-ray sources (Greiner, 1996).

AG Dra is especially known for undergoing irregular phases of activity, which are characterized by a sequence of light maxima (outbursts) separated by ~ 350 d. Many X-ray observations were made by the EXOSAT and ROSAT satellites during quiescence and outburst. These observations showed an intriguing peculiarity of AG Dra: its X-ray flux weakens when the star brightens. This is best illustrated

*Based on X-ray observations collected with the XMM–Newton Observatory, on INES data from the IUE satellite, and on optical spectra collected with the Asiago–Cima Ekar, Bologna–Loiano and La Palma–Galileo Italian telescopes.

Astrophysics and Space Science **296**: 435–439, 2005.
© Springer 2005

by Figure 1, where the X-ray to visual energy distribution of AG Dra is presented for two extreme cases of activity, covered by both X-ray (ROSAT) and UV-space (IUE) observations. During quiescence, the red giant spectrum dominates the visual to red range, while the 'nebular' f–f, f–b emission emerges shortward of ~4000 Å. The WD spectrum comes out in the far UV and contributes with its Wien tail to the soft X-ray range (dashed curve in Figure 1, see Greiner et al., 1997). During the 1994 outburst, the blue-to-UV flux rose dramatically, due to the much-increased contribution of the nebular emission. Conversely, in September 1994, the X-ray flux was much lower than previously observed. The quiescent flux was recovered a few months later, but a new flux minimum was detected by ROSAT during the following 1995 outburst (Greiner et al., 1997). Previously, AG Dra was pointed by EXOSAT during the minor 1985 and 1986 outbursts. A large X-ray fading was again found near the time of the two light maxima (see Viotti et al., 2003).

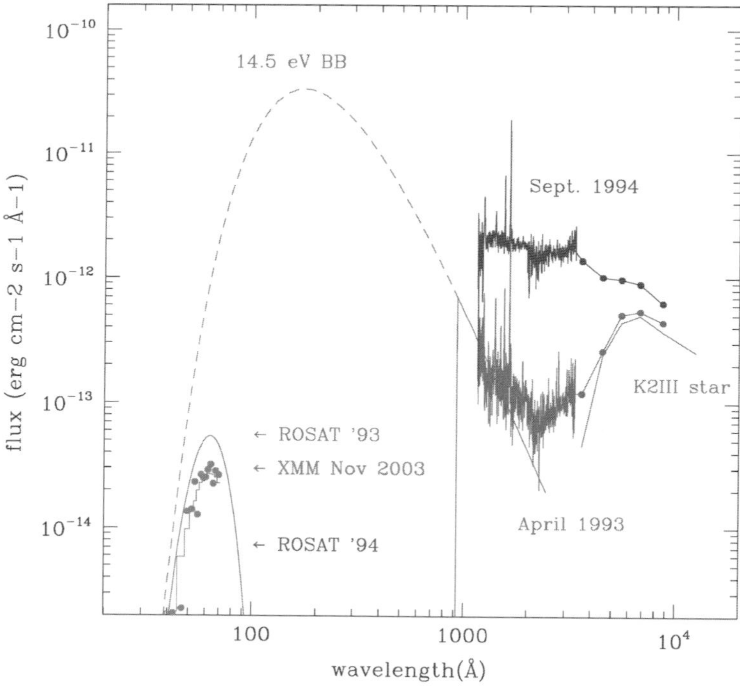

Figure 1. The energy distribution of AG Dra during two different activity phases: quiescence (April 1993) and outburst (September 1994), based on optical photometry, ultraviolet spectra (IUE), and ROSAT observations. The ROSAT maximum X-ray flux at the two times is indicated. The dashed and continuum curves are the unabsorbed and absorbed 14.5 eV black-body spectra of the accreting WD during quiescence, respectively. The XMM observation of November 2003 (dots and histogram) is shown below the ROSAT quiescent spectrum. The energy spectrum of the red component is also shown.

The analysis of the IUE observations led González-Riestra et al. (1999) to identify in AG Dra *cool* and *hot* outbursts. During the cool outbursts of 1994 and 1995, the He II Zanstra temperature of the WD decreased, as the result of a large expansion of the stellar *effective* radius. This resulted in a general shift of the WD energy distribution to lower frequencies, out of the sensitivity range of the X-ray satellites. As for the decrease of the X-ray flux during the 1985 and 1986 hot outbursts, it was attributed by González-Riestra et al. to an increase of the opacity of the WD envelope and wind to the extreme–UV photons, shortward of the N^{+4} ionisation limit.

We are monitoring the optical spectrum of AG Dra at high resolution with the Asiago 1.8 m Cima Ekar and the La Palma 3.5 m Galileo Italian telescopes. We found that the radial velocity of the K-star absorption lines is clearly modulated with the $\sim$550 d orbital period. As for the emission lines, their intensity is strongly variable also during quiescence. The behaviour of the equivalent width of the Hβ and He II 4686 Å emission lines during 1993–2003 is illustrated in Figure 2. Note the relative weakness of the He II line during the *cool* outbursts of 1994 and 1995, while it was stronger during the c4 and c5 outbursts of the previous active phase, and especially at the time of the October 2003 outburst.

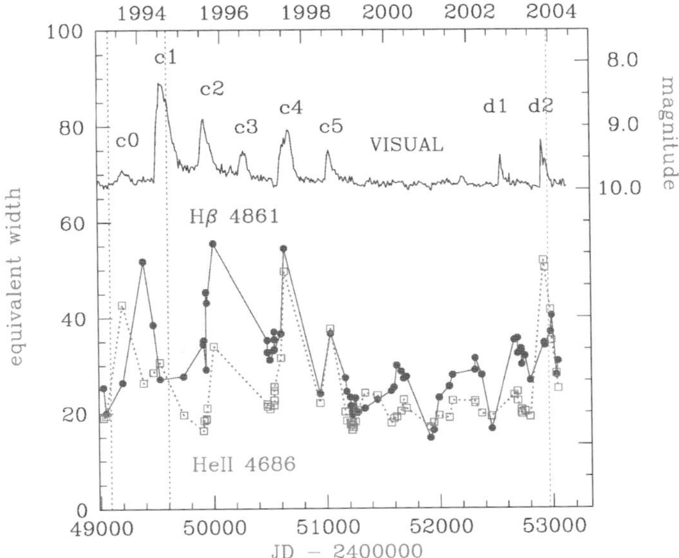

Figure 2. Variations of Hβ (filled circles) and He II 4686 Å (open squares) during 1993–2003 (Asiago Observatory). The visual light curve (from AAVSO) is shown for comparison at the top of the figure. The outbursts of the last two activity phases are marked following the denomination of González-Riestra et al. Vertical lines indicate the times of ROSAT and XMM observations reported in Figure 1.

2. The October 2003 Outburst

By the end of September 2003, AG Dra began a new luminosity outburst (d2) and reached $V \approx 9.2$ on October 1. The spectrum of AG Dra was obtained on October 6 at the Bologna–Loiano Observatory and is shown in Figure 3. Besides the prominent emission lines of hydrogen (Balmer and Paschen series), He I and He II, the spectrum displayed a remarkable Balmer continuum emission and strong and broad red O VI Raman scattering lines at 6825 and 7082 Å. The large He II 4686 Å/Hβ ratio (see Figure 2) seems to indicate that near the peak of the 2003 outburst the WD component of the system was hotter than during quiescence, a behaviour similar to that of the hot outbursts of 1985 and 1986.

X-ray observations were performed with the XMM–Newton Observatory on November 1, 2003, 5 weeks after the outburst maximum, when AG Dra had faded to $V = 9.6$. AG Dra has been clearly detected with XMM–EPIC in the 0.1–0.7 keV range. As shown in Figure 1, the flux of AG Dra was very soft, with an energy distribution similar to, but weaker than that observed by ROSAT during quiescence (see Greiner et al., 1997), and is well represented by a black-body temperature of $kT = 14.5\,\mathrm{eV}$ (170 000 K) and $\log N(H) = 20.5$ (the histogram in the figure). Although our XMM observations were carried out more than one month after the

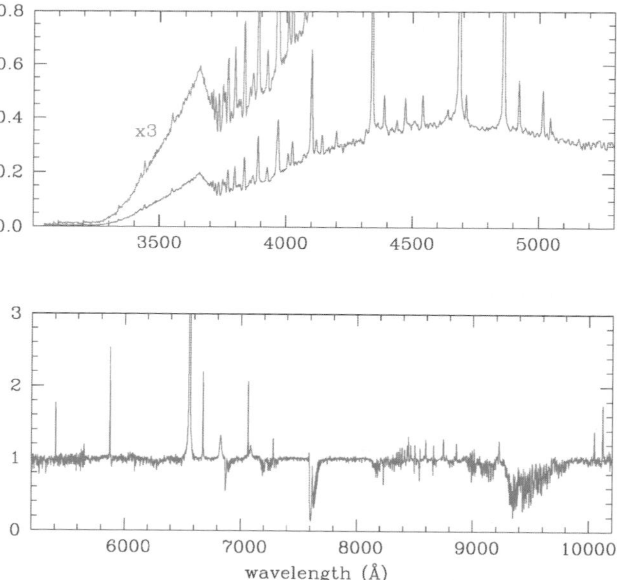

Figure 3. The near-UV–to–near-IR spectrum of AG Dra during the October 2003 outburst (1.5 m Cassini telescope, Loiano). Note the strong Balmer continuum emission. He II 4686 Å is about 60% stronger than Hβ.

outburst peak, the X-ray flux was still much weaker than during quiescence, which confirms the visual/X-ray anti-correlation observed during the previous outbursts.

3. Discussion and Conclusions

During quiescence, the WD luminosity is very high, around that of the K-giant companion (e.g. Greiner et al., 1997). This high luminosity is explained by thermonuclear burning, on the WD's surface, of H-rich matter accreted from the K-giant wind at quite a high rate (around 3×10^{-8} $M_\odot$ yr^{-1} for a distance of 2.5 kpc, Greiner et al., 1997). During all the five outbursts so far observed with X-ray satellites, the soft X-rays weakened independently of the outburst's peak luminosity. The mechanism producing the outbursts is still unknown, but it could be associated with the second $\sim$350 d variability cycle of AG Dra, which is sometimes attributed to pulsations of the K star. According to González-Riestra et al. (1999), the decrease of the X-ray flux during outbursts can be attributed to two mechanisms depending on the outburst type. During the weaker 1985, 1986 and 2003 *hot* outbursts, the WD was surrounded by a hot envelope with a large EUV brightness, but opaque for soft X-rays. During the strong 1994 and 1995 outbursts, the envelope has extended to several stellar radii, with a lower effective temperature and EUV brightness. However, we cannot exclude that, during the earlier phases of the stronger outbursts, AG Dra passed through a *hot* phase, as it is suggested, for instance, by the IUE observations of the rising phase of the 1980 and 1994 outbursts. Indeed, our knowledge of the nature of AG Dra largely benefitted from the very extended monitoring of the ultraviolet spectrum of AG Dra with the IUE satellite. An important step forward would certainly be provided by new far-UV observations, especially during the outburst rising phases.

Acknowledgements

We acknowledge with thanks the variable star observations from the AAVSO International Database contributed by observers worldwide and used in this research.

References

Anderson, C.M., Cassinelli, J.-P. and Sanders, W.T.: 1981, *Astrophys. J.* **247**, L127–L130.
González-Riestra, R., Viotti, R., Iijima, T. and Greiner, J.: 1999, *A&A* **347**, 478–493.
Greiner, J.: 1996, in: J. Greiner (ed.), *Supersoft X-Ray Sources, Lecture Notes in Physics* **472**, 299.
Greiner, J. et al.: 1997, *A&A* **322**, 576–590.
Viotti, R.F., Montagni, F., Maesano, M., et al.: 2003, in: R.L.M. Corradi, J. Mikolajewska and T.J. Mahoney (eds.), Symbiotic Stars Probing Stellar Evolution, *ASP Conf. Ser.* **303**, 298–302.

TOWARDS AN UNDERSTANDING OF RADIAL VELOCITY SHIFT IN UV SPECTRA OF SYMBIOTIC STARS

JOANNA MIKOŁAJEWSKA[1] and MICHAEL FRIEDJUNG[2]

[1]*N. Copernicus Astronomical Center, Bartycka, Warsaw, Poland; E-mail: mikolaj@camk.edu.pl*
[2]*Institut d'Astrophysique, Paris, France*

(accepted April 2004)

Abstract. A systematic redshift of high ionization resonance emission lines relative to the intercombination lines was found by Friedjung et al. (1883, *A & A* **126**, 4071983) in the UV spectra of a number of symbiotic binaries. The interpretation was not then clear. We present a study of archival IUE and GHRS/HST spectra of the symbiotic binary CI Cyg. The shift varies during the orbital cycle which can be understood in terms of the presence of strong circum-binary line absorption formed in an outer expanding region.

Keywords: symbiotic binaries, circumstellar matter, mass loss, CI Cyg

1. Introduction

As already mentioned by other speakers, symbiotic binaries contain a cool giant component, transfering mass to a much hotter compact component, believed to be usually a white dwarf. Spectra of these objects also show the presence of a third "nebular" conponent, produced by the ionized wind of the cool giant, while in some cases other components can also contribute to this nebular emission. The light curves and spectra go through epochs of activity with outbursts, whose nature is badly understood for most objects.

In a study of the ultraviolet IUE spectra of a number of symbiotic binaries, Friedjung et al. (1983) found a systematic redshift of the zero-volt optically very thick emission lines of highly ionized atoms ($C\,IV$, $N\,V$ and $Si\,IV$) with respect to the intercombination lines, which should be optically thin. The interpretation was not then clear. There appeared to be two possibilities, involving either the presence of blueshifted absorption components, absorbing the blue wings of the zero-volt lines, or a radiative transfer effect which occurs when photons of an optically very thick line suffer multiple scattering in an expanding medium.

In order to better understand the shift, we have examined the high resolution IUE spectra of the eclipsing symbiotic binary CI Cyg at different orbital phases from the INES archive, as well as GHRS/HRS archival spectra of this object.

Astrophysics and Space Science **296**: 441–444, 2005.
© Springer 2005

2. Results

We have examined the orbital profile and radial velocity variations. The profiles of different lines are plotted in Figure 1 and radial velocity variations in Figure 2. The C IV line plotted is the stronger 1548 Å line of the doublet, the weaker line behaves similarly, but has a much more noisy profile.

Figure 2 shows that He II traces the hot component, but the systemic velocity is blueshifted by about 18 km/s with respect to the solution for the M giant of Kenyon et al. (1991). Combining its curve with that of the cool component, suggests a mass ratio of 2.45, consistent with other estimates. The intercombination line radial velocities are also in phase with the hot component, but the amplitude, $K \sim 7$ km/s, is much lower, while the systemic radial velocity is also ~ 7 km/s.

The most striking aspect of Figure 2 is however the phase variation of the velocity difference resonance – intercombination lines. It traces the velocity variation of the cool component, with a systemic velocity difference of ~ 23 km/s^{-1}. In order to understand better what is going on, we can look at Figure 1. It is clear that the blue part of the C IV 1548 Å profile is strongly absorbed and, as the underlying

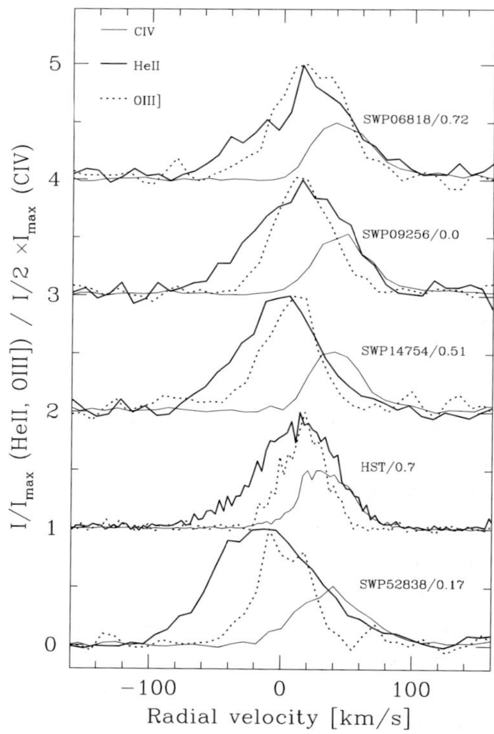

Figure 1. 1548 Å C IV, 1640 Å He II and 1666 Å O III emission line profiles at different orbital phases.

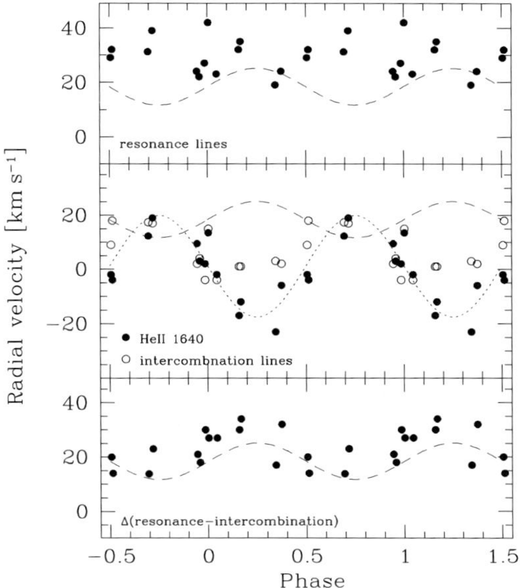

Figure 2. Orbital radial velocity variations with phase of the orbital cycle. The dashed line repeats the orbit of the cool component and the dotted line – the hot component solution.

continuum is extremely weak, we conclude that it is the line radiation which is absorbed.

3. Discussion

Absorption of the C IV doublet must occur in an expanding region with a maximum velocity of the order of at least 100 km/s. Estimating very approximately an optical thickness of not less than 3 for the weaker 1551 Å line and assuming that almost all C^{+2} is in the ground state, we obtain a minimum column density of 7.7×10^{14} cm^{-2} for C^{+2}. Supposing all carbon three times ionized, leads for solar abundances to a minimum hydrogen column density of 2×10^{18} cm^{-2}.

The phase variation of the velocity difference is understandable if the expanding region, responsible for the absorption is circum-binary and does not move with either stellar component. Then if the region emitting the highly ionized atom zero-volt emission lines is not the same, but moves with the hot component, the blue wings of the emission lines will be more absorbed, when the hot component is approaching. The cool component will be receding at that time and the emission lines will be most redshifted. At the opposite phase the emission lines will be less absorbed and less redshifted and the cool component will be approaching.

It is interesting to estimate the size of the C IV emitting region. The ionization energy for C IV is lower than for He II. The latter is formed in a photoionized region according to Kenyon et al. (1991), so we may suppose that the resonance lines of C IV are produced in such a region. The sum of emissivities of these collisionally excited lines is given in Table VI of Nussbaumer and Schmutz (1983), and the radius of a spherical region emitting the total flux of this zero-volt multiplet in Eq. (11) of the same paper. In the present calculation we take $d = 2$ kpc, $E(B - V) = 0.4$ (Mikołajewska and Ivison 2001, and references therein) with the reddening curve of Seaton (1979), while following Nussbaumer and Schmutz, we take the relative abundance of C^{+3}/C equal to 0.5 for a radiation temperature of more than 80 000 K. We measure a total flux of the C IV multiplet of 6×10^{-12} erg cm^{-2} s^{-1} on MJD 44831 (phase 0.506), so we obtain for $N_e = 10^{10}$ cm^{-3} (Kenyon et al., 1991), $r = 1.2 \times 10^{13}$, 4.6×10^{12} and 2.9×10^{12} cm assuming electron temperatures of 10 000, 15 000 and 20 000 K, respectively. This radius can be compared with the radius of the cool giant of 1.6×10^{13} cm (Kenyon et al., 1991).

The nature of the expanding region is not clear. Its velocity seems to be too high to be that of the wind of the cool component. Blueshifted absorption components were seen during activity (Gravina, 1975). His velocity of 20 km/s is also too low; excited atoms of the cool component wind may then have been detected. CI Cyg is an eclipsing system with an inclination of 73 ° (Kenyon et al., 1991), so it appears unlikely that an accretion disk wind is involved.

Acknowledgements

This research was partly supported by KBN Research Grant No. 5 P0 019 20, and by the Polish–French program LEA Astro-PF. It also made use of the IUE-INES and HST archives.

References

Friedjung, M., Stencel, R.E. and Viotti, R.A.: 1883, *A&A* **126**, 407.
Gravina, R.: 1975, *C.R. Acad. Sci. Paris, Série B* **280**, 115.
Kenyon, S.J., Oliverson, N.A., Mikołajewska, J., Mikołajewski, M., Stencel, R.E., Garcia, M.R. and Anderson, C.M.: 1991, *AJ* **101**, 637.
Mikołajewska, J. and Ivison, R.J.: 2001, *MNRAS* **324**, 1023.
Nussbaumer, H. and Schmutz, W.: 1983, *A&A* **126**, 59.
Seaton, M.J.: 1979, *MNRAS* **187**, 73.

PRELIMINARY PHOTOMETRIC RESULTS FOR THE 2003 ECLIPSE OF EE CEP

M. MIKOLAJEWSKI[1], C. GALAN[1], K. GAZEAS[2], P. NIARCHOS[2], S. ZOLA[3],
M. KURPINSKA-WINIARSKA[3], M. WINIARSKI[3], A. MAJEWSKA[3], M. SIWAK[3],
M. DRAHUS[3], W. WANIAK[3], A. PIGULSKI[4], G. MICHALSKA[4],
Z. KOLACZKOWSKI[4], T. TOMOV[1], M. GROMADZKI[1], D. GRACZYK[1],
J. OSIWALA[1], A. MAJCHER[1], M. HAJDUK[1], M. CIKALA[1], A. ZAJCZYK[1],
D. KOLEV[5], D. DIMITROV[5], E. SEMKOV[5], B. BILKINA[5], A. DAPERGOLAS[6],
L. BELLAS-VELIDIS[6], B. CSAK[7], B. GERE[7], P. NEMETH[7] and G. APOSTOLOVSKA[8]

[1]*Uniwersytet Mikolaja Kopernika, Centrum Astronomii, ul. Gagarina 11, 87-100 Torun, Poland;*
E-mail: mamiko@astri.uni.torun.pl
[2]*Department of Astrophysics, Astronomy and Mechanics, Faculty of Physics, University of Athens,*
Panepistimiopolis, GR-15784 Zografos, Athens, Greece
[3]*Astronomical Observatory, Jagiellonian University, ul. Orla 171, 30-244 Krakow, Poland*
[4]*Astronomical Institute, Wroclaw University, ul. Kopernika 11, 51-622 Wroclaw, Poland*
[5]*National Astronomical Observatory Rozhen, Institute of Astronomy, BAS, PO Box 136,*
4700 Smolyan, Bulgaria
[6]*Institute of Astronomy and Astrophysics, National Observatory of Athens, PO Box 20048,*
11810 Athens, Greece
[7]*Department of Experimental Physics and Astronomical Observatory, University of Szeged,*
Dom ter 9, H-6720 Szeged, Hungary
[8]*Institute of Physics, Faculty of Natural Sciences, PO Box 162, 1000 Skopje, Macedonia*

(accepted April 2004)

Abstract. We report multicolour photometric observations of the 2003 eclipse of the long-period (5.6 yr) eclipsing binary EE Cep. Measurements were obtained with ten telescopes at eight observatories in four countries. In most cases, $UBV(RI)_C$ broad band filters have been used. The light curve shape shows that the obscuring body is an almost dark disk around a low-luminosity central object. However, variations of the colour indices during the eclipse indicate that the obscuring body emits a considerable amount of radiation in the near infrared.

Keywords: binaries, eclipsing-stars, individual, EE Cep

1. Introduction and Observations

EE Cep is an unique, long-period (5.6 years) binary system, in which the primary component is a B5 III star and the eclipsing body is most probably a dark opaque disk around an invisible companion (Mikolajewski and Graczyk, 1999). The secondary eclipse is not observed (Meinunger, 1973), while the primary minimum demonstrates very large variations in duration ($30 \div 60$ days) and depth ($0\overset{m}{.}6 \div 2\overset{m}{.}1$) from epoch to epoch.

Astrophysics and Space Science **296**: 445–449, 2005.
© Springer 2005

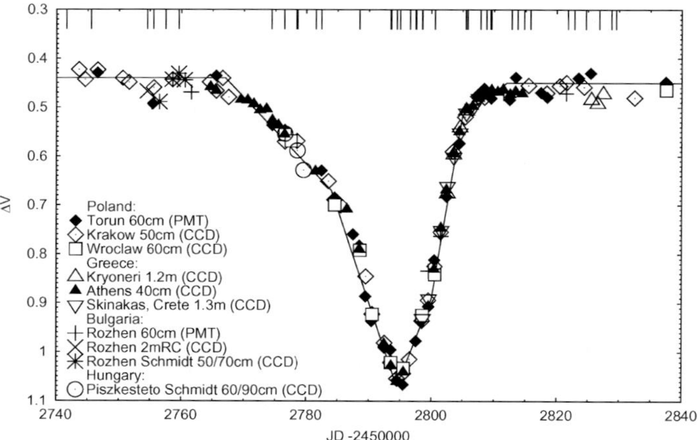

Figure 1. The *V* light curve and the schematic shape of the primary minimum of EE Cep in 2003. Marks at the top denote all collected spectroscopic observations (see Mikolajewski et al., this volume).

The recent minimum of EE Cep took place during May–June 2003. The appeal of Mikolajewski et al. (2003) aimed at a precise covering of the eclipse with photometric and spectral observations. As a result, we have collected the most numerous $UBV(RI)_C$ photometric data so far, from ten telescopes in four countries (Figure 1). Most of the telescopes were equipped with CCD cameras, two of them (Torun and Rozhen 60 cm) were operating with single-channel aperture photometers equipped with photomultipliers (PMTs). The photometric data were shifted to eliminate the systematic off-set between particular instrumental systems. The shifts were calculated using data obtained on the same night. As a zero point, we have adopted the most numerous and the best-quality data from Krakow and Athens. Low-resolution, flux-calibrated spectra were obtained with the Canadian Copernicus Spectrograph (CCS) at the 90 cm telescope of the Torun observatory.

2. Discussion

The *V*-band data collected during the 2003 eclipse are presented in Figure 1. Despite the fact that the minimum was about 1^{m}0 shallower than most of the previous ones, the same six characteristic contacts, dividing the eclipse into five different phases, can be distinguished (compare with Figure 2 in Mikolajewski et al., 2003). These contacts can be easily understood if the obscuring body is a highly inclined opaque disk inside an elongated semi-transparent envelope (Graczyk et al., 2003). The projection of this disk forms a characteristic oblong shape tilted to the direction of its motion during the eclipse.

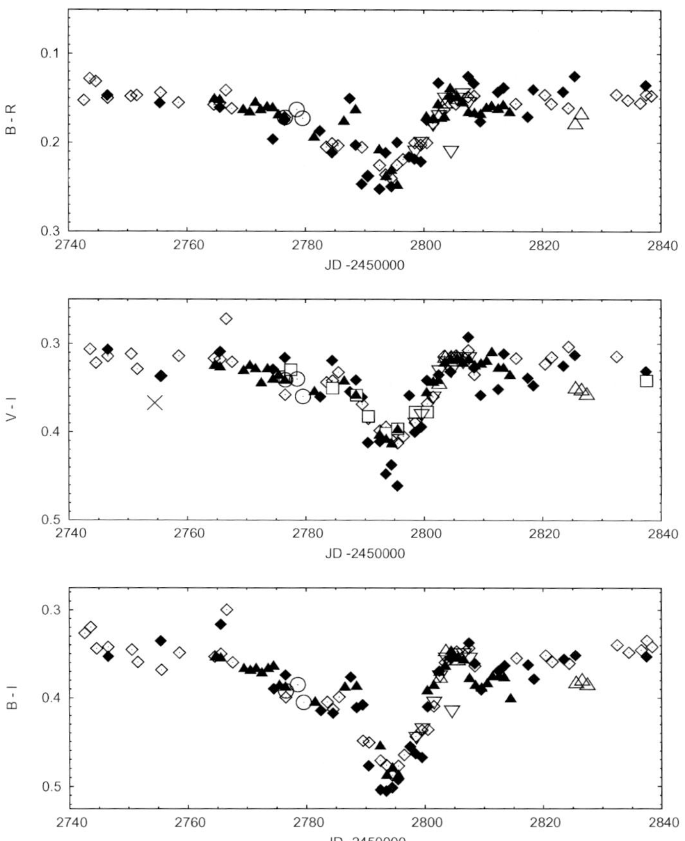

Figure 2. The variations of $B - R$, $V - I$ and $B - I$ colour indices during the 2003 eclipse. Different symbols are the same as in Figure 1.

The large asymmetry between the duration of ingress and egress is caused by the tilt of the obscuring body provided that the eclipse is not central. During a non-central transit, the eclipsed surface of the B star varies significantly, causing the characteristic slope-bottom phase. Changes of the tilt angle produce variations in the duration and slope of the ingress, egress and transit phases, from eclipse to eclipse. The effective thickness of the projected disk also changes during different eclipses, causing variations in the depth of minimum. The precession of the disk can change both the inclination of the disk to the line of sight and the tilt of its cross-section to the transit direction. According to the model described above, the phases of the 2003 eclipse distinguished in Figure 1 can be interpreted as follows:

– the relatively extended atmospheric part of ingress (17 days);
– the ingress phase – a partial eclipse by an opaque oblong body (9 days);

TABLE I

The $UBV(RI)_C$ amplitudes of the 1997 and 2003 eclipses

Year	ΔB	$\Delta(U-B)$	$\Delta(B-V)$	$\Delta(V-R)$	$\Delta(R-I)$
1997	1.65	0.08	0.07	0.10	0.03
2003	0.68	0.04	0.05	0.05	0.05

– the transit of the opaque body – the slope-bottom phase (6 days);
– the sharp egress – a partial eclipse again (5 days);
– the short atmospheric egress (7 days).

The previous eclipse occurred in 1997. During this eclipse, multi-colour $UBVRI$ observations of EE Cep were obtained for the first time (Mikolajewski and Graczyk, 1999). Both the 1997 and the 2003 minimum appeared to be almost grey, however a slight reddening of about $0.^{m}03$–$0.^{m}10$ between the consecutive photometric bands can be noticed (Table I). The selective extinction (with a reddening law similar to that of interstellar matter) in semi-transparent parts of the obscuring cloud can explain the reddening only during the atmospheric phases of the eclipse, while the largest reddening is observed during the slope-bottom phase, i.e. the transit of the opaque body (Figure 2).

An alternative explanation of the observed reddening can be the domination of the limb-darkened regions of the B star photosphere during transit of an oblong body. We have estimated the influence of the limb darkening effect during both minima using nonlinear logarithmic and square-root limb-darkening laws. The limb-darkening coefficients for all passbands were taken from the tables published by Van Hamme (1993). Adopting $T_{\text{eff}} = 15\,000$ K for the temperature of the eclipsed B5 III star and $\log g = 3.5$ for its surface gravity, the calculated influences of the limb darkening are very small: $0.^{m}007$, $0.^{m}014$, and $0.^{m}021$ for the 2003 eclipse; and $0.^{m}019$, $0.^{m}038$, and $0.^{m}055$ for the 1997 eclipse, in V, R, and I, respectively.

Another possible interpretation of the decreasing amplitude towards red wavelengths (Table I) is a remarkable amount of radiation from the obscuring body, detected mostly in the R_C and I_C bands. Neglecting both selective extinction and limb darkening, we assumed the same obscuring factor for all photometric fluxes estimating it from the eclipse depth in U light. For the amplitude $\Delta U = 0.^{m}72$ (Table I), there are noticeable differences between the observed and calculated $BVRI$ fluxes in mid-eclipse (Figure 3). Also our low resolution spectrum F_{mid} has a similar excess in comparison to the calculated one, using that observed close to the end of minimum and $\Delta U = 0.^{m}55$. The obtained residual spectrum (F_{res} in Figure 3) can be interpreted as a G–K photosphere of the disk edge or its partly obscured central star.

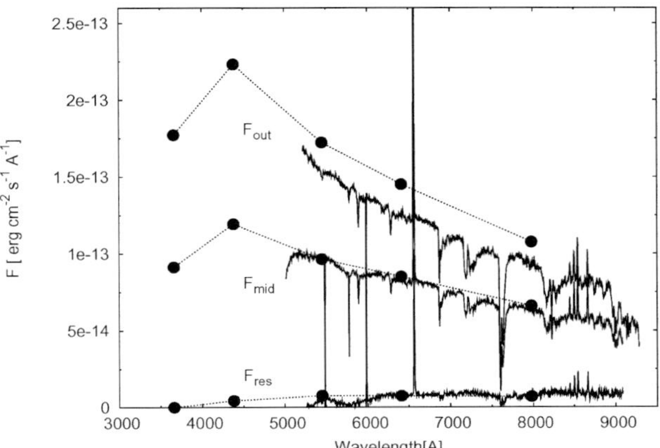

Figure 3. Photometric energy distributions (dots) and flux calibrated spectra (solid line) of EE Cep out of eclipse (F_{out}) and in the mid-eclipse (F_{mid}). The residual fluxes give a possible spectrum of the eclipsing body: $F_{res} = F_{mid} - 10^{-0.4\Delta U} F_{out}$, where ΔU denotes the eclipse depth corresponding to the obscuring factor.

Acknowledgements

M. Cikala, M. Hajduk and C. Galan are grateful to the Copernicus Foundation for Polish Astronomy for the financial support. This poster was supported by KBN Grant No. 5 P03D 003 20 and Grant No. 2 P03D 019 25. E. Semkov would like to thank the Director of Skinakas Observatory, Prof. I. Papamastorakis, and Dr. I. Papadakis for the telescope time.

References

Graczyk, D., Mikolajewski, M., Tomov, T., Kolev, D. and Iliev, I.: 2003, *A&A* **403**, 1089.
Meinunger, L.: 1973, *Mitt. Veränd. Sterne* **6**, 89.
Mikolajewski, M. and Graczyk, D.: 1999, *MNRAS* **303**, 521.
Mikolajewski, M., Tomov, T., Graczyk, D., Kolev, D., Galan, C. and Galazutdinov, G.: 2003, *IBVS*, No. 5412.
van Hamme, W.: 1993, *AJ* **106**, 5.

SPECTROSCOPIC OBSERVATIONS OF THE EE CEP ECLIPSE IN 2003*

M. MIKOLAJEWSKI[1], T. TOMOV[1], M. HAJDUK[1], M. CIKALA[1], J. OSIWALA[1],
C. GALAN[1], A. ZAJCZYK[1], D. KOLEV[2], I. ILIEV[2], P. MARRESE[3], U. MUNARI[3],
T. ZWITTER[4], G. GALAZUTDINOV[5], F. MUSAEV[6], A. BONDAR[7], L. GEORGIEV[8],
C.T. BOLTON[9], R.M. BLAKE[9] and W. PYCH[9]

[1]*Uniwersytet Mikolaja Kopernika, Centrum Astronomii, ul. Gagarina 11, 87-100 Torun, Poland*
E-mail: mamiko@astri.uni.torun.pl
[2]*Institute of Astronomy, National Astronomical Observatory, P.O.B. 136, BG-4700,*
Smolyan, Bulgaria
[3]*Osservatorio Astronomico di Padova, Sede di Asiago, 36012 Asiago (VI), Italy*
[4]*University of Ljubljana, Department of Physics, Jadranska 18, 1000 Ljubljana, Slovenia*
[5]*Bohyunsan Optical Astronomy Observatory, Jacheon P.O.B. 1, YoungChun, KyungPook, 770-820,*
South Korea
[6]*Special Astrophysical Observatory, 369167 Nizhnij Arhyz, Russia*
[7]*International Center for Astronomical, Medical and Ecological Researches, 361605 Terskol, Russia*
[8]*Instituto de Astronomie, UNAM, Apartado Postal 70-264, Cindad Universitaria, 04510 Mexico,*
D.F. Mexico
[9]*David Dunlap Observatory, University of Toronto, P.O. Box 360, Richmond Hill, ON L4C 4Y6,*
Canada

(accepted April 2004)

Abstract. High-resolution spectroscopy during the eclipse of EE Cep was obtained and presented for the first time. The star's spectroscopic behaviour can be roughly interpreted as a partial eclipse of the high luminosity Be primary and its emitting gaseous ring by the semi-transparent gaseous envelope around an invisible, opaque secondary, most probably a dark disk.

Keywords: binaries, eclipsing-stars, individual, EE Cep

1. Introduction and Observations

EE Cep is an extraordinary long-period ($P_{orb} \sim 5.6$ yr) system showing only primary eclipses with remarkable and complex changes of the shape and the depth of the minima (Graczyk et al., 2003 and references therein). The eclipses during the last several decades were relatively well covered with photometric observations, while the changes in the spectrum of EE Cep outside and during eclipses were poorly studied. Here we present preliminary results of an international campaign (Mikolajewski et al., 2003), which was aimed to follow for the first time the spectral variations during the recent eclipse. Basic information concerning

*This paper is based on spectroscopic observations from Asiago (Italy), DDO (Canada), Rozhen (Bulgaria), SPM (Mexico), Terskol (Russia) and Torun (Poland) observatories.

Astrophysics and Space Science **296**: 451–455, 2005.
© Springer 2005

TABLE I

Summary of the EE Cep spectroscopic observations

Observatory	Telescope	Spectrograph	Resolving power	Spectral region
Rozhen (Bulgaria)	2 m	Coude	15000, 30000	Hα, Hβ, NaI
Asiago (Italy)	1.8 m	Echelle	20000	4600–9200 Å
SPM (Mexico)	2.1 m	Echelle	18000	3700–6800 Å
DDO (Canada)	1.9 m	Cassegrain	16000	Hα, NaI
Terskol (Russia)	2 m	Echelle	13500	4200–6700 Å
Asiago (Italy)	1.8 m	AFOSC/echelle	3600	3600–8800 Å
Torun (Poland)	0.9 m	CCS	1200, 700	4800–9000 Å

our spectroscopic observations is given in Table I. Their dates are shown together with the collected light curve in Figure 1 of Mikolajewski et al. (this volume).

2. Results and Preliminary Interpretation

The model of EE Cep assumes the existence of a normal B5-type primary eclipsed by a flat, opaque companion, most probably a disk visible almost edge-on (Mikolajewski and Graczyk, 1999; Graczyk et al., 2003). Small differences between its inclination during successive conjunctions change its projection on the visible companion and lead to extreme variations in the depth and duration of the particular eclipses. The oblong shape of the projected disk during eclipses causes an asymmetry between the ingress and egress phases and the characteristic slope bottom during the transit (see Figure 6 in Graczyk et al., 2003).

The depth of the minimum depends on the projected height of the disk, H_d. During the 2003 eclipse, about half of the stellar disk was screened by the eclipsing body, giving at least $H_\mathrm{d} \approx R_*$, where R_* is the radius of the eclipsed B5 primary. In such geometry, the longer of the two phases – ingress or egress – should be proportional to the stellar diameter. The durations of particular photometric phases of the 2003 eclipse are roughly estimated by Mikolajewski et al. (this volume). The ingress was the longer one and lasted about 9 days. The time interval including the ingress and the slope-bottom phase should be roughly proportional to the eclipsing body length (diameter of the dark disk's projection), thus $R_\mathrm{d}/R_* \approx 15/9$. The full duration of the eclipse in continuum, together with the atmospheric phases, is about 45 days. This indicates that the dimensions of the semi-transparent parts of the disk are at least 5 times larger than the eclipsed star.

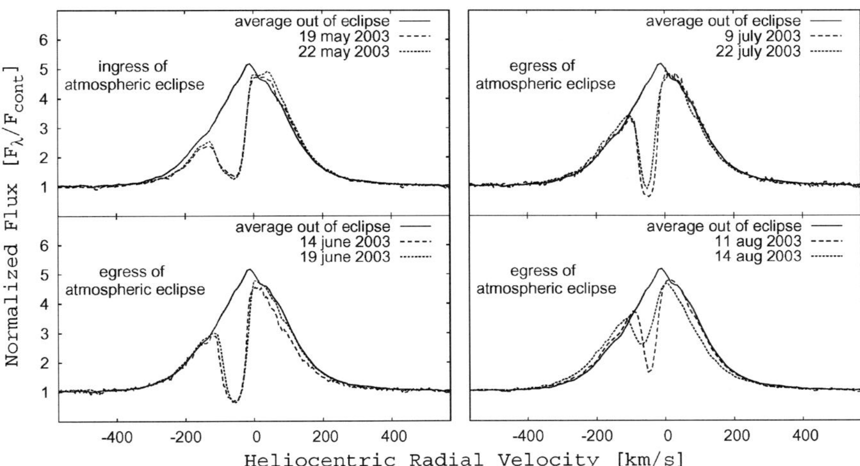

Figure 1. The average out-of-eclipse Hα profiles compared to those during the atmospheric phases of the eclipse.

Our spectroscopic observations show that the eclipse effects can be detected even 2.5 months after the mid-eclipse. Two profiles of the Hα emission, obtained on November 20, 2002 and October 12, 2003, are very similar and free from eclipse effects. Three profiles obtained on June 2 and 6 during the slope-bottom phase (mid-eclipse) are also very similar, however, their intensity obviously exceeds emission present out of eclipse. Only one profile obtained during the egress phase of the photospheric eclipse (June 9) presents an intermediate level of emission. This excess can be easily explained by the rapid decrease of the continuum level due to the eclipse by the opaque parts of the disk.

During the atmospheric phases of the 2003 eclipse, all Hα profiles demonstrate practically the same emission component as out of eclipse, with a mean heliocentric velocity about −15 km/s (Figure 1). Superimposed on the Hα emission, there appears a narrow shell absorption slowly evolving between May 19 and August 14. The depth of the absorption with respect to the local continuum is at least 2–3 times larger than the continuum level of the B5 star. Therefore, it must be real absorption of the photons produced in the Hα emission region, which is occulted by the outer region of the eclipsing companion. The same shell absorptions with heliocentric velocities varying between −40 and −65 km/s are visible also in the Na I doublet and Fe II lines (Figure 2). No measurable eclipse effects were noticed in the well-visible diffuse interstellar bands (5780, 5797, 6196, 6203, 6270, 6284, 6614 Å).

The eclipse effects in Hα prove that the line emission region is associated with the eclipsed star. Hence, the primary in EE Cep belongs to the group of Be stars.

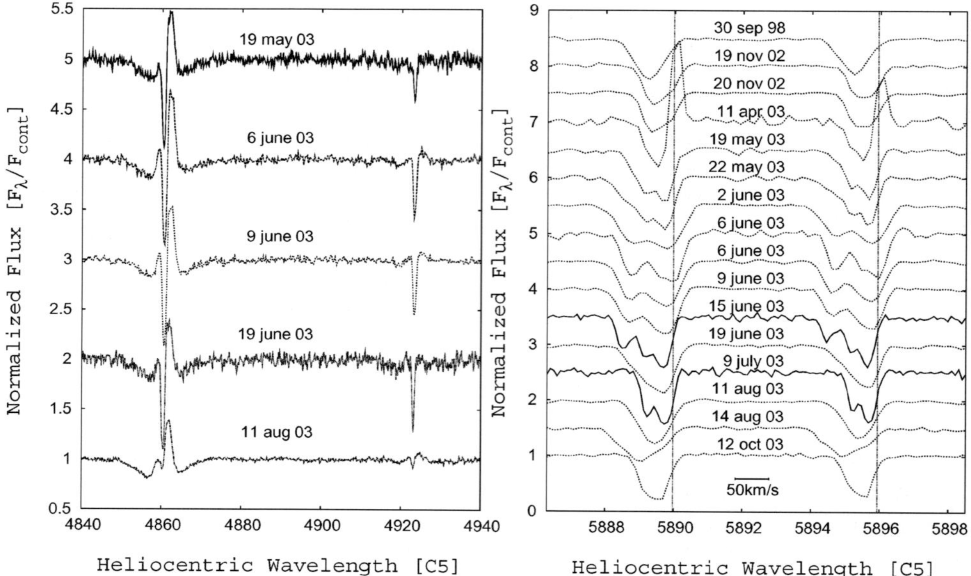

Figure 2. The sample spectra showing the line profile variations in Na I doublet (right), Hβ and Fe II 4924 Å (left). The profiles are shifted vertically respectively by 0.5 and 1.0 for better display.

The shape of the Hα profile observed out of eclipse (Figure 1) together with the earlier observations (Graczyk et al., 2003) allows to classify this star as an H2-type Be star (Hanuschik et al., 1996), i.e. with highly inclined emitting disk and asymmetric profiles. The profiles of Fe II lines close to the end of the eclipse (in August) confirm this classification (Figure 2).

The profiles of the higher Balmer lines (see Hβ in Figure 2) are dominated by broad absorptions of a rapidly rotating star. The best fit to H8–H11 lines was obtained for temperatures and log g characteristic of a B5 III star and $v \sin i \approx 350$ km/s. The mean heliocentric radial velocity of the photospheric Balmer absorptions Hγ – H11 is about −50 km/s, i.e. close to the velocity of the shell absorption.

The eclipse effects visible in Hα are apparent at least to mid-August, while the bluest component of the Na I doublet vanishes between June 19 and 25 (Figure 2). This suggests that the radius of the eclipsing cloud producing the Na I lines is at least $6R_*$, which is about $1R_*$ larger than its radius producing the eclipse in the continuum. The rapid decay of the shell absorption in Hα between August 11 and 14 (Figure 1) suggests that the gaseous ring around the B star producing the Hα emission has an almost twice larger size than the eclipsing cloud, i.e. $\sim 10R_*$. A summary of these estimates is given in Table II.

TABLE II

The main ingredients of the EE Cep system

Binary component	Primary		Secondary		
Nature of the object	Be5III star		Dark disk around the invisible component		
Element	Photo-sphere	Gaseous ring emitting HI, FeII lines (H2 type)	Opaque inner part	Semitransparent part in continuum	Outer envelope (optically thick in HI, FeII, NaI)
Radius	5–10 $R_\odot$	$\sim 10 R_*$	1.6–1.9 R_* $\sim 5 R_*$		$\sim 6 R_*$
Height	–	$\sim 1 R_*$	$\gtrsim 1 R_*$ $\gtrsim 2 R_*$		$\gtrsim 2 R_*$

Acknowledgements

M. Hajduk, C. Galan and M. Cikala are grateful to the Copernicus Foundation for Polish Astronomy for the financial support. We thank T. Karmo, Stefan Mochnacki, and G. Coinidis for contributed data. This project was supported by the Polish KBN grants Nos. 5 P03D 003 20 and 2 P03D 019 25. We acknowledge support of the work of R.M. Blake and W. Pych through Discovery Grants from the Canadian Natural Sciences and Engineering Council to C. T. Bolton and S. Rucinski, respectively.

References

Graczyk, D., Mikolajewski, M., Tomov, T., Kolev, D. and Iliev, I.: 2003, *A&A* **403**, 1089.

Hanuschik, R.W., Hummel, W., Sutorius, E., Dietle, O. and Thimm, G.: 1996, *A&AS* **116**, 309.

Mikolajewski, M. and Graczyk, D.: 1999, *MNRAS* **303**, 521.

Mikolajewski, M., Tomov, T., Graczyk, D., Kolev, D., Galan, C. and Galazutdinov, G.: 2003, *IBVS*, No. 5412.

BM CAS: A LONG-PERIOD ECLIPSING BINARY WITH A SUPERGIANT AND A COMMON ENVELOPE

P. KALV[1], V. HARVIG[1] and I.B. PUSTYLNIK[2]

[1]*Tallinn Technical University, Estonia; E-mail: izold@aai.ee*
[2]*Tartu Observatory, Estonia*

(accepted April 2004)

Abstract. New results of *UBVR* photometry of the long-period eclipsing binary BM Cas obtained at Tallinn Observatory between 1979 and 1996 are discussed.

Keywords: stars, binaries, close-stars, individual, BM Cas

We report some results of *UBVR* photometry of the long-period eclipsing binary BM Cas ($P_{\mathrm{orb}} = 197\overset{\mathrm{d}}{.}275$) conducted between 1979 and 1996 at Tallinn Observatory.

New epochs of primary minimum in *UBV* colours have been determined (see data in Table I). No significant period variations have been detected during this time span (see the *O–C* diagram, Figure 2, as well as similar data in Kreiner et al. (2001). Scattering of data points in the *O–C* diagram is caused basically by pronounced intrinsic variability seen in all colours (Figure 1). Sporadic displacements of the primary minimum epochs from the predicted ephemerides amounting to $0.02 P_{\mathrm{orb}}$ have been observed. The observed $\Delta(B - V)$ changes in both minima of the light curve are consistent with the presence of a hot region located between the components. A similar pattern was found for another long-period eclipsing binary, UU Cnc (Kalv and Oja, unpublished).

To our knowledge, no satisfactory solution of the light curves of BM Cas exists. The existing estimates of the masses and the radii of the components range between $M_1 = 1.7 M_{\odot}$, $M_2 = 7.3 M_{\odot}$, $R_1 = 74 R_{\odot}$, $R_2 = 18 R_{\odot}$ (Barwig, 1976) and $M_1 = 47 M_{\odot}$, $M_2 = 32 M_{\odot}$, $R_1 = 100 R_{\odot}$, $R_2 = 57 R_{\odot}$ (Kalv, 1980). The colour excess for BM Cas is $1\overset{\mathrm{m}}{.}0$, and probably the binary, located slightly below the Perseus arm, belongs to Association III of Cassiopeia. Assuming $A_{\mathrm{v}} = 3.2 E_{\mathrm{B-V}}$ and $d = 2.5$ kpc, Kalv (1980) finds for the primary component $M_{\mathrm{V}} = -5\overset{\mathrm{m}}{.}7$.

A close resemblance between the observed parameters of BM Cas and those of $h\&\chi$ Per, namely the distance to the object (2.5 kpc) and the mean radial velocity $-43\,\mathrm{km\,s}^{-1}$ has been noted by several authors. BM Cas could be an example of a system at the common envelope stage. Available observational data do not contradict this idea. The mechanism responsible for a putative hot region between the components could be the "drag luminosity" as the secondary spirals towards the primary component (Iben and Livio, 1993).

Astrophysics and Space Science **296**: 457–460, 2005.
© Springer 2005

TABLE I

List of primary minima of BM Cas

Minimum JD	E	$O{-}C$, d	Minimum JD	E	$O{-}C$, d
2419255.400	−151	−6.529	2441553.500	−38	−1.162
2425775.400	−118	3.128	2441554.500	−38	−0.162
2426957.400	−112	1.436	2441554.500	−38	−0.162
2427152.700	−111	−0.545	2441750.400	−37	−1.540
2427154.700	−111	1.454	2441751.100	−37	−0.840
2427352.200	−110	1.672	2441947.800	−36	−1.417
2427550.800	−109	2.990	2441948.100	−36	−1.117
2427746.400	−108	1.309	2442144.500	−35	−1.995
2427941.500	−107	−0.872	2442144.600	−35	−1.895
2428139.800	−106	0.146	2442736.700	−32	−1.627
2428337.500	−105	0.564	2442736.700	−32	−1.627
2428532.400	−104	−1.816	2443130.100	−30	−2.781
2428731.200	−103	−0.297	2445299.080	−19	−3.846
2428927.800	−102	−0.979	2445299.130	−19	−3.796
2429321.900	−100	−1.441	2445302.600	−19	−0.326
2429720.933	−98	3.029	2445302.700	−19	−0.226
2432081.000	−86	−4.271	2445303.400	−19	0.473
2432282.600	−85	0.047	2445304.800	−19	1.873
2432480.800	−84	0.967	2445308.300	−19	5.373
2432677.800	−83	0.687	2445498.700	−18	−1.503
2432875.300	−82	0.907	2446883.000	−11	1.862
2433074.700	−81	3.027	2446883.400	−11	2.262
2433270.600	−80	1.647	2446884.100	−11	2.962
2433466.500	−79	0.267	2447078.500	−10	0.086
2433665.300	−78	1.787	2447078.700	−10	0.286
2433859.900	−77	−0.892	2447078.700	−10	0.286
2434057.500	−76	−0.572	2447078.960	−10	0.546
2434457.500	−74	4.868	2447274.000	−9	−1.689
2434648.000	−73	−1.911	2447274.300	−9	−1.389
2434846.500	−72	−0.691	2447274.600	−9	−1.089
2434848.500	−72	1.308	2447471.500	−8	−1.465
2435047.300	−71	2.829	2447471.800	−8	−1.165
2437017.400	−61	0.137	2447472.100	−8	−0.865
2439384.270	−49	−0.335	2447472.812	−8	−0.153
2439974.300	−46	−2.139	2448263.474	−4	1.405
2439974.300	−46	−2.139	2448657.680	−2	1.060
2440173.200	−45	−0.517	2448657.680	−2	1.060
2440173.800	−45	0.082	2449050.720	0	−0.450
2440963.400	−41	0.570	2449051.300	0	0.129
2440963.800	−41	0.970	2449051.700	0	0.529
2441158.590	−40	−1.517	2449052.500	0	1.329
2441356.300	−39	−1.084	2450037.500	5	−0.046
2441356.400	−39	−0.984	2450828.800	9	2.152
2441366.780	−39	9.395	2451620.500	13	4.753
2441546.800	−38	−7.862	2451811.100	14	−1.921

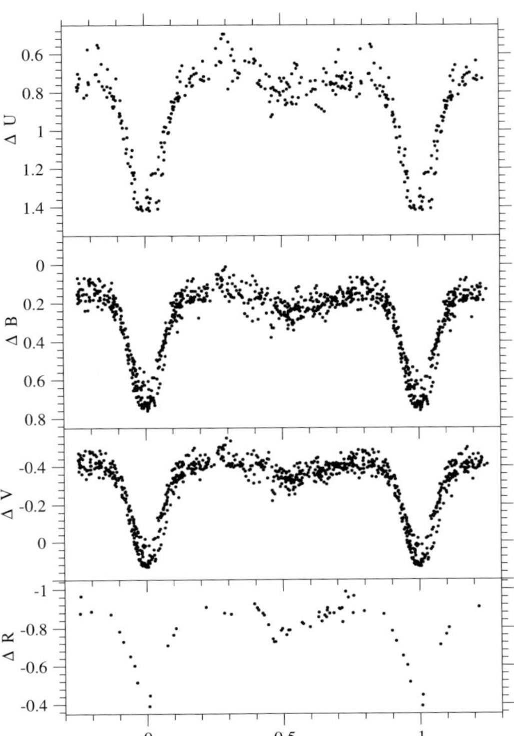

Figure 1. The UBVR light curves of BM Cas obtained in Tallinn observatory in 1979–1996.

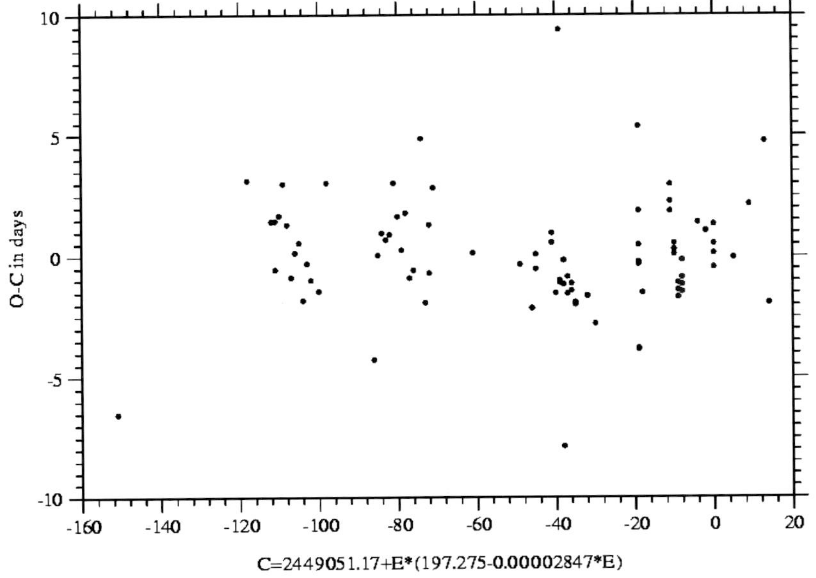

Figure 2. The *O–C* diagram for BM Cas.

P. KALV ET AL.

Acknowledgements

Supporting Grant 5760 from Estonian Science Foundation is gratefully acknowledged.

References

Barwig, H.D.: 1976, *Spektroskopische Untersuchung des Bedeckungsveränderlichen BM Cassiopeiae*. Inaugural-Dissertation zur Erlangung des Doktorgrades des Fachbereiches Physik der Ludwig-Maximilians-Universität München.

Iben, I., Jr. and Livio, M.: 1993, *PASP* **105**, 1373.

Kalv, P.: 1980, *Investigation of the Selected Long-Period Eclipsing Stars Based on Photoelectric Observations*. Ph.D. Thesis, Tartu (unpublished).

Kalv, P. and Oja, T.: private communication.

Kreiner, J., Kim, Chun-Hwey and Nha, Il-Seong: 2001, *An Atlas of O–C Diagrams of Eclipsing Binary Stars*, Krakow.

MV Lyr: TRANSITION FROM LOW TO HIGH BRIGHTNESS STATE

ELENA PAVLENKO[1], KIRILL ANTONIUK[1], SERGEI YU. SHUGAROV[2],
NATALY A. KATYSHEVA[2], LENIE MUZHDABAEVA[3]
and VYACHESLAV MIKHAJLOV[4]

[1]*Crimean Astrophysical Observatory, Nauchny, Crimea 98409, Ukraine;*
E-mail: pavlenko@crao.crimea.ua
[2]*Sternberg State Astronomical Institute, Universitetskij Prosp. 13, Moscow 119992, Russia*
[3]*Tavrichesky National University, Crimea 95007, Ukraine*
[4]*Simferopol Society of Amator Astronomers, Ukraine*

(accepted April 2004)

Abstract. We present results of multicolour photometry of the nova-like binary MV Lyr in the years 2002–2003, corresponding to the transition of the star from its low ($V = 17.8$) to high ($V = 12.3$) brightness state. The transition lasted at most 200 days. During the first $\sim$50 days MV Lyr brightened at the rate 0.06 mag/day, and during the next $\sim$150 days at the rate 0.01 mag/day. The brightening was accompanied by a blue shift from 0.45 to 0.05 in $V - R$, interpreted as an increase of the accretion disk contribution to the total light. During the transition MV Lyr displayed well-known 'quasi-orbital' light variations and fast quasi-periodic oscillations with a typical time scale of tens of minutes.

Keywords: stars, binaries, MV Lyr

1. Introduction

For long MV Lyr was thought to belong to the VY Scl-type stars or "anti-dwarf novae", which mostly remain in a high brightness state and abruptly fall to a faint state for a short time. However, our idea of this binary changed as the sporadic observations were replaced by systematic ones during the last 25 years. In 1979 MV Lyr entered a minimum and spent there 10 years. In that state it showed relatively short outbursts of amplitude 1^m–4^m and duration of 7–100 days. In 1989–1995 MV Lyr was in a high state (the 25-year light curve is shown in Katysheva et al., 2000). The next low state lasted nearly 7 years. No orbital light variations have been found in the faint state, while in the high and intermediate states some observers detected light modulations close to the orbital period, but systematically different by several minutes (Borisov, 1992; Skillman et al., 1995; Pavlenko and Shugarov, 1999). In 2002 MV Lyr entered the high state. Here we present the results of multicolour observations of this star in 2002–2003.

2. Observations

Photometric observations of MV Lyr in 2002–2003 have been carried out at the Crimean Astrophysical Observatory (CrAO) and at the Crimean laboratory of the

 *Astrophysics and Space Science* **296**: 461–464, 2005.
© Springer 2005

Sternberg Astronomical Institute (SAI) with different telescopes and equipment. $UBVRI$ (Johnson) observations were obtained with the 1.25 m telescope AZT-11 (CrAO) using the double beam chopping photometer/polarimeter designed by Korhonen and Piirola (1984); BVR (Johnson) observations were obtained using the CCD SBIG ST7 camera at the K-380 telescope (CrAO) and the Zeiss-600 telescope (SAI).

3. Results

We collected observations of MV Lyr covering the transition from low (faint) state, when the variable was at $V = 17.8$, to high (bright) brightness state ($V = 12.3$). The light curve and $V - R$ behaviour are shown in Figure 1. The photoelectric data are marked by open symbols. One could see that the brightening of the star occurred in two steps: the fast brightening lasted at most 50 days and had a rate of $\sim$0.06 mag/day; it was followed by a 6-times slower rate of 0.01 mag/day, which lasted $\sim$150 days. The star became more blue during its brightening. Figure 2 displays the behaviour of MV Lyr in the magnitude-colour diagram. $V - R$ changed from $\sim$0.45 in minimum to 0.05 in maximum with a ratio of $(V - R)/R = 0.03$. The V, $V - R$ dependence looks roughly linear, but there are some additional short-term light and colour variations (Andronov and Antonyuk, 2004). An example of two nightly light curves is displayed in Figure 3. They show variability on a scale of several hours – so-called "quasi-orbital" variations (Andronov et al., 1992) and decades of minutes.

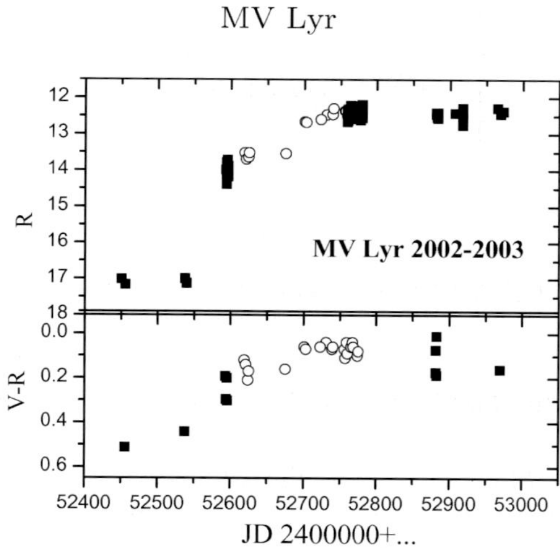

Figure 1. Light and $V - R$ curve of MV Lyr during transition from low to high brightness state.

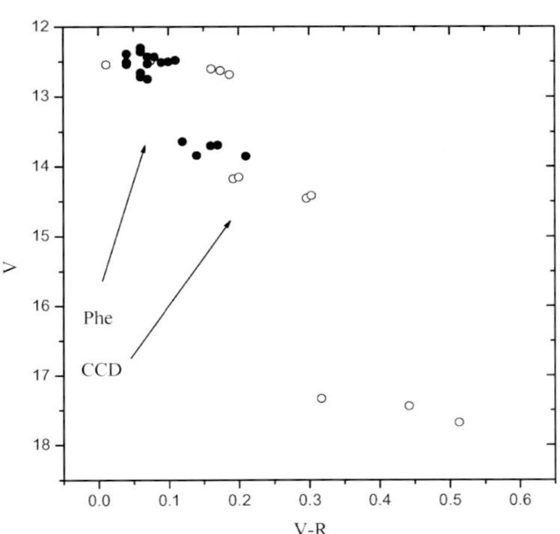

Figure 2. $V, V - R$ diagram for transition from low to high brightness state.

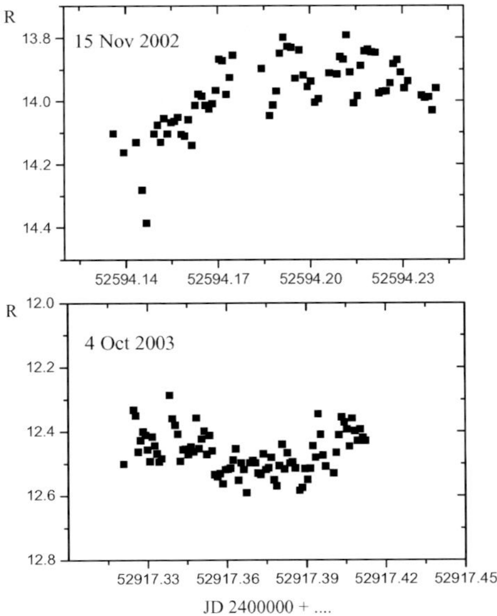

Figure 3. Two nightly light curves for intermediate and high brightness level.

4. Discussion: The Problem of the Orbital Light Modulation

The orbital period of MV Lyr was obtained by spectroscopy as 0.1336(17) days (Schneider et al., 1981) or 0.1329(4) days (Skillman et al., 1995). The search for a photometric orbital modulation was unsuccessful, perhaps due to the low binary inclination (10–13°). Only in the low state, when the disk contribution is less than in the high state, a possible light modulation caused by the ellipsoidal effect or reflection effect could be expected. Recently Hoard et al., hereafter HLS (2004), reported on the absence of an accretion disk and evidence for a hot white dwarf (47 000 K) in the deep low state. Our $V - R$ measurement showed that in the faint state we observe light from both a white dwarf and a red dwarf in agreement with the HLS result. The blue shifting of $V - R$ simultaneously with brightening means an increased contribution of the accretion disk and a rise of its temperature. HLS predicted a high reflection effect in this close binary. At least two attempts were undertaken to search for and to confirm the orbital period by photometry during the last faint state (1995–2002). The first one was made at the beginning of the faint state in 1995–1996 in a broad "BV" system (Pavlenko, 1996), and the second one just prior to the high state in 2004 in V (HLS). The first attempt was unsuccessful, no indication of the orbital period was found. Very close to the orbital period a low-amplitude (less than 0.1 mag) modulation was detected at a period of 0.142 day. Meanwhile HLS found the photometric 0.1 mag modulation with an orbital period of ~ 0.133 d, but its amplitude is much less than the predicted one. The authors explain this by the presence of star spots on the secondary component near the L_1 point and possible shadowing in the equatorial region of the secondary by the nascent accretion disk. Future observations in the faint state are necessary to study the orbital modulation appearance and disappearance and its connection with star spots and/or shadowing by the disk.

Acknowledgements

This work was partially supported by the Ukrainian Fund For Fundamental Researches 02/07/00451 and by the grants 02-02-16235, 02-02-17524 of Russian Foundation of Basic Researches and Russian President Grant NS-388.2003.3.

References

Andronov, I.L. and Antonyuk, K.: 2004, in astrophysics of Cataclysmic Variables and Related Objects, Hameury, H.M. and Lasota, J.P. (eds.), in press.
Andronov, I.L., Borodina, I.G., Chinarova L.L. et al.: 1992, *Soobsch. Spets. Astroph. Obs.* **69**, 125.
Borisov, G.V.: 1992, *A&A* **261**, 154.
Hoard, D.W., Linnell, A.P., Szkody, P. et al.: 2004, *ApJ* **604**, 346 .
Katysheva, N.A., et al.: 2000, *Kinemat. Phys. Celes. Bodies* **3**, 393.
Korhonen, T. and Piirola V.: 1984, ESO Messenger no.38, p. 20.
Piirola, V.: 1988, in Polarized Radiation of Circumstellar Origin, Coyne, G.V., Magalhaes, A.M., Moffat, A.F.J., Schulte-Ladbeck, R.E., Tapia, S., Wickramasinghe, D.T. (eds.), Vatican Observatory, p. 261.
Pavlenko, E.P.: 1996, *Odessa Astronom. Publ.* **9**, 38.
Pavlenko, E.P. and Shugarov S.Yu.: 1999, *A&A* **343**, 909.
Schneider, D.P., Young, P. and Shestman, S.A.: 1981, *ApJ* **245**, 644.
Skillman, D.R., Patterson, J. and Thorstensen, J.R.: 1995, *PASP* **107**, 545.

PRELIMINARY ANALYSIS OF PHOTOMETRIC VARIATIONS OF THE CENTRAL STAR OF THE PLANETARY NEBULA Sh 2-71

ZDENĚK MIKULÁŠEK[1,3], LUBOŠ KOHOUTEK[2], MILOSLAV ZEJDA[3]
and ONDŘEJ PEJCHA[3]

[1]*Institute of Theoretical Physics and Astrophysics, Masaryk University, Kotlářská 2, Brno, Czech Republic; E-mail: mikulas@ics.muni.cz*
[2]*Hamburg Observatory, Hamburg, Germany*
[3]*N. Copernicus Observatory and Planetarium, Kraví hora 2, Brno, Czech Republic*

(accepted April 2004)

Abstract. We confirm the presence of regular $UBV(RI)_C$ light variations of the object in the center of the planetary nebula Sh 2-71, with an improved period of $P = 68.132 \pm 0.005$ days. The shapes and amplitudes of light curves, in particular colours, are briefly discussed.

Keywords: $UBV(RI)_C$ photometry, planetary nebula, Sh 2-71, period, PCA

1. Introduction

The nebula Sh 2-71 (=PK 036-01 1) was discovered by Minkowski (1946), who classified it as a diffuse or peculiar nebulosity. Sharpless (1959) included this object in his catalogue of H II regions and considered it as a possible planetary nebula.

Kohoutek (1979) reported light variations (>0.7 mag) of the star in the center of planetary nebula Sh 2-71, and preliminarily classified the central star as of B8 spectral type. Sabbadin et al. (1985, 1987) showed the nebula being a high excitation PN, with a strong stratification effect. The very high effective temperature of the ionizing source of 129 000 K was determined by Preite-Martinez et al. (1989) and confirmed by (Kaler et al., 1990; Bohigas, 2001). Feibelman (1999) showed a variable Mg II λ 2800 emission line superposed on a variable stellar continuum. Jurcsik (1993) revealed periodic light variations of the central star with a period of 68.064 days, while colour variations in $UBV(RI)_C$ were found to be insignificant.

2. Observations

We have processed a large set of 3370 observations from three sources (see the Table I):

1. Partly unpublished *UBV* observations made by Kohoutek at Wise Observatory at Mitzpe Ramon, Israel, in 1977–1979, at the European Southern Observatory

Astrophysics and Space Science **296**: 465–468, 2005.
© Springer 2005

TABLE I

Journal of photometric observations: name of author or group, interval of observations, number of measurements in UBV$(RI)_C$

Author	Time interval	U	B	V	R_C	I_C	Sum
Kohoutek	08.1977–11.1979	81	82	100	0	0	263
Jurcsik et al.	07.1990–07.1993	392	401	403	403	402	2001
Brno	08.1999–11.2002	0	0	196	201	159	556
Vyškov	05.2001–09.2002	0	0	267	283	0	550
N_{tot}		473	483	966	887	561	**3370**
obs. error [mag]		0.12	0.04	0.04	0.03	0.05	0.06
scatter [mag]		0.16	0.09	0.11	0.10	0.13	0.12
eff. ampl. [mag]		0.84	0.70	0.71	0.73	0.81	

at La Silla, Chile, and at the Hamburg-Bergedorf Observatory, Germany, in 1979.

2. Unpublished $UBV(RI)_C$ observations made by Jurcsik (2003) at Konkoly Observatory in 1990–1993.

3. Unpublished *VRI* CCD photometry made by M. Zejda, P. Hájek, O. Pejcha, J. Šafář, and P. Sobotka at Brno in 1999–2002 and Vyškov Observatories in 2001–2002.

Stars *a* and *b* (Kohoutek, 1979) were measured as comparison stars and HD 175 544 (B3 V) as photometric standard (Menzies et al., 1991).

3. Ephemeris of Light Variations

Analysis of photometric data confirmed the presence of strong quasi-periodic variations of the object in all photometric colours, with a period of about 68 days. 966 individual V observations of the star span 135 cycles of light variations, which is quite sufficient for an adequately accurate determination of linear ephemeris. Using an own version of the non-linear robust regression (Mikulášek et al., 2003) we derived the following expression for the moment of the light maximum:

$$JD_{\max} = (2449795.85 \pm 0.22) + (68.132 \pm 0.005)(E - 95).$$

The beginning of the epoch counting was chosen so that the maximum in $E = 0$ corresponds to the first light maximum preceding the first photometric observation of the star.

This linear ephemeris should be considered only as a first approximation. There are strong indications that both the period and the form of light curves are not stable. Consequently, the above given uncertainties of the ephemeris are only formal.

All other checked beat periods give a markedly worse fit than the 68 d period.

4. Shapes of Light Curves

The comparison of light curves in various colours was done with an own method combining principal component analysis and robust regression. We have found that periodically repeating parts of light curves in all colours can be satisfactorily expressed as a linear combination of only two principal light curves (see Figure 2a), where the first one contains the basic features of stellar variations and the second one corresponds to the differences of individual light curves. The fit based on a superposition of both basic curves is displayed in Figure 1. The dependencies of semi-amplitudes A_1 and A_2 corresponding to both principal curves on the effective wavelength are plotted in Figure 2b.

A detailed analysis of the photometric behaviour of the object will be published elsewhere.

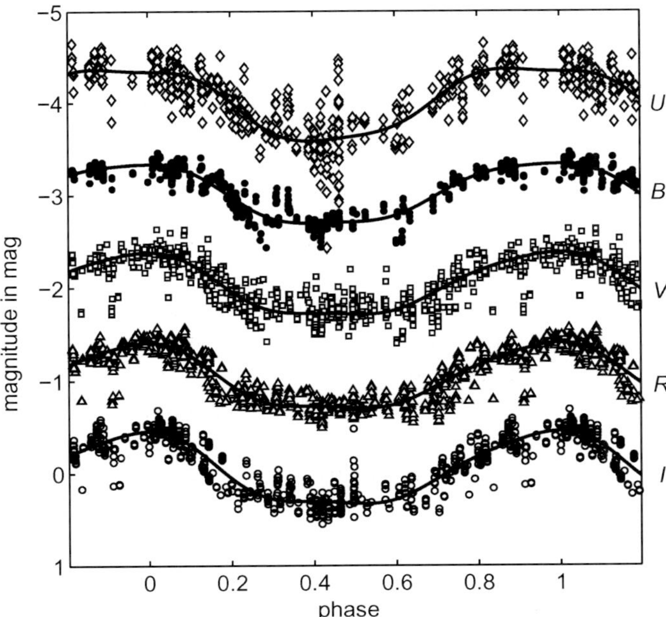

Figure 1. Phase diagram of light variations of Sh 2-71 in $UBV(RI)_{\rm C}$ colours folded with period $P = 68.132$ d. Plots of individual light curves are offset for clarity. In U and to lesser extent in I a specific type of stellar activity is evident. The observed scatter is significantly larger than observational errors (see Table I).

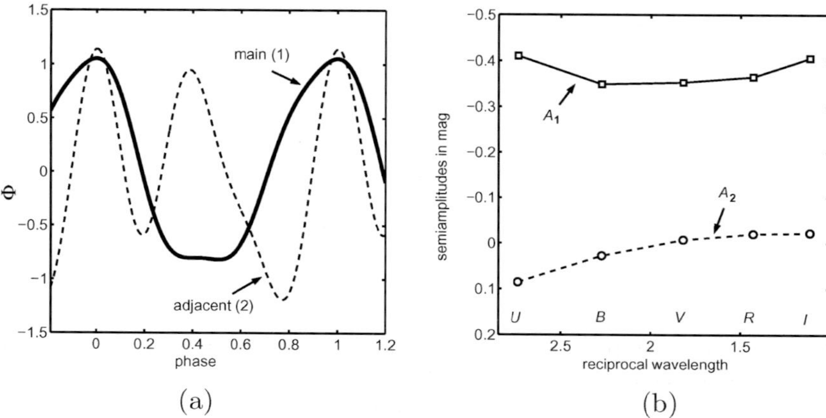

Figure 2. (a) Phase graph of the first and second mutually orthogonal principal functions. (b) Dependence of semi-amplitudes A_1 and A_2 of the decomposition of light curves into the first two principal functions on reciprocal effective wavelengths expressed in microns.

Acknowledgements

The work was supported by the projects 205/02/0445, 205/04/1267 and 205/04/2063 of the grant Agency of the Czech Republic.

References

Bohigas, J.: 2001, *Revista MxAA* **37**, 237.

Feibelman, W.A.: 1999, *PASP* **111**, 719.

Jurcsik, J.: 1993, in: R. Weinberger and A. Acker (eds.), *Planetary Nebulae: Proceedings of the 155 Symposium of the IAU, Innsbruck; Austria; 13–17 July 1992*, Kluwer Academic Publishers; Dordrecht, p.399.

Kaler, J.B., Shaw, R.A. and Kwitter, K.B.: 1990, *Astrophys. J.* **359**, 392.

Kohoutek, L.: 1979, *IBVS* No. 1672.

Menzies, J.W., Marang, F., Laing, J.D., Coulson, I.M. and Engelbrecht, C.A.: 1991, *Mon. Not. Roy. Astron. Soc.* **248**, 642.

Mikulášek, Z., Žižňovský, J., Zverko, J. and Polosukhina, N.S.: 2003, *Contr. Astron. Obs. Sk. Pleso* **33**, 29.

Minkowski, R.: 1946, *Publ. Astron. Soc. Pacif.* **58**, 305.

Preite-Martinez, A., Acker, A., Koeppen, J. and Stenholm, B.: 1989, *Astron. Astrophys. Suppl. Ser.* **67**, 541.

Sabbadin, F., Ortolani, S. and Bianchini, A.: 1985, *Mon. Not. Roy. Astr. Soc.* **213**, 563.

Sabbadin, F., Falomo, R. and Ortolani, S.: 1987, *Astron. Astrophys. Suppl. Ser.* **81**, 309.

Sharpless, S.: 1959, *Astrophys. J. Suppl.* **4**, 257.

OBSERVATIONS OF THE DEEPLY ECLIPSING DWARF NOVA GY Cnc

NATALY A. KATYSHEVA and SERGEI YU. SHUGAROV

Sternberg State Astronomical Institute, Universitetskij Prosp., Moscow, Russia;
E-mail: nk@sai.msu.ru

(accepted April 2004)

Abstract. We present photometric measurements of the eclipsing dwarf nova and X-ray source GY Cnc. The observations were collected during outbursts and in quiescence. The investigation of plates from the Sonneberg archive showed that the mean outburst interval is about 210–270 days, that the outburst is very fast, and lasts for about 5 days.

Keywords: binary stars, cataclysmic variables, X-ray source

1. Introduction

Cataclysmic variables (CVs) are binary stars of very short orbital period, in which a low-mass red K–M dwarf star (the secondary) transfers mass to a white dwarf (the primary). CVs are subdivided into some sub-classes including dwarf novae and nova-likes variables. RX J0909.8 + 1849 = GY Cnc ($\alpha_{2000} = 09^h09^m50.56^s$, $\delta_{2000} = 18°49'47.2''$) was identified as an X-ray source by *ROSAT* and as a possible CV by Bade et al. (1998). The first outburst of GY Cnc with an amplitude of up to 12 mag was independently detected by Gänsicke et al. (2000) and the VSNET-team (Kato et al., 2000) in February 2000. The photometric observations during the outburst were provided by the VSNET team (Kato et al., 2000) and Gänsicke et al. (2000) immediately, and they classified GY Cnc as a dwarf nova. Both groups observed the object during several February nights. They found the eclipse with a depth of 2–3 mag and pointed out that there are no humps in the light curve during the outburst. In April 2000 Thorstensen (2000) and Shafter et al. (2000) studied this star spectroscopically during quiescence. The next outbursts occurred in October 2000 and November 2001. The November 2001 outburst was described by Kato et al. (2002), who gave the ephemeris

$$BJD\,(\text{mid} - \text{ecl}) = 2451586.21271(8) + 0.17544251(5) \cdot E.$$

The observations during the fourth outburst were carried out by Goranskij and Barsukova (private communication), on December 5 and 9, 2002. Only two sets of observations have been obtained. The latest outburst happened in April 2004.

GY Cnc is one of a few eclipsing dwarf novae with a period above the "period gap": EX Dra, IP Peg, BD Pav, and the possibly new dwarf nova V367 Peg (see

Katysheva and Pavlenko, 2003). Also only a few dwarf novae show X-ray emission, e.g. EX Dra. So GY Cnc can be an intermediate polar (Kato et al., 2002).

2. Photographic Archive Data

Shugarov et al. (2003) analyzed 730 plates of the Sonneberg plate collection from 1930 to 1990 to verify any outburst activity of GY Cnc. The star was seen only during its outbursts (above 13–14 mag) and was not visible in quiescence ($\sim$17 mag). At least 24 outbursts of this star have been observed (18 are certain and 6 uncertain). In maximum it reached $12.^m5$. The magnitudes of the comparison stars in the blue photographic system pg (close to B) are given in Shugarov et al. (2003). It is very difficult to estimate the outburst cycle, because observations are scarce. But the minimum outburst interval we found is about 43–60 days. In Table I, we present Julian Dates of the bright states and the intervals between outbursts (Δt). Of course, the epochs of archive observations were sporadic. But the mean outburst interval may be about 210–270 days. A coadded outburst light curve was obtained by a displacement of the individual light curves in time. It is very fast and lasts for about 5 days (Shugarov et al., 2003). The outburst curve shows a very sharp rise (about 1 day), a plateau (2 days) and a decline (2 days). Gänsicke et al. (2000) and Kato et al. (2000) also noted that outbursts of GY Cnc are shorter than the typical ones of IP Peg, for example.

3. CCD Observations

We began our CCD observations of GY Cnc after having received the VSNET notification in February 2000 and kept on our monitoring since then. One run (in R) was obtained during the second outburst on 21 October 2000. Last photometry

TABLE I

Dates of outbursts, JD 2400000+

Date	Δt (days)	Date	Δt (days)	Date	Δt (days)	Date	Δt (days)
27924		39024	522	47238	367	51584	1862
33305	5381	39579	555	47852:	614	51839	255
34120	815	42149	2570	48361	509	52238	399
35209	1089	45017	2868	48691:	330:	52618	380
35870	661	45760	743	48986:	295:	53098	480
36246	376	46079	319	49029*	43*		
36631:	385	46437	358	49090*	61*		
38502:	1871	46871	434	49722	632		

Note. ":" denotes uncertain value, "*" denotes the minimum interval between two outbursts.

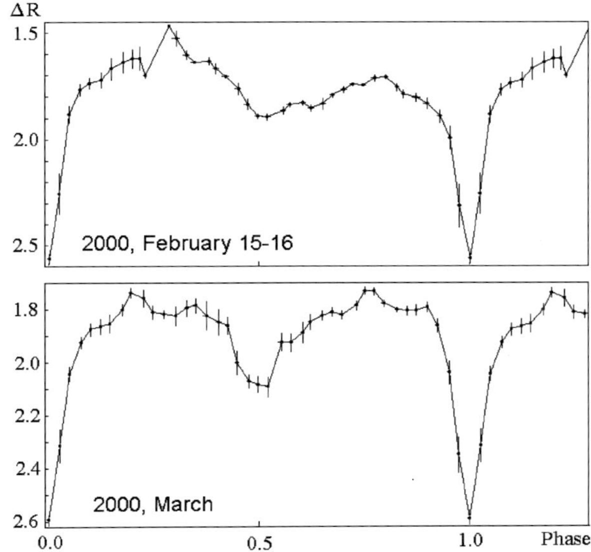

Figure 1. Mean light curves of GY Cnc in *R* band (relative units).

was obtained in January 2004. In general, our runs were made at the Crimean Laboratory of Sternberg Astronomical Institute. We used the 38, 60 and 125 cm telescopes equipped with the SBIG ST–7, ST–8 and AP–7 CCD cameras.

Figure 1 presents the mean *R* light curves at the end of the outburst in February 2000 and during quiescence (March 2000) folded with the period $0.^d175446$. In contradiction to the very smooth outburst light curves (Gänsicke et al., 2000; Kato et al., 2002; Shugarov et al., 2003) we can see the eclipse of the secondary in *R* ($\Delta R \sim 0.4$). It is noteworthy that there is also a secondary eclipse in *I* of about 0.25 mag and a possible hump before the main eclipse.

Unfortunately, we got only one light curve in the *R* band during the outburst on 21 October, 2000, and we do not know what outburst day it is, but probably close to an outburst maximum. The eclipse light curve of this day (Figure 2) is very smooth, without any humps. The absence of humps during the outburst can be explained by an enlarged disk giving rise to the main part of radiation. The minimum is asymmetric, with the egress part being steeper than the ingress (16.5 min versus 6.6 min). The observations show how the disk is becoming more homogeneous during the rise to this outburst.

For the mass and radius of the secondary we got $M_2/M_\odot = 0.48$ and $R_2/R_\odot = 0.46$, using the relations by Warner (Warner, 1995): $M_2/M_\odot = 3.18 \cdot 10^{-5} P_{orb}$ (sec) and $R_2/R_\odot = 0.959 M_2/M_\odot$. These values are similar to the ones given by other authors.

The archive and modern data provide evidence for very short outbursts and a large outburst interval. The hot spot is weaker than in most dwarf novae. There

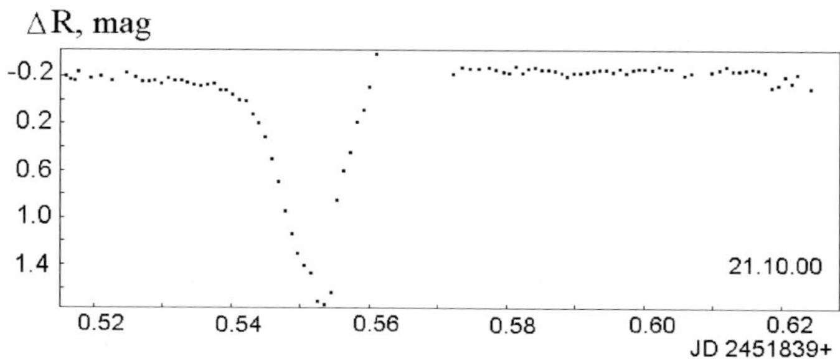

Figure 2. The outburst *R* light curve of GY Cnc (relative units).

were two other outbursts observed besides those described by the above-mentioned authors.

Acknowledgements

The authors are grateful to RFBR 02-02-16235, 02-02-17524 and NS-388.2003.3 – grants for partial support of this work.

References

Bade, N., Engels, D., Voges, W., et al.: 1998, *A&AS* **127**, 145.
Gänsicke, B.T., et al.: 2000, *A&A* 356, L79.
Goranskij, V.P. and Barsukova, E.A. (private communication).
Kato, T., et al.: 2000, *IBVS* No. **4873**, 1.
Kato, T., Ishioka, R. and Uemura, M.: 2002, *PASJ* **54**, 1023.
Katysheva, N.A. and Pavlenko, E.P.: 2003, *Ap* **46**, 114 (*Astrofizika* **46**, 147).
Shafter, A.W., Clark, L. Lee, Holland, J., Williams, S.J..: 2000, *PASP* **112**, 1467.
Shugarov, S., Katysheva, N. and Kroll, P.: 2003, in: D. Proust, M. Verdenet and J. Minois (eds.), Stellar Variability, Intern. conf. var. stars, Editions Buriller, p. 95.
Thorstensen, J.R.: 2000, *PASP* **112**, 1276.
Warner, B.: 1995, Cataclysmic Variable Stars, Cambridge University Press, Cambridge, UK.

WAVELENGTH DEPENDENCE OF THE ORBITAL VARIABILITY OF THE ECLIPSING NOVA-LIKE OBJECT DW UMa

N.I. OSTROVA[1], S.YU. SHUGAROV[2] and I.L. ANDRONOV[1]

[1]*Astronomical Observatory and Department of Astronomy, Odessa State University
T.G. Shevchenko Park, Odessa Ukraine; E-mails: natali_ostrova@ukr.net, il-a@mail.od.ua*
[2]*Sternberg Astronomical Institute, Moscow State University, Moscow, Russia*

(accepted April 2004)

Abstract. Results of three-colour VRI CCD photometry are presented obtained during 11 nights in 2001. Twelve new minima timings are derived. The eclipse depth decreases with wavelength, being equal to 1.28, 1.13 and 1.02 for V, R and I, respectively. The instrumental magnitudes and colours are tabulated for the mid-eclipse, out-of-eclipse part and the eclipsed component. The $V - R$ colour of the eclipsed component of emission is by 0.22 mag bluer than that for mid-eclipse, indicating a much higher temperature of the eclipsed region.

Keywords: stars, nova-like individual, DW UMa

1. Introduction

Cataclysmic variables (CVs) are close binary systems consisting of a late-type, typically main-sequence star (the secondary) that fills its critical Roche lobe and transfers material to a white dwarf companion (the primary) onto an accretion disk. DW UMa (=PG 1030 + 590) is a relatively well-studied eclipsing nova-like CV with a period of 3.27 h, an inclination of $80°$, and a deep $(1.^{m}5)$ eclipse (Kopylov and Somov, 1987, Shafter et al., 1988; Kazennova and Shugarov, 1992). Eclipse mapping (Bíró, 2000) inferred flat disk temperature profiles as compared to the standard disk model, a feature not uncommon among nova-like objects.

This paper presents results of an analysis of CCD observations of this interesting star.

2. Observations

Our CCD observations of DW UMa were carried out using the 1.2 m ZTE, Zeiss-300, and AZT-2 telescopes of the Sternberg Astronomical Institute and K-380 of the Crimean Astrophysical Observatory. A total of 1140 brightness estimates have been obtained during 11 nights in the filters V, R and I. The start and the end of observations, exposure time, filter and number of measurements are given in Table I.

Astrophysics and Space Science **296**: 473–476, 2005.
© Springer 2005

TABLE I

Journal of observations

Telescope	JD, 24...	JD, 24...	t_{exp} (sec)	N	Filter
Zeiss-300	51994.3185	51994.3867	90	50	R
Zeiss-300	51996.2969	51996.4622	70	150	R
Zeiss-300	51998.2778	51998.4579	60	190	V
Zeiss-300	52000.3126	52000.4657	90	48	I
AZT-2	52003.2872	52003.3879	60	89	R
ZTE	52015.4628	52015.4804	90	15	I
K-380	52023.3115	52023.3801	60	67	R
K-380	52026.2642	52026.3259	120	31	R
K-380	52027.2475	52027.5035	90	114	I
K-380	52029.2728	52029.5178	90	185	V
K-380	52030.2558	52030.5245	90	201	R

The instrumental magnitudes for all stars in the field were measured using the program by V.P. Goranskiy, 2002 (Program WINFITS, private communication) for aperture photometry.

3. Orbital Variations

The star was in its normal high state. The light curves are shown in Figure 1, folded into orbital phase according to the ephemeris published by Stanishev et al. (2004):

$$T_{\min}[HJD] = 2446229.00687(9) + 0.136606527(3) \cdot E, \tag{1}$$

To obtain a smoothed light curve, we applied the trigonometric polynomial method using a program published by Andronov (1994). The statistically significant degrees of the trigonometric polynomial are equal to 9, 9 and 7 for V, R and I, respectively. The fits with a "1σ corridor" are shown in Figure 1. We have separated the data into "eclipse" and "out-of-eclipse" parts to estimate parameters for the different phase ranges. They are listed in Tables II and III. All brightness differences were computed with respect to the comparison star ($\alpha = 10^{\mathrm{h}}33^{\mathrm{m}}17^{\mathrm{s}}$, $\delta = +58°44'43''$). The flux from the "eclipsed source" was computed as the difference between the fluxes "out-of-eclipse" and at mid-eclipse. Note that this source has a much larger temperature as compared to other sources of emission.

TABLE II

Brightness relative to the comparison star for the minimum, out-of-eclipse part, eclipsed light, and the eclipse depth

Filter	$\Delta m_{\min}$	Δm_{out}	m_{ecl}	Amplitude
V	3.161 ± 0.025	1.881 ± 0.005	2.280 ± 0.014	1.280 ± 0.026
R	3.410 ± 0.018	2.277 ± 0.004	2.749 ± 0.010	1.133 ± 0.018
I	3.497 ± 0.047	2.473 ± 0.014	3.009 ± 0.038	1.024 ± 0.045

TABLE III

Mean colours relative to the comparison star for the minimum, out-of-eclipse part, eclipsed light

Colour	Minimum	Out-of-eclipse	Eclipsed part	Amplitude
$V - R$	-0.248 ± 0.031	-0.396 ± 0.006	-0.468 ± 0.017	0.147 ± 0.032
$R - I$	-0.088 ± 0.051	-0.196 ± 0.015	-0.260 ± 0.040	0.108 ± 0.053

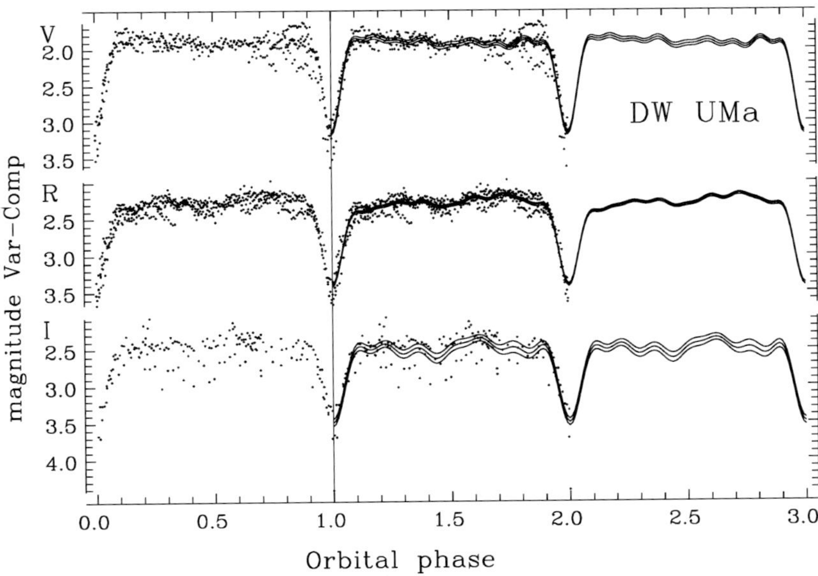

Figure 1. Light curves of DW UMa of March–April, 2001 in V, R and I colours. Vertical line marks the zero phase. The solid curves are trigonometric polynomial fits with $\pm 1\,\sigma$ corridor.

4. Minima Timings

To determine individual times of minima, we have used the method of "asymptotic parabola" (Marsakova and Andronov, 1996). The descending branch, minimum and ascending branch are approximated with a spline of variable order $n = 1+2+1$ (Andronov, 2003), i.e. by asymptotic straight lines connected with a piece of

TABLE IV

Minima timings for DW UMa in the VRI filters

JD 24...	JD 24...	JD 24...
51996.39750 ± 0.00033 R	52003.36416 ± 0.00010 R	52029.31932 ± 0.00039 V
51998.31014 ± 0.00044 V	52026.31416 ± 0.00002 R	52029.45707 ± 0.00045 V
51998.44642 ± 0.00022 V	52027.27128 ± 0.00050 I	52030.27599 ± 0.00017 R
52000.35904 ± 0.00198 I	52027.40751 ± 0.00053 I	52030.41258 ± 0.00038 R

parabola. The defect $k = 1$ means that the derivatives of this function are continuous up to the order $(n - k)$, i.e. the smoothing function and its derivative are continuous. The unknown parameters determined using the least-squares fit are the slopes of asymptotic lines and the signal value at their point of intersection. The additional free parameters are the "basic" times of switching to/from the parabola. Thus the shape ranges from a pure parabola (like that used for the same star by Bíró (2000)), where "basic" times coincide with the start and end of the data, to a broken line, where both "basic" times coincide, with no intermediate parabola. All these 5 parameters are determined using non-linear least-squares fitting. Such a fit is preferred for signals with possibly different slopes of the ascending and descending branches. The main parameters of the fits are listed in Table II. The phases are very close to zero, despite brightness at minimum varies strongly from cycle to cycle.

References

Andronov, I.L.: 1994, *Odessa Astron. Publ.* **7**, 49.
Andronov, I.L.: 2003, *ASP Conf. Ser.* **292**, 391.
Bíró, I.B.: 2000, *A&A* **364**, 573.
Kazennova, E.A. and Shugarov, S.Yu.: 1992, *ASP Conf. Ser.* **29**, 390.
Kopylov, I.M. and Somov, N.N.: 1987, *SoSAO* **56**, 51.
Marsakova, V.I. and Andronov, I.L.: 1996, *Odessa Astron. Publ.* **9**, 127.
Shafter, A.W., Hessman, F.V. and Zhang, E.H.: 1988, *ApJ* **327**, 248.
Stanishev, V., Kraicheva, Z., Boffin, H.M.J., et al.: 2004, *A&A*, **416**, 1057.

FLICKERING IN THE MAGNETIC CV STAR AM HERCULIS

BELINDA KALOMENI, E. RENNAN PEKÜNLÜ and KADRI YAKUT

Department of Astronomy and Space Sciences, Faculty of Science, University of EGE, Bornova, İzmir, Turkey

(accepted April 2004)

Abstract. In this study, we present a photometric study of AM Her, a prototype of a class of magnetic CVs. Optical photometry of AM Her was obtained using the Russian–Turkish 1.5 m telescope at TÜBİTAK National Observatory (TUG) in August 2003. The R band light curve of the system shows two maxima and two minima during one orbital cycle. In both observing nights the star showed flickering at a significant level. The measured flickering time scale is about 5 min.

Keywords: stars: binaries close, stars: magnetic field, stars: individual: AM Her

1. Introduction

AM Herculis is the prototype of a subclass of cataclysmic variable (CV) stars, also known as "polars", with strongly magnetic white dwarf primaries (10–230 MG) accreting matter from a low-mass secondary. As a result of the intense magnetic fields, the rotational and orbital motion of the white dwarf is synchronized, hence the emission of polars is strongly modulated with the orbital period, and no accretion disks can form due to the magnetic interaction.

AM Her, classified as a CV by Berg and Duthie in 1977, is the brightest known polar. The orbital period is 3.1 h (Cowley et al., 1976; Tapia, 1976). The system has been observed many times by numerous researchers in U, B, V, and R bands (e.g. Olson, 1977; Szkody and Brownlee, 1977; Priedhorsky and Krzeminski, 1978; Gilmozzi et al., 1978; Szkody and Margon, 1980; Crosa et al., 1981; Bonnet-Bidaud, 1991; Kjurkchieva and Marchev, 1995; Bonnet-Bidaud, 2000; de Martino et al., 2002). Since most observable properties of CVs depend on the instantaneous accretion rate, the photometric, spectroscopic and polarimetric behaviour of AM Her may change with respect to the observation time.

2. Observations and Reduction

Optical photometry of AM Her was obtained using the Russian–Turkish 1.5 m telescope (RTT150) at TÜBİTAK National Observatory (TUG) on 6–7 August 2003 in the R filter. The exposure times were set to 15 s. GSC 3533 1026 and GSC 3533 1021 were chosen as comparison and check stars, respectively. Observational data of both nights cover more than one orbital cycle.

Astrophysics and Space Science **296**: 477–480, 2005.
© Springer 2005

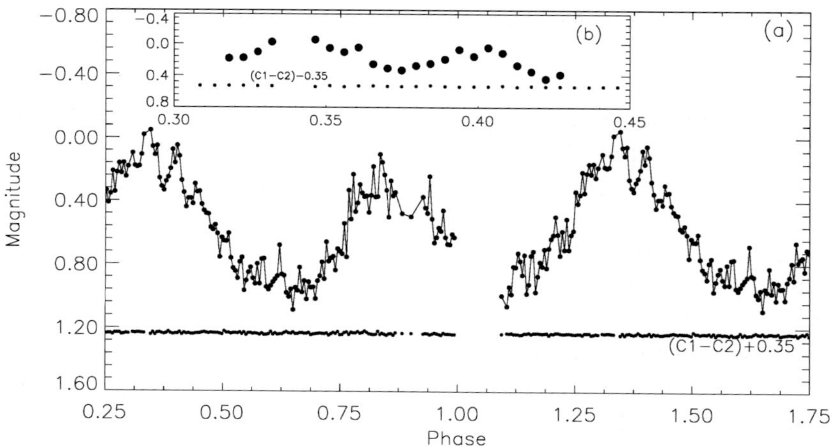

Figure 1. (a) Light curve of AM Her on 6 August 2003; (b) The flickering between phases 0.315 and 0.43.

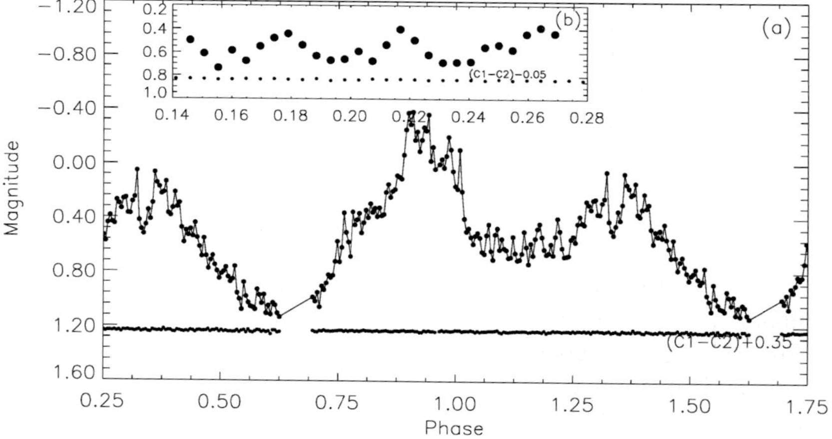

Figure 2. (a) Light curve of AM Her on 7 August 2003; (b) The flickering between phases 0.15 and 0.27.

All CCD reductions were done with the IRAF[1] (DAOPHOT/PHOT package). The standard deviations of single differential observations are 0.0050 and 0.0049 for 6 and 7 August, respectively. The timings were converted to orbital phases using the ephemeris of Heise and Verbunt (1988).

The light curves of AM Her are shown in Figures 1a and 2a. Figures 1b and 2b show the flickering variability on an enhanced scale in the phase ranges 0.315–0.43 and 0.15–0.27, respectively.

---

[1]IRAF is distributed by National Optical Astronomy Observatories.

3. The Flickering Effect in AM Her

A general feature of AM Her-type systems are brightness variations characterized as flickering on time scales of seconds to minutes (e.g., Middleditch, 1982; Larsson, 1987, 1989; Watson et al., 1987; Middleditch et al., 1991, 1997; Bonnet-Bidaud et al., 1996), which are commonly observed in X-ray and/or optical bandpasses (e.g., Szkody and Margon, 1980; Watson et al., 1987; Bonnet–Bidaud et al., 1991; Middleditch et al., 1997). There is observational evidence that the flickering in AM Her occurs on two different time scales: ~4.5 min (Bonnet-Bidaud et al., 1991) and ~8 min (King, 1989).

In our two-night observations, we detected non-periodic changes, with 3–10-min duration. In principle, the period of flickering can be determined by subtracting the 3.1 h variation. An application of the autocorrelation function (ACF) (see Szkody et al., 1980; Andronov, 1994) to the residuals relative to the best fit of the light curves yields a flickering time scale of about 5 min. In Figures 1b and 2b, the most significant flickering features of 6 and 7 August are shown, with brightness variations of 0.4 and 0.23 mag, respectively. For comparison, the difference between comparison and check stars, C1–C2, is shown at the bottom of Figures 1b and 2b. From this it is clear that the comparison star does not have any effect on the observed flickering. To explain the flickering properties different models have been proposed (i.e., Kuijper and Pringle, 1982; Frank et al., 1988; Szkody and Margon, 1980; King, 1989; Tuohy et al., 1981).

4. Summary

The optical light curves of AM Her-type systems are different from other well-known binary systems. There are three main types of variability in the observed light curves. First, long period changes called high/low state, which are non-periodic, with more than 2 magnitudes amplitude. Second, there are periodic, orbital phase-dependent variations with a period of 3.1 h in the case of AM Her; the amplitude of this variation changes with time. Finally, there appear non-periodic variations, which are known as flickering. Our observations of 6–7 August 2003 cover the transition period, when AM Her passed from high to low state. In these observations, along with the 3.1 h period, we detected flickering with a time scale of about 5 min. The reason for this flickering is still unknown. Proposed mechanisms discussed in the literature are inhomogeneities in the accretion flow or ionizing radiation and changes in the mass accretion rate.

Acknowledgements

The authors would like to thank TUG (TÜBİTAK National Observatory) and TÜBİTAK-BAYG for their supports. This study has been partly supported by Ege University Research Project (2002/FEN/002).

References

Andronov, I.: 1994, *AN* **315**, 353.

Berg, R. and Duthie, J.: 1977, *AJ* **211**, 859.

Bonnet-Bidaud, J.M., Somova, T.A. and Somov, N.N.: 1991, *A&A* **251**, 27.

Bonnet-Bidaud, J.M., et al.: 2000, *A&A* **354**, 1003.

Bonnet-Bidaud, J.M., Mouchet, M., Somova, T.A. and Somov, N.N.: 1996, *A&A* **306**, 199.

Cowley, A., Crampton, D., Szkody, P. and Brownlee, D.: 1976, IAU Circ., No. 2984.

Crosa, L., Szkoda, P., Stokes, G., Swank, J. and Wallerstein, G.: 1981, *ApJ* **247**, 984.

de Martino, D., et al.: 2002, *A&A* **396**, 213.

Frank, J., King, A.R. and Lasota, J.-P.: 1988, *A&A* **193**, 113.

Gilmozzi, R., Messi, R. and Natali, G.: 1978, *A&A* **68**, L1.

Heise, J. and Verbunt, F.: 1988, *A&A* **189**, 112.

Olson, E.C.: 1977, *ApJ* **215**, 166.

Priedhorsky, W. and Krzeminski, W.: 1978, *AJ* **219**, 597.

King, A.R.: 1989, *MNRAS* **241**, 365K.

Kjurkchieva, D. and Marchev, D.: 1995, *Ap&SS* **234**, 207.

Kuijpers, J. and Pringle, J.E.: 1982, *A&A* **114L**, 4K.

Larsson, S.: 1987, *Ap&SS* **130**, 187.

Larsson, S.: 1989, *A&A* **217**, 146.

Middleditch, J.: 1982, *ApJ* **257**, L71.

Middleditch, J., Imura, J.N., Wolff, M.T. and Steiman-Cameron, T.Y.: 1991, *ApJ* **382**, 315.

Middleditch, J., Imura, J.N. and Steiman-Cameron, T.Y.: 1997, *ApJ* **489**, 912.

Tapia, S.: 1976, IAU Circ., No. 2987.

Szkody, P. and Brownlee, D.E.: 1977, *ApJ* **212**, L113.

Szkody, P. and Margon, B.: 1980, *ApJ* **236**, 862.

Tuohy, I.R., Mason, K.O., Garmire, G.P. and Lamb, F.K.: 1981, *ApJ* **245**, 183.

Watson, M.G., King, A.R. and Williams, G.A.: 1987, *MNRAS* **226**, 867.

POST-COMMON-ENVELOPE BINARIES WITH DISTORTED SECONDARIES

ADELA KAWKA[1] and STÉPHANE VENNES[2]

[1]*Stellar Department, Astronomický ústav AV ČR, Fričova 298, CZ-251 65 Ondřejov, Czech Republic; E-mail: kawka@sunstel.asu.cas.cz*
[2]*Department of Physics and Astronomy, 3400 North Charles Street, Johns Hopkins University, Baltimore, MD, USA*

(accepted April 2004)

Abstract. We will examine the properties of binary systems which have Roche lobe filling secondaries and white dwarf primaries as well as systems where the secondary is only partially filling its Roche lobe. We will also discuss observational properties such as ellipsoidal variations, light reprocessing and radial velocity measurements of the close binary systems BPM 71214, EC 13471-1258 and GD 245. The properties of BPM 71214 and EC 13471-1258 show that these systems may be pre-cataclysmic variables just prior to the onset of mass transfer or they may be hibernating novae. GD 245 is a pre-cataclysmic variable.

Keywords: close binary stars, white dwarfs

1. Introduction

Short-period binaries containing a white dwarf primary and a low-mass main-sequence secondary are the outcome of common-envelope evolution (Paczynski, 1976, and references therein). They are also the progenitors of cataclysmic variables, systems where the secondary is transferring mass onto the white dwarf. Schreiber and Gänsicke (2003) have recently published a list of 30 well-observed post-common-envelope binaries and determined their evolutionary status. We will discuss three of these systems. Two of these systems (BPM 71214 and EC 13471-1258) show the secondary star filling its Roche lobe but with no observable mass transfer. Spectroscopic and photometric studies of BPM 71214 were carried out by Kawka et al. (2002) and Kawka and Vennes (2003). Similar studies of EC 13471-1258 were presented by Kawka et al. (2002) and O'Donoghue et al. (2003). The properties of the third system, GD 245, suggests that the secondary fills in over 40% of the volume of its Roche lobe (Schmidt et al., 1995).

2. Roche Lobe Filling Secondary Stars

Ellipsoidal variations are observed when the secondary is distorted by the gravitational field of the primary. BPM 71214 and EC 13471-1258 both display ellipsoidal

Astrophysics and Space Science **296**: 481–484, 2005.
© Springer 2005

TABLE I

Parameters for BPM 71214 and EC 13471-1258

Parameter	BPM 71214[a,b]	EC 13471-1258[a]
Orbital period	0.201626 ± 0.000004 days	0.15074 ± 0.00004 days
WD mass	$0.77 \pm 0.06\,M_\odot$	$0.77 \pm 0.04\,M_\odot$
WD temperature	17200 ± 1000 K	14085 ± 100 K
Secondary mass	$\sim 0.54\,M_\odot$	$0.55 \pm 0.11\,M_\odot$
Secondary radius	$\sim 0.56\,R_\odot$	$\sim 0.44\,R_\odot$
Inclination	$30 \pm 2°$	$74 \pm 2°$

[a]Kawka et al. (2002).
[b]Kawka and Vennes (2003).

variations. Kawka et al. (2002) and Kawka and Vennes (2003) presented spectroscopic and photometric observations that suggested that the secondary stars of BPM 71214 and EC 13471-1258 are just underfilling their Roche lobe.

2.1. BPM 71214

The parameters of BPM 71214 are summarized in Table I. The orbital parameters were determined from spectroscopic observations as cited, and the secondary star properties were constrained by comparing model light curves to R and I photometry.

Since Hα emission is likely to trace an orbit that is inferior to the orbit traced by the true center of the secondary mass if the secondary is irradiated by the hot white dwarf, the radial velocities were remeasured by cross-correlating the absorption features in the range 6240–6540 Å against the M2 dwarf GL 250B using FXCOR in IRAF. Absorption features are more likely to trace the center of mass of the secondary. The velocities were phased with the orbital period to obtain a secondary velocity semi-amplitude of $K_{\rm sec} = 116.2 \pm 2.5\,{\rm km\ s^{-1}}$.

2.2. EC 13471-1258

The parameters of EC 13471-1258 are summarized in Table I, where the secondary mass is significantly larger than the mass ($0.41 \pm 0.09\,M_\odot$) that O'Donoghue et al. (2003) obtained. The secondary mass and inclination in Table I were calculated using $K_{\rm sec} = 245 \pm 6\,{\rm km\ s^{-1}}$, white dwarf mass, orbital period and the duration of the eclipse (15.4 ± 1.0 min) as inputs into an iterative code where an initial estimate of the inclination $i = 80°$ was made. The radius of the red dwarf was assumed to be the Roche lobe radius, which was checked against the radius obtained from the FWHM of Hα (Kawka et al., 2002).

The secondary mass was recalculated by replacing $K_{\rm sec} = 245\,{\rm km\ s^{-1}}$ with $K_{\rm sec} = 266 \pm 5\,{\rm km\ s^{-1}}$ (O'Donoghue et al., 2003) to obtain $M_{\rm sec} = 0.42 \pm 0.09\,M_\odot$, which is consistent with O'Donoghue et al., and $i = 76 \pm 2°$ which is consistent with the earlier calculation. Kawka et al. (2002) used the Hα emission line to obtain their $K_{sec} = 241 \pm 8$ km s^{-1}, where as O'Donoghue et al. (2003) used absorption features of the secondary. However, Hα emission is likely to favor an orbit that is inferior to the true center of mass of the secondary. Therefore, the radial velocities were remeasured using absorption features in the range of 6240–6540 Å against the spectra of GL 190 (M3.5) and GL 250B (M2) which resulted in $K_{sec} = 246 \pm 12$ km s^{-1} and $K_{sec} = 251 \pm 12$ km s^{-1}, respectively. These values are still marginally lower than those of O'Donoghue et al. Therefore, the mass of the secondary is probably between 0.41 $M_\odot$ and 0.55 $M_\odot$.

Figure 1 shows the R and B photometry compared to synthetic light curves which were computed using the Wilson–Devinney (WD) code (Wilson, 1979, 1990). The fixed inputs into the code were the parameters in Table I and the eclipse duration of 15.4 min. The binary was assumed to be detached, with black-body source spectra and no spots present on either star. The radius of the secondary was varied until the amplitude of the ellipsoidal variations was satisfied, then the effective temperature of the secondary star was varied until the depth of the eclipse was satisfied. First, the R light curve was fitted, for which $T_{eff} = 2790$ K and $R_{sec} = 0.44\ R_\odot$ provided the best fit. These were then adopted to the B light curve, but $T_{eff} = 2980$ K was required. $T_{eff} = 2900 \pm 150$ K is only an estimate since a black-body spectrum does not represent a M-dwarf spectrum well.

Synthetic light curves were also calculated for $M_{sec} = 0.42\ M_\odot$, however, similar values of secondary radius and temperature were required. Note, that if the secondary mass is decreased, then the separation must also decrease, and therefore the mass of the secondary has little effect on the resulting light-curve unless strict mass–radius relations are kept.

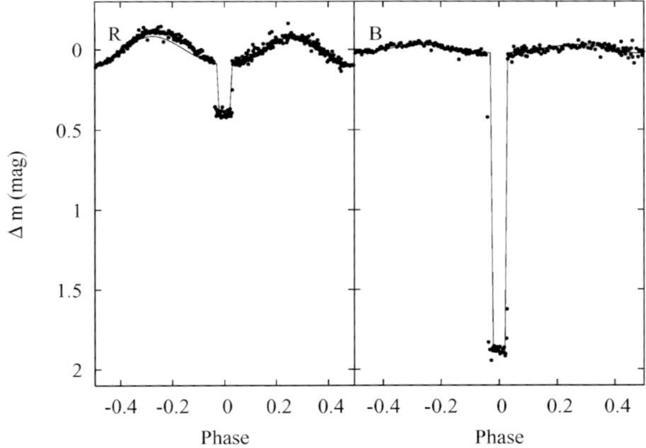

Figure 1. Photometry of EC 13471-1258 in R and B showing ellipsoidal variations and the eclipse. The light curves are compared to model light curves (black line).

The orbital parameters of these systems indicate that they are pre-cataclysmic variables just prior to the onset of mass transfer or they may be hibernating novae (Shara et al., 1986).

3. Pre-Cataclysmic Variable Stars

Most of the known white dwarf plus dMe binaries are not close enough for the secondary to be detectably distorted by the primary white dwarf. Therefore, ellipsoidal variations are not going to be observed in these systems, however, some of these systems may be variable as a result of light reprocessing. Irradiation of the secondary star is observed if the binary components are close enough and the white dwarf is hot enough to produce a sufficient amount of EUV/FUV flux.

GD 245 is an example of such a system. Schmidt et al. (1995) obtained spectroscopic and photometric observations of GD 245 and determined the system parameters. They found the orbital period to be 4.17 h, with $M_{\mathrm{WD}} = 0.48\,M_{\odot}$ and $T_{\mathrm{eff}} = 22170 \pm 130$ K, and a mass ratio of 0.46. They also observed photometric variations in U, B and V, and variations in the Hα equivalent width, which are the result of the secondary being irradiated by the white dwarf.

These parameters indicate that the secondary main-sequence star fills in over 40% of the volume of its Roche lobe. The secondary is not close enough to the primary to be detectably distorted by the white dwarf, and according to (Schreiber and Gänsicke, 2003), the binary has completed only $\sim$2% of its post-common-envelope binary life-time, and it will come into contact within a Hubble time.

References

Kawka, A., Vennes, S., Koch, R. and Williams, A.: 2002, *AJ* **124**, 2853.
Kawka, A. and Vennes, S.: 2003, *AJ* **125**, 1444.
O'Donoghue, D., et al.: 2003, *MNRAS* **345**, 506.
Paczynski, B.: 1976, in: P. Eggleton, S. Milton and J. Whelan (eds.), IAU Symposium 73: Structure and Evolution of Close Binary Systems, Dordrecht: Reidel, p. 75.
Schmidt, G.D., Smith, P.S., Harvey, D.A. and Grauer, A.D.: 1995, *AJ* **110**, 398.
Schreiber, M.R. and Gänsicke, B.T.: 2003, *A&A* **406**, 305.
Shara, M.M., Livio, M., Moffat, A.F.J. and Orio, M.: 1986, *ApJ* **311**, 163.
Wilson, R.E.: 1979, *ApJ* **234**, 1054.
Wilson, R.E.: 1990, *ApJ* **356**, 613.

SEARCH FOR CORRELATIONS BETWEEN BATSE GAMMA-RAY BURSTS AND SUPERNOVAE

JIŘÍ POLCAR[1], MARTIN TOPINKA[2], GRAZIELLA PIZZICHINI[3],
ELIANA PALAZZI[3], NICOLA MASETTI[3], RENÉ HUDEC[4] and VĚRA HUDCOVÁ[4]

[1]*Institute of Theoretical Physics & Astrophysics, Faculty of Science, Masaryk University, Brno, Czech Republic; E-mail: polcar@physics.muni.cz*
[2]*Faculty of Mathematics and Physics, Charles University, Prague, Czech Republic*
[3]*C.N.R. Instituto di Astrofisica Spaziale e Fisica Cosmica – I.A.S.F./CNR, Bologna, Italy*
[4]*Astronomical Institute of the Academy of Sciences of Czech Republic, Ondřejov, Czech Republic*

(accepted April 2004)

Abstract. We report on our statistical research of space–time correlated supernovae and CGRO-BATSE gamma-ray bursts (GRBs). There exists a significantly higher abundance of core-collapse supernovae among the correlated supernovae, but the subset of all correlated objects does not seem to be physically different from the whole set.

Keywords: gamma-ray bursts, supernovae, correlations

1. Introduction

The origin and source of gamma-ray bursts (GRBs) still remain a puzzle. This tremendous energy could be released on a short time scale during a collapse of a massive star in a supernova-like explosion. There are several pieces of observational evidence supporting the connection between GRBs and supernovae (SNe): some GRBs reveal an underlying SN in an optical afterglow light curve (e.g. GRB980326), sometimes supported by spectral and colour signatures (e.g. GRB030329 and SN2003dh), and there are a few objects with space–time coincidence (e.g. GRB980425 and SN1998bw).

We used the current CGRO-BATSE GRB catalogue (up to date version to April 2003) (BATSE) and combined data from Asiago Padova (Padova), Harvard (CFA) and Sternberg (SAI) catalogues of SNe.

2. Search for Space–Time Correlation

We developed a fast matching engine for matching databased data from two selected catalogues written in `perl` under Linux OS. We looked for space and time coincidences. A SN and a GRB make up a pair, if the position of the SN falls into the GRB errorbox. In the time domain, it is not theoretically clear what is the time delay between an eventual gamma-ray emission of a SN (which could be observed

Astrophysics and Space Science **296**: 485–488, 2005.
© Springer 2005

as a GRB) and its optical emission. We simplified our problem and divided SNe into type Ia SNe and core-collapse SNe. The time delays between the time of the explosion and the time of the maximum of the SN are assumed to be 0 ± 30 days for core-collapsed SNe and 20 ± 7 days for type Ia SNe for the purpose of our analysis (Petschek, 1996; MacFayeden, 1999; Vietri, 1998); the errors given denote the uncertainty in the width of the time interval.

Only a fraction of the SNe provide information on the date of the maximum, and also the type is not defined or at least doubtful for part of these objects. We use here a statistical approach and assume the time delay between the time of the maximum and the time of the discovery to be the median (-4 days) of all known time delays. We used the weighted average of all types and all time delays for SNe of unknown type. The uncertainty increases with increasing size of the time interval window.

We ended up with 92 possibly connected GRB/SN pairs (see Figure 1). Note that we do not know how many of these coincidences are real and how many just incidental.

We rotated coordinates of the SNe before matching with GRB to get rid of any possible GRB/SN correlation. So we get random pairs of GRB/SN for each rotation to compare with real match pairs. Surprisingly, there is a significantly larger fraction of core-collapse SNe among the possibly correlated pairs in the real match than in the random scheme (see Figures 2 and 3).

3. Search for Correlations with Physical Properties

The number of all events and the number of the pairs is too small to allow for any significant conclusions. We suppose that not only a fraction of GRBs and SNe is space–time correlated, but also that this fraction is intrinsically physically different from the rest of the sample. We analyzed all available physical quantities of both

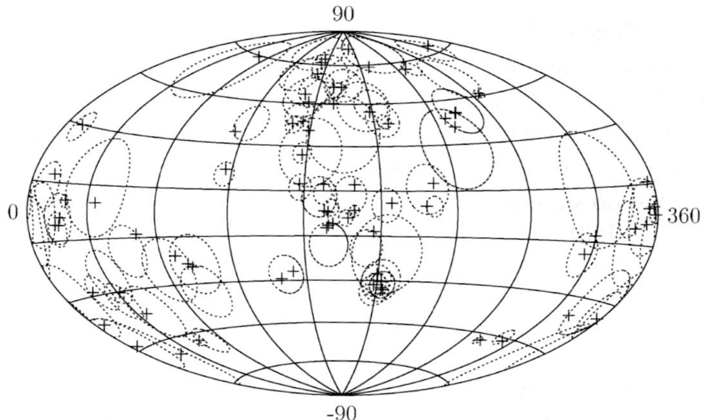

Figure 1. Final match – GRB/SN pairs.

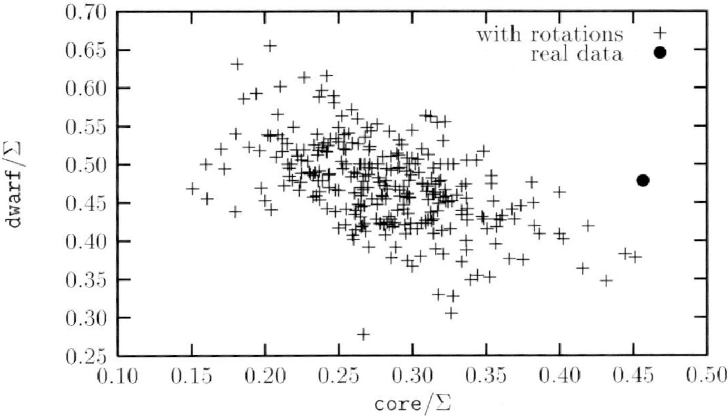

Figure 2. Fraction of core-collapsed SNe and Ia SNe in matched pairs.

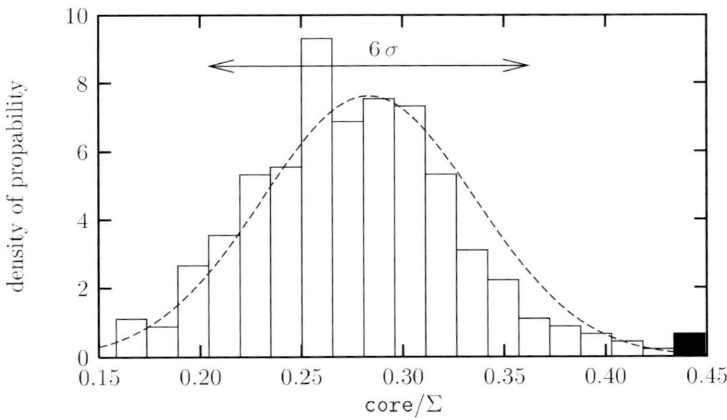

Figure 3. Probability function of fraction of core-collapsed SNe in matched pairs.

GRBs and SNe. None of the GRBs detected by BATSE has a redshift measured. In the analysis below we assume that the redshift of the SN in the pair (if known) is also the redshift of the corresponding GRB.

We searched for possible correlations between physical properties of SNe and GRBs as well as for possible deviations of the matched pairs from the whole sample and found no significant correlations according to a Kolmogorov–Smirnov test (Press et al., 1968).

4. Conclusions

At this moment, we are not able to conclude if there is any positive or negative or any correlation at all between GRBs and SNe. We plan to rebuild the database and check the results with more robust methods, which we have already developed (e.g., by means of a Monte Carlo simulation and by setting the upper limit for the number of the correlated pairs in the sample). We also plan to provide a more complete statistical background to maximize the amount of information one can extract from the GRB and SN catalogues.

Acknowledgements

We acknowledge the support by the Grant Agency of the Academy of Sciences of the Czech Republic, grant A3003206.

References

BATSE: The Burst and Transient Source Experiment, catalogue of GRBs http://cossc.gsfc.nasa.gov/batse/index.html.

CFA: Center for Astrophysics, Harvard a Smithsonian Observatory, catalogue of supernovae http://cfa-www.harvard.edu/iau/lists/Supernovae.html.

MacFayeden, A.I. and Woosley, S.E.: 1999, *ApJ* **524**, 262.

The Asiago Supernova Catalogue http://merlino.pd.astro.it/~supern/snean.txt.

Petschek, A.G.: 1996, *Supernovae*, Springer Verlag, New York.

Press, W.H., Flanney, P., Teukolsky, S.A.N. and Vetterling, W.T.: 1968, Numerical Recipes, Cambridge University Press.

SAI: Sternberg Astronomical Institute, catalogue of supernovae http://www.sai.msu.su/sn/.

Vietri, M. and Stella, L.: 1998, *ApJ* **507**, L45.